FORTSCHRITTE DER CHEMIE ORGANISCHER NATURSTOFFE

PROGRESS IN THE CHEMISTRY OF ORGANIC NATURAL PRODUCTS

HERAUSGEGEBEN VON · EDITED BY

L. ZECHMEISTER
CALIFORNIA INSTITUTE OF TECHNOLOGY, PASADENA

SECHSUNDZWANZIGSTER BAND
TWENTY-SIXTH VOLUME

VERFASSER · AUTHORS

K. BERNAUER · J. W. BUCHLER · R. B. COREY · D. L. DREYER
D. DÜTTING · H. GERLACH · W. HOFHEINZ · H. H. INHOFFEN
P. JAGER · W. KELLER-SCHIERLEIN · K. LÜBKE · R. E. MARSH
E. SCHRÖDER · G. P. SCHWARTZ · A. C. TRAKATELLIS

MIT 97 ABBILDUNGEN · WITH 97 FIGURES

1968

WIEN · SPRINGER-VERLAG · NEW YORK

ISBN-13: 978-3-7091-7135-6   e-ISBN-13: 978-3-7091-7134-9
DOI: 10.1007/978-3-7091-7134-9

Titel Nr. 8233

Inhaltsverzeichnis
Contents

Limonoid Bitter Principles. By DAVID L. DREYER, U. S. Dept. of Agriculture, Fruit and Vegetable Chemistry Laboratory, Pasadena, California 190

Methoden und Ergebnisse der Sequenzanalyse von Ribonucleinsäuren.

Von DIETER DÜTTING, Max-Planck-Institut für Virusforschung, Mole-
kularbiologische Abteilung, Tübingen

X-Ray Diffraction Studies of Crystalline Amino Acids, Peptides and Proteins

By **R. B. COREY** and **R. E. MARSH**, Pasadena, California

With 14 Figures

Contents

Acknowledgement. The preparation of this paper was supported in part by a Public Health Service Research Grant No. HE-02143 from the National Heart Institute of the National Institutes of Health, Public Health Service. We should like to thank Miss Lillian Casler for making most of the drawings, and Miss Allison Kimball for help in preparing the manuscript.

Introduction

Three previous reviews have appeared in this Series describing the use of X-ray diffraction methods for the investigation of molecular structure: one by Kratky and Mark (*4*) on proteins and other natural products, one by Corey (*1*) on amino acids and peptides, and one by Pauling and Corey (*11*) on the configuration of polypeptide chains. During the 14 years that have passed since the last of these reviews, great progress has been made in X-ray diffraction techniques. As a result of these improved techniques, much more accurate and reliable data are now available concerning the dimensions of small biological molecules; in addition, detailed information concerning the structure of vastly more complex molecules is now being obtained. Whereas in 1954 structural information concerning proteins had to be deduced from knowledge of the structures of, at best, simple dipeptides, today such information is being obtained directly from crystalline proteins themselves.

It is the purpose of this review to discuss the results and significance of recent X-ray diffraction investigations of, first, amino acids and peptides and, later, of crystalline proteins.

I. Outline of the X-Ray Method

The determination of the structure of a crystal is based on two fundamental equations. The first is Bragg's Law,

$$\lambda = 2\,\mathrm{d}(h\,k\,l)\,\sin\Theta\,(h\,k\,l), \tag{1}$$

which defines the conditions for which a diffraction maximum may be observed when a crystal is irradiated. Here, λ is the wave length of the radiation, Θ is the angle of incidence (and reflection) between the X-rays and the diffracting set of crystallographic planes, h, k, and l are the Miller indices defining that set of planes, and d is the spacing between successive, parallel planes. By measuring values of Θ for a number of sets of planes, using a radiation of known wave length, it is immediately possible to determine the size and shape of the crystallographic unit cell—the unit of structure which, when repeated translationally in three dimensions, generates the crystal. Typically, unit cells of crystals of moderately complex compounds have dimensions in the range 10 to 20 Å; for crystalline proteins these dimensions may be as large as 100 Å. The wave lengths of the X-rays used to examine the crystals are about 1 to 2 Å.

The determination of the structure of a crystalline compound involves knowing not only the size and shape of the unit cell but also the locations of the various atoms within the cell. This knowledge can be obtained from measurements of the intensities I $(h\,k\,l)$ of the X-ray beam after it has been diffracted by a set of crystal planes $h\,k\,l$. The intensity of the diffracted beam is proportional to the square of the magnitude of the quantity $F_{h\,k\,l}$—the so-called "structure factor". The structure factor is the basis of the second fundamental equation:

$$F_{h\,k\,l} = \sum_j f_j\, T_j\, \exp.\ 2\,\pi\,i\,(h\,x_j + k\,y_j + l\,z_j). \tag{2}$$

Here, $x_j\,y_j\,z_j$ are the fractional coordinates, relative to the unit cell axes a, b and c, defining the position of the jth atom of the structure, f_j is the scattering power of that atom (approximately proportional to its atomic number), and T_j is a term describing its vibrational motion within the crystal lattice; the summation is over all atoms in the unit cell.

Thus, if one knows the positions, identities and patterns of motion of all atoms within the unit cell, one can calculate the structure factors and, from them, the intensities of all diffraction maxima. The inverse is not true; for in order to determine the positions of the atoms from a knowledge of the intensities, one must know not only the magnitudes but also the phase angles of the structure factors. Determination of these phase angles—the "phase problem"—is therefore the crucial factor in a successful X-ray diffraction investigation.

Various methods are used to solve the phase problem. One of the most powerful tools is the Patterson function

$$P_{u\,v\,w} = \sum_{h,\,k,\,l\,=-\infty}^{\infty\ \ \infty\ \ \infty} F_{h\,k\,l}^2 \cos 2\,\pi\,(h\,u + k\,v + l\,w). \tag{3}$$

This function involves a triple Fourier summation of the observed values of F^2 (which are readily derived from intensity measurements) for all crystal planes. The result of this summation is a three-dimensional map in which maximum values of $P_{u\,v\,w}$ occur at positions—defined by the coordinates u, v, w—corresponding to vectors between pairs of atoms in the structure. Since a structure with n atoms will generate n^2 such vectors, Patterson maps are difficult to interpret except for small compounds or for molecules containing one or two very heavy atoms, in which case vectors involving these atoms will be prominent. Nevertheless, a majority of successful crystal-structure determinations have been based on the interpretation of Patterson maps.

Other methods used for solving the phase problem include: (i) attempting to guess the structure from the size and symmetry of the unit cell, making use of known features of molecular geometry, packing and hydrogen bonding; (ii) preparing isomorphous, heavy-atom derivatives, for which the heavy atom can be readily located from Patterson maps and then used, in conjunction with the differences in intensities between the two derivatives, to assign phase angles; (iii) using statistical methods of phasing, in which the trigonometric nature of the structure factor is made the basis of probability relationships between the phase angles of related reflections [a recent application of this method is given by Karle

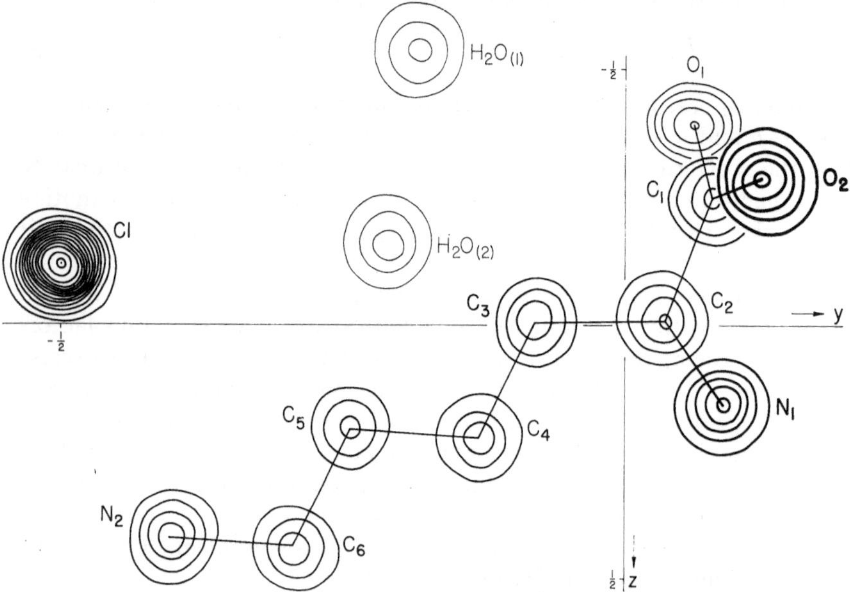

Fig. 1. A representation of the electron density in crystals of L-lysine hydrochloride dihydrate (38), as viewed down the a axis of the unit cell. Contours are drawn at equal intervals of electron density. The electron density associated with the hydrogen atoms is too small to be apparent in maps of this type. The complete structure of the crystal is shown in Figure 3 [Acta Crystallogr. 15, 54 (1962)]

and Karle (22, 45)]. In general, method (i) is applicable only to very simple molecules; method (iii) has been used on relatively complex structures—up to about 50 atoms; while method (ii) can be used on very complex structures and is the method usually employed in protein crystallography.

Once an approximate structure, or set of phase angles, has been derived, an improved picture of the structure can be obtained by calculating the three-dimensional Fourier series

$$\varrho\,(x\,y\,z) = \sum_{h,\,k,\,l\,=\,-\infty}^{\infty}\sum^{\infty}\sum^{\infty} F\,(h\,k\,l)\,\exp. - 2\,\pi\,i\,(h\,x + k\,y + l\,z). \qquad (4)$$

A plot of this function depicts the value of the electron density ϱ at each point $(x\,y\,z)$ in the unit cell. Regions of high electron density correspond to the positions of atoms, the electron density being approximately proportional to the atomic number of the atom. Typical electron density maps are shown in *Figures 1 and 2*.

The final step in a crystal-structure investigation is the refinement of the positional coordinates x, y, z and vibrational parameters T of all the atoms in the unit cell, resulting in a set of calculated structure factors

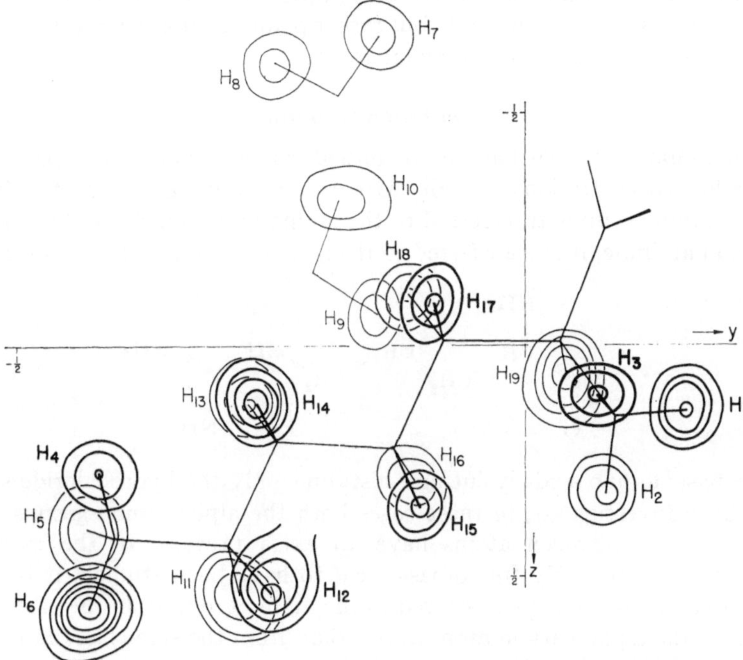

Fig. 2. An electron density map of L-lysine hydrochloride, showing the locations of the hydrogen atoms. The terms entering into this Fourier summation (Equation 4, p. 4) were values of $F\,(h\,k\,l)$ from which the calculated contributions of the C, N, O and Cl atoms had been subtracted [Acta Crystallogr. 15, 54 (1962)]

$F\,(h\,k\,l)$ in optimum agreement in magnitude with the observed values. This refinement is now carried out almost exclusively by the method of least-squares, and makes full use of the speed and storage capabilities of modern digital computers. The intensities $I\,(h\,k\,l)$ calculated from these coordinates should be in satisfactory agreement with the observed intensities. The accuracy of the resulting set of atomic parameters is often such that interatomic distances are accurate to less than 0.01 Å and angles to about 0.1°.

II. Crystal Structures of the Amino Acids

The crystal structures of nearly all of the amino acids—or, in some cases, of one or more of their hydrohalide salts—have by now been determined (12–39). The accuracy of these determinations varies widely, the standard deviations in the interatomic distances ranging from perhaps 0.1 Å for early investigations based on partial, two-dimensional intensity data to less than 0.005 Å in a few recent instances where exceptionally careful three-dimensional refinements were carried out. In all cases, however, the accuracy has been sufficient to permit the fundamental structural and configurational features to be seen. We shall not discuss each of these structures in detail, but rather point out the general structural features which seem to be common to all.

1. Zwitterion Structures

All amino acids that have been studied (as well as all simple peptides) crystallize as zwitterions. In all cases except arginine, the proton from the carboxyl group is transferred to the amino group on the alpha carbon atom; in arginine, it is transferred to the guanidinium group. In the cases

$$O=C(O^{\ominus})-CH(NH_2)-CH_2-CH_2-CH_2-CH_2-NH-C(NH_2)(=NH_2^{\oplus})$$

of the basic amino acids lysine and histidine, only the hydrochloride salts have been investigated; in these cases both the alpha amino groups and the side-chain nitrogen atoms have an extra proton. In the case of glutamic acid, the only free dibasic acid to have been studied, it is probable that the proton is transferred to the amino group from the carboxyl group on the alpha carbon atom rather than from the side-chain carboxyl group.

2. Hydrogen Bonding

Intermolecular hydrogen bonding is an important feature of the crystal structures of all the amino acids. The ammonium group ($-NH_3^+$) of the zwitterion is an excellent hydrogen-bond donor and the carboxylate ion ($-CO_2^-$) is an excellent acceptor; as a result, strong hydrogen bonds are invariably formed involving these two groups. The length of the hydrogen bond, the N ... O distance, is usually in the range 2.8 to 2.9 Å, but occasionally is as long as 3.1 Å.

Typical hydrogen-bond arrangements, as found in crystals of L-lysine hydrochloride, glycine, and of L- and DL-alanine, are shown in *Figures 3–6*. The relationship between the structures of the two forms of alanine is particularly interesting, and points up the importance of hydrogen bonding in determining the arrangement of molecules in crystals. Crystals of DL-alanine contain planes of symmetry which relate molecules of

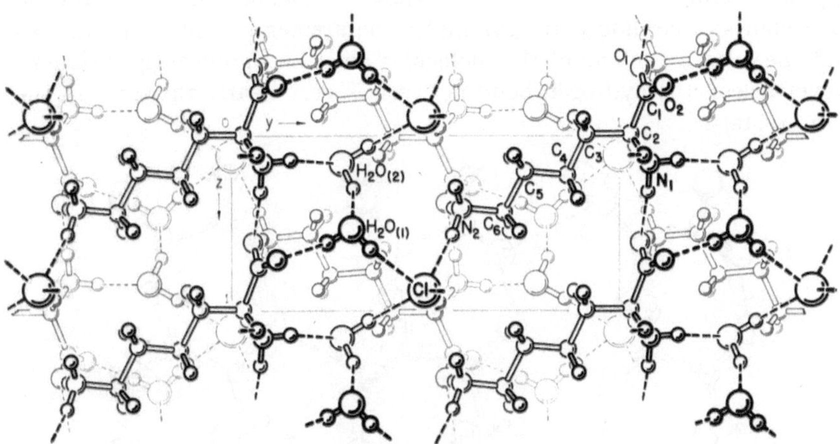

Fig. 3. The crystal structure of L-lysine hydrochloride dihydrate, as determined by X-ray diffraction procedures (*38*). The view is along the *a* axis. All ten acidic protons—three from each of the ammonium groups and two from each of the water molecules—form hydrogen bonds, indicated by dashed lines; the chloride ion and the oxygen atoms of the carboxylate groups and water molecules serve as acceptors. The molecule shown in Figures 1 and 2 is at the extreme left [Acta Crystallogr. 15, 54 (1962)]

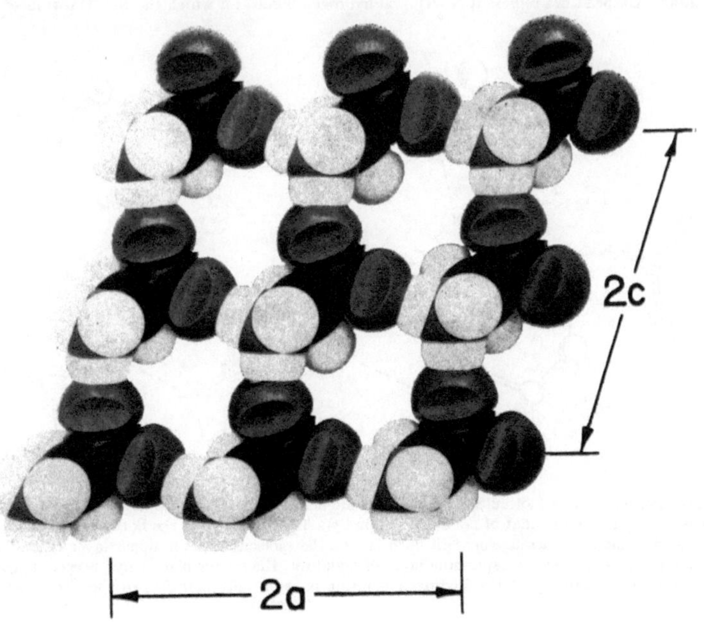

Fig. 4. A photograph of space-filling atomic models representing a portion of the crystal structure of glycine (*12, 24*). Only two of the three hydrogen bonds are shown; the remaining hydrogen atoms (white) of the ammonium groups are bonded to oxygen atoms in a layer above

opposite configuration, whereas crystals of L-alanine can contain no such planes; accordingly, the symmetry requirements of the space groups and the overall packing of the molecules must be significantly different. Nevertheless, the hydrogen bond network is very nearly the same in the two crystals.

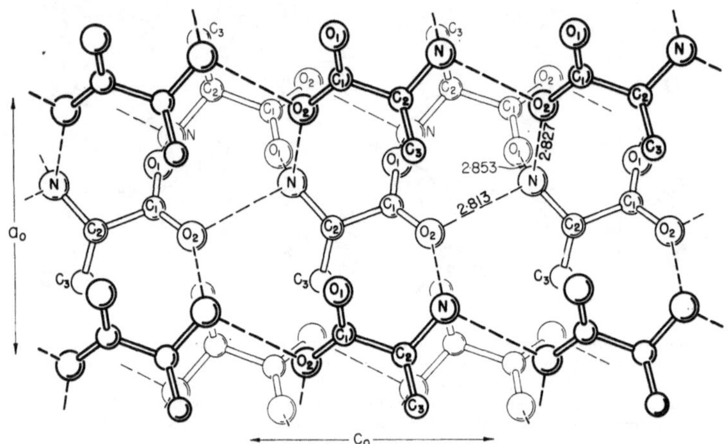

Fig. 5. The arrangement of molecules in crystals of L-alanine, as viewed down the b axis (34). Hydrogen atoms are not shown. Dashed lines represent N—H...O hydrogen bonds, for which the N...O distances are given

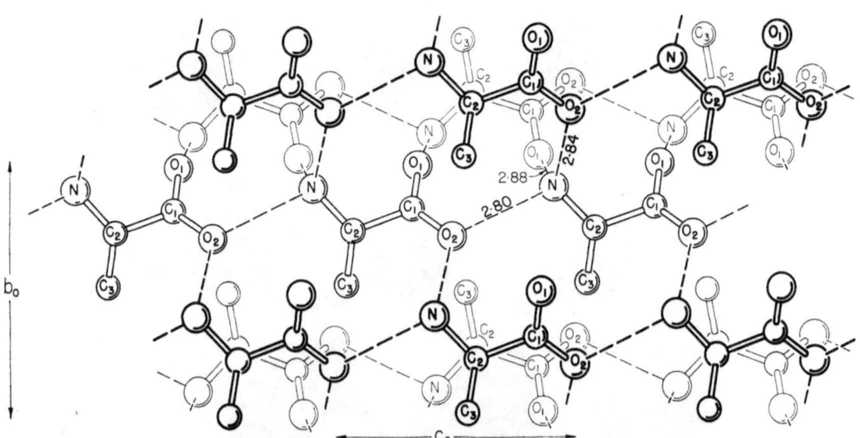

Fig. 6. The arrangement of molecules in crystals of DL-alanine, as viewed down the a axis (15, 23). This structure is closely related to that of L-alanine (Figure 5). The molecules in the bottom layers (farthest from the viewer) have the same arrangement in both cases; the molecules in the upper layer of DL-alanine are mirror images of those in the corresponding layer of L-alanine, the mirror plane being perpendicular to the c axis. The geometry of the hydrogen bonding is essentially identical in the two cases

An interesting situation also obtains in crystals of arginine dihydrate (22). As discussed above, the zwitterion is formed by transfer of a proton to the side-chain guanidinium group. The resulting positive charge

References, pp. 40—47

causes this group to become an excellent hydrogen-bond donor with all five protons participating. However, the alpha amino group is uncharged and hence is a much poorer donor; in fact, it forsakes its role as a donor and instead *accepts* a hydrogen bond from a neighboring guanidinium group.

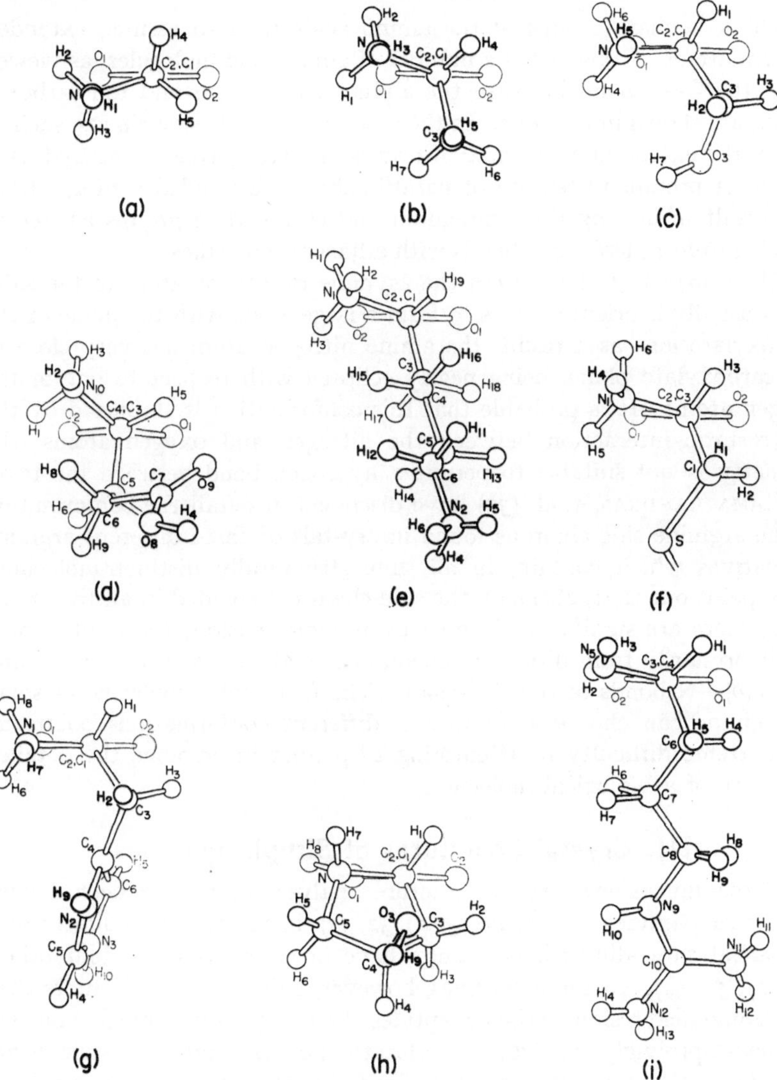

Fig. 7. The conformation of various amino acid molecules, as found in crystal structure investigations. The view is down the C(α)—C(carboxyl) bond. In each case the numbering of the atoms is taken from the original paper. (a) glycine (*12, 24*). (b) L-alanine (*34*). (c) L-serine (*32*). (d) L-glutamic acid (*19*). (e) L-lysine (hydrochloride) (*38*). (f) L-cystine (*27*). (g) L-histidine (hydrochloride) (*16*). (h) L-hydroxyproline (*17*) and (i) L-arginine (*22*) [In part: Adv. Protein Chem. **22**, 235 (1967)]

In crystals of amino acids which crystallize as hydrates, the water molecules can act as both acceptors and donors in hydrogen bond formation (see Fig. 3, p. 7). Halide ions of hydrohalide salts are also excellent hydrogen-bond acceptors.

3. Conformation of Side-Chains

The side-chain groups of the amino acids tend to assume extended conformations. Drawings of a number of amino-acid molecules, as viewed down the C—C bond between the alpha carbon atom and the carboxyl group, are shown in *Figure 7*. In the cases of aliphatic side chains, such as lysine, the extended conformation results in a staggering of the hydrogen atoms—a prominent feature of paraffin chains. Extended conformations also result in leaving the ammonium and carboxylate groups as free as possible to form hydrogen bonds with adjacent molecules.

The C (α)—C (β) bond from the asymmetric carbon atom to the side-chain usually is oriented so as to form a large angle with the plane of the carboxylate ion; as a result, the amino nitrogen atom lies very close to the carboxylate plane, being nearly eclipsed with respect to one of the oxygen atoms. It is probable that this conformation is stabilized by the electrostatic interaction between the nitrogen and oxygen atoms (the geometry is not suitable to permit a hydrogen bond between the two).

Ramachandran et al. (*30*) have discussed in detail the conformation of the arginine side chain as found in crystals of three different arginine derivatives which contain, in all, five structurally distinct molecules. They point out that, although the side-chain is extended in all five molecules, there are significant differences in conformation; these differences result primarily from different torsion angles about the C (α)—C (β) and the C (δ)—N bonds of the side-chain. The fact that a molecule as small as arginine can choose a number of different conformations points up the extreme difficulty in attempting to predict in advance the detailed structure of a biological molecule.

III. Crystal Structures of Simple Peptides

Three-dimensional crystal-structure analyses have now been carried out for ten peptides (*40, 41, 43, 45–49, 52, 53*); in addition, two-dimensional or partial three-dimensional results have been reported for four others (*42, 44, 50, 51*). Of these peptides, however, only four contain more than two amino-acid residues: the tripeptides glycyl-L-phenylalanylglycine (*47*), L-leucyl-L-prolylglycine (*46*) and glutathione (*52*) and the cyclic hexapeptide cyclohexaglycyl (*45*). But even though information is available for only these simple peptides, it serves to establish certain common structural features which relate to the configurations of polypeptide chains in proteins.

1. Hydrogen Bonding in Peptides

As in the case of the amino acids, all of the peptides so far studied exist as zwitterions, the proton on the carboxyl group of the C-terminal residue being transferred to the nitrogen atom of the N-terminal residue*. As a result, the N-terminal nitrogen atom, with three protons and a formal positive charge, is an excellent hydrogen-bond donor. Similarly, the C-terminal carboxylate group, with two oxygen atoms and a formal negative charge, is an excellent acceptor. In contrast, the nitrogen and oxygen atoms of the intermediate peptide groups have no formal charge and only a single hydrogen atom, and hence are much poorer donors and acceptors. Thus, the hydrogen bonding potential of small peptides is concentrated in the terminal groups. Accordingly, the inter-molecular arrangements in crystals of these compounds—including, no doubt, many details of the side-chain conformations—are dictated primarily by the arrangement of hydrogen bonds involving the end groups.

A typical hydrogen-bond arrangement, as found in crystals of α-glycyl-glycine (*40*), is shown in *Figure 8* (p. 12). The N-terminal nitrogen atom, N_1, forms three relatively strong hydrogen bonds, of average length 2.74 Å, to carboxylate oxygen atoms of three neighboring molecules. On the other hand, although the nitrogen atom of the peptide group, N_5, forms a single hydrogen bond to the carbonyl oxygen atom of a neighboring molecule, the length of this bond (2.966 Å) suggests that it is quite weak.

It appears that the formation of N—H...O hydrogen bonds between nitrogen and oxygen atoms of peptide groups, which is generally considered to be of primary importance in determining the configuration of polypeptide chains in proteins, is of only secondary importance in the crystallization of small peptides.

---

* In glutathione, γ-L-glutamyl-L-cysteinylglycine, the fusion between the glutamic acid and cystein residues involves the side-chain carboxyl group of the glutamic

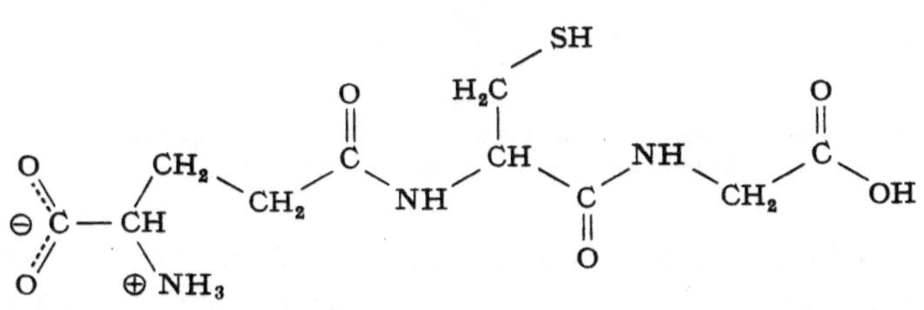

acid molecule. In this case, the zwitterion is formed entirely within the glutamic acid residue, a proton being transferred from the main-chain carboxyl group to the alpha amino group. The carboxyl group of the glycine residue retains its proton.

1 a*

2. The Geometry of the Peptide Group

A second feature of the structure of simple peptides is the coplanarity of the atoms of the peptide (or amide) group:

$$O{=}C_1{-}N \begin{matrix} C_\alpha \\ \ \\ H \end{matrix} \quad \begin{matrix} \\ C_\alpha \end{matrix}$$

This feature was first pointed out a number of years ago by Pauling and Corey (*1, 2*), and it was one of the fundamental assumptions which

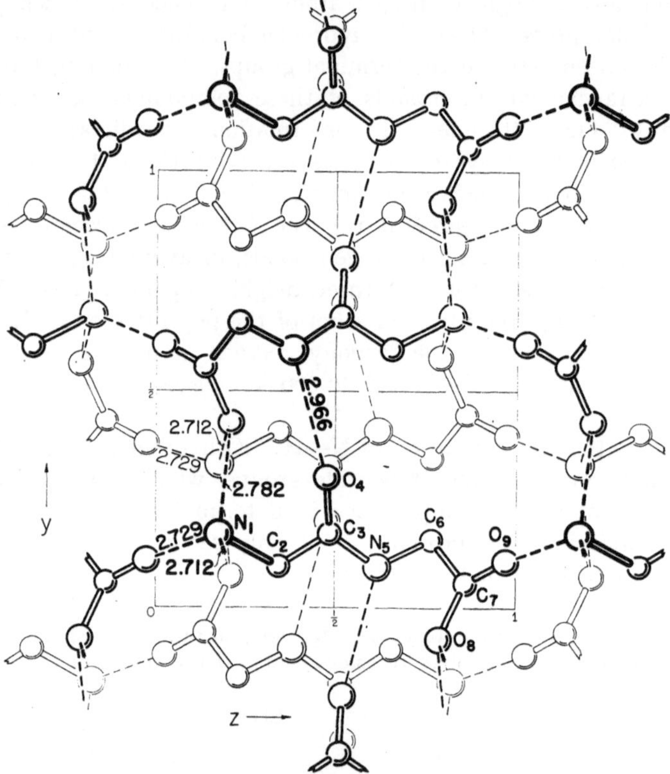

Fig. 8. The structure of α-glycylglycine, viewed down the *a* axis (*40*). Hydrogen bonds are indicated by dashed lines [Acta Crystallogr. B **24**, 40 (1968)]

led to their formulation of various stable configurations of the poly-peptide chain in proteins (*6–11*). From time to time, however, other workers have proposed structures for polypeptide chains which have required that the atoms of the peptide group be non-planar. We should like to reaffirm here that, for every peptide whose detailed structure has been determined, the peptide group has been found to be planar

References, pp. 40—47

within experimental error. In several of the more recent determinations the uncertainties in the atomic positions are 0.02 Å or less, so that it seems reasonable to believe that in general the atoms of the peptide group are coplanar within 0.02 Å.

In a paper describing the determination of the crystal structure of tosyl-L-prolyl-L-hydroxyproline (*43*), the authors state that the peptide group "is not strictly coplanar". However, as pointed out by MARSH and DONOHUE (*5*), the six atoms of the peptide group in this structure are coplanar within 0.11 Å, an amount which is comparable with the experimental uncertainties in the atomic positions. DYER (*42*) also reports a non-planar peptide group for the complex between cysteinyl glycine and sodium iodide; however, this determination was based on limited experimental data and hence details of the structure are of doubtful significance.

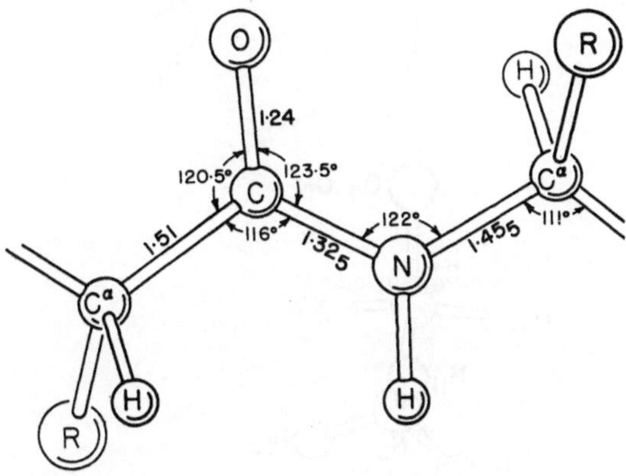

Fig. 9. Interatomic distances and angles within the peptide group. The various dimensions are weighted averages of the individual values reported in recent three-dimensional investigations of the crystal structures of simple peptides [Adv. Protein Chem. **22**, 235 (1967)]

The bond distances and angles in the peptide group are shown in *Figure 9*. These dimensions were proposed by MARSH and DONOHUE (*5*), and are weighted averages of the distances found in three-dimensional crystal-structure investigations. As pointed out by these authors, the dimensions are very close to the values originally selected by PAULING and COREY (*2, 11*).

3. Conformation of the Peptide Chain

A third structural feature of the small peptides that have so far been studied is that the polypeptide chain tends to assume a relatively extended conformation.

Typical conformations are shown in *Figures 10 and 11* (pp. 14, 15,) which are drawings of glycyl-L-asparagine (*49*) and glycyl-L-tryptophan (*48*) as viewed down the C—O bond of the peptide link, and in *Figure 12* (p. 16), which is a drawing of the crystal structure of N,N'-diglycyl-L-cystine dihydrate (*53*) viewed down the *b* axis of the unit cell.

In constructing the backbone of a polypeptide chain, the planarity of the peptide group limits the number of degrees of freedom to two per residue; these are defined by EDSALL et al. (3) as the angle Φ, describing the amount of rotation about the $N-C_\alpha$ bond, and the angle ψ, describing the amount of rotation about the $C_\alpha-C'$ bond. In a fully extended

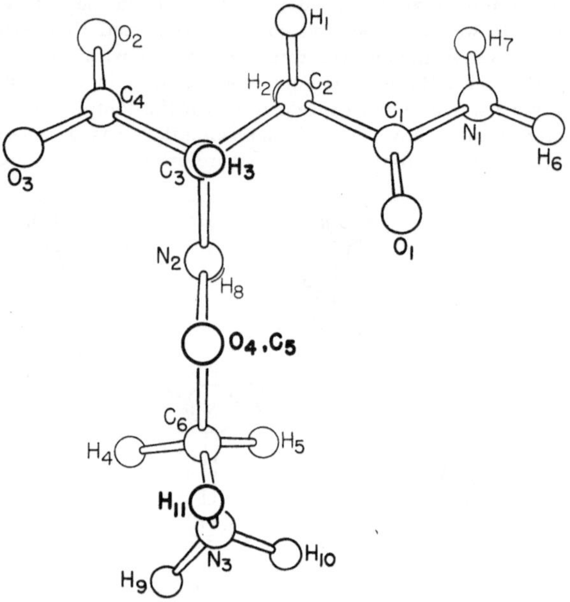

Fig. 10. The structure of glycyl-L-asparagine as viewed down the O_4-C_5 bond of the peptide group. The side-chain of the L-asparagine residue is at the upper right. The numbering of atoms is taken from (49)

conformation, both of these angles are 0°. In crystalline peptides the angle ψ has invariably been found to be quite small, so that the nitrogen and oxygen atoms of the same residue are approximately eclipsed. (This

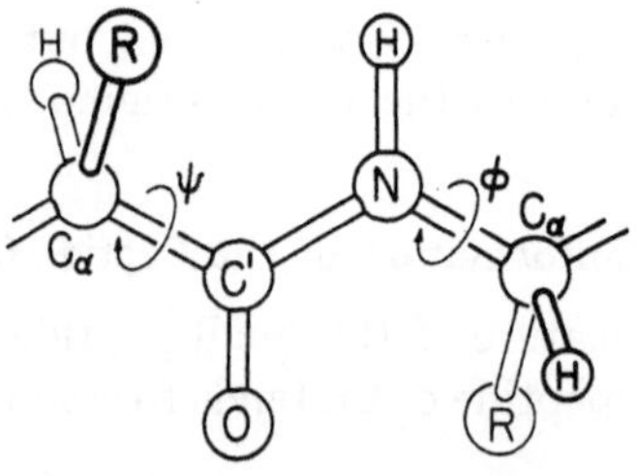

is analogous to the situation in the amino acids themselves, as discussed earlier.) The angle Φ takes on a wider range of values. It is small (around 0°) for the glycyl peptides, near 90° for peptides involving bulky side-chains (glycyltryptophan, glycylasparagine, glycyltyrosine), and very close to 120° for prolyl or hydroxyprolyl peptides, as required by packing constraints imposed by the five-membered pyrrolidine ring. *Figure 13* (p. 17) is a stereoscopic view of the L-leucyl-L-prolylglycine molecule showing the conformation about the two peptide groups.

The extended conformation of the polypeptide chain, as found in the crystal structures of simple peptides, is similar to that found in the

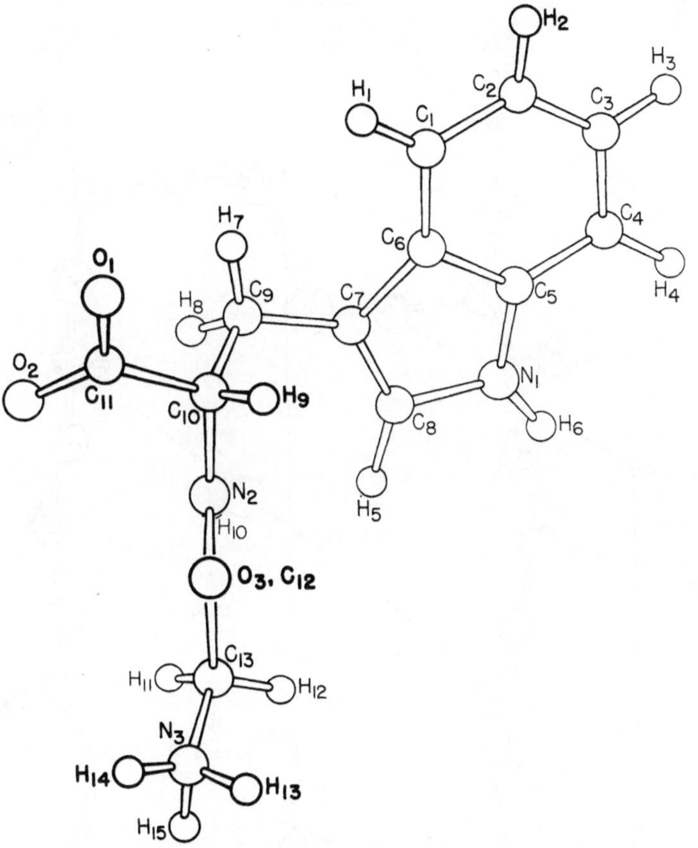

Fig. 11. The structure of glycyl-L-tryptophan as viewed down the O_3—C_{12} bond of the peptide group. The numbering of atoms is taken from (*48*)

pleated-sheet structures of the β proteins. Indeed, the arrangement of molecules in crystals of glycyl-L-phenylalanylglycine (*Figure 14*, p. 17) is very similar to the parallel-chain pleated sheet structure originally proposed by PAULING and COREY (9). Conspicuous by its absence in structures of small peptides is any evidence of the alpha helix, which is so prominent a structural feature of many fibrous and crystalline proteins. However, the formation of an alpha helix involves a hydrogen bond from the nitrogen atom of one peptide group to an oxygen atom of a peptide group

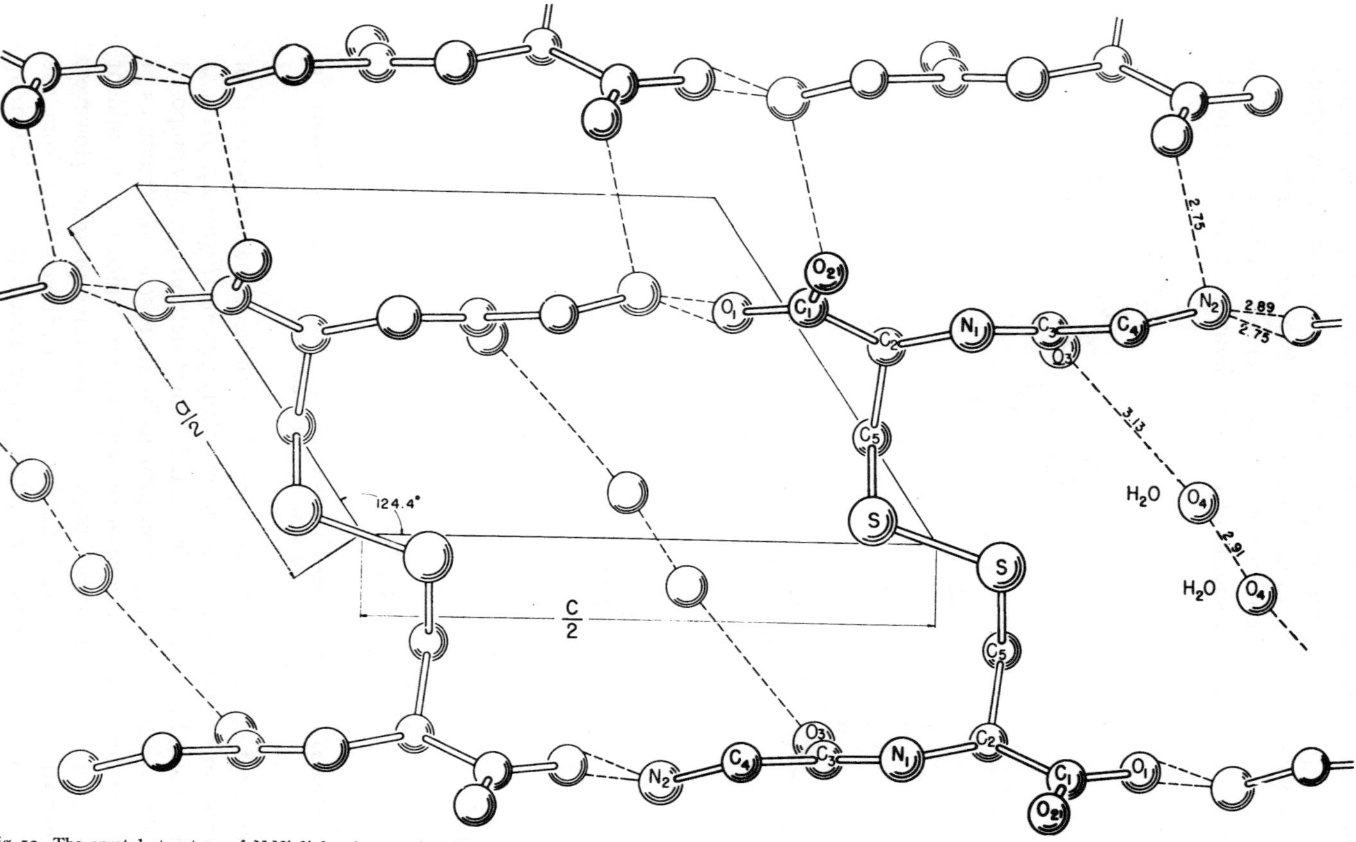

Fig. 12. The crystal structure of N,N'-diglycyl-L-cysteine dihydrate (53), as viewed down the b axis of the unit cell. Hydrogen atoms are not shown. Dashed lines represent hydrogen bonds, for which the O...O or N...O distances are given [Fortschr. Chem. organ. Naturstoffe 11, 180 (1954)]

3 groups removed along the chain; no crystalline peptides having chains of sufficient length to permit this arrangement have yet been studied. Moreover, as pointed out earlier, hydrogen bonding involving the charged

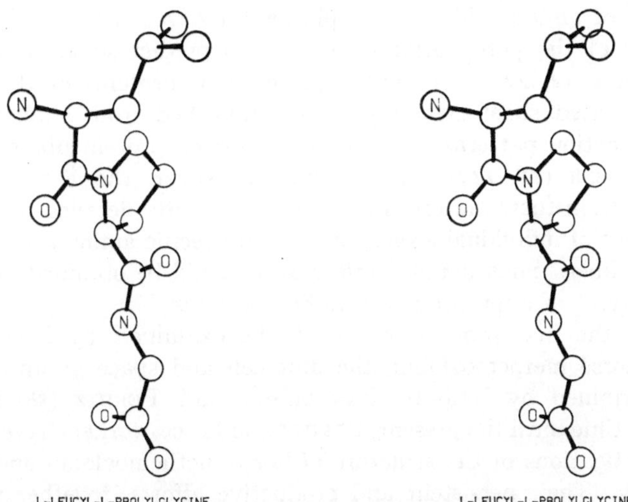

Fig. 13. A stereoscopic drawing of L-leucyl-L-prolylglycine (46). The side chain of the leucyl residue is at top right. Hydrogen atoms are not shown; carbon atoms are not labelled. (We are indebted to Drs. H. Levy, G. Brown and C. Johnson of the Chemistry Division, Oak Ridge National Laboratory, Oak Ridge, Tennessee, for assistance in preparing this figure.) [Adv. Protein Chem. 22, 235 (1967)]

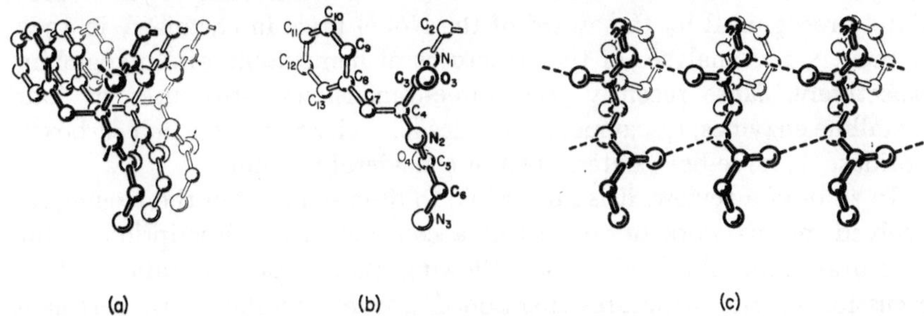

Fig. 14. Three views of the structure of glycyl-L-phenylalanylglycine (47), showing the parallel-chain pleated sheet motif of hydrogen bonding. (a) Portions of three molecules, viewed at an angle of 15° from the c axis. Hydrogen atoms are not shown. (b) The same portion of a single molecule, viewed down the c axis. (c) Three molecules viewed along a. Dashed lines represent N—H...O hydrogen bonds between peptide groups of adjacent molecules [Acta Crystallogr. 14, 1110 (1961)]

terminal groups would be expected to be far more important, in determining the conformation of a short peptide, than the hydrogen bonds formed between peptide groups. Accordingly, it is perhaps unwarranted to assume that the conformations of either the peptide chain or of the side chains of a particular protein can be deduced from a knowledge of the sequence of the constituent amino acids.

IV. Crystalline Proteins

1. Introduction

About 15 years ago, the interatomic distances and bond angles found in crystals of amino acids and simple peptides were used to derive the dimensions of the polypeptide chain and to predict some of its stable configurations (*2, 11*). The actual presence in proteins of the α-helix and the pleated sheet structures was established from studies of the X-ray diffraction patterns of naturally occurring protein fibers, such as hair, wool, silk (*11, 111*), and the like. However, the X-ray patterns obtainable from these materials were not sufficiently detailed to establish the positions of individual atoms, or even of specific amino-acid residues, within the fiber. Such detailed information could be obtained only from X-ray analyses of appropriate crystalline proteins.

Among the first protein crystals to be examined by X-rays were those of horse methemoglobin, the unit cell and space group of which were determined by Bernal, Fankuchen and Perutz (*58*) in 1938. From that time until the present, Perutz and his co-workers have pursued their investigations of the structure of horse methemoglobin and related compounds. Their persistent and productive efforts, together with the interest and collaboration of Sir Lawrence Bragg, were largely responsible for the development of the methods and techniques which are the basis of current protein crystallography. The outstanding work of Perutz and of Kendrew, both at Cambridge University, was appropriately recognized by the award of the Nobel prize in chemistry in 1962 for their X-ray analyses of the structures of hemoglobin and myoglobin, respectively. More recently, the three-dimensional structures of four crystalline enzymes, lysozyme, ribonuclease, α-chymotrypsin and carboxypeptidase A, have been determined in considerable detail.

In so brief a review, it is impossible to discuss in detail the techniques involved in this work or to present a comprehensive description of the structural data obtained. The following paragraphs contain a short discussion of the structures mentioned above, together with pertinent references to sources of more detailed information. Some of the work now in progress on the structures of other proteins is also included.

2. Myoglobin

The protein myoglobin, which has a molecular weight of about 17 000, occurs in muscle tissue. It contains one heme group per molecule and is involved in the storage of oxygen and its transfer between the blood and the tissue cells. In 1948 Kendrew (*97*) determined the unit cells and space groups of crystals of myoglobin from horse and whale. In a later publication (*103*) he listed similar data for two crystal types of sperm whale myoglobin, a monoclinic form, type A, having the space group $P2_1$ with two molecules per unit cell, and an orthorhombic form, type B, with space group $P2_12_12_1$ and four molecules per unit cell. Intensity data from the three principal zones of both types of crystals were collected and used for the preparation of Patterson vector plots from which tentative conclusions were drawn concerning the general arrangement of the myoglobin molecules in the crystals, and speculations were made regarding the orientation and packing of the polypeptide chains (*102*). Specific information concerning the shape and internal structure of the molecule could be derived only by the preparation of a detailed plot of the electron density throughout the crystal. The necessary prerequisite for the preparation of such a plot was the determination of the phase angle associated with each of the X-ray reflections. At that time, no method had been devised by which this might be achieved.

A major break-through in the determination of protein structure was made in 1954 by PERUTZ and his group, who had been working for many years on crystals of horse methemoglobin. This break-through consisted in the discovery that the method of isomorphous replacement, which had already been used for the determination of crystal structures of simpler compounds, could also be applied successfully to proteins (*90*). Two molecules of p-chloromercuribenzoate were found to react with one molecule of hemoglobin, giving p-mercuribenzoate-hemoglobin; and although the unit-cell dimensions of crystals of this complex were the same as those of crystals of hemoglobin itself, the intensities of the X-ray reflections were changed. Crystals of hemoglobin which contained two silver ions per hemoglobin molecule were also prepared and were likewise found to be isomorphous with those of the original protein.

The changes in the intensities of the reflections for both of these isomorphous derivatives were used to determine the positions in the unit cell of the heavy mercury and silver ions. In turn, knowledge of the positions of the heavy atoms was used to assign probable phases to the X-ray reflections from the h0l planes. Thus, the technique, which had been used for many years in the determination of the structures of less complex crystals, was shown by PERUTZ and his colleagues to be applicable to protein crystals also.

Type A monoclinic crystals of sperm whale myoglobin were selected for application of the isomorphous replacement technique (*65*). When these crystals were grown from solutions containing potassium mercuric iodide, K_2HgI_4, X-ray diffraction data showed that the heavy mercuric iodide ion had diffused into the crystals, occupying specific sites without changing seriously the arrangement of the protein molecule. Similar results were also obtained by the use of p-chloromercuribenzene sulfonate, $Cl—Hg—C_6H_4—SO_3^-$ (PCMBS), the mercury atom occupying a different site from that occupied by the mercuric iodide ion. Data from these two derivatives were used to determine the signs of the h$0l$ reflections from the myoglobin crystals, which in turn were used to calculate two two-dimensional Fourier projections of the electron density, one from data extending to 6.6 Å and the other to 4 Å resolution. Unfortunately, both projections were essentially uninterpretable because of the overlapping of the structural features. It was clear that a three-dimensional plot was necessary.

Such a plot was soon obtained by KENDREW and his colleagues (*100, 66*). They prepared crystals of five different heavy-atom derivatives, which proved to be isomorphous with crystals of the native protein. Intensity data collected from these five kinds of crystals showed that each heavy-atom reagent (one of which—a substituted phenylhydroxylamine–had been designed to combine specifically with the iron atom of the heme group) occupied a different site. These data were then combined in the calculation of phase angles for each of the 400 reflections extending to a resolution of about 6 Å, and a three-dimensional electron density map was calculated.

This map showed the myoglobin molecule to consist of a number of prominent rods of high electron density lying at distances of 8 to 10 Å from their neighbors, their dimensions and distribution strongly suggesting α-helical sections of the polypeptide chain. Because of the sharp corners connecting these rods of high density, the chain could not be α-helical throughout, a reasonable estimate of its helical content being about 70%. A single disk-shaped region clearly represented the heme group. Although this three-dimensional model of the myoglobin molecule represented great progress, it was clear that information concerning further details of the structure must await a corresponding study based on X-ray data extending to much greater resolution.

In 1960, KENDREW and his colleagues (*101*) published a description of such a study based on data extending to 2 Å resolution. These data were obtained from unsubstituted sperm whale myoglobin crystals and from isomorphous derivatives containing the following heavy-atom compounds: *p*-chloromercuribenzene sulfonate, mercury diamine, and aurichloride, together with a double derivative containing the first two substituents simultaneously. The number of reflections with spacing greater than 2 Å was 9600, all of which were measured for the unsubstituted protein and for each of the derivatives. After the positions of the heavy atoms had been determined, the phases of all reflections were computed and used, together with the observed amplitudes, for the calculation of the three-dimensional plot of electron density.

This plot was the first successful demonstration of the power of the X-ray diffraction technique in revealing the structural details of a protein molecule. The course and configuration of the polypeptide chain was clearly evident, especially in the straight sections which were found to be hollow cylindrical tubes of high density. Closer examination of the density contours projected onto this cylinder revealed an α-helix having the dimensions given by PAULING and COREY (8); moreover, the side-chains were found to emerge at intervals of 100° around the helix and at intervals of 1.5 Å in the direction of its axis. In all of the helical sections, the β-carbon atom was found to be on the side of the main chain opposite the oxygen atoms of the carboxyl groups, which showed that the α-helix is right-handed. Where the chain turned a corner, its configuration was less certain; the identification of side-chains was also somewhat uncertain. The iron atom of the heme group was resolved from the nitrogens of the porphyrin ring and the structure of this group as a whole corresponded closely with expectation.

In a subsequent publication KENDREW and his colleagues (104) compared the amino-acid sequences of sperm whale myoglobin as partially determined by X-ray techniques with incomplete results of chemical analyses (86). The problem of identifying the side-chains from the contours of the electron density plot involved locating first the main chain atoms and then the atoms of the side-chains themselves. In all, 634 atoms were located in the helical regions and the heme group and 262 atoms in the side-chains, or about 75% of the atoms in the structure. In general, the composition of the sixteen tryptic peptides identified chemically agreed with the composition determined from the model.

As discussed in Section 3 below, PERUTZ and his group (123) showed that the tertiary structures of the α- and β-chains of horse hemoglobin closely resemble that of sperm whale myoglobin. In view of the crystallographic evidence that the tertiary structures of hemoglobins and of myoglobins of different species are very similar, it seemed appropriate to compare the amino-acid sequences of sperm whale myoglobin with the known sequences in the α- and β-chains of human hemoglobin, the sequences in horse hemoglobin being still undetermined. This comparison was made by WATSON and KENDREW (136). Although several correspondences (identical residues in corresponding positions in all three chains) were found, the main impression resulting from this study was that, in spite of the close resemblances in the tertiary structures, the correspondences were remarkably few.

In his Nobel lecture in 1962 KENDREW (98) outlined briefly the further refinement of the structure of sperm whale myoglobin already in progress at that time. This refinement, which abandoned the use of heavy-atom derivatives, was carried on by more conventional methods. Intensity data from about 25000 reflections extending to a resolution of 1.4 Å were collected. The phases of all these reflections were calculated from the coordinates of the atoms which had already been located, and an electron density plot was then computed from the observed amplitudes and these calculated phases. In addition to the atoms already used in the phasing, this plot revealed additional atoms and also indicated small shifts in the positions of those previously located. In the next cycle, the phases were calculated from the previous set of atoms, with corrected coordinates, and the additional atoms located by the first cycle. The improvement already attained is indicated in some of the illustrations accompanying the lecture and is discussed in a later publication (*99*).

3. Hemoglobin

Since the first X-ray diffraction study of crystals of horse methemoglobin (*58*) in 1938, investigations of the structure of this protein have been vigorously pursued by Perutz and his collaborators. In the earlier studies (*67, 68, 114*) data were collected from crystals representing several shrinkage stages and also from those in which the density of the liquid surrounding the molecules was varied by altering the salt concentration. From these data phase angles could be assigned to low order h0l reflections from which very low-resolution projections of the electron density could be calculated. These investigations provided a basis for speculations concerning the dimensions and arrangement of the protein molecules in the crystal but gave no indication of their internal structure.

The next step in the study of horse methemoglobin crystals (*115*) consisted in the estimation of the intensities of some 7000 X-ray reflections extending to a spacing of 2.8 Å. These data were used for the calculation of a three-dimensional Patterson vector diagram, a plot in which the high-density regions corresponded to concentrations of interatomic vectors in the crystal (see Chapter I, p. 2). The interpretation of this plot suggested that the hemoglobin molecule was a cylinder, 57 Å in diameter and 34 Å in height, and consisted of polypeptide chains running parallel to the base of the cylinder. The chains apparently contained a short-range fold with a prominent vector of 5 Å parallel to the chain direction; they appeared to be arranged in four layers about 9 Å apart. Additional information concerning the packing of the protein molecules in crystals of horse methemoglobin and other hemoglobins was presented in a subsequent series of papers by Bragg and Perutz (*69, 71, 72*). The first showed that the absolute intensities of 0kl reflections were only one-third of what would be expected if the polypeptide chains were straight and parallel throughout the molecule. This information suggested that the molecule might be made up of structures other than straight chains running parallel to the a-axis. In the remaining investigations of this series, the molecules in crystals of horse methemoglobin were surrounded by solutions of $(NH_4)_2SO_4$ of varying concentrations and X-ray data were collected from hO*l* and O*kl* reflections. The changes in intensity with changing salt concentration were used to arrive at a more definite picture of the outer form and dimensions of the hydrated hemoglobin molecule, about $55 \times 55 \times 65 - 80$ Å. Studies of other hemoglobins from horse and from man indicated that the dimensions of these molecules are remarkably similar.

The next series of papers by PERUTZ and his collaborators described the progressive steps which led them to the successful determination of the three-dimensional structure of the hemoglobin molecule at 5.5 Å resolution. In the first three papers of this series (*70, 73, 116*) a more accurate picture of the external shape and dimensions of the hemoglobin molecule was obtained by an extension of the methods already described in the preceding series. The next paper (*90*) described the preparation of isomorphous crystals of hemoglobin containing the heavy atoms mercury and silver and their use in determining the phase angles of the hO*l* reflections (see above, Myoglobin, p. 8). These phase angles were in turn used to calculate a projection of electron density on the (010) plane (*74*). From this projection an approximate outline of the molecule was derived. The only unexpected feature was a hollow at the center indicating that the thickness of the molecule at this point was only about 32 Å.

It was now obvious that a three-dimensional analysis would be required to provide any detailed information about the structure of the hemoglobin molecule. A primary step in this direction was taken by BLOW (*63*) in an extensive study of procedures for the determination of the phase angles of the non-centrosymmetric reflections. By the use of three mercury derivatives he was able to assign values to the phase angles of the O*kl* reflections extending to about 6 Å resolution. These phase angles were used for the calculation of projections of electron density on the (100) plane. Although these projections did not indicate specific structural features, the study as a whole showed that if the polypeptide chains were in the α-helix configuration, their paths should show up clearly if an analysis at 6 Å resolution were made in three dimensions. The results of such a three-dimensional analysis were first published by PERUTZ and his colleagues (*123*) in 1960; details of the procedure and a more thorough discussion of the structure appeared somewhat later (*76, 77*). A brief account of this work and some of its implications are presented in the following paragraphs.

Horse oxyhemoglobin or methemoglobin (molecular weight, 67000) forms monoclinic crystals having the space group C2 with two molecules per unit cell. A two-fold symmetry axis passes through the center of each molecule, which therefore consists of two identical parts. X-ray intensities from 1200 reflections and extending to a resolution of 5.5 Å

were collected from crystals of the native protein and from each of six
isomorphous heavy-atom derivatives. X-ray intensities were measured
photographically, and also by the use of a spectrometer. A phase angle
for each reflection from the native protein was derived by considering
the changes of intensity resulting from the introduction of the heavy atoms.
Finally, a three-dimensional plot of the electron density was prepared.
The absolute configuration of the molecule was determined from anomalous dispersion effects.

The four heme groups containing the iron atoms were clearly represented by four conspicuous peaks; their orientations, which had already
been determined exactly by electron spin resonance (*94*), corresponded
approximately to the flattening of the peaks. The prominent features
of the plot were the more or less cylindrical clouds of electron density
which curved about to form four structural units arranged in two identical
pairs, the members of each pair being related by the two-fold axis which
passes through the center of the molecule. In a plastic model of the
structure, these two pairs were designated as "white" and "black",
respectively. Except for minor details, the individual black unit closely
resembled the white one, and the configurations of both units were very
similar to that of the molecule in a corresponding plot of the myoglobin
structure. Clearly, these four units, which make up the hemoglobin
molecule, represent four polypeptide chains the configurations of which
correspond closely to the configuration of the polypeptide chain in sperm
whale myoglobin.

In the hemoglobin molecule, there is comparatively little contact
between the two members which make up an identical pair, but the
surface contours of the white chains exactly fit those of the black, so
that there is a large area of contact between them. The general arrangement of the four groups in the molecule is tetrahedral, with the four
heme groups lying in separate pockets on the surface of the molecule.
From the more detailed analysis of myoglobin, it was clear that the
straight segments of the chains are α-helices. The similarity of the configuration of the polypeptide chains in sperm whale myoglobin, seal
myoglobin (*127*) and horse hemoglobin strongly suggested that all hemoglobins and myoglobins of vertebrates follow the same pattern.

Comparisons of the structures of myoglobin and hemoglobin with
their corresponding amino-acid sequence data have led to some significant
conclusions. The "black" chain in the hemoglobin model has been shown
to correspond chemically to the β-chain in horse and human hemoglobins (*130*). A preliminary comparison (*77, 117*) was made of the com-
positions of the α- and β-chains of horse and human hemoglobins with
that of sperm whale myoglobin. The sequences of the amino-acid residues
in the α- and β-chains of hemoglobin were found to be similar, but by

no means identical; they could be brought into register by leaving appropriate gaps in one or the other of the chains. Similarity between hemoglobin and myoglobin was much less marked. However, 21 residues appeared to occupy identical positions in all chains and to play a vital part in the configuration of the structure. For example, all prolyl residues occurred at corners or in non-helical regions, although their presence was not necessary for turning corners.

The preceding work on the structure of hemoglobin was discussed by PERUTZ (*118*) in his Nobel prize lecture of 1962.

The reaction of horse hemoglobin with oxygen produces a change in crystal form, suggesting that it causes a major change in the structure of the molecule. Unfortunately, those crystal forms of oxygen-free or reduced horse hemoglobin which were first available were unsuitable for detailed X-ray analysis. Reduced human hemoglobin was therefore chosen for study (*113*). The investigation involved the use of three isomorphous heavy-atom compounds and led to a three-dimensional electron-density plot at 5.5 Å resolution. Although the α-chains showed no significant differences, the β-chains were displaced relative to one another by a distance of 7 Å, so that the molecule was cut across by a groove separating the two β-chains. Although it seemed likely that these structural differences were due to the oxygenation reaction, the point could be definitely settled only by solving the structure of the oxygenated and reduced forms of the same species.

A new form of reduced horse hemoglobin was discovered in which the four molecules in the unit cell are situated on two-fold axes, as in crystals of horse oxyhemoglobin (*120*). A single isomorphous mercury derivative was prepared and the positions of the mercury atoms determined in two projections along the two-fold axes of the crystal. When the distance between two mercury atoms was determined it was found to agree with the value 37.7 Å found in reduced human hemoglobin, and to differ from the 30.0 Å found in horse oxyhemoglobin. The dimensions of the hole surrounding the two-fold axis at the center of the molecule were also in agreement with the former and different from the latter structure. These results suggest strongly that the reduced forms of horse and human hemoglobin have very similar structures, which differ from the oxyhemoglobins in the arrangement of the β-chains.

Hemoglobin H is one of the abnormal human hemoglobins. Its molecule consists of four β-chains and its heme groups react with oxygen independently of each other. Crystals of oxyhemoglobin H are closely similar to those of normal human reduced hemoglobin, which suggested that crystals of the oxygenated and reduced forms of hemoglobin H have the same or very similar structures. This conclusion was substantiated by the following observation (*122*). When crystals of oxyhemoglobin H were immersed in a solution of ferrous citrate until their absorption spectrum changed to that of reduced hemoglobin, no other change in the crystals was observed; sharp reflections extended to 3 Å spacing and the unit-cell dimensions were unchanged. On the other hand, when normal hemoglobin was reduced, the crystals developed numerous cracks, and reflections from spacings less than 8 Å faded out altogether. These observations confirmed the earlier conclusions that the heme-heme interaction in normal hemoglobin is linked to the structural changes which accompany oxygenation.

In the first of a recent series of papers, Perutz (*119*) has described a model of horse oxyhemoglobin constructed by combining the three-dimensional electron-density synthesis at 5.5 Å with that of sperm whale myoglobin at 1.4 Å resolution, together with the complete sequence data for both compounds. The positions of the heme groups and of the amino acid residues in oxyhemoglobin were thus located within narrow limits. The distribution of the amino acid residues, the nature of the intra- and intermolecular forces, and the dissociation properties of hemoglobin are discussed in the light of this model. A second paper in this series (*121*) discusses the relations between the configuration of the polypeptide chain and the sequence of the amino acid residues. The most prominent common feature is the almost total exclusion of polar residues from interior sites on the globin molecule. Polar amino acids tend to repeat at regular intervals of about 3.6 residues, and thus are concentrated along one side of the surface of the α-helix. The opposite surface, which faces towards the center of the molecules therefore consists primarily of non-polar residues. At the external surface many different kinds of replacements seem to occur without affecting the tertiary structure. These include replacement of polar by non-polar residues and vice versa.

The collection of X-ray data from crystals of horse hemoglobin to a resolution of 3 Å has been begun.

4. Lysozyme

The protein lysozyme (molecular weight, 14600) is an enzyme which breaks down the cell walls of certain bacteria. It is widely distributed in the fluids and tissues of the human body, and in some animals and plants. The best source of lysozyme is egg white.

An X-ray crystallographic study of the structure of lysozyme was begun several years ago at the Royal Institution in London by Phillips and his collaborators. Their first publication (*59*) described a three-dimensional Fourier plot of electron density distribution in the crystal. This plot was calculated from diffractometric measurements of the intensities of reflections extending to 6 Å resolution and included data from three isomorphous heavy-atom derivatives. Measurements of anomalous dispersion were used to distinguish between two enantiomorphous space groups. Although the boundaries of the molecule could not be definitely established at this resolution, it appeared to be roughly ellipsoidal in shape. Rod-like features representing helical configurations were far less prominent than in corresponding plots of myoglobin and hemoglobin.

A subsequent paper (61) describes the results of a three-dimensional analysis of lysozyme at 2 Å resolution. Three isomorphous heavy-atom derivatives were used, one taken from the preceding analysis, the other two being new derivatives which yielded improved intensity data. Each set of data comprised more than 9000 reflections. In each of the derivatives, the heavy atoms occupied more than one site. The three-dimensional plot showed a continuous ribbon of high electron density representing the main polypeptide chain together with protruding side-chains. About six portions of the main chain appeared to be α-helical. The side-chains were most easily identified in these helical regions, but were also clearly recognizable in some other portions of the chain. For the complete indentification of side-chains, the chemical sequence data were most helpful; no significant discrepancies were observed.

The determination of the structure of the lysozyme molecule has thrown considerable light on the mechanism of its enzymatic action. In its attack on bacteria, lysozyme hydrolyzes the mucopolysaccharide component of the cell walls, releasing N-acetyl amino sugars. Preliminary clues to the mechanism of this action were obtained by JOHNSON and PHILLIPS (95) through the preparation of a three-dimensional difference Fourier plot calculated from 6-Å X-ray data obtained from crystals of lysozyme chloride and of lysozyme chloride inhibited by N-acetylglucosamine and related molecules. The resulting electron-density maps contained peaks which indicated that the inhibitor molecule was embedded in a crevice in the surface of the enzyme, which was thus indentified as the location of the active site.

Additional details concerning the structure of the lysozyme molecule as determined from the 2 Å data have recently been published by PHILLIPS and his collaborators (62). The construction of a molecular model helped to clarify the interpretation of the three-dimensional electron-density plot. The helical regions are restricted to three which contain about 10 residues each and one which is shorter. At one point in the molecule the chain doubles back on itself to form an antiparallel-chain pleated sheet. As in other protein structures, the acidic and basic side-chains and most other polar side-chains are distributed over the surface of the molecule, whereas the great majority of the non-polar side-chains are in the interior. Each protein molecule is in contact with its neighbor over only a small fraction of its surface. It appears probable that the conformation of the molecule in the crystal is essentially uninfluenced by its neighbors and is therefore the same as its conformation in solution.

In a related paper (*60*) a correspondingly detailed discussion is also given of recent structural studies of the enzymic activity of lysozyme. It appears that important constituents of the cell walls of bacteria which are lysed by lysozyme are long-chain mucopolysaccharides composed of alternate units of N-acetylglucosamine (NAG) and N-acetylmuramic acid (NAM) joined by β-(1-4) linkages, and that the enzyme attacks the linkage between NAM and NAG. The investigations of the mechanism of enzymatic activity by the use of difference-Fourier methods at 6 Å resolution described above were extended by the use of other inhibitors. The resulting plot of electron density showed clearly that all inhibitor molecules were bound within the cleft, and that they occupied a number of different sites. In particular, tri-NAG was bound stably, covering three apparent binding sites. Plots at 2 Å resolution showed that the interactions between NAG molecules and the enzyme consist of hydrogen bonds between the NH and carbonyl oxygen atoms of the N-acetyl side groups and the CO and NH groups of the main polypeptide chain. There were also clear indications that the conformation of the enzyme changes slightly when the inhibitor is bound. When a fourth NAG molecule is added to the trimer, it also makes reasonable contacts with the enzyme, and even if a fifth and sixth molecule is added, good interactions with the enzyme suggest themselves. There are indications that the enzyme promotes hydrolysis between the fourth and fifth residue. This conclusion is also in agreement with some chemical observations.

The work described above is also discussed in a recent paper by PHILLIPS (*124*). Further crystallographic and model-building experiments are in progress.

Crystals of lysozyme chloride are also being used in an investigation of the possible usefulness of heavy-atom complex ions such as $Ta_6Cl_{12}^{++}$, $Nb_6Cl_{12}^{++}$, PtI_6^{--}, and the like, for the determination of the detailed structure of a protein molecule (*75, 131, 132*). The ions $Ta_6Cl_{12}^{++}$ and $Nb_6Cl_{12}^{++}$ are practically identical in dimensions (*135*), so that they might be expected to form completely isomorphous derivatives when incorporated into protein crystals. This has indeed been shown to be the case (*75*). The outstanding characteristic of these complex ions is their very strong scattering power for X-rays, so that there should be little difficulty in determining their positions, even in crystals of biological molecules having relatively large molecular weights—100000 or more.

At a resolution of 5 Å the orientation of the ions has little if any effect on their scattering powers (*75, 132*). Consequently, at such resolution, the positions of these ions in crystals of lysozyme chloride can be determined with relative ease and definiteness. On the other hand,

References, pp. 40—47

early experiments showed that the presence of these large ions produced a significant change in the arrangement of the protein molecule within the crystals.

In later investigations (*131*), it became clear that even at a resolution of 4 Å, information concerning the orientation of these large ions would be essential in order to arrive at a realistic estimate of the X-ray scattering by the protein molecule itself. This information was derived by calculating the X-ray scattering by the ions corresponding to a systematic series of orientations and comparing the differences between these calculated values obtained for the different ions with the observed differences in the X-ray intensities obtained from the corresponding crystal derivatives. When these calculations were extended to a resolution of 2 Å, conspicuous minima in these differences were found to be associated with quite definite orientations of the ions. Assuming these orientations of the ions, a three-dimensional Fourier plot of the electron density in the crystal was calculated from data extending to 1.8 Å. Although the resulting plot strongly suggested the course of a polypeptide chain, no unambiguous picture of the detailed structure of the lysozyme molecule could be derived from it.

All of the X-ray intensity data upon which the final plot was based were derived from visual estimation of the intensities of spots on Weissenberg X-ray photographs. Inaccuracies in the contours of the plot doubtless reflect the inaccuracies in the X-ray data, and the consequent errors in the estimation of the orientations of the ions. An automatic X-ray diffractometer is now being used for the collection of a completely new set of intensities. It is hoped that the use of these new and more accurate data will either establish the usefulness of the heavy-atom complex ions in protein structure analysis or show quite definitely that they are without value for this purpose.

5. Ribonuclease

The structure of the enzyme ribonuclease (molecular weight, 13 683) has been under investigation by X-ray crystallographic methods for many years. Two papers have recently been published in which the structure of the ribonuclease molecule has been discussed in some detail, one by CARLISLE and his colleagues (*57*) at Birkbeck College, London, and the other by HARKER and collaborators (*96*) at the Roswell Park Memorial Institute, Buffalo, New York. Like those of myoglobin, crystals of ribonuclease are monoclinic, space group $P2_1$, with two molecules in the unit cell. Each molecule consists of a single polypeptide chain composed of 124 amino-acid residues which include eight cysteines, the latter being cross-linked to form four S—S bridges between different portions of the molecule.

The paper by Carlisle and his group (57) describes the preparation of a three-dimensional electron density map of the ribonuclease molecule obtained from X-ray data collected with a diffractometer and extending to 5.5 Å resolution. Five isomorphous heavy-atom derivatives were used. In all but one derivative the heavy atoms occupied multiple sites. This map shows the general outline of the molecule and indicates the presence of a cleft in its surface. A difference Fourier map calculated from data obtained from crystals of the enzyme inhibited with 2'-cytidylic acid showed that the inhibitor was situated in the cleft. One of the cystine bridges was identified by means of a chemically modified protein, and likely positions were found for the other three bridges. An approximate arrangement of the polypeptide chain in the ribonuclease molecule was suggested.

In a note added in proof, Carlisle (57) stated that the model described above differs from the one described by Harker (see below) in two respects. First, there are differences in the electron density maps themselves, and, second, there are differences in interpretation of the parts of the maps which are the same. The resolution of these differences must await further progress in the determination of the detailed structure of ribonuclease.

In the investigation carried out by Harker and his group (96), crystals of ribonuclease were grown in the presence of a phosphate buffer. An X-ray diffractometer was used for the measurement of intensities of reflections from the native protein and from seven isomorphous heavy-atom derivatives. A three-dimensional Fourier synthesis at 3 Å resolution showed the position of the amino end of the polypeptide chain, a depression on the surface of the molecule where the phosphate ion is located, and the positions of three of the four S—S bridges. The 2 Å map was based on more than 7200 reflections from each of the seven derivatives. In this map it was possible to trace the main polypeptide chain from the amino end to the carboxyl end. The positions of bulky side-groups and the disulfide bridges served as good checks when adjustments were needed. No attempt has yet been made to check the correctness of the chemical sequence by identifying the side-groups from X-ray data alone.

The molecule is roughly kidney-shaped with a deep depression in one side. The helical content is extremely low, the only obvious helical segment consisting of about two turns near the amino end of the chain. As in other proteins, polar side-chains are on the outside of the molecule.

There is much chemical evidence indicating that the active site is near the location of the phosphate ion. Possible information about the position of the active site was therefore obtained from the location of the phosphate group. This was derived from a difference map computed from a phosphate and an arsenate derivative. It showed clearly that the phosphate group is embedded in the depression of the kidney-shaped surface of the molecule. The amino-acid residues closest to the phosphate group are two histidines; slightly further away are two lysines and a third histidine. A more detailed description of the ribonuclease molecule is in preparation.

The structure of ribonuclease-S has recently been determined by Wyckoff, Richards and their collaborators at Yale University. This enzyme is formed by the hydrolysis of the peptide bond between residues 20 and 21 or 21 and 22 of bovine pancreatic ribonuclease-A. The determination of its structure to a resolution of 3.5 Å is described in a series of three articles.

The first article (*137*) describes the design of a diffractometer and flow cell system which was used for the collection of the X-ray intensity data and discusses some of its applications to the crystal chemistry of ribonuclease-S. The cell consists of a small polyethylene tube held in a brass yoke which is firmly attached to a standard goniometer head. The crystal is embedded in the tube in cotton linters or particles of Sephadex to keep it from moving, and is surrounded by a changeable liquid medium during the X-ray diffraction studies. In addition to the collection of X-ray intensity data from heavy-atom derivatives used in phasing the reflections, this flow cell system was also used for studies of diffusion rates of solvents and inhibitors into and out of the crystal, the measurement of pH effects on cell constants, the determination of curve of binding of iodinated inhibitors as a function of external concentration, the comparison of binding sites of various inhibitors, and other significant observations.

The second aricle (*139*) describes the determination of the structure of ribonuclease-S at a resolution of 6 Å. The crystals used were trigonal, space group $P3_121$, with 6 molecules per unit cell. Three-dimensional data were collected to 6 Å resolution from crystals of (a) the enzyme, (b) the enzyme plus a uridine phosphate inhibitor, and (c) the enzyme plus an iodinated form of the inhibitor. These data were used for calculations which culminated in electron-density projections, down the two-fold axis, for all three types of crystals. The projections showed clearly the positions of the uridylate moiety, the iodouridylate moiety and the iodine atom. By the use of data from a uranyl derivative and anomalous scattering, three-dimensional electron density maps were obtained at 6 Å and at 3 Å resolution. The 3 Å map showed much of the backbone of the polypeptide chain and some side-chains, but in other areas the chain was lost and over-all interpretation was impossible. When data from two additional platinum derivatives were incorporated, the 6 Å map was considerably cleared up so that the molecule was clearly outlined over most of its surface and appeared to contain three short lengths of helix. Comparison indicated that this 6 Å structure was in essential agreement with the structure of ribonuclease-A derived by HARKER and his group. The models were very similar and the phosphate binding site in ribonuclease-A was very close to the nucleotide inhibitor site in ribonuclease-S.

The third article (*138*) describes the extension of the work to 3.5 Å resolution. The necessary additional data were collected from the same three heavy-atom derivatives. The resulting three-dimensional electron density map was clearly interpretable in terms of the backbone of the polypeptide chain. All of the disulfide bridges were seen, as were all of the methionine residues. The large side-chains, known from the sequence, fell into appropriate regions of high density. All ring structures were visible, but in most cases no flattening was apparent to reveal the orientation. In all, about 70 to 80% of all side-chains could be seen. The three sections of helix were found to include about 15% of the residues. As in other crystalline proteins, the charged residues were on the surface or in the liquid of crystallization, and a hydrophobic core included about 15% of the molecule. The S-peptide (residues 1 to 20) occupied about one-third of the surface of the hydrophobic core. The terminal residues 20 and 21 were separated by about 10 to 15 Å. A long twisted and bent antiparallel pair of chains was also present in the structure. The location of the active center was implied by the binding site of uridylic acid which was situated in a groove near two histidine residues. Many other details of the structure were evident and are discussed in the publication. The only marked difference in the polypeptide chains in ribonuclease-A and ribonuclease-S was the separation of residues 20 and 21 in the latter. The positions of the bound phosphate and arsenate in ribonuclease-A appeared to be consistent with the location and orientation of the inhibitor in ribonuclease-S.

6. α-Chymotrypsin

The alpha form of the enzyme chymotrypsin (molecular weight, about 25 000) forms monoclinic crystals, space group $P2_1$, with four molecules in the unit cell (*58*), two per asymmetric unit. The structure of α-chymotrypsin has been under investigation by BLOW and his collaborators at Oxford University for several years.

First, a low-resolution (5.8 Å) study (*64*) was made of crystals of α-chymotrypsin containing planar complex haloplatinite ions. Application of the Rossmann-Blow rotation function (*125*) showed that the two molecules in each asymmetric unit were related by a non-crystallographic two-fold rotation axis perpendicular to the $b c$ plane of the crystal.

In a subsequent investigation (*129*), crystals of α-chymotrypsin were treated with various permanent inhibitors containing sulfur, mercury and iodine atoms and analogous inhibitors containing light atoms only. Phases of the reflections were determined by the use of isomorphous derivatives containing $(PtCl_4)^{2-}$ ions. Difference Fourier projections of electron density calculated from data extending to 2.8 Å resolution were used to locate the positions of the heavy atoms in the different inhibitor molecules. All the inhibitors were found to occupy the same sites. These sites, which were presumably the regions of catalytic activity, lay close to the non-crystallographic two-fold axis which related the two molecules in the asymmetric unit in the crystal.

In a recent paper, BLOW and his colleagues (*112*) have described a determination of the detailed structure of an inhibited derivative of α-chymotrypsin based on X-ray data extending to 2 Å resolution. Tosyl-α-chymotrypsin was used in this investigation rather than the native enzyme in order to take advantage of its close isomorphism with the corresponding pipsyl derivative and also because of its better binding to phenylmercuric acetate (PMA). In addition, it provided an unambiguous identification of the serine residue in the active site. The X-ray diffraction data were recorded on precession photographs of α-chymotrypsin and of each of four derivatives. About 66500 reflections were measured for each derivative, representing about 90% of all reflections with a spacing of more than 2 Å. (The PMA-tosyl data were limited to 2.8 Å resolution.) After averaging the values for equivalent reflections and including anomalous scattering measurements, the number of independent intensities derived from each of the derivatives containing a heavy atom was about 35000. Positions of the heavy-atom sites were derived from difference Patterson and Fourier projections and the phase angles of 24500 structure factors were determined. A three-dimensional Fourier synthesis of electron density was then calculated from which a three-dimensional map of the structure was plotted.

This map confirmed the general arrangement of the molecule previously deduced from the low-resolution studies. The structure can be visualized as a series of long chains of molecules running parallel to the *c*-axis. In the vicinity of the non-crystallographic two-fold axis the two molecules in the asymmetric unit are very nearly identical, but the correspondence deteriorates at greater distances. In the interpretation of the map, four cystine bridges could be readily recognized and chains of density could be followed which corresponded in length to the A, B and C chains of chymotrypsin. With the exception of a short section of α-helix, the chains tend to be fully extended, and often run parallel to one another. The four terminal groups which arise in the enzyme from the hydrolysis of the zymogen are in positions consistent with a splitting off of peptides from the surface of a zymogen molecule having the same general structure. Although many of the bulky side-chains

could be clearly identified, there were many others which required the
aid of the chemical sequence data. Like other protein molecules whose
structures have been determined, almost all of the charged side-groups
are situated on the surface and the interior is composed largely of interacting hydrophobic groups attached to different chains.

Unlike lysozyme or ribonuclease, there is no pronounced cleft in
the surface of the molecule. However, in the vicinity of the active serine
there are open regions on the surface of the molecule where the structure
is not closely packed. Both serine 195 and histidine 57 extend into one
of these open regions, their locations giving support to the accepted view
that they exert a concerted action in the catalytic mechanism.

A more complete study of the results of this investigation is in progress.

As a part of their investigation of the structure of chymotrypsinogen,
Kraut and his collaborators (*107*) have collected X-ray data to 5 Å resolution from crystals of π-, δ- and γ-chymotrypsin. These data were used
by them to prepare difference Fourier plots containing structural information regarding the transformation of the zymogen into the different
active forms of the enzyme. They are discussed in Section 8 c.

7. Carboxypeptidase A

During the last few years, excellent progress has been made by
Lipscomb and his coworkers at Harvard University on the determination
of the structure of bovine pancreatic carboxypeptidase A (molecular
weight, 34600), an enzyme which catalyzes the hydrolysis of both peptide
and ester bonds. In its native state it is complexed with a single zinc
atom which can be replaced by atoms of other metals. Crystals of carboxypeptidase A are monoclinic, space group $P2_1$, with two molecules in the
unit cell.

The work of Lipscomb and his collaborators is presented in a series
of four papers (*93, 108–110*). The first three papers of this series (*93,
108, 110*) describe work which culminated in the derivation of a three-
dimensional map showing the structure of the enzyme molecule at 6 Å
resolution. The electron density map described in the last of these three
papers (*108*) was prepared from data derived from the native protein
and four heavy-atom derivatives, in three of which the heavy atom occupied multiple sites. Anomalous scattering was used to determine the
correct enantiomorph. The boundaries of the molecule were unusually
clear. All heavy atoms in the derivatives appeared to be near the surface
of the molecule. The different portions of the molecule were made up
of either long, straight, dense sections about 5 to 6 Å in diameter or
more tortuous and complicated regions. The former were interpreted as
representing six or seven helix-like regions containing about 75 residues,
or about 25% of the molecule.

References, pp. 40—47

The zinc atom, which has been considered to be situated at the active site, was located from a 6 Å three-dimensional difference electron-density map derived from crystals of the native enzyme and the apoenzyme (the enzyme without the zinc atom). It was found to lie in a depression in the surface of the molecule and adjacent to a pocket which extends into the molecular interior. A scale model of the typical substrate glycyl-L-phenylalanine could be placed on the model of the enzyme so that the aromatic group was in the pocket and the carboxyl oxygen of the peptide bond to be split was near the zinc. This enzyme-substrate model is consistent with the observation that carboxypeptidase A efficiently hydrolyzes substrates having hydrophobic side-chains.

In the fourth paper LIPSCOMB and his colleagues (*109*) give a preliminary report of the results obtained from two new three-dimensional electron-density maps prepared from data extending to 2.8 Å resolution. In the first of these maps, two of the conspicuously dense regions which suggested helical structures on the 6 Å map were readily identified as right-handed helices in which the number of residues per turn and the pitch of the helix agreed well with the parameters expected for an α-helix. The amino-to-carboxyl direction of the polypeptide chain was also clearly in agreement with the conclusions drawn from the 6 Å map. The second 2.8 Å map included data from a new mercury derivative which considerably increased its clarity and ease of interpretation. The continuity of electron density along the polypeptide chain, especially in non-helical regions, and also the connection of the side-chains to the main chain were improved over the same features in the earlier 2.8 Å map.

In the same paper (*109*), three-dimensional X-ray studies of crystals of the apoenzyme complexed with the substrate glycyl-L-tyrosine have also been reported. A difference map at 6 Å resolution was calculated from data from the apoenzyme and from the apoenzyme-substrate complex. The resulting peak which corresponds to the position of the substrate has its maximum about 4–5 Å from the zinc position and extends into the adjacent pocket in the molecule. The relatively low enzymic activity of crystals of carboxypeptidase A permitted the preparation of crystals of the native enzyme complexed with the substrate glycyl-L-tyrosine. These crystals were isomorphous with those of the native enzyme itself. The position of the substrate in these enzyme crystals was found to be only a few Ångstroms from that of the same substrate in crystals of the apoenzyme.

All of these investigations are being continued.

8. Some Other Proteins

Work on the structure of many other crystalline proteins is in progress in laboratories throughout the world. Much of this work has been described in publications; several examples are cited in the following brief discussion.

a. Insulin

X-ray diffraction data from rhombohedral crystals of insulin were first published by Dorothy Crowfoot (*78*) in 1938. More recently, Low and her coworkers have carried out several studies of crystals of insulin sulfate in which attempts were made to derive structural information about the arrangement of the molecules, primarily from Patterson functions (*87, 128*). Within the last year, two papers (*82, 92*) have been published by Dorothy Crowfoot Hodgkin and her collaborators at Oxford University. These papers describe a continuation of the earlier study (*78*) of rhombohedral insulin crystals. The crystals, space group R3, are of two types, designated as 2 Zn and 4 Zn to indicate the ratio of the number of zinc atoms to the six insulin monomers (molecular weight, 5778 each) in the rhombohedral unit cell. Wet crystals of both types have been used for the collection of X-ray data to a spacing limit of 2.2 Å, and the corresponding three-dimensional Patterson functions have been calculated. An interpretation of these functions indicates the presence of a two-fold axis perpendicular to the crystallographic c-axis of the hexagonal unit cell. The molecular arrangement in 2 Zn and 4 Zn insulin crystals has been shown to be very similar. Additional work, including the preparation and study of heavy-atom derivatives, is already well advanced and will be described in subsequent publications.

b. Cytochrome c

Considerable progress has already been made in an X-ray diffraction study of the structure of horse heart ferricytochrome c (molecular weight, 12400) which is currently being carried on by Dickerson and his collaborators at the California Institute of Technology. A preliminary paper (*79*) describes the preparation and interpretation of a two-dimensional electron-density projection derived from data extending to 4 Å resolution for two isomorphous heavy-atom derivatives. The position and general dimensions of the molecule are indicated and some conclusions regarding the distribution of hydrophobic and hydrophilic groups are drawn. Much more information has now been derived from a 4 Å three-dimensional electron density plot based on data from the same derivatives (*80, 81*). A cleft crevice along one side of the molecule contains the heme group, the plane of which is normal to the surface of the molecule with only one of its edges exposed to the solvent. One of the iron ligands has been identified as the imidazole side-chain of a histidyl residue. The molecule contains little or no α-helix, but is composed of a shell of extended polypeptide chain surrounding a core of hydrophobic side-chains. Further refinement of this structure is now in progress.

References, pp. 40—47

c. Chymotrypsinogen

The structure of chymotrypsinogen (molecular weight, 25000) has been under investigation by KRAUT and his collaborators for several years, first at the University of Washington and more recently at the University of California in San Diego. In a preliminary paper (*106*) they described the preparation of a three-dimensional electron-density plot of the molecule of bovine chymotrypsinogen A from X-ray data extending to 5 Å resolution. These data were obtained from six heavy-atom isomorphous derivatives, each heavy atom occupying more than one site. The plot indicated that the conformation of the polypeptide chain was complex, but included little if any α-helix. A similar analysis extending to 4 Å resolution (*105*) yielded no more detailed information concerning the structure of the molecule with the exception of its absolute configuration, which was determined from anomalous scattering effects.

In a recent publication, KRAUT and his coworkers (*107*) have used three-dimensional Fourier plots at 5 Å resolution to investigate the structural changes involved in the transformation of chymotrypsinogen into π-, δ- and γ-chymotrypsin. First, a new 5-Å electron-density map of chymotrypsinogen was calculated from selected portions of the data used in the preceding study (*105*) and the corresponding molecular model was constructed. A similar electron-density map was also calculated from data from inhibited crystals of δ-chymotrypsin. A comparison of the molecular models of the zymogen and the enzyme indicates a striking similarity in their general configuration, together with two conspicuous differences. A continuous portion of the zymogen chain shows a distinct break at one point in the model of the δ enzyme, doubtless representing the severing of the polypeptide chain which is responsible for the enzymic activation. In addition, the activation to the δ enzyme appears to cause a portion of the polypeptide chain to swing outwards, forming a pocket within which certain inhibitor molecules are bound, as shown by subsequent experiments. Difference-Fourier maps prepared from isomorphous crystals of the inhibited π, δ and γ enzymes and also of the active γ enzyme locate the activation dipeptide serine-arginine, the second dipeptide threonine-asparagine associated with the transition from the δ to the α and γ enzymes, and the serine residue at the active site. This work is now being extended in an effort to obtain a complete and unambiguous structure for the entire polypeptide chain.

d. Papain

Work on the structure of crystals of the enzyme papain (molecular weight, 22000) has been carried on by DRENTH and JANSONIUS and their collaborators (*83, 84*) at the University of Groningen. A preliminary study indicated that mercuripapain crystallizes in the monoclinic space group P2 with two molecules in the unit cell. More recently (*84*), work has been started on the determination of the structure of orthorhombic crystals of papain C. Four isomorphous heavy-atom derivatives have been prepared and the coordinates of the heavy atoms determined. In three of these derivatives the atoms occupy more than one site. Photographic X-ray intensity data to 5 Å resolution collected around the a and c axes were reduced to the absolute scale and used for the preparation of two-dimensional Fourier projections of the electron density. These projections provided little if any structural information.

Work on the structure of papain C has now been extended to three dimensions (*85*) by the use of data to 4.5 Å resolution obtained from the native protein and three heavy-atom derivatives. Measurements of anomalous dispersion also provided information concerning the absolute configuration of the structure. On the three-dimensional plot, individual molecules can be separated quite easily. The molecular structure appears to be very complex and to include only a small number of short pieces of α-helix. Two parts of the polypeptide chain come close together at many places, some of which must represent disulfide bridges. Since it is impossible to select its direction in the vicinity of these intersections, the course of the polypeptide chain cannot be followed throughout the molecule. The —SH group, known to be part of the active site, has been located in a slight depression on the surface of the molecule. Work on this structure is now being extended to 2.8 Å resolution.

e. Carbonic Anhydrase

Preliminary investigations of the structure of crystals of carbonic anhydrase (molecular weight, 30000) have been made by Strandberg and his collaborators at the University of Uppsala. The crystals are monoclinic, space group $P2_1$, with two molecules in the unit cell (*133*). Each molecule of human carbonic anhydrase contains a single sulfhydryl group and is associated with one atom of zinc. Several isomorphous derivatives, containing primarily mercury atoms, were prepared from the native and the zinc-free enzyme (*134*). X-ray intensity data extending to 5.5 Å resolution were used to determine the positions of the mercury atoms and the degrees of occupancy of their single sites. The positions of the zinc atom and the sulfhydryl group were indicated. Data from the native enzyme and from four different heavy-atom derivatives were used to prepare a two-dimensional projection of the electron density.

Work on this structure has now been extended (*88*) to the preparation of a three-dimensional electron density plot at 5.5 Å resolution of the enzyme itself and of an enzyme-inhibitor complex. X-Ray intensity data from crystals of the native enzyme and of four mercury-containing derivatives were used. These derivatives included mercury compounds of the zinc-free enzyme and of the enzyme complexed with two different inhibitors. Only in the single derivative of the native zinc-containing enzyme did the mercury atom occupy a single site; the three other derivatives involved multiple sites. The electron density map of the native enzyme lead to a plausible model of the enzyme molecule. The zinc atom is situated near the center of the molecule in a large cavity. Interpretation of the model resulted in a proposed folding of the polypeptide chain and a definite identification of one of its ends. Several straight sections involving about 30% of the residues might be interpreted as α-helices. The single —SH group is situated on the surface of the molecule near the identified end of the polypeptide chain. The electron density map of the enzyme-inhibitor complex showed no change in the folding of the polypeptide chain. The inhibitor molecule occupies a rather narrow slit in the surface of the enzyme. This work is being continued in an effort to push the resolution as far as possible.

f. Lactic Dehydrogenase

Crystals of dogfish muscle lactic dehydrogenase are being investigated by ROSSMANN and his coworkers (*126*) at Purdue University. The molecule is composed of four polypeptide chains, each of about 35 000 molecular weight. The crystals are tetragonal, space group I 422, with one polypeptide chain per asymmetric unit. The four chains which compose the molecule are therefore related to each other either by a four-fold rotation axis or by three mutually perpendicular two-fold axes. The results of model building strongly favor the latter. Two useful heavy-atom derivatives have been obtained. X-ray intensity data extending to 4.5 Å resolution were used to derive the coordinates of the heavy atoms. The general distribution of the intensities with scattering angle suggests that the molecule of dogfish lactic dehydrogenase contains a very significant amount of α-helix. This observation is confirmed by the results of optical rotatory dispersion measurements.

g. β-Lactoglobulin

At the National Institute for Research in Dairying, University of Reading, England, ASCHAFFENBURG and DREWRY (*54*) showed in 1955 that individual cows produce either one or the other of two electrophoretically distinct β-lactoglobulins (then termed β_1- and β_2-lactoglobulins) or a mixture of the two. Crystals grown from mixed milk, and therefore composed of both forms, had been examined by X-rays by several workers (*91*). They were found to be orthorhombic, space group P$2_1 2_1 2_1$, with eight molecules (molecular weight, about 35 000) per unit cell. Investigations of the individual types, β_1 and β_2, have been made by Green and Aschaffenburg and their collaborators at the Royal Institution, London. In their first study (*91*) monoclinic crystals of β_2-lactoglobulin were obtained which were transformed on standing into rectangular crystals similar in appearance to those of β_1-lactoglobulin. Comparison of cell dimensions indicated that both types were very nearly isomorphous with those obtained from mixed milk. Although the intensity patterns of both types of crystals showed distinct differences, they were still so closely similar as to indicate that β_1- and β_2-lactoglobulis were almost identical proteins. In a succeeding publication (*55*) the nomenclature of these proteins was changed so that the β_1- and β_2-lactoglobulins are now designated β-lactoglobulins A and B, respectively.

A cadmium derivative of β-lactoglobulin B and a p-chloromercuribenzoate derivative of β-lactoglobulin A were prepared and investigated by X-ray diffraction (*89*). This study showed that in each type the molecule of molecular weight 35000 was composed of two similar subunits. More recently, several additional heavy-atom derivatives of the β-lactoglobulins have been prepared and their characteristics determined (*56*). The symmetry of the new crystal forms strongly supports the view that the two polypeptide chains of β-lactoglobulin are identical and that the molecule has two-fold symmetry. Work is being continued with these new derivatives.

References

Nos. 1—11 refer to general literature, nos. 12—39 to amino acids, nos. 40—53 to peptides and nos. 54—139 to proteins.

General

1. Corey, R. B.: X-Ray Diffraction Studies of Crystalline Amino Acids and Peptides. Fortschr. Chem. organ. Naturstoffe **8**, 310 (1951).

2. Corey, R. B. and L. Pauling: Fundamental Dimensions of Polypeptide Chains. Proc. Roy. Soc. (London) **B 141**, 10 (1953).

3. Edsall, J. T., P. J. Flory, J. C. Kendrew, A. M. Liquori, G. Nemethy, G. N. Ramachandran and H. A. Scheraga: A Proposal of Standard Conventions and Nomenclature for the Description of Polypeptide Conformations. J. Mol. Biol. **15**, 399 (1966); J. Biol. Chem. **241**, 1004 (1966); Biopolymers **4**, 121 (1966).

4. Kratky, O. und H. Mark: Anwendung physikalischer Methoden zur Erforschung von Naturstoffen: Form und Größe dispergierter Moleküle. Röntgenographie. Fortschr. Chem. organ. Naturstoffe **1**, 255 (1938).

5. Marsh, R. E. and J. Donohue: Crystal Structure Studies of Amino Acids and Peptides. Adv. Protein Chem. **22**, 235 (1967).

6. Pauling, L. and R. B. Corey: Stable Configurations of Polypeptide Chains. Proc. Roy. Soc. (London) **B 141**, 21 (1953).

7. — — Two Hydrogen-Bonded Spiral Configurations of the Polypeptide Chain. J. Amer. Chem. Soc. **72**, 5349 (1950).

8. — — Atomic Coordinates and Structure Factors for Two Helical Configurations of Polypeptide Chains. Proc. Nat. Acad. Sci. (USA) **37**, 235 (1951).

9. PAULING, L. and R. B. COREY: The Pleated Sheet; A New Layer Configuration of Polypeptide Chains. Proc. Nat. Acad. Sci. (USA) **37**, 251 (1951).

10. — — Configurations of Polypeptide Chains with Favored Orientations Around Single Bonds: Two New Pleated Sheets. Proc. Nat. Acad. Sci. (USA) **37**, 729 (1951).

11. — — The Configuration of Polypeptide Chains in Proteins. Fortschr. Chem. organ. Naturstoffe **11**, 180 (1954).

Amino Acids

12. ALBRECHT, G. and R. B. COREY: The Crystal Structure of Glycine. J. Amer. Chem. Soc. **61**, 1087 (1939).

13. BUERGER, M. J., E. BARNEY and T. HAHN: The Crystal Structure of Diglycine Hydrobromide. Z. Kristallogr. **108**, 130 (1956).

14. DAWSON, B.: The Crystal Structure of DL-Glutamic Acid Hydrochloride. Acta Crystallogr. **6**, 31 (1953).

15. DONOHUE, J.: The Crystal Structure of DL-Alanine. II. Revision of Parameters by Three-Dimensional Fourier Analysis. J. Amer. Chem. Soc. **72**, 949 (1950).

16. DONOHUE, J., L. R. LAVINE and J. S. ROLLETT: The Crystal Structure of Histidine Hydrochloride Monohydrate. Acta Crystallogr. **9**, 655 (1956).

17. DONOHUE, J. and K. N. TRUEBLOOD: The Crystal Structure of Hydroxy-L-Proline. II. Determination and Description of the Structure. Acta Crystallogr. **5**, 419 (1952).

18. HAHN, T. and M. J. BUERGER: The Crystal Structure of Diglycine Hydrochloride, $2(C_2H_5O_2N) \cdot HCl$. Z. Kristallogr. **108**, 419 (1957).

19. HIROKAWA, S.: A New Modification of L-Glutamic Acid and its Crystal Structure. Acta Crystallogr. **8**, 637 (1955).

20. IITAKA, Y.: The Crystal Structure of β-Glycine. Acta Crystallogr. **13**, 35 (1960).

21. — The Crystal Structure of γ-Glycine. Acta Crystallogr. **14**, 1 (1961).

22. KARLE, I. L. and J. KARLE: An Application of the Symbolic Addition Method to the Structure of L-Arginine Dihydrate. Acta Crystallogr. **17**, 835 (1964).

23. LEVY, H. A. and R. B. COREY: The Crystal Structure of DL-Alanine. J. Amer. Chem. Soc. **63**, 2095 (1941).

24. MARSH, R. E.: Refinement of the Crystal Structure of Glycine. Acta Crystallogr. **11**, 654 (1958).

25. MATHIESON, A. McL.: The Crystal Structures of the Dimorphs of DL-Methionine. Acta Crystallogr. **5**, 332 (1952).

26. — Polymorphism of DL-Norleucine. Acta Crystallogr. **6**, 399 (1953).

27. OUGHTON, B. M. and P. M. HARRISON: The Crystal Structure of Hexagonal L-Cystine. Acta Crystallogr. **12**, 396 (1959).

28. PARTHASARATHY, R.: The Structure of L-Valine Hydrochloride. Acta Crystallogr. **21**, 422 (1966).

29. PETERSON, J., L. K. STEINRAUF and L. H. JENSEN: Direct Determination of the Structure of L-Cystine Dihydrobromide. Acta Crystallogr. **13**, 104 (1960).

30. RAMACHANDRAN, G. N., S. K. MAZUMDAR, K. VENKATESAN and A. V. LAKSHIMINARAYANAN: Conformation of the Arginine Side-Group and its Variations. J. Mol. Biol. **15**, 232 (1966).

31. RAMAN, S.: Determination of the Structure and Absolute Configuration of L(+)-Lysine Hydrochloride Dihydrate by the Anomalous-Dispersion Method. Z. Kristallogr. **111**, 301 (1959).

32. SHOEMAKER, D. P., R. E. BARIEAU, J. DONOHUE and C.-S. LU: The Crystal Structure of DL-Serine. Acta Crystallogr. **6**, 241 (1953).

33. Shoemaker, D. P., J. Donohue, V. Schomaker and R. B. Corey: The Crystal Structure of L₈-Threonine. J. Amer. Chem. Soc. **72**, 2328 (1950).

34. Simpson, H. J., Jr. and R. E. Marsh: The Crystal Structure of L-Alanine. Acta Crystallogr. **20**, 550 (1966).

35. Steinrauf, L. K., J. Peterson and L. H. Jensen: The Crystal Structure of L-Cystine Hydrochloride. J. Amer. Chem. Soc. **80**, 3835 (1958).

36. Subramanian, E.: The Crystal Structure of L-Leucine Hydrobromide. Acta Crystallogr. **22**, 910 (1967).

37. Trommel, J. and J. M. Bijvouet: Crystal Structure and Absolute Configuration of the Hydrochloride and Hydrobromide of D(—)-Isoleucine. Acta Crystallogr. **7**, 703 (1954).

38. Wright, D. A. and R. E. Marsh: The Crystal Structure of L-Lysine Monohydrochloride Dihydrate. Acta Crystallogr. **15**, 54 (1962).

39. Zussman, J.: The Structure of Hydroxyproline. Acta Crystallogr. **4**, 493 (1951).

Peptides

40. Biswas, A. B., E. W. Hughes, B. D. Sharma and J. W. Wilson: The Crystal Structure of α-Glycylglycine. Acta Crystallogr. **B 24**, 40 (1968).

41. Degeilh, R. and R. E. Marsh: A Refinement of the Crystal Structure of Diketopiperazine (2,5-Piperazinedione). Acta Crystallogr. **12**, 1007 (1959).

42. Dyer, H. B.: The Crystal Structure of Cysteylglycine-Sodium Iodide. Acta Crystallogr. **4**, 42 (1951).

43. Fridrichsons, J. and A. McL. Mathieson: The Crystal Structure of Tosyl-L-prolyl-L-hydroxyproline Monohydrate. Acta Crystallogr. **15**, 569 (1962).

44. Hughes, E. W. and W. J. Moore: The Crystal Structure of β-Glycylglycine. J. Amer. Chem. Soc. **71**, 2618 (1949).

45. Karle, I. L. and J. Karle: An Application of a New Phase Determination Procedure to the Structure of Cyclo(hexaglycyl) Hemihydrate. Acta Crystallogr. **16**, 969 (1963).

46. Leung, Y. C. and R. E. Marsh: The Crystal Structure of L-Leucyl-L-Prolyl-Glycine. Acta Crystallogr. **11**, 17 (1958).

47. Marsh, R. E. and J. P. Glusker: The Crystal Structure of Glycylphenyl-alanyl-glycine. Acta Crystallogr. **14**, 1110 (1961).

48. Pasternak, R. A.: The Crystal Structure of Glycyl-L-Tryptophan Dihydrate. Acta Crystallogr. **9**, 341 (1956).

49. Pasternak, R. A., L. Katz and R. B. Corey: The Crystal Structure of Glycyl-L-Asparagine. Acta Crystallogr. **7**, 225 (1954).

50. Smits, D. W. and E. H. Wiebenga: The Crystal Structure of Glycyl-L-Tyrosine Hydrochloride. Acta Crystallogr. **6**, 531 (1953).

51. Tranter, T. C.: Crystal Structure of Glycyl-L-Alanine Hydrochloride. Nature **177**, 37 (1956).

52. Wright, W. B.: The Crystal Structure of Glutathione. Acta Crystallogr. **11**, 632 (1958).

53. Yakel, H. L., Jr. and E. W. Hughes: The Crystal Structure of N,N'-Diglycyl-L-Cystine Dihydrate. Acta Crystallogr. **7**, 291 (1954).

Proteins

54. Aschaffenburg, R. and J. Drewry: Occurrence of Different β-Lactoglobulins in Cow's Milk. Nature **176**, 218 (1955).

55. — — Genetics of the β-Lactoglobulins of Cow's Milk. Nature **180**, 376 (1957).

56. Aschaffenburg, R., D. W. Green and R. M. Simmons: Crystal Forms of β-Lactoglobulin. J. Mol. Biol. **13**, 194 (1965).

57. AVEY, H. P., M. O. BOLES, C. H. CARLISLE, S. A. EVANS, S. J. MORRIS, R. A. PALMER, B. A. WOOLHOUSE and S. SHALL: Structure of Ribonuclease. Nature 213, 557 (1967).

58. BERNAL, J. D., I. FANKUCHEN and M. F. PERUTZ: An X-Ray Study of Chymotrypsin and Hemoglobin. Nature 141, 523 (1938).

59. BLAKE, C. C. F., R. H. FENN, A. C. T. NORTH, D. C. PHILLIPS and R. J. POLJAK: Structure of Lysozyme. A Fourier Map of the Electron Density at 6 Å Resolution Obtained by X-Ray Diffraction. Nature 196, 1173 (1962).

60. BLAKE, C. C. F., L. N. JOHNSON, G. A. MAIR, A. C. T. NORTH, D. C. PHILLIPS and V. R. SARMA: Crystallographic Studies of the Activity of Hen Egg-white Lysozyme. Proc. Roy. Soc. (London) B 167, 378 (1967).

61. BLAKE, C. C. F., D. F. KOENIG, G. A. MAIR, A. C. T. NORTH, D. C. PHILLIPS and V. R. SARMA: Structure of Hen Egg-white Lysozyme. A Three-dimensional Fourier Synthesis at 2 Å Resolution. Nature 206, 757 (1965).

62. BLAKE, C. C. F., G. A. MAIR, A. C. T. NORTH, D. C. PHILLIPS and V. R. SARMA: On the Conformation of the Egg-white Lysozyme Molecule. Proc. Roy. Soc. (London) B 167, 365 (1967).

63. BLOW, D. M.: The Structure of Hemoglobin. VII. Determination of Phase Angles in the Non-centrosymmetric [100] Zone. Proc. Roy. Soc. (London) A 247, 302 (1958).

64. BLOW, D. M., M. G. ROSSMANN and B. A. JEFFERY: The Arrangement of α-Chymotrypsin Molecules in the Monoclinic Crystal Form. J. Mol. Biol. 8, 65 (1964).

65. BLUHM, M. M., G. BODO, H. M. DINTZIS and J. C. KENDREW: The Crystal Structure of Myoglobin. IV. A Fourier Projection of Sperm-Whale Myoglobin by the Method of Isomorphous Replacement. Proc. Roy. Soc. (London) A 246, 369 (1958).

66. BODO, G., H. M. DINTZIS, J. C. KENDREW and H. W. WYCKOFF: The Crystal Structure of Myoglobin. V. A Low-Resolution Three-Dimensional Fourier Synthesis of Sperm-Whale Myoglobin Crystals. Proc. Roy. Soc. (London) A 253, 70 (1959).

67. BOYES-WATSON, J., E. DAVIDSON and M. F. PERUTZ: An X-Ray Study of Horse Methemoglobin. I. Proc. Roy. Soc. (London) A 191, 83 (1947).

68. BOYES-WATSON, J. and M. F. PERUTZ: X-Ray Analysis of Hemoglobin. Nature 151, 714 (1943).

69. BRAGG, W. L., E. R. HOWELLS and M. F. PERUTZ: Arrangement of Polypeptide Chains in Horse Methemoglobin. Acta Crystallogr. 5, 136 (1952).

70. — — — The Structure of Hemoglobin. I. Proc. Roy. Soc. (London) A 222, 33 (1954).

71. BRAGG, W. L. and M. F. PERUTZ: The External Form of the Hemoglobin Molecule. I. Acta Crystallogr. 5, 227 (1952).

72. — — The External Form of the Hemoglobin Molecule. II. Acta Crystallogr. 5, 323 (1952).

73. — — The Structure of Hemoglobin. Proc. Roy. Soc. (London) A 213, 425 (1952).

74. — — The Structure of Hemoglobin. VI. Fourier Projection on the 010 Plane. Proc. Roy. Soc. (London) A 225, 315 (1954).

75. COREY, R. B., R. H. STANFORD, Jr., R. E. MARSH, Y. C. LEUNG and L. M. KAY: An X-Ray Investigation of Wet Lysozyme Chloride Crystals. Preliminary Report on Crystals Containing Complex Ions of Niobium and Tantalum. Acta Crystallogr. 15, 1157 (1962).

76. Cullis, A. F., H. Muirhead, M. F. Perutz, M. G. Rossmann and A. C. T.
    North: The Structure of Hemoglobin. VIII. A Three-Dimensional Fourier
    Synthesis at 5.5 Å Resolution; Determination of the Phase Angles. Proc. Roy.
    Soc. (London) **A 265**, 15 (1961).
77. — — — — — The Structure of Hemoglobin. IX. A Three-Dimensional
    Fourier Synthesis at 5.5 Å Resolution; Description of the Structure. Proc.
    Roy. Soc. (London) **A 265**, 161 (1962).
78. Crowfoot, D.: The Crystal Structure of Insulin. I. The Investigation of
    Air-Dried Insulin Crystals. Proc. Roy. Soc. (London) **A 164**, 580 (1938).
79. Dickerson, R. E., M. L. Kopka, C. L. Borders, Jr., J. (C.) Varnum, J. E.
    Weinzierl and E. Margoliash: A Centrosymmetric Projection at 4 Å of
    Horse Heart Oxidized Cytochrome c. J. Mol. Biol. **29**, 77 (1967).
80. Dickerson, R. E., M. L. Kopka, J. (E.) Weinzierl, J. (C.) Varnum, D. Eisen-
    berg and E. Margoliash: Location of the Heme in Horse Heart Ferricyto-
    chrome *c* by X-Ray Diffraction. J. Biol. Chem. **242**, 3015 (1967).
81. — — — — — — An Interpretation of a Two-Derivative 4 Å Resolution
    Electron Density Map of Horse Heart Ferricytochrome c. Sympos. Cyto-
    chromes, Osaka, Japan, 1967.
82. Dodson, E., M. M. Harding, D. C. Hodgkin and M. G. Rossmann: The
    Crystal Structure of Insulin. III. Evidence for a 2-Fold Axis in Rhombohedral
    Zinc Insulin. J. Mol. Biol. **16**, 227 (1966).
83. Drenth, J. and J. N. Jansonius: The Unit Cell of Mercuripapain Crystals.
    Nature **184**, 1718 (1959).
84. Drenth, J., J. N. Jansonius, R. Koekoek, J. Marrink, J. Munnik and
    B. G. Wolthers: The Crystal Structure of Papain C. I. Two-Dimensional
    Fourier Syntheses. J. Mol. Biol. **5**, 398 (1962).
85. Drenth, J., J. N. Jansonius and B. G. Wolthers: The Crystal Structure of
    Papain. II. A Three-Dimensional Fourier Synthesis at 4.5 Å Resolution.
    J. Mol. Biol. **24**, 449 (1967).
86. Edmundson, A. B. and C. H. W. Hirs: The Amino-Acid Sequence of Sperm
    Whale Myoglobin. Chemical Studies. Nature **190**, 663 (1961).
87. Einstein, J. R., A. S. McGavin and B. W. Low: Insulin. A Probable Gross
    Molecular Structure. Proc. Nat. Acad. Sci. (USA) **49**, 74 (1963).
88. Fridborg, K., K. K. Kannan, A. Liljas, J. Lundin, B. Strandberg, R.
    Strandberg, B. Tilander and G. Wiren: Crystal Structure of Human
    Erythrocyte Carbonic Anhydrase C. III. Molecular Structure of the Enzyme
    and of One Enzyme-Inhibitor Complex at 5.5 Å Resolution. J. Mol. Biol. **25**,
    505 (1967).
89. Green, D. W. and R. Aschaffenburg: Twofold Symmetry of the β-Lacto-
    globulin Molecule in Crystals. J. Mol. Biol. **1**, 54 (1959).
90. Green, D. W., V. M. Ingram and M. F. Perutz: The Structure of Hemo-
    globin. IV. Sign Determination by the Isomorphous Replacement Method.
    Proc. Roy. Soc. (London) **A 225**, 287 (1954).
91. Green, D. W., A. C. T. North and R. Aschaffenburg: Crystallography of
    the β-Lactoglobulins of Cows' Milk. Biochim. Biophys. Acta **21**, 583 (1956).
92. Harding, M. M., D. C. Hodgkin, A. F. Kennedy, A. O'Connor and P. D. J.
    Weitzmann: The Crystal Structure of Insulin. II. An Investigation of Rhombo-
    hedral Zinc Insulin Crystals and a Report of Other Crystalline Forms. J. Mol.
    Biol. **16**, 212 (1966).
93. Hartsuck, J. A., M. L. Ludwig, H. Muirhead, T. A. Steitz and W. N. Lips-
    comb: Carboxypeptidase A. II. The Three-Dimensional Electron Density
    Map at 6 Å Resolution. Proc. Nat. Acad. Sci. (USA) **53**, 396 (1965).

94. INGRAM, D. J. E., J. F. GIBSON and M. F. PERUTZ: Electron Spin Resonance in Myoglobin and Hemoglobin. Orientation of the Four Heme Groups in Hemoglobin. Nature **178**, 905 (1956).

95. JOHNSON, L. N. and D. C. PHILLIPS: Structure of Some Crystalline Lysozyme-Inhibitor Complexes Determined by X-Ray Analysis at 6 Å Resolution. Nature **206**, 761 (1965).

96. KARTHA, G., J. BELLO and D. HARKER: Tertiary Structure of Ribonuclease. Nature **213**, 862 (1967).

97. KENDREW, J. C.: Preliminary X-Ray Data for Horse and Whale Myoglobins. Acta Crystallogr. **1**, 336 (1948).

98. — Myoglobin and the Structure of Proteins. Science **139**, 1259 (1963).

99. — The Molecular Structure of Myoglobin and Hemoglobin. In: M. Sala (Ed.), New Perspectives in Biology, p. 18. Amsterdam-New York-London: Elsevier Publ. Co. 1964.

100. KENDREW, J. C., G. BODO, H. M. DINTZIS, R. G. PARRISH, H. WYCKOFF and D. C. PHILLIPS: A Three-Dimensional Model of the Myoglobin Molecule Obtained by X-Ray Analysis. Nature **181**, 662 (1958).

101. KENDREW, J. C., R. E. DICKERSON, B. E. STRANDBERG, R. G. HART, D. R. DAVIES, D. C. PHILLIPS and V. C. SHORE: Structure of Myoglobin. A Three-Dimensional Fourier Synthesis at 2 Å Resolution. Nature **185**, 422 (1960).

102. KENDREW, J. C. and R. G. PARRISH: The Crystal Structure of Myoglobin. III. Sperm Whale Myoglobin. Proc. Roy. Soc. (London) **A 238**, 305 (1956).

103. KENDREW, J. C., R. G. PARRISH, J. R. MARRACK and E. S. ORLANS: The Species Specificity of Myoglobin. Nature **174**, 946 (1954).

104. KENDREW, J. C., H. C. WATSON, B. E. STRANDBERG, R. E. DICKERSON, D. C. PHILLIPS and V. C. SHORE: The Amino-Acid Sequence of Sperm Whale Myoglobin. A Partial Determination by X-Ray Methods, and Its Correlation with Chemical Data. Nature **190**, 666 (1961).

105. KRAUT, J., D. F. HIGH and L. C. SIEKER: Chymotrypsinogen: Increased Resolution and Absolute Configuration. Proc. Nat. Acad. Sci. (USA) **51**, 839 (1964).

106. KRAUT, J., L. C. SIEKER, D. F. HIGH and S. T. FREER: Chymotrypsinogen: A Three-Dimensional Fourier Synthesis at 5 Å Resolution. Proc. Nat. Acad. Sci. (USA) **48**, 1417 (1962).

107. KRAUT, J., H. T. WRIGHT, M. KELLERMAN and S. T. FREER: π-, δ-, and γ-Chymotrypsin: Three-Dimensional Electron Density and Difference Maps at 5 Å Resolution, and Comparison with Chymotrypsinogen. Proc. Nat. Acad. Sci. (USA) **58**, 304 (1967).

108. LIPSCOMB, W. N., J. C. COPPOLA, J. A. HARTSUCK, M. L. LUDWIG, H. MUIRHEAD, J. SEARL and T. A. STEITZ: The Structure of Carboxypeptidase A. III. Molecular Structure at 6 Å Resolution. J. Mol. Biol. **19**, 423 (1966).

109. LUDWIG, M. L., J. A. HARTSUCK, T. A. STEITZ, H. MUIRHEAD, J. C. COPPOLA, G. N. REEKE and W. N. LIPSCOMB: The Structure of Carboxypeptidase A. IV. Preliminary Results at 2.8 Å Resolution, and a Substrate Complex at 6 Å Resolution. Proc. Nat. Acad. Sci. (USA) **57**, 511 (1967).

110. LUDWIG, M. L., I. C. PAUL, G. S. PAWLEY and W. N. LIPSCOMB: The Structure of Carboxypeptidase A. I. A Two-Dimensional Superposition Function. Proc. Nat. Acad. Sci. (USA) **50**, 282 (1963).

111. MARSH, R. E., R. B. COREY and L. PAULING: An Investigation of the Structure of Silk Fibroin. Biochim. Biophys. Acta **16**, 1 (1955).

112. MATTHEWS, B. W., P. B. SIGLER, R. HENDERSON and D. M. BLOW: Three-Dimensional Structure of Tosyl-α-Chymotrypsin. Nature **214**, 652 (1967).

113. Muirhead, H. and M. F. Perutz: Structure of Hemoglobin. A Three-Dimensional Fourier Synthesis of Reduced Human Hemoglobin at 5.5 Å Resolution. Nature **199**, 633 (1963).

114. Perutz, M. F.: X-Ray Analysis of Hemoglobin. Nature **149**, 491 (1942).

115. — An X-Ray Study of Horse Methemoglobin. II. Proc. Roy. Soc. (London) **A 195**, 474 (1949).

116. — The Structure of Hemoglobin. III. Direct Determination of the Molecular Transform. Proc. Roy. Soc. (London) **A 225**, 264 (1954).

117. — Relation Between Structure and Sequence of Hemoglobin. Nature **194**, 914 (1962).

118. — X-Ray Analysis of Hemoglobin. Nobel prize lecture, 1962; Science **140**, 863 (1963).

119. — Structure and Function of Hemoglobin. I. A Tentative Atomic Model of Horse Oxyhemoglobin. J. Mol. Biol. **13**, 646 (1965).

120. Perutz, M. F., W. Bolton, R. Diamond, H. Muirhead and H. C. Watson: Structure of Hemoglobin. An X-Ray Examination of Reduced Horse Hemoglobin. Nature **203**, 687 (1964).

121. Perutz, M. F., J. C. Kendrew and H. C. Watson: Structure and Function of Hemoglobin. II. Some Relations between Polypeptide Chain Configuration and Amino Acid Sequence. J. Mol. Biol. **13**, 669 (1965).

122. Perutz, M. F. and L. Mazzarella: A Preliminary X-Ray Analysis of Hemoglobin H. Nature **199**, 639 (1963).

123. Perutz, M. F., M. G. Rossmann, A. F. Cullis, H. Muirhead, G. Will and A. C. T. North: Structure of Hemoglobin. A Three-Dimensional Fourier Synthesis at 5.5 Å Resolution, Obtained by X-Ray Analysis. Nature **185**, 416 (1960).

124. Phillips, D. C.: The Hen Egg-white Lysozyme Molecule. Proc. Nat. Acad. Sci. (USA) **57**, 484 (1967).

125. Rossmann, M. G. and D. M. Blow: The Detection of Sub-Units Within the Crystallographic Asymmetric Unit. Acta Crystallogr. **15**, 24 (1962).

126. Rossmann, M. G., B. A. Jeffery, P. Main and S. Warren: The Crystal Structure of Lactic Dehydrogenase. Proc. Nat. Acad. Sci. (USA) **57**, 515 (1967).

127. Scouloudi, H.: The Myoglobin Molecule. Nature **183**, 374 (1959).

128. Shoemaker, C. B., J. R. Einstein and B. W. Low: Insulin. The Three-Dimensional Patterson Function for Insulin Sulfate Type A Crystals. Acta Crystallogr. **14**, 459 (1961).

129. Sigler, P. B., B. A. Jeffery, B. W. Matthews and D. M. Blow: An X-Ray Diffraction Study of Inhibited Derivatives of α-Chymotrypsin. J. Mol. Biol. **15**, 175 (1966).

130. Smith, D. B. and M. F. Perutz: Identification of the Black Sub-Unit of the Crystallographic Model of Horse Hemoglobin with the Valyl-Glutaminyl Polypeptide Chain. Nature **188**, 406 (1960).

131. Stanford, R. H., Jr. and R. B. Corey: Determination of the Structure of Proteins by X-Ray Diffraction: Possible Use of Large Heavy Ions in Phase Determination. In: A. Rich and N. Davidson (Ed.), Structural Chemistry and Molecular Biology. San Francisco and London: Freeman and Co. 1968.

132. Stanford, R. H., Jr., R. E. Marsh and R. B. Corey: An X-Ray Investigation of Lysozyme Chloride Crystals Containing Complex Ions of Niobium and Tantalum: Three-Dimensional Fourier Plot Obtained from Data Extending to a Minimum Spacing of 5 Å. Nature **196**, 1176 (1962).

133. Strandberg, B., B. Tilander, K. Fridborg, S. Lindskog and P. O. Nyman: The Crystallization and X-Ray Investigation of One Form of Human Carbonic Anhydrase. J. Mol. Biol. **5**, 583 (1962).

134. TILANDER, B., B. STRANDBERG and K. FRIDBORG: Crystal Structure Studies on Human Erythrocyte Carbonic Anhydrase C. II. J. Mol. Biol. **12**, 740 (1965).

135. VAUGHAN, P. A., J. H. STURDIVANT and L. PAULING: The Determination of the Structure of Complex Molecules and Ions from X-Ray Diffraction by Their Solutions: The Structures of the Groups $PtBr_6^{--}$, $PtCl_6^{--}$, $Nb_6Cl_{12}^{++}$, $Ta_6Br_{12}^{++}$, and $Ta_6Cl_{12}^{++}$. J. Amer. Chem. Soc. **73**, 5477 (1950).

136. WATSON, H. C. and J. C. KENDREW: Comparison between the Amino-Acid Sequences of Sperm Whale Myoglobin and of Human Hemoglobin. Nature **190**, 670 (1961).

137. WYCKOFF, H. W., M. DOSCHER, D. TSERNOGLOU, T. INAGAMI, L. N. JOHNSON, K. D. HARDMAN, N. M. ALLEWELL, D. M. KELLY and F. M. RICHARDS: Design of Diffractometer and Flow Cell System for X-Ray Analysis of Crystalline Proteins with Applications to the Crystal Chemistry of Ribonuclease-S. J. Mol. Biol. **27**, 563 (1967).

138. WYCKOFF, H. W., K. D. HARDMAN, N. M. ALLEWELL, T. INAGAMI, L. N. JOHNSON and F. M. RICHARDS: The Structure of Ribonuclease-S at 3.5 Å Resolution. J. Biol. Chem. **242**, 3984 (1967).

139. WYCKOFF, H. W., K. D. HARDMAN, N. M. ALLEWELL, T. INAGAMI, D. TSERNOGLOU, L. N. JOHNSON and F. M. RICHARDS: The Structure of Ribonuclease-S at 6 Å Resolution. J. Biol. Chem. **242**, 3749 (1967).

(Received, November 9, 1967)

Synthese von Peptiden und Peptidwirkstoffen

Von **E. Schröder** und **K. Lübke**, Berlin

Mit 35 Abbildungen

Inhaltsübersicht

Abkürzungen

Die in dieser Arbeit verwendeten Abkürzungen richten sich im wesentlichen nach Empfehlungen in:

"Peptides", Proceedings 5th European Symposium, Oxford 1962, G. T. Young (ed.), Pergamon Press 1963, p. 262 und

IUPAC-IUB-Kommission über Biochemische Nomenklatur, Abkürzungen für Aminosäurederivate und Peptide, Tentative Rules, J. Biol. Chem. **241**, 527, 2491 (1966); Biochim. Biophys. Acta **121**, 1 (1966).

Es werden verwendet:

1. Aminosäurereste:

Ala	Alanin,	Me · Ile	N-Methyl-Isoleucin,
Arg	Arginin,	Me · Leu	N-Methyl-Leucin,
Asp	Asparaginsäure,	Me · Val	N-Methyl-Valin,
Asn	Asparagin [auch Asp(NH$_2$)],	Nle	Norleucin,
Cys	Cystein,	Orn	Ornithin,
Dab	α,γ-Diaminobuttersäure,	Phe	Phenylalanin,
Eti	Äthionin,	Pro	Prolin,
Glu	Glutaminsäure,	⌐Glu = Pyroglu	Pyroglutaminsäure,
Gln	Glutamin [auch Glu(NH$_2$)],	Sar	Sarkosin (N-Methyl-Glycin),
Gly	Glycin,	Ser	Serin,
His	Histidin,	Thr	Threonin,
Ile	Isoleucin,	Try	Tryptophan,
Leu	Leucin,	Tyr	Tyrosin,
Lys	Lysin,	Val	Valin.
Met	Methionin,		

2. Hydroxysäuren:

Glyc	Glycolsäure,	Oxisoval	α-Hydroxyisovaleriansäure,
Lac	Milchsäure,	PheLac	Phenylmilchsäure.

3. Schutzgruppen, aktivierte Ester etc., vgl. auch Abb. 12—15, SS. 64—71:

Ac	Acetyl,	NP	p-Nitrophenyl,
BOC	tert. Butyloxycarbonyl,	NPS	o-Nitrophenylsulfenyl,
tBu	tert. Butyl,	Pht	Phthalyl,
Bz	Benzoyl,	Pip	Piperidyl,
Bzl	Benzyl,	PZ	Phenylazobenzyloxycarbonyl,
CP	2,4,5-Trichlorphenyl,	Q	8-Chinolyl,
EC	Äthylcarbamoyl,	Su	N-Succinimido,
Et	Äthyl,	Tos	Toluolsulfonyl,
Form	Formyl,	Trit	Trityl,
Me	Methyl,	Z	Benzyloxycarbonyl,
MOB	p-Methoxybenzyl,	Z(NO$_2$)	p-Nitrobenzyloxycarbonyl,
NB	p-Nitrobenzyl,	Z(OMe)	p-Methoxybenzyloxycarbonyl.
NO$_2$	Nitro,		

Vorwort

Die Zeit nach der ersten Synthese eines Peptidwirkstoffes, des Oxytocins, durch DU VIGNEAUD im Jahre 1953, ist mit mannigfaltigen Bemühungen angefüllt, die präparativen Methoden der Peptidchemie zu vervollkommnen. Stimulierend für diese Bemühungen waren die Fortschritte bei der Isolierung und Strukturaufklärung einer Vielzahl von Naturstoffen mit Peptidstruktur und der Wunsch nach eindeutigen und rationellen Synthesen der neuen biologisch aktiven Substanzen. Mit der Zugänglichkeit von Peptidwirkstoffen durch Synthese wurden eingehende pharmakologische und medizinische Untersuchungen ermöglicht, die viel zum heutigen Verständnis über die physiologische Bedeutung der verschiedensten Polypeptidwirkstoffe beigetragen haben. In einigen Fällen konnten mit synthetischen Produkten sogar therapeutische Anwendungsmöglichkeiten erschlossen werden.

Bei den vielen Berührungspunkten, die die Peptidchemie heute mit anderen Fachdisziplinen aufweist, wird es selbst für den Spezialisten immer schwieriger, auf allen Teilgebieten gleich gut informiert zu bleiben. Die vorliegende Arbeit soll in erster Linie dem Uneingeweihten eine Übersicht über die wichtigsten Peptidwirkstoffe geben*. Eine Zusammenstellung der Methoden der Peptidsynthese beschränkt sich auf eine Auswahl, die sich für die Synthese höher molekularer Peptide in der Praxis durchgesetzt hat. Charakteristische Synthesebeispiele für jeden Peptidwirkstofftyp sollen schließlich die Fortschritte zeigen, die mit einer verbesserten Methodik auf diesem präparativ so mühevollen Gebiet erreicht werden konnten. Trotz vieler Erfolge, die sich in der Synthese komplizierter Moleküle, wie dem adrenocorticotropen Hormon, dem Insulin, den Polymyxinen und etwa 1000 Wirkstoffanaloga wiederspiegeln, bleibt die Notwendigkeit, bessere Synthesemöglichkeiten zu finden, eine der wichtigsten Aufgaben in der Peptidchemie.

---

* Ausführliche Monographien (*36, 63, 162, 163*).

Literaturverzeichnis: SS. 107—119

I. Biologisch aktive Peptide

Die Peptidwirkstoffe, die im folgenden besprochen werden, lassen sich in fünf Gruppen einteilen: die von endokrinen Drüsen sezernierten Peptidhormone (adrenocorticotropes Hormon, α- und β-melanocytenstimulierende Hormone, Oxytocin, Vasopressin, Insulin, Glukagon), die Peptidhormone des Intestinaltraktes (Gastrin, Secretin), die sogenannten Gewebshormone (Angiotensine, Bradykinin), Peptide unterschiedlicher Herkunft (z. B. aus Amphibienhäuten, Bienengift usw.) und die Peptidantibiotika.

Das adrenocorticotrope Hormon (ACTH) gehört neben den drei Gonadotropinen, dem thyreotropen Hormon und dem Wachstumshormon zu den Hormonen des Hypophysenvorderlappens. Es ist über die Sekretion der Nebennieren-Steroid-Hormone am Gesamtstoffwechsel beteiligt *(131)*. Therapeutisch verwendet wird das ACTH zur Entzündungshemmung, bei Allergien und in einigen Fällen auch bei bestimmten Formen von Hypophyseninsuffizienz. Die im Handel befindlichen Präparate sind bevorzugt tierischen Ursprungs. Seit einiger Zeit ist eine aus den ersten 24 Aminosäuren bestehende Teilsequenz des ACTH mit praktisch voller Wirksamkeit als synthetisches Präparat im Handel *(28)*. Mit 39 Aminosäureresten ist das ACTH zur Zeit der längste synthetisierte lineare Peptidwirkstoff. Die speziesbedingten Strukturunterschiede sind in der Teilsequenz 25—33 lokalisiert *(163)* *(Abb. 1)*.

Vom Hypophysenmittellappen werden die melanocytenstimulierenden Hormone α- und β-MSH sezerniert *(102)*. Die physiologische Rolle dieser Hormone beim Menschen ist nicht bekannt. Bei niederen Tieren sind sie für die Anpassung der Hautfärbung an die Umgebung verantwortlich. Die Aminosäuresequenz des α-MSH ist mit der Sequenz 1—13 des ACTH identisch und enthält zusätzlich eine N-terminale Acetyl- und eine C-terminale Säureamidgruppe. Speziesunterschiede sind nicht bekannt *(Abb. 2)*. Auch das β-MSH mit 18 Aminosäuren stimmt über längere Bereiche mit der Sequenz des ACTH überein. Die aus verschiedenen Spezies isolierten Hormone unterscheiden sich in den Positionen 2, 6 und 16. Das humane β-MSH mit 22 Aminosäuren (Abb. 2) weicht stärker von den anderen β-MSH-Typen ab. α-MSH und β-MSH (Rind) sind synthetisiert worden; eine therapeutische Verwendung haben die melanocytenstimulierenden Hormone nicht gefunden.

Ein aus der Hypophyse isolierbares Hormon mit lipolytischer Wirkung, das β-Lipotropin (β-LPH) zeigt in seiner Primärstruktur (90 Aminosäuren) ebenfalls große Übereinstimmungen mit dem ACTH und vor allem mit dem humanen β-MSH *(105)* (Abb. 2).

Von den Hormonen des Hypophysenhinterlappens *(77)* wirkt das Oxytocin auf die Entleerung der lactierenden Milchdrüse. Ob der Wirkung auf den Uterus eine physiologische Bedeutung zukommt, ist noch um-

H-Ser-Tyr-Ser-Met-Glu-His-Phe-Arg-Try-Gly-Lys-Pro-Val-Gly-Lys-Lys-Arg-Arg-Pro-Val-Lys-Val-Tyr-Pro-
1   2   3   4   5   6   7   8   9   10 11 12 13 14 15 16 17 18 19 20 21 22 23 24

Rind: -Asp-Gly-Glu-*Ala-Glu-Asp*-Ser-Ala-Gln-Ala-Phe-Pro-Leu-Glu-Phe-OH

Schaf: -*Ala*-Gly-Glu-*Asp*-Asp-*Glu-Ala-Ser*-Gln-Ala-Phe-Pro-Leu-Glu-Phe-OH

Schwein: -Asp-Gly-*Ala*-Glu-Asp-Gln-*Leu*-Ala-*Glu*-Ala-Phe-Pro-Leu-Glu-Phe-OH

Mensch: -Asp-*Ala-Gly*-Glu-Asp-Gln-Ser-Ala-*Glu*-Ala-Phe-Pro-Leu-Glu-Phe-OH
            25 26 27 28 29 30 31 32 33 34 35 36 37 38 39

Abb. 1. Adrenocorticotrope Hormone

α-MSH: CH$_3$CO-Ser-Tyr-Ser-Met-Glu-His-Phe-Arg-Try-Gly-Lys-Pro-Val-NH$_2$
            1   2   3   4   5   6   7   8   9   10 11 12 13

β-MSH: H-Asp-*Ser*-Gly-Pro-Tyr-Lys-Met-Glu-His-Phe-Arg-Try-Gly-Ser-Pro-Pro-Lys-Asp-OH
(Rind, Schaf) 1   2   3   4   5   6   7   8   9   10 11 12 13 14 15 16 17 18

β-MSH: H-Asp-Glu-Gly-Pro-Tyr-Lys-Met-Glu-His-Phe-Arg-Try-Gly-Ser-Pro-Pro-Lys-Asp-OH
(Schwein, Schaf)

β-MSH: H-Asp-Glu-Gly-Pro-Tyr-Lys-Met-Glu-His-Phe-Arg-Try-Gly-Ser-Pro-*Arg*-Lys-Asp-OH
(Pferd)

β-MSH: H-Asp-Glu-Gly-Pro-Tyr-*Arg*-Met-Glu-His-Phe-Arg-Try-Gly-Ser-Pro-Pro-Lys-Asp-OH
(Affe)

β-MSH: H-Ala-Glu-Lys-Lys-Asp-Glu-Gly-Pro-Tyr-Arg-Met-Glu-His-Phe-Arg-Try-Gly-Ser-Pro-Pro-Lys-Asp-OH
(Mensch)   1   2   3   4   5   6   7   8   9   10 11 12 13 14 15 16 17 18 19 20 21 22

β-LPH: ----Ala-Glu-Lys-Lys-Asp-Ser-Gly-Pro-Tyr-Lys-Met-Glu-His-Phe-Arg-Try-Gly-Ser-Pro-Pro-Lys-Asp----
(Schaf)   37 38 39 40 41 42 43 44 45 46 47 48 49 50 51 52 53 54 55 56 57 58

Abb. 2. Melanocytenstimulierende Hormone

stritten, dennoch wird Oxytocin zur Geburtseinleitung verwendet. Ferner hat es sich in Form von Nasenspray als milchaustreibendes Mittel bewährt. Da die aus natürlichen Quellen zugänglichen Oxytocin-Präparate nur schwer vasopressinfrei zu erhalten sind, bedeutete die Einführung von synthetischen Produkten einen wesentlichen Fortschritt. Das Vasopressin ist als antidiuretisches Hormon für eine physiologische Salzkonzentration des Harns verantwortlich (*166*). Bei ungenügendem Vasopressinspiegel werden sehr große Harnmengen mit niedrigem Salzgehalt

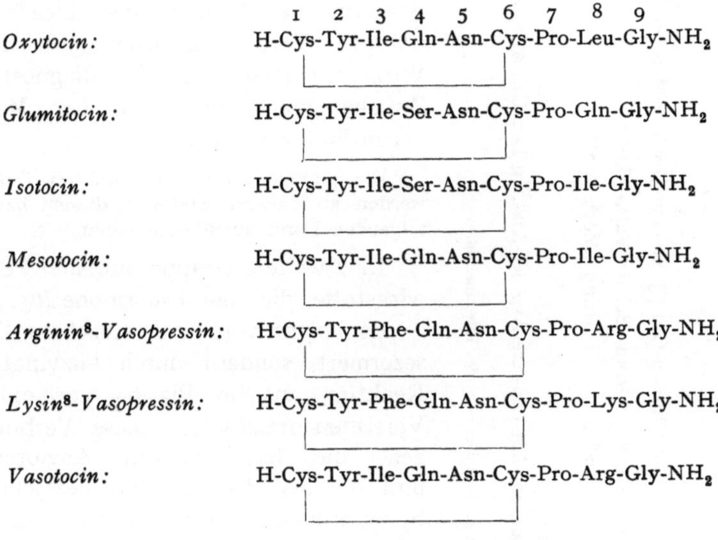

Abb. 3. Peptidhormone des Hypophysenhinterlappens

ausgeschieden. Diese als Diabetes insipidus bezeichnete Krankheit ist die wichtigste Indikation des Vasopressins. Die Nonapeptide Oxytocin und Vasopressin enthalten als besonderes Strukturmerkmal einen cyclischen Disulfidring *(Abb. 3)*. Beim Vasopressin unterscheidet man je nach Herkunft das Arg⁸- und das Lys⁸-Vasopressin. Weitere Hinterlappenhormone mit analoger Struktur konnten aus verschiedenen Spezies isoliert werden. Während das Isotocin, das Mesotocin und das Glumitocin typische Oxytocinanaloga sind, enthält das Vasotocin den Disulfidring des Oxytocins und die C-terminale Seitenkette des Arg⁸-Vasopressins (Abb. 3). Die Vasopressine sind nur in Säugetieren (Lys⁸-Vasopressin nur beim Schwein), das Vasotocin als antidiuretisches Hormon bei Vögeln, Amphibien und Fischen aufgefunden worden. Als oxytocisches Prinzip wurde Oxytocin in Säugetieren und Vögeln, Mesotocin in Amphibien, Isotocin in Knochenfischen und Glumitocin in Knorpelfischen erkannt (*1*).

Die Peptidhormone des Intestinaltraktes, das Gastrin und das Sekretin sind lineare Peptide mit 17 bzw. 27 Aminosäureresten *(Abb. 4)*. Vom Gastrin, das als Gastrin I und II vorkommt, sind Speciesunterschiede bekannt *(19)*. Sekretin und die Gastrine sind synthetisiert worden. Gastrin stimuliert die Sekretion von Säure und Pepsin *(179)*, Sekretin die Bildung eines bicarbonathaltigen Verdauungssaftes *(90)*. Beide Wirkstoffe lassen sich für diagnostische Zwecke zur Überprüfung der Magen-Darm-Funktion verwenden.

Die Pankreashormone Insulin und Glukagon werden an anderer Stelle in diesem bzw. im folgenden Band ausführlich behandelt.

Eine weitere Gruppe humaner Peptidwirkstoffe, die Gewebshormone *(65, 199)*, werden nicht von einer einzelnen Drüse sezerniert, sondern durch enzymatische Reaktion aus im Plasma vorhandenen Vorstufen freigesetzt. Diese Verbindungen, die hypertensiven Angiotensine und die als Plasmakinine bezeichneten hypotensiven Peptide, Bradykinin, Kallidin und Methionyl-lysyl-bradykinin *(Abb. 5)* sind kreislaufwirksam und stimulieren die glatte Muskulatur. Bisher hat nur das Angiotensin bei postoperativem Kreislaufkollaps zur Wiederherstellung des normalen Blutdruckes eine therapeutische Anwendung gefunden. Über die physiologische Rolle der Gewebshormone ist nur wenig bekannt. Das Angiotensin wird mit dem renalen Hochdruck in Zusammenhang gebracht, ferner wird der Einfluß des Angiotensins auf die Aldosteronsekretion diskutiert. Den Plasmakininen wird eine Beteiligung bei verschiedenen pathologischen und physiologischen Prozessen zugesprochen *(217)*.

$$\text{1 \quad 2 \quad 3 \quad 4 \quad 5 \quad 6 \quad 7 \quad 8 \quad 9 \quad 10 \quad 11 \quad 12 \quad 13 \quad 14 \quad 15 \quad 16 \quad 17}$$

Gastrin I: Pyroglu-Gly-Pro-Try-☐-Glu-Glu-Glu-☐-Ala-Tyr-Gly-Try-Met-Asp-Phe-NH$_2$

$$\overset{\displaystyle SO_3}{\underset{\textstyle |}{}}$$

Gastrin II: Pyroglu-Gly-Pro-Try-☐-Glu-Glu-Glu-☐-Ala-Tyr-Gly-Try-Met-Asp-Phe-NH$_2$

Mensch: Leu^5, Glu^{10} Rind, Schaf: Val^5, Ala^{10}
Schwein: Met^5, Glu^{10} Hund: Met^5, Ala^{10}

$$\text{1 \qquad 5 \qquad 10 \qquad 15 \qquad 20 \qquad 25 \quad 27}$$

Sekretin: H-His-Ser-Asp-Gly-Thr-Phe-Thr-Ser-Glu-Leu-Ser-Arg-Leu-Arg-Asp-Ser-Ala-Arg-Leu-Gln-Arg-Leu-Leu-Gln-Gly-Leu-Val-NH$_2$

Abb. 4. Peptidhormone des Intestinaltraktes

Literaturverzeichnis: SS. 107—119

Relativ gut bekannt sind die Freisetzungsmechanismen der Gewebshormone. Aus einer im Plasma vorhandenen Proteinvorstufe, dem Angiotensinogen wird durch Einwirkung von Renin unter Spaltung der Leu^{10}-Leu^{11}-Bindung zunächst das inaktive Ile^5-Angiotensin I freigesetzt. Aus diesem entsteht unter Einfluß eines „converting enzyme" unter Abspaltung des C-terminalen Dipeptids H-His-Leu-OH der biologisch aktive Wirkstoff Ile^5-Angiotensin II. Durch tryptischen Abbau einer Globulinfraktion konnte eine Partialsequenz des Angiotensinogens, das „Polypeptid-Renin-Substrat" (ein Tetradecapeptid, vgl. Abb. 5) isoliert, in seiner Struktur aufgeklärt und synthetisiert werden (*193*). Ile^5-Angiotensin I und II wurden aus Pferde- und Schweineplasma, ein Val^5-Angiotensin I aus Rinderplasma isoliert. Aus menschlichem Plasma konnte Ile^5-Angiotensin I erhalten werden (*13*). Die Plasmakinine werden ebenfalls aus im Serum vorhandenen inaktiven Vorstufen [Kininogene (Bradykininogen, Kallidinogen)] durch proteolytische Einwirkung von

Angiotensinogen:

1 2 3 4 5 6 7 8 9 10 11 12 13 14
H-Asp-Arg-Val-Tyr-Ile-His-Pro-Phe-His-Leu-Leu-Val-Tyr-Ser ...

Ile^5-Angiotensin I:

1 2 3 4 5 6 7 8 9 10
H-Asp-Arg-Val-Tyr-Ile-His-Pro-Phe-His-Leu-OH

Ile^5-Angiotensin II:

1 2 3 4 5 6 7 8
H-Asp-Arg-Val-Tyr-Ile-His-Pro-Phe-OH

Val^5-Angiotensin II:

1 2 3 4 5 6 7 8
H-Asp-Arg-Val-Tyr-Val-His-Pro-Phe-OH

Bradykinin:

1 2 3 4 5 6 7 8 9
H-Arg-Pro-Pro-Gly-Phe-Ser-Pro-Phe-Arg-OH

Kallidin:

1 2 3 4 5 6 7 8 9 10
H-Lys-Arg-Pro-Pro-Gly-Phe-Ser-Pro-Phe-Arg-OH

Methionyl-Lysyl-Bradykinin:

1 2 3 4 5 6 7 8 9 10 11
H-Met-Lys-Arg-Pro-Pro-Gly-Phe-Ser-Pro-Phe-Arg-OH

Abb. 5. Gewebshormone

Kallikrein, Plasmin und Trypsin sowie von einigen Schlangen- und Bakteriengiften freigesetzt. Kürzlich gelang es, durch peptischen Abbau der α_2-Globulin-Fraktion das Tetradecapeptid H-Met-Lys-Arg-Pro-Pro-Gly-Phe-Ser-Pro-Phe-Arg-Ser-Val-Gln-OH zu isolieren und damit die Lage der Kinine in der Proteinvorstufe genauer zu lokalisieren (*72*).

Neben den bisher genannten Wirkstoffen sind verschiedene Polypeptide nur aus tierischem Material isoliert worden: das in der Speicheldrüse von Tintenfischen vorkommende Eledoisin (*200*) und die aus Amphibienhäuten isolierbaren Wirkstoffe Physalaemin (*27*) und Phyllokinin (*3*). Diese Undekapeptide haben beim Säuger kininähnliche biologische Eigenschaften. Während das Phyllokinin ein C-terminal verlängertes Bradykinin ist, zeigen Eledoisin und Physalaemin untereinander trotz unterschiedlicher Herkunft strukturelle Ähnlichkeiten (*Abb. 6*). Ferner müssen die Bienengiftpeptide Melittin (*73*) und Apamin (*75, 192*) genannt werden. Das Melittin ist das haemolysierende Prinzip des Bienengiftes, das Apamin zeigt Wirkungen auf das Zentralnervensystem. Die Aminosäuresequenzen beider Peptide sind bekannt *(Abb. 7)*. Für das Apamin ist jedoch die Lage der Disulfidbrücken, die die vier Cysteinreste verbinden, noch nicht festgelegt.

$$
\begin{array}{l}
\quad 1 \quad 2 \quad 3 \quad 4 \quad 5 \quad 6 \quad 7 \quad 8 \quad 9 \quad 10 \quad 11 \\
Eledoisin:\quad \text{Pyroglu-Pro-Ser-Lys-Asp-Ala-Phe-Ile-Gly-Leu-Met-NH}_2
\end{array}
$$

Eledoisin:

    1 2 3 4 5 6 7 8 9 10 11

Pyroglu-Pro-Ser-Lys-Asp-Ala-Phe-Ile-Gly-Leu-Met-NH$_2$

Physalaemin:

    1 2 3 4 5 6 7 8 9 10 11

Pyroglu-Ala-Asp-Pro-Asn-Lys-Phe-Tyr-Gly-Leu-Met-NH$_2$

OSO$_3$H
|
Phyllokinin: H-Arg-Pro-Pro-Gly-Phe-Ser-Pro-Phe-Arg-Ile-Tyr-OH

    1 2 3 4 5 6 7 8 9 10 11

Abb. 6. Kinin-ähnliche Peptidwirkstoffe

Zahlreiche Peptidwirkstoffe sind aus Mikroorganismen zugänglich. Die Mehrzahl zeigt antibiotische Eigenschaften. Von den meisten ist die Struktur noch nicht bekannt und nur einige sind bisher synthetisiert worden. Fast alle diese Wirkstoffe sind cyclischer Natur und zeichnen sich durch besondere strukturelle Eigenarten aus. Sie enthalten häufig D-Aminosäuren und zum Teil unnatürliche Aminosäuren. [Vgl. Bodanszky

Melittine:

 1 2 3 4 5 6 7 8 9 10 11 12 13 14 15 16 17 18 19 20

H-Gly-Ile-Gly-Ala-Val-Leu-Lys-Val-Leu-Thr-Thr-Gly-Leu-Pro-Ala-Leu-Ile-Ser-Try-Ile

            21 22 23 24 25 26

Melittin I: Lys-Arg-Lys-Arg-Gln-Gln-NH

            21 22 23 24 25 26 27

Melittin II: Ser-Arg-Lys-Lys-Arg-Gln-Gln-NH

Apamin:

 1 2 3 4 5 6 7 8 9 10 11 12 13 14 15 16 17 18

H-Cys-Asn-Cys-Lys-Ala-Pro-Glu-Tyr-Ala-Leu-Cys-Ala-Arg-Arg-Cys-Gln-Gln-His-OH
  | | | |

Abb. 7. Peptidwirkstoffe des Bienengiftes

und Perlman *(39)*.] Im folgenden sollen nur einige charakteristische Vertreter genannt werden. So werden z. B. von *Bacillus brevis* in ihrer Struktur unterschiedliche Peptidantibiotika produziert: die linearen Gramicidine A, B und C *(153, 154)*, das cyclische Gramicidin S *(53)* und die ebenfalls cyclischen Tyrocidine A, B und C *(146)*. Während die linearen Gramicidine keine strukturellen Beziehungen zu den cyclischen Verbindungen zeigen, haben Gramicidin S und die Tyrocidine eine Pentapeptidsequenz gemeinsam *(Abb. 8)*. Eine in ihrer Struktur sehr ähnliche Gruppe cyclischer Peptidantibiotika sind die Polymyxine, Colistine und Circuline *(208)*. Es sind Dekapeptide mit einem Heptapeptidring, der

Literaturverzeichnis: SS. 107—119

Gramicidin A, B und C:

$$\text{H-CO-}\boxed{X}\text{-Gly-L-Ala-D-Leu-L-Ala-D-Val-L-Val-D-Val-L-Try-D-Leu-}\boxed{Y}\text{-D-Leu-L-Tyr-D-Leu-L-Try-NH-CH}_2\text{-CH}_2\text{-OH}$$

positions: 1 2 3 4 5 6 7 8 9 10 11 12 13 14 15

	X	Y
Valin-Gramicidin A:	L-Val	L-Try
Valin-Gramicidin B:	L-Val	L-Phe
Valin-Gramicidin C:	L-Val	L-Tyr

	X	Y
Isoleucin-Gramicidin A:	L-Ile	L-Try
Isoleucin-Gramicidin B:	L-Ile	L-Phe
Isoleucin-Gramicidin C:	L-Ile	L-Tyr

Gramicidin S:

→ L-Val → L-Orn → L-Leu → D-Phe → L-Pro—
└ L-Pro ← D-Phe ← L-Leu ← L-Orn ← L-Val ←

Tyrocidin A, B und C:

→ L-Val → L-Orn → L-Leu → D-Phe → L-Pro—
└ L-Tyr ← L-Gln ← L-Asn ← $\boxed{X}$ ← $\boxed{Y}$ ←

	X	Y
Tyrocidin A:	D-Phe	L-Phe
Tyrocidin B:	D-Phe	L-Try
Tyrocidin C:	D-Try	L-Try

Abb. 8. Peptidantibiotika der Gramicidin/Tyrocidingruppe

über die γ-Aminogruppe eines α,γ-Diaminobuttersäurerestes geschlossen
ist, und einer Aminosäureseitenkette. Außerdem enthalten sie als N-terminalen Heterobestandteil (d. h. als Nichtaminosäurebestandteil) einen

```
                                          ┌─L-Thr ← L-Dab ← L-Dab ←
                                          │
                                          ↓
         R → L-Dab → L-Thr →│ X │→ L-Dab → L-Dab →│ Y │→│ Z │─
         ┆                      ┆                     ┆     ┆
```

	R	X	Y	Z
Polymyxin B₁:	MOA*	L-Dab	D-Phe	L-Leu
Polymyxin B₂:	IOA**	L-Dab	D-Phe	L-Leu
Polymyxin D₁:	MOA	D-Ser	D-Leu	L-Thr
Polymyxin D₂:	IOA	D-Ser	D-Leu	L-Thr
Colistin A: (= Polymyxin E₁)	MOA	L-Dab	D-Leu	L-Leu
Colistin B: (= Polymyxin E₂)	IOA	L-Dab	D-Leu	L-Leu
Circulin A:	MOA	L-Dab	D-Leu	L-Ile

* (+)-6-Methyloctansäure [(+)-Isopelargonsäure].
** 6-Methylheptansäure [Isooctansäure].

Abb. 9. Peptidantibiotika der Polymyxingruppe

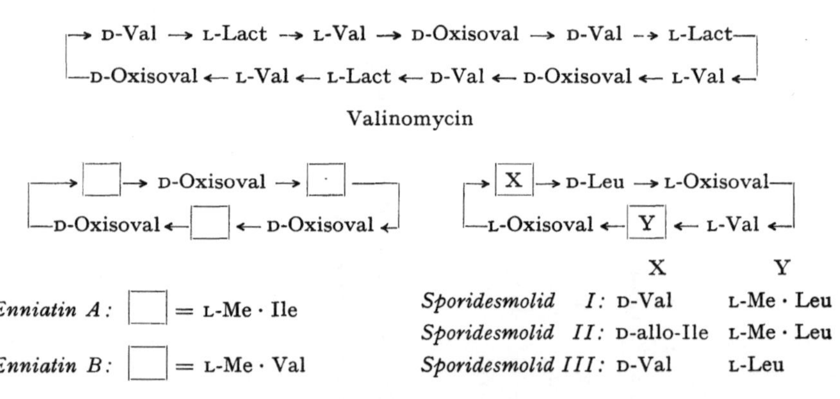

Abb. 10. Peptidwirkstoffe mit Depsipeptidstruktur

verzweigten Fettsäurerest [(+)-6-Methyloctansäure (MOA) bzw. 6-Methyl-
heptansäure (Isooctansäure, IOA)] *(Abb. 9)*. Die Mehrzahl der hier aufgeführten Peptidantibiotika ist synthetisiert worden. Im allgemeinen
steht die starke Toxizität aller Peptidantibiotika einer breiteren prak-
tischen Anwendung entgegen.

Literaturverzeichnis: SS. 107—119

Peptide, die neben den Peptidbindungen Esterbindungen enthalten, werden als Depsipeptide bezeichnet. Wird die Esterbindung aus einer Hydroxyaminosäure und der Carboxylgruppe der C-terminalen Aminosäure gebildet, so liegen Peptidlactone vor, z. B. die Actinomycine, Etamycin, Staphylomycin Faktor S, Ostreogrycin und Mikamycin. Zu einer zweiten Gruppe von Depsipeptiden gehören Verbindungen, die an Stelle einer Aminosäure Hydroxysäuren innerhalb der Peptidkette enthalten. Vorzugsweise wurden α-Hydroxyisovaleriansäure, Milchsäure und α-Hydroxyisocapronsäure als „Heterobestandteile" gefunden. Diese in der Natur vorkommenden und auch als Peptolide bezeichneten Verbindungen sind cyclisch und können einen symmetrischen Aufbau (Enniatine, Valinomycin) oder einen unsymmetrischen Aufbau (Sporidesmolide) haben *(Abb. 10) (109)*.

II. Methoden der Peptidsynthese

A. Prinzip der Peptidsynthese

Eine Peptidsynthese ist charakterisiert durch die Bildung einer Amidbindung zwischen der Carboxylgruppe einer und der Aminogruppe einer zweiten Aminosäure. Um einheitliche Produkte zu erhalten, ist es erforderlich, alle zusätzlichen funktionellen Gruppen, sofern sie an der Reaktion teilnehmen können, mit Schutzgruppen zu blockieren. Nach der Einführung geeigneter Schutzgruppen in die Aminosäuren werden die Aminosäurederivate zu einem Dipeptid gekuppelt. Die weitere Kettenverlängerung erfolgt nach selektiver Abspaltung der N-terminalen Amino- oder der C-terminalen Carboxylschutzgruppe. Die Peptidsynthese endet mit der vollständigen Abspaltung aller Schutzgruppen. Fast immer wird sich eine Synthese jedoch nicht ohne Kompromisse durchführen lassen, da die Zahl der zur Verfügung stehenden Schutzgruppen und Kupplungsmethoden begrenzt ist. Auch lassen sich nicht alle Schutzgruppen untereinander und nicht alle Schutzgruppen und Kupplungsmethoden miteinander beliebig kombinieren. Darüber hinaus sind weitere Einschränkungen durch die besonderen Eigenschaften bestimmter Aminosäuren gegeben. Die Synthese eines Peptids erfordert daher eine sehr sorgfältige Planung, die nach BODANSZKY und ONDETTI (36) in „Strategie" und „Taktik" zu unterteilen ist. Während die Strategie die prinzipiellen Möglichkeiten der Verknüpfung von Aminosäuren zu Peptiden umfaßt, behandelt die Taktik Probleme der Schutzgruppen und Kupplungsmethoden.

An prinzipiellen Synthesemöglichkeiten stehen zur Zeit die Methode der Fragmentkondensation, die schrittweise Methode und die Peptidsynthese an ,fester Phase (Merrifield-Methode) zur Verfügung.

1. Konventionelle Synthesen

a) Fragmentkondensation: Die zu synthetisierende Aminosäure‑sequenz wird schematisch an geeigneten „Schnittstellen" in einzelne Teilsequenzen zerlegt. Diese Fragmente werden zunächst unabhängig voneinander synthetisiert und anschließend zur gewünschten Verbindung kondensiert. Als geeignete Schnittstellen sind diejenigen anzusehen, bei denen die Kupplung der Fragmente racemisierungsfrei verläuft (d. i. nach Glycin oder Prolin) und bei denen im Verlauf der Kupplung weder Nebenreaktionen noch eine unvollständige Umsetzung z. B. durch sterische Hinderung zu erwarten sind.

b) Schrittweise Kondensation: Die zu synthetisierende Sequenz wird, beginnend mit der C-terminalen Aminosäure, durch nacheinanderfolgende Ankondensation von jeweils nur einer Aminosäure aufgebaut. Analog kann eine Synthese auch mit der N-terminalen Aminosäure begonnen werden. Dieses Verfahren wird aber nur in Ausnahmefällen verwendet.

Selbstverständlich ist es möglich, die für eine Fragmentkondensation erforderlichen Teilsequenzen schrittweise aufzubauen oder zunächst eine Teilsequenz durch Fragmentkondensation herzustellen und diese dann schrittweise zum Endprodukt zu verlängern.

Es ist nicht möglich, eines dieser Syntheseprinzipien als das bessere zu bezeichnen. Die Fragmentkondensation ist in vieler Hinsicht rationeller, und vor allem ist die Auswahl der Schutzgruppen einfacher. Es gelingt meistens, die Fragmente so zu wählen, daß sich störende Schutzgruppen oder Reaktionen nicht in ein und demselben Fragment vorkommen. Nachteilig sind der größere Reinigungsaufwand und die meist geringeren Ausbeuten bei der Kondensation längerer Teilsequenzen. Die schrittweise Synthese ist in Hinblick auf Ausbeute und Reinigungsaufwand am günstigsten. Sie ist jedoch zeitraubender und stellt sehr hohe Anforderungen an die verwendeten Schutzgruppen und ihre selektive Abspaltbarkeit bzw. Stabilität.

2. Synthese an fester Phase (Merrifield-Methode)

In sehr kurzer Zeit hat die Peptidsynthese an fester Phase (*115*) ein allgemeines Interesse gefunden. Die Synthese beginnt mit der Fixierung der C-terminalen Aminosäure an einem unlöslichen hochpolymeren Träger (mit Divinylbenzol vernetztes Polystyrol) *(Abb. 11)*. Zur Einführung der ersten Aminosäure wird das Harz am Phenylring chlormethyliert und mit einer N-geschützten Aminosäure unter Bildung eines Benzylesters umgesetzt. Nach Abspaltung der N-Schutzgruppe erfolgt dann der Aufbau der Peptidkette durch Ankondensation der folgenden Aminosäuren, vorzugsweise nach der Carbodiimidmethode. Zur Erzielung eines praktisch quantitativen Umsatzes ist ein 3—5facher Überschuß erforderlich. Dieser vollständige Umsatz ist die Voraussetzung für eine erfolgreiche Synthese, da eine Abtrennung von am Harz fixierten Produkten, die durch Nebenreaktionen oder unvollständige Umsetzung entstanden sind, nicht möglich ist. Das Hauptproblem der Merrifield-

Synthese besteht im Fehlen ausreichender analytischer Methoden, die den sicheren Beweis einer quantitativen Kupplung liefern. Wie YOUNG et al. (225) gezeigt haben, bedeutet eine korrekte Aminosäureanalyse keineswegs, daß einheitliche Produkte vorliegen. BODANSZKY und SHEEHAN (40) haben die Methode der aktivierten Ester, WEYGAND und RAGNARSSON (216) erfolgreich eine Fragmentkondensation unter intermediärer Esterbildung mit N,N'-Dicyclohexylcarbodiimid/N-Hydroxysuccinimid eingesetzt.

$$
\begin{array}{c}
R \\
|\\
\mathrm{BOC-NH-CH-COOH} + \mathrm{Cl-CH_2-} \textcircled{P} \\
\downarrow\ \substack{Et_3N\\(-HCl)} \\
\mathrm{BOC-NH-CH-CO-O-CH_2-}\textcircled{P} \\
\downarrow\ \substack{1.\,HCl/AcOH\ oder\ Dioxan\\2.\,Et_3N/DMF} \\
\mathrm{H-NH-CH-CO-O-CH_2-}\textcircled{P} \\
\downarrow\ \substack{+BOC-NH-CH-COOH\\(DCCI)} \\
\mathrm{BOC-NH-CH-CONH-CH-CO-O-CH_2-}\textcircled{P} \\
\downarrow\ etc. \\
\mathrm{BOC-NH-CH-CO-[NH-CH-CO]_n-NH-CH-CO-O-CH_2-}\textcircled{P} \\
\downarrow\ HBr/Trifluoressigsäure \\
\mathrm{H_2N-CH-CO-[NH-CH-CO]_n-NH-CH-COOH} + \mathrm{Br-CH_2-}\textcircled{P}
\end{array}
$$

Abb. 11. Schema einer Peptidsynthese nach der Merrifield-Methode

Der große Überschuß an aktivierter Carboxylkomponente macht eine Blockierung aller funktionellen Gruppen weitgehend erforderlich. So werden Arginin als N^G-Nitro- oder N^G-Tosylderivat*, Serin und Tyrosin als O-Benzyl- und Histidin als N^{Im}-Benzylderivat* eingesetzt. Synthesen mit blockierter Hydroxylfunktion des Threonins sind bisher nicht bekannt. ω-Carboxylgruppen werden als Benzylester, ω-Aminogruppen mit dem Benzyloxycarbonyl- oder Tosylrest geschützt. Die Verwendung von Histidin mit ungeschützter Imidazolfunktion und Arginin in der protonisierten Form ist ebenfalls möglich (107). Zur Kettenverlängerung wird als α-Aminoschutz bevorzugt die tert.-Butyloxycarbonylgruppe, gelegentlich auch die o-Nitrophenylsulfenylgruppe (121) verwendet.

Nach Beendigung der Kettensynthese wird das Peptid vom Harz abgespalten. Da es in Form einer Benzylesterbindung vorliegt, ist eine Reaktion mit Bromwasserstoff in Trifluoressigsäure notwendig. Die

---

* G = Guanidin, Im = Imidazol.

relativ energischen Reaktionsbedingungen (90 Min., Raumtemperatur) können zur Spaltung von Peptidbindungen und zur Zerstörung einiger Aminosäuren führen. Auch eine Abspaltung vom Harz durch alkalische Verseifung ist beschrieben (*107*), ebenso eine Ammonolyse unter Amidbildung (*40*). Eine weitere Variation ist die von Sakakibara et al. (*149*) entwickelte und von Lenard und Robinson (*104*) bereits erfolgreich an einer substituierten Bradykininsequenz eingesetzte Abspaltung vom Harz mit flüssigem Fluorwasserstoff bei 0°, wobei neben der Abspaltung auch andere Schutzgruppen entfernt werden.

Mit der Synthese von Bradykinin (*116*), Bradykinylbradykinin (*121*), Met-Lys-Bradykinin (*117*), Ile^5-Angiotensin II (*113*), Ala^3-Ile^5-Angiotensin II C (*96*) und einiger Sequenzen des Tabakmosaikvirusproteins (*225*) wurde die Brauchbarkeit dieser Methode auch an Peptidwirkstoffen bewiesen. Nicht voll befriedigend war bisher der Aufbau der Insulinketten (*118, 226*). Zwar lieferten diese synthetischen Ketten bei der Kombination mit natürlichen Ketten oder untereinander insulinaktives Material, die geringen Ausbeuten im Vergleich zu der Kombination natürlicher Ketten lassen sich aber nicht nur auf die ungünstigen Bedingungen bei der Schutzgruppenabspaltung (Natrium in flüssigem Ammoniak) zurückführen.

In einer interessanten Variation der Merrifield-Methode wurde von Shemyakin et al. (*189*) ein in organischen Lösungsmitteln lösliches polymeres Trägermaterial benutzt. Ausgehend vom Aminoacylpolystyrol konnte die nachfolgende Kupplungsreaktion mit tert.-Butyloxycarbonyl-aminosäurehydroxysuccinimidester in Lösung ausgeführt werden. Der Überschuß des Aminosäurederivates und Nebenprodukte lassen sich nach Abspaltung der N-terminalen Schutzgruppe durch Eingießen der Reaktionsmischung in Wasser entfernen. Die Verwendbarkeit der Methode für Peptidwirkstoffe konnte am Beispiel des Gly^5-Gly^{10}-Gramicidin S demonstriert werden (*97*). In einem anderen Prinzip wurde von Fridkin et al. (*60*) mit 4% Divinylbenzol vernetztes Poly-4-hydroxy-3-nitrostyrol zur Bildung aktivierter Ester von z. B. Benzyloxycarbonyl- oder *o*-Nitrophenyl-sulfenyl-aminosäuren verwendet. In diesem Fall erfolgt die Kettenverlängerung durch Aminolyse der am Polymeren fixierten aktivierten Esterbindung mit einem zugesetzten Aminosäure- oder Peptidester. Nach diesem neuen Verfahren beschrieben Patchornik et al. (*129*) die Synthese des Bradykinins. Eine aussichtsreiche Anwendung dieser Technik liegt, wie weiterhin gezeigt werden konnte, in der Synthese von Cyclopeptiden. Eine an der Aminogruppe geschützte Peptidkette, die über die aktivierte Esterbindung mit dem polymeren Träger verknüpft ist, cyclisiert nach selektiver Entfernung der Aminoschutzgruppe durch intramolekulare Aminolyse.

B. Aminosäuren und die Blockierung ihrer funktionellen Gruppen

Nur wenige Probleme entstehen bei der Synthese eines Peptids aus einfachen Monoamino-monocarbonsäuren (Glycin, Alanin, Valin, Leucin, Isoleucin, Phenylalanin). Höhere Peptide aus ausschließlich oder bevorzugt diesen Aminosäuren sind sowohl in Wasser als auch in organischen Lösungsmitteln extrem unlöslich. Reinigungsoperationen sind dadurch meistens sehr erschwert. Ferner zeigen speziell die Aminosäuren mit verzweigter Seitenkette (Valin, Leucin, Isoleucin) häufig eine so starke sterische Hinderung, daß Kupplungen nur schlecht verlaufen. Die für eine Peptidsynthese mit nur diesen Aminosäuren zur Verfügung stehende Anzahl von Schutzgruppen ist sehr groß, da nur eine Selektivität zwischen der Amino- und der Carboxylschutzgruppe erforderlich ist. Die heterocyclischen Aminosäuren Tryptophan und Prolin sind ebenfalls Aminosäuren ohne zusätzliche funktionelle Gruppe. Allerdings verlaufen hier Peptidkupplungen und speziell Schutzgruppenabspaltungen nicht immer ohne Komplikationen. Tryptophanhaltige Peptide zeigen bei acidolytischen Schutzgruppenabspaltungen häufig Zersetzungen unter Violettfärbung. Bei prolinhaltigen Peptiden wird unter alkalischen oder sauren Bedingungen gelegentlich die Spaltung der Prolyl-Aminosäure-Bindung beobachtet. Enthält das zu synthetisierende Peptid Aminosäuren mit einer zusätzlichen funktionellen Gruppe, so sind höhere Anforderungen an die Schutzgruppenkombination zu stellen. Ist diese zusätzliche Funktion eine Aminogruppe (Lysin, Ornithin, α,γ-Diaminobuttersäure) oder eine Carboxylgruppe (Glutaminsäure, Asparaginsäure), so werden im Prinzip zur Blockierung dieser Funktionen dieselben Schutzgruppen verwendet, die auch zur Blockierung der α-Aminogruppe bzw. der α-Carboxylgruppe eingesetzt werden können.

Aus der Vielzahl der zur Blockierung der Aminofunktion vorgeschlagenen Gruppen (*42*) (*Abb. 12*, S. 64) haben sich für die Synthese von Peptidwirkstoffen nur wenige bewährt. An erster Stelle sind die bereits 1932 von BERGMANN und ZERVAS (*21*) eingeführte Benzyloxycarbonyl-(Carbobenzoxy-) und die 1957 von ANDERSON und McGREGOR (*6*) beschriebene tert.-Butyloxycarbonylgruppe zu nennen.

Die Benzyloxycarbonylgruppe kann in alle Aminosäuren über das Benzyloxycarbonylchlorid (Chlorameisensäurebenzylester) eingeführt werden. Die tert.-Butyloxycarbonyl-aminosäuren werden entweder durch Aminolyse des tert.-Butyloxycarbonyl-*p*-nitrophenylesters (tert.-Butyl-*p*-nitrophenyl-carbonat) (*6*) oder über das tert.-Butyloxycarbonylazid (*178*) erhalten. Eine Herstellung der tert.-Butyloxycarbonyl-aminosäuren mit wesentlich besseren Ausbeuten gelingt nach SCHNABEL (*155*) unter sorgfältiger pH-Kontrolle.

Die große Bedeutung dieser beiden Schutzgruppen liegt in ihrer Kombinationsmöglichkeit, wobei eine fast ideale Selektivität erhalten wird. Der Benzyloxycarbonylrest ist selektiv neben der tert.-Butyloxycarbonylgruppe durch katalytische Hydrierung abspaltbar. Anderseits ist die tert.-Butyloxycarbonylgruppe mit Chlorwasserstoff in einem geeigneten Lösungsmittel oder mit Trifluoressigsäure unter so milden Bedingungen acidolytisch spaltbar, daß die Benzyloxycarbonylgruppe nicht angegriffen wird. Zu ihrer acidolytischen Spaltung sind stärkere Reaktionsbedingungen notwendig, z. B. die Umsetzung mit Bromwasserstoff in einem organischen Lösungsmittel, bevorzugt in Eisessig. Auf diesem Wege gelingt die gemeinsame Abspaltung beider Gruppen nach beendeter Synthese oder, in Abwesenheit der tert.-Butyloxycarbonylgruppe, die Abspaltung der Benzyloxycarbonylgruppe, wenn z. B. in Gegenwart schwefelhaltiger Aminosäuren eine katalytische Hydrierung nicht durchführbar ist.

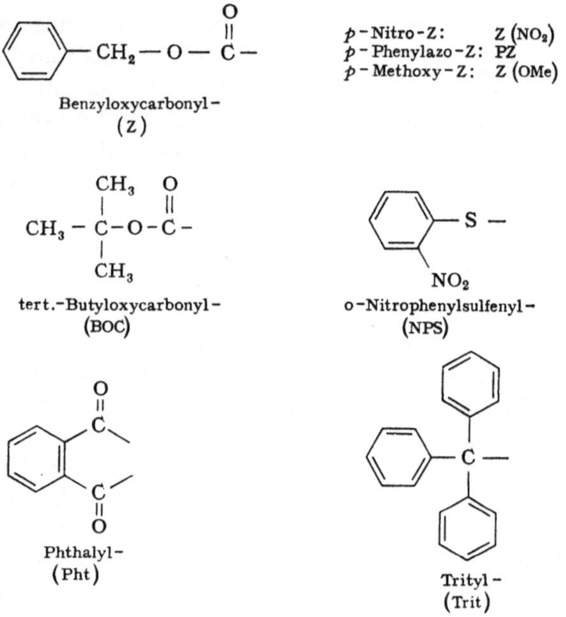

Abb. 12. Aminoschutzgruppen

Neben der Benzyloxycarbonylgruppe finden gelegentlich *p*-substituierte Derivate Anwendung. Der *p*-Nitro- (*50*) und *p*-Chlorbenzyloxycarbonylrest (*98*) werden in erster Linie wegen der besseren Kristallisation der Derivate verwendet. Der *p*-Methoxybenzyloxycarbonyl- (*215*) und der tert.-Amyloxycarbonylrest (*150*) sind ähnlich leicht acidolytisch spaltbar, wie die tert.-Butyloxycarbonylgruppe.

Die Kombination Benzyloxycarbonyl- und tert.-Butyloxycarbonylgruppe reicht jedoch in vielen Fällen nicht aus, speziell dann nicht, wenn im Molekül weitere hydrierbare Schutzgruppen verwendet werden müssen oder wenn durch schwefelhaltige Aminosäuren die Hydrierung verhindert oder erschwert ist. In solchen Fällen haben sich drei weitere Schutzgruppen bewährt (Abb. 12): Bedingt geeignet ist der durch Hydrazinolyse spaltbare Phthalylrest, der allerdings nicht die Anwesenheit hydrazinolytisch spaltbarer Esterbindungen gestattet. Günstig ist eine Kombination mit dem hydrazinolytisch stabilen tert.-Butylester (*4, 30*).

Literaturverzeichnis: SS. 107—119

Bei längeren Peptiden wird jedoch ein vorteilhafter Einsatz des Phthalylrestes als Schutz der α-Aminogruppe fragwürdig. Der häufiger verwendete Tritylrest ist auf Grund seiner leichten Acidolyse durch heiße Essigsäure eine geeignete Schutzgruppe neben dem Benzyloxycarbonylund dem tert.-Butyloxycarbonylrest (*173*). Nachteilig ist die unbefriedigende Einführung der Tritylgruppe in freie Aminosäuren bzw. die durch sterische Hinderung oft erschwerte alkalische Verseifung von N-Trityl-aminosäureestern. Als Ausweg wird eine Teilsequenz zunächst mit einer anderen Schutzgruppe aufgebaut und nach deren Abspaltung trityliert (*175*). Eine schnelle Verbreitung hat seit 1963 der o-Nitrophenylsulfenylrest gefunden (*229*). Wenn auch seine acidolytische Abspaltung neben dem tert.-Butyloxycarbonylrest nur bedingt durchführbar ist (vgl. *137*) und z. B. bei tryptophanhaltigen Peptiden nur unter Bildung

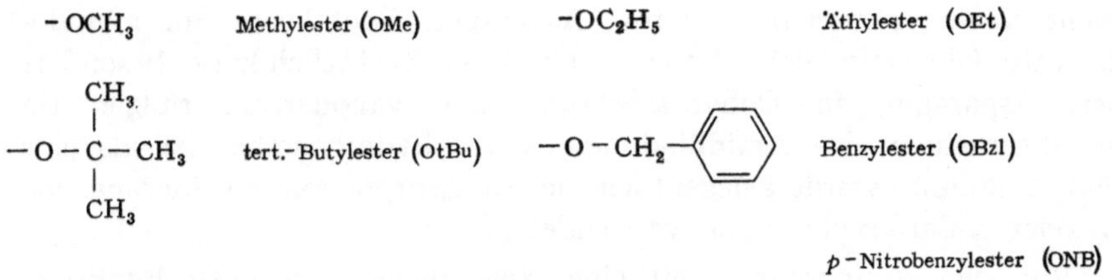

$-\mathrm{OCH_3}$ Methylester (OMe) $-\mathrm{OC_2H_5}$ Äthylester (OEt)

$$-\mathrm{O}-\underset{\underset{\mathrm{CH_3}}{|}}{\overset{\overset{\mathrm{CH_3}}{|}}{\mathrm{C}}}-\mathrm{CH_3}$$ tert.-Butylester (OtBu) $-\mathrm{O}-\mathrm{CH_2}-\langle\!\bigcirc\!\rangle$ Benzylester (OBzl)

p-Nitrobenzylester (ONB)

Abb. 13. Carboxylschutzgruppen

von Nebenprodukten gelingt (*219*), läßt er sich mit Sicherheit selektiv durch geeignete Thiole unter Bildung des symmetrischen Disulfids abspalten (*59, 95*). Die o-Nitrophenylsulfenylgruppe hat bereits bei Naturstoffsynthesen gute Dienste geleistet.

Der bevorzugte Schutz der Carboxylfunktion *(Abb. 13)* ist die Esterbildung. Bei kürzeren Sequenzen stehen der Methyl- oder Äthylester immer noch an erster Stelle. Bei längeren Sequenzen ist jedoch die notwendige alkalische Verseifung erschwert und kann zu einer unerwünschten Racemisierung führen. Häufig hat sich daher der durch katalytische Hydrierung abspaltbare Benzylester oder noch besser der gegenüber acidolytischen Bedingungen stabilere p-Nitrobenzylester bewährt (*44, 167*). Eine selektive katalytische Hydrierung der Aminoschutzgruppen ist dann natürlich nicht mehr möglich. Als dritter Estertyp zum Schutz der Carboxylfunktion steht der tert.-Butylester zur Verfügung (*4*). Auf Grund seiner leichten sauren Verseifung spielt er etwa die gleiche Rolle zum Schutz von ω-Carboxylgruppen wie die tert.-Butyloxycarbonylgruppe zum Schutz von ω-Aminogruppen. Die Veresterung an der freien oder an der N-blockierten Aminosäure gelingt leicht durch Umsetzung mit Isobuten oder durch Umesterung mit tert.-Butylacetat (*203*).

Eine Carboxylblockierung kann auch durch eine gleichzeitige oder vorbereitete Aktivierung erreicht werden. Unter bestimmten Voraussetzungen können die zur Peptidkupplung verwendeten aktivierten Ester und die zur Azidkupplung erforderlichen Hydrazide, sofern sie an der freien NH_2-Gruppe des Hydrazins reversibel blockiert sind, als Carboxylschutz eingesetzt werden (*44*).

Ein sehr einfacher Weg, die Carboxylfunktion zu blockieren ist die Salzbildung. Sie legt allerdings hinsichtlich der verwendbaren Kupplungsmethoden gewisse Beschränkungen auf. So ist es erforderlich, die für die Peptidbindung vorgesehene Carboxylgruppe getrennt zu aktivieren und erst dann die Komponente mit der salzblockierten Carboxylgruppe zuzugeben. Andernfalls würde eine Konkurrenzreaktion an der salzblockierten Carboxylgruppe zu unerwünschten Nebenreaktionen führen.

Aminodicarbonsäuren mit amidierter ω-Carboxylfunktion (Glutamin, Asparagin) sind häufig Bestandteile von Peptidwirkstoffen. Bei Synthesen mit diesen Aminosäuren können an der ω-Säureamidgruppierung leicht Nebenreaktionen auftreten: Asparagin bildet leicht Succinimidoderivate, Glutamin leicht Pyroglutaminsäure. Zusätzlich kann, besonders beim Asparagin, eine Dehydratisierung zum Cyanoderivat erfolgen. Da im allgemeinen eine Amidbindung leichter hydrolysierbar ist als eine Peptidbindung, wurde gelegentlich die Amidgruppe zur Blockierung von ω- oder α-Carboxylgruppen verwendet.

Von den Aminosäuren mit einer zusätzlichen basischen Funktion, wird Histidin meistens mit ungeschützter Imidazolfunktion zur Peptidsynthese eingesetzt. Als brauchbarer Schutz (Abb. 14) kann der Benzylrest (*56*) verwendet werden, der jedoch zu seiner Abspaltung, speziell an längeren Peptidketten, drastische Bedingungen erfordert. Seltener haben N^{α},N^{Im}-Dibenzyloxycarbonyl-histidin (*128*) bzw. der entsprechende *p*-Nitrophenylester (*122*) und N^{Im}-trityliertes Histidin für Naturstoffsynthesen Anwendung gefunden (*38, 71*).

Für die Guanidofunktion der Arginins sind im Laufe der Zeit eine große Zahl von Schutzgruppen beschrieben worden. Von diesen hat sich als umfassend anwendbar bisher nur die Nitrogruppe durchsetzen können. Sie läßt sich durch katalytische Hydrierung abspalten. Allerdings sind hier bei längeren Peptiden, besonders wenn der Nitroargininrest nicht terminal steht, lange Hydrierzeiten und große Katalysatormengen oder Druckhydrierungen erforderlich, wobei trotzdem nicht immer eine vollständige Reaktion erreicht wird. Von Sakakibara et al. (*149*) wird die Entfernung der Nitrogruppe mit flüssigem Fluorwasserstoff beschrieben. Größere Erfahrungen mit dieser Methode stehen noch aus. Da die Nitrogruppe als Blockierung der ω-Funktion bei einem weiteren Kettenaufbau die Kombination mit dem ebenfalls hydrierbaren Benzyloxycarbonylrest nicht gestattet, werden häufig kleinere Nitroarginin enthaltende Fragmente mit anderen Aminoschutzgruppen, z. B. der Trityl- oder tert.-

Literaturverzeichnis: SS. 107—119

Butyloxycarbonylgruppe hergestellt. Nach Einführung der Benzyl-oxycarbonylgruppe wird diese gemeinsam mit der Nitrogruppe durch Hydrierung abgespalten. Im Verlauf der weiteren Synthese wird die Guanidofunktion des Arginins durch Protonisierung geschützt.

Imidazol – Stickstoff:

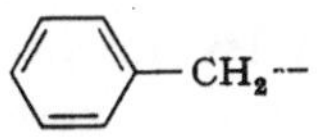

Benzyl (Bzl)

Guanidogruppe:

NO_2

Nitro (NO_2)

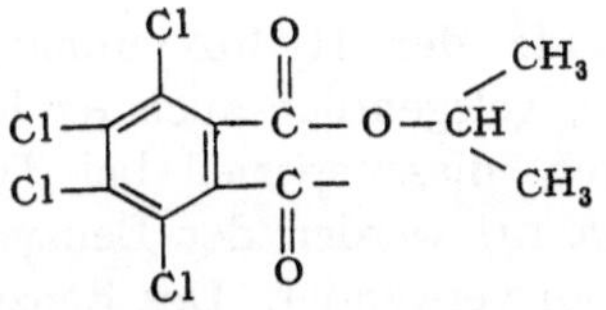

2 – lsopropyloxycarbonyl – 3,4,5,6 – tetrachlorbenzoyl (TIP)

Hydroxylgruppen:

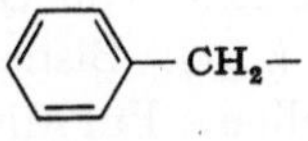

Benzyl (Bzl)

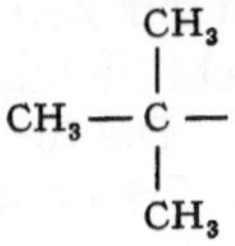

tert.-Butyl (tBu)

Thiolgruppen:

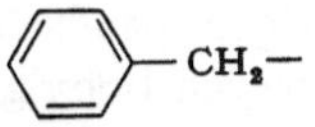

Benzyl (Bzl)

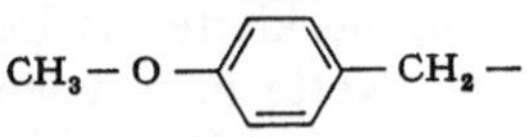

$CH_3 - O$ ... $CH_2 -$ p – Methoxybenzyl (MOB)

$C_2H_5 - NH - CO -$ Äthylcarbamoyl (EC)

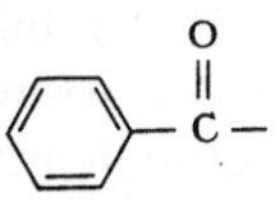

Benzoyl (Bz)

Abb. 14. Schutzgruppen für spezielle funktionelle Gruppen

Als Guanidoschutz läßt sich auch der Benzyloxycarbonylrest oder der Tosylrest verwenden. Als nachteilig haben sich die ungünstigen Kombinationsmöglichkeiten mit anderen Schutzgruppen im Falle des N^G,N^G-Dibenzyloxycarbonylschutzes und die nur mit Natrium in flüssigem Ammoniak durchführbare Abspaltung des Tosylrestes erwiesen. Als Schutz für die Guanidofunktion wurde von GUTTMANN und PLESS (69) der N^G-p-Nitrocarbobenzoxyrest vorgeschlagen, der im Vergleich zu den N^G,N^G-Dibenzyloxycarbonylgruppen acidolytisch stabil ist, aber leicht durch Hydrierung entfernt werden kann. Ein ideales Argininderivat stellt das N^α-Benzyloxycarbonyl-N^G-tert.-butyloxycarbonylarginin dar, das jedoch infolge einer nur schwierigen Herstellung noch keine breitere Anwendung gefunden hat (*130*). Ebenfalls wenig Erfahrung liegt mit der 2-Isopropyloxy-3,4,5,6-tetrachlorbenzoylgruppe vor, die gegenüber einer katalytischen Hydrierung und HBr/Eisessig stabil ist, aber mit Eisessig bei 100° entfernt werden kann (*69*).

Aminosäuren mit einer zusätzlichen Hydroxylfunktion (alkoholisch: Serin, Threonin; phenolisch: Tyrosin) werden gewöhnlich ohne Schutz der Hydroxylfunktion zur Peptidsynthese eingesetzt. In den Fällen, wo ein größerer Überschuß an carboxylaktivierter Verbindung zur Kupplung herangezogen wird, ist ein O-Schutz wünschenswert oder sogar erforderlich (schrittweise Methode, Merrifield-Methode). Auch einige Kupplungsmethoden erfordern den Schutz der Hydroxylfunktion (Imidazolidmethode, Säurechloridmethode, gelegentlich auch Azidmethode und die Methode der aktivierten Ester, diese speziell bei Tyrosin). Als geeignete O-Schutzgruppen *(Abb. 14)* werden der Benzyläther (*125, 220*) und der tert.-Butyläther (*29, 49*) verwendet. Der Benzyläther ist in die aliphatischen Hydroxyaminosäuren schwer einführbar. L-Serin-O-benzyläther ist z. B. nur durch eine Totalsynthese von O-Benzyl-DL-serin mit folgender Racematspaltung zugänglich (*221*). Bisher sehr unrationell ist der L-Threonin-O-benzyläther herzustellen. Für die Synthese nach der Merrifield-Methode ist dieser Schutz trotzdem zur Zeit die Methode der Wahl. Die tert.-Butyläther sind relativ leicht durch Umsetzung der Hydroxyaminosäure mit Isobuten zugänglich. Die leichte acidolytische Spaltung des tert.-Butyläthers erlaubt unter Erhalt der Ätherbindung eine Kombination mit der Benzyloxycarbonyl- bzw., in Anwesenheit von Nitroarginin, mit der o-Nitrophenylsulfenyl- oder Methionin mit der Phthalyl- oder der o-Nitrophenylsulfenylgruppe als Aminoschutz (*30, 136*). Eine Kombination mit der tert.-Butyloxycarbonylgruppe ermöglicht nur die Einführung der Hydroxyaminosäure mit blockierter Hydroxyfunktion, nicht aber eine weitere Verlängerung der Kette mit diesem O-Schutz.

Von den schwefelhaltigen Aminosäuren wird Methionin stets ohne Schutz der Thioätherfunktion eingesetzt. Als Nebenreaktion sind die leichte Oxydierbarkeit zum Methioninsulfoxid und bei acidolytischer Spaltung der Benzyloxycarbonylgruppe der Austausch der S-Methylgruppe durch den Benzylrest störend. Cystein wird ausschließlich mit geschützter Thiolfunktion (Abb. 14, S. 67) zur Peptidsynthese eingesetzt. Trotz vieler cysteinhaltiger Peptidwirkstoffe, die bereits synthetisiert worden sind, ist das Problem der S-Schutzgruppe noch nicht befriedigend gelöst. Die Mehrzahl aller Synthesen ist mit dem S-Benzylschutz durchgeführt worden (*218*), der sehr stabil ist unter den Bedingungen, die zur Abspaltung anderer Schutzgruppen erforderlich sind. Als einzige befriedigende Methode zur Abspaltung des S-Benzylrestes wird die Reaktion mit Natrium in flüssigem Ammoniak verwendet. Diese Reaktion, die sehr häufig zu unerwünschten Nebenreaktionen führt, war Anlaß für eine Reihe von systematischen Untersuchungen (*15, 66*). Unvorteilhaft ist diese Schutzgruppe bei Insulinsynthesen. Die wesentlich schlechteren Ausbeuten bei der Kombination synthetischer Ketten im Vergleich zu der Kombination natürlicher Ketten konnte auf Nebenreaktionen bei der Behandlung mit Natrium in flüssigem Ammoniak zurückgeführt werden (*205*). Es ist daher verständlich, daß eine größere Zahl weiterer S-Schutzgruppen vorgeschlagen worden ist. Von diesen sind der S-Tritylrest (*206*), der S-p-Methoxybenzylrest (*149*), der S-Carbobenzoxyrest und der S-Benzoylrest (*132*), sowie der S-Äthylcarbamoylrest (*67*) bereits bei Synthesen auf dem Oxytocin-Vasopressin-Gebiet verwendet worden. Der Beweis ihrer wirklichen Überlegenheit gegenüber dem S-Benzylrest, z. B. bei der Synthese der Insulinketten, steht jedoch noch aus. Die S-Tritylgruppe ist relativ leicht acidolytisch spaltbar. Zum Aufbau der Peptidkette kommen daher zusätzlich nur der N^{α}-Tritylrest bzw. die N^{α}-o-Nitrophenylsulfenylgruppe in Frage. Der S-Carbobenzoxyrest ist acidolytisch relativ stabil, so daß der Aufbau der Peptidkette über tert.-Butyloxycarbonylderivate erfolgen kann. Eine Entfernung gelingt unter stärker sauren Bedingungen und durch Einwirkung von wäßrigem Ammoniak (*20*). Der S-Benzoylrest ist acidolytisch stabil, er ermöglicht also ohne Schwierigkeiten die Synthese eines cysteinhaltigen Peptides. Die Abspaltung erfolgt unter alkalischen Bedingungen (*132*). Die S-Carbobenzoxy- und die S-Benzoylgruppe unterliegen als Thioester leicht einem S → N-Shift und geben dadurch Anlaß zu Nebenreaktionen. Auch der S-Äthylcarbamoylrest ist gegen Säuren stabil. Die Abspaltung erfordert alkalische Bedingungen, ist aber auch mit flüssigem Ammoniak oder mit Hydrazinhydrat möglich (*67*).

C. Bildung der Peptidbindung

In der Literatur sind seit 1950 eine Vielzahl von Methoden zur Aktivierung der Carboxylgruppe beschrieben worden (vgl. *2*). Nur wenige haben sich als „Kupplungsmethoden" zur Synthese von Peptidwirkstoffen geeignet erwiesen, da sie die nachstehend aufgeführten Voraussetzungen wenigstens im weiten Maße erfüllen konnten.

1. Die Aktivierung der Carboxylgruppe muß so stark sein, daß auch im Falle längerer Peptide oder sterisch gehinderter Aminosäuren die Peptidbindung in guter Ausbeute gebildet wird.

2. Die Aktivierung sollte weder durch Umlagerungen oder Reaktionen mit anderen Teilen des Moleküls zu unerwünschten Nebenprodukten führen. Auch müssen die auf Grund der Reaktionsgleichung zwangsläufig neben dem gewünschten Peptid entstehenden Produkte leicht zu entfernen sein.

3. Sowohl die Aktivierung als auch die Kupplung muß so verlaufen, daß die optisch aktiven Zentren der Aminosäuren nicht angegriffen werden.

Zu den bewährten Methoden gehört auch heute noch die von Curtius 1904 eingeführte Azidmethode (vgl. *Abb. 15*). Die zur Azidbildung erforderliche Vorstufe, die Säurehydrazide, wird durch Hydrazinolyse von N-geschützten Aminosäure- oder Peptidestern erhalten. Bei längeren oder sterisch gehinderten Peptiden ist diese Reaktion häufig erschwert. Als Ausweg kann die Hydrazidgruppe auf einer früheren Synthesestufe, vorzugsweise bereits am Aminosäurederivat eingeführt werden. Um eine Beteiligung der Hydrazidfunktion an den folgenden Kupplungsreaktionen zu vermeiden, muß ihre NH_2-Gruppe blockiert sein. Zum Schutz werden die gleichen Gruppen eingesetzt, die üblicherweise zur Blockierung von Aminogruppen verwendet werden (Benzyloxycarbonyl-, tert.-Butyloxycarbonyl- und Tritylgruppen). Zu einem gewünschten Zeitpunkt ist es möglich, die Hydrazidfunktion freizulegen und eine Azidkupplung durchzuführen. Erschwert ist in diesen Fällen selbstverständlich der Schutz von ω-Aminogruppen innerhalb des Moleküls, für die eine weitere selektive Aminoschutzgruppe erforderlich ist. Die Azidbildung wird durch Umsetzung des Säurehydrazids mit salpetriger Säure oder einem Derivat dieser Säure, bevorzugt tert.-Butylnitrit (*83*), vorgenommen. Ein wesentlicher Vorteil der Azidmethode ist der racemisierungsfreie Verlauf der Kupplungsreaktion. Eine Umlagerung des Azids zum Isocyanat kann zu Nebenreaktionen führen. Als Folge dieser Umlagerung bilden sich mit der Aminokomponente Harnstoffderivate oder mit ungeschützten Hydroxylgruppen Urethane.

Eine seit ihrer Einführung im Jahre 1951 aktuell gebliebene Kupplungsmethode ist die der gemischten Anhydride. Von den vielen zur Anhydridbildung vorgeschlagenen „Hilfssäuren" werden fast ausschließlich die Kohlensäuremonester verwendet (Abb. 15). Die Aktivierung der Carboxylgruppe erfolgt im allgemeinen intermediär, sie ist leicht durchzuführen und als weitere Reaktionsprodukte entstehen nur CO_2 und der entsprechende Alkohol. Ein entscheidender Nachteil liegt in der großen Racemisierungstendenz, die immer dann vorhanden ist, wenn eine Acylaminosäure in das gemischte Anhydrid übergeführt wird. Da auch Peptide Acylaminosäuren sind, können N-geschützte Peptidsäuren nur racemisierungssicher gekuppelt werden, wenn die C-terminale Aminosäure Glycin oder Prolin ist. Durch eine Reihe von Untersuchungen ist sichergestellt, daß der wesentliche, wenn auch nicht der einzige Grund für eine Racemisierung die Bildung eines Azlactons ist. Bei einer Urethangruppierung, z. B. der Benzyloxycarbonyl- oder tert.-Butyloxycarbonylschutzgruppe, kann eine Azlactonbildung und damit eine Racemisierung nicht auftreten. Neben der Kupplung von Peptidsäuren mit C-terminalem

Glycin oder Prolin wird daher die Methode der gemischten Anhydride bevorzugt zur Kupplung von N-geschützten Aminosäuren verwendet.

Bei der von SHEEHAN und HESS (*180*) eingeführten Carbodiimid-Methode werden zur Bildung der Peptidbindung N,N'-dialkylierte Carbo-

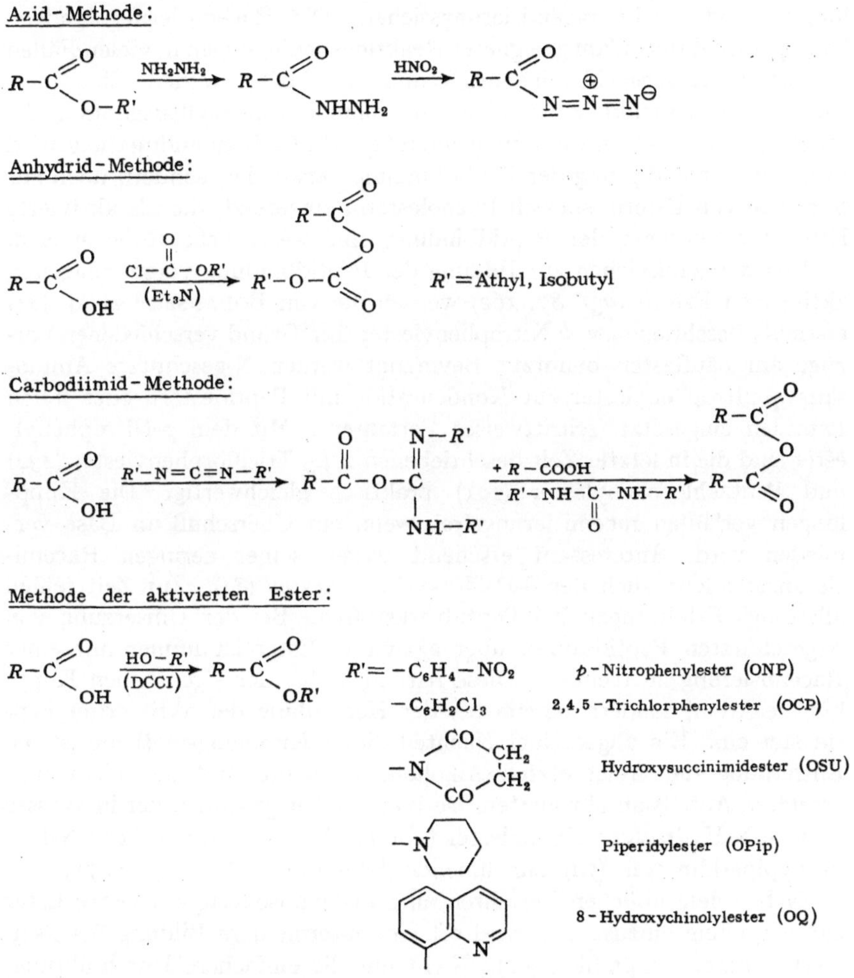

Azid-Methode:

Anhydrid-Methode:

$R' = $ Äthyl, Isobutyl

Carbodiimid-Methode:

Methode der aktivierten Ester:

$R' = -C_6H_4-NO_2$ *p*-Nitrophenylester (ONP)

$-C_6H_2Cl_3$ 2,4,5-Trichlorphenylester (OCP)

Hydroxysuccinimidester (OSU)

Piperidylester (OPip)

8-Hydroxychinolylester (OQ)

Abb. 15. Methoden zur Aktivierung der Carboxylgruppe

diimide, bevorzugt N,N'-Dicyclohexylcarbodiimid, in Sonderfällen wasserlösliche Carbodiimide, z. B. N-Cyclohexyl-N'-(2-morpholinyl-(4)-äthyl)-carbodiimid, benutzt. Die Carbodiimidmethode (Abb. 15) verläuft ebenfalls über eine anhydridartige Zwischenstufe, die durch Anlagerung der Carboxylgruppe an die C=N-Doppelbindung entsteht. Der weitere Verlauf der Kupplungsreaktion ist noch nicht bis ins einzelne geklärt. Diskutiert werden die direkte Umsetzung dieses „gemischten Anhydrides" mit der Aminokomponente oder eine weitere Umsetzung zum symmetrischen Anhydrid, das dann die Peptidbindung bildet. Durch den Verlauf über ein anhydridartiges Zwischenprodukt ist auch diese Kupplungsmethode nicht racemisierungssicher. Die Racemisierungstendenz kann jedoch durch Wahl geeigneter Reaktionsbedingungen in vielen Fällen bis auf ein zu vernachlässigendes Minimum reduziert werden. Eine unerwünschte Nebenreaktion ist die leichte Bildung von Acylharnstoffen, die durch eine O → N-Acylwanderung entstehen. Die Carbodiimidmethode wird nicht nur zur Knüpfung der Peptidbindung verwendet, sondern auch zur Synthese von Estern, speziell Phenolestern, eingesetzt, die als aktivierte Ester zur Synthese der Peptidbindung eine wesentliche Rolle spielen.

Von den zahlreichen zur Bildung der Peptidbindung vorgeschlagenen aktivierten Estern (vgl. *87, 162*) werden die von Bodanszky et al. (*41*) erstmals beschriebenen *p*-Nitrophenylester auf Grund verschiedener Vorzüge am häufigsten benutzt. Bevorzugt werden N-geschützte Aminosäure-*p*-nitrophenylester zur Kondensation mit Peptidestern oder freien Peptiden eingesetzt (schrittweises Verfahren). Mit dem *p*-Nitrophenylester sind die in letzter Zeit beschriebenen 2,4,5-Trichlorphenylester (*134*) und Pentachlorphenylester (*101*) praktisch gleichwertig. Die Kupplungen verlaufen racemisierungsfrei, wenn ein Überschuß an Base vermieden wird. Interessant erscheint wegen seiner geringen Racemisierungstendenz auch der 8-Hydroxychinolinester (*88*). Zur Zeit fehlen allerdings Erfahrungen bei Peptidwirkstoffen. Bei der Umsetzung von N-geschützten Peptidsäuren über aktivierte Ester ist immer mit einer Racemisierung zu rechnen. Diese tritt nicht bei der eigentlichen Kupplungsreaktion, sondern bereits bei der Herstellung der aktivierten Peptidester ein. Ein allgemeiner Nachteil dieser Kupplungsmethode ist die Entfernung der freigesetzten Alkoholkomponente und des nicht umgesetzten Anteils an aktiviertem Derivat. Daher gewinnen der in Wasser lösliche N-Hydroxysuccinimidester (*7*) oder der in Säuren lösliche N-Hydroxypiperidinester (*74*) zunehmende Bedeutung (Abb. 15, S. 71).

Neben dem üblichen Verfahren, in Substanz isolierte aktivierte Ester zur Kupplung einzusetzen, wird oft eine intermediäre Bildung des aktivierten Esters empfohlen (*61*). Nicht nur die einfachere Durchführung, sondern auch eine geringere Racemisierungstendenz sind wesentliche Vorteile. Besonders günstig scheint die Verwendung von Carbodiimid/N-Hydroxysuccinimid zu sein (*214*). Auch die Kombination gemischte Anhydrid-Methode/N-Hydroxysuccinimid wurde untersucht (*5*). Nach den letzten beiden Variationen der aktivierten Estermethode konnte bisher selbst bei N-geschützten Peptidsäuren keine Racemisierung nachgewiesen werden.

Literaturverzeichnis: SS. 107—119

Bei den im voranstehenden beschriebenen Kupplungsmethoden handelt es sich um die Verfahren, die nach den auf Grund systematischer Untersuchungen gewonnenen Erkenntnissen die geringste Wahrscheinlichkeit für eine Racemisierung bieten, bzw. bei denen unter bestimmten präparativen Vorsichtsmaßnahmen eine prinzipiell mögliche Racemisierung auf ein Minimum reduziert werden kann. Welche Bedeutung die Racemisierung für die Peptidchemie und speziell für die Synthese von Peptidwirkstoffen hat, zeigt der Umfang an Arbeiten, der jedes Jahr auf dem europäischen Peptidsymposium (*32*, *48*, *228*) diesem Thema gewidmet wird.

Da eine auch nur oberflächliche Darstellung der vielen bisher erarbeiteten Befunde über die Theorie der Racemisierung, physikalische, biologische, chemische und enzymatische Nachweismethoden und schließlich das Ausmaß der Racemisierung bei den einzelnen Kupplungsmethoden den Rahmen dieser Abhandlung sprengen würde, sei auf eine weitere Zusammenfassung verwiesen (*162*).

D. Reinigungsmethoden und Analytik

Reinigungsmethoden: Verglichen mit anderen Gebieten der organisch-präparativen Chemie nimmt im Verlauf der Synthese eines Peptidwirkstoffes die Reinigung von Zwischenprodukten und Endprodukt einen unverhältnismäßig breiten Raum ein. Abgesehen von Aminosäurederivaten und einfachen Peptiden gelingt eine Kristallisation nur in Ausnahmefällen. Durch ein Umfällen der Produkte wird meistens keine vollständige Reinigung erreicht. Es ist also erforderlich, moderne Methoden wie Gegenstromverteilung, Säulenchromatographie oder präparative Elektrophorese anzuwenden. Welches Verfahren im jeweiligen Falle geeignet ist, richtet sich in starkem Maße nach der Löslichkeit der Substanz.

Für wasserlösliche Produkte ist die Zahl der zur Verfügung stehenden Methoden relativ groß. Unterscheiden sich die zu trennenden Substanzen in ihrer Molekülgröße, so kann die Gelfiltration mit Erfolg angewendet werden (*54*). Der Molekularsiebeffekt kann gelegentlich von einer Adsorptionschromatographie überlagert werden. Die zur Verfügung stehende Vielzahl verschiedenartiger Gele gestattet es, den Molekularsiebeffekt zur Trennung auch kleinerer Moleküle mit einem Molekulargewicht von unter 1000 einzusetzen. Eine Verteilungschromatographie an Gelen wurde von YAMASHIRO (*223*) beschrieben. Unterscheiden sich die zu trennenden Verbindungen in ihrer elektrischen Ladung, so können die verschiedenartigsten Typen von Harzaustauschern und Austauschercellulosen eingesetzt werden. Durch die Einführung von sauren und basischen Gruppen in die typischen Träger der Gelfiltration ist es auch möglich geworden, den Molekularsiebeffekt mit einer Austauscherchromatographie zu kombinieren. Gegenüber diesen Methoden tritt die reine Adsorptionschromatographie an z. B. Cellulosepulver oder Kieselgel immer mehr in den Hintergrund. Ebenfalls geeignet für die Reinigung von Substanzen, die sich in ihrer elektrischen Ladung unterscheiden, ist die präparative Elektrophorese. Ihr etwas größerer apparativer Aufwand wird im allgemeinen dadurch kompensiert, daß es fast immer gelingt, das Ergebnis der analytischen Papierelektrophorese auf die präparative Trennung zu übertragen. Von den verschiedenen Prinzipien der präparativen Elektrophorese ist die trägerfreie, auf Grund ihrer Möglichkeit, größere

Mengen durchzusetzen, den anderen Methoden überlegen (*165*). Kombinierte Methoden, wie eine Gelelektrophorese oder eine Dünnschichtelektrophorese, werden in Spezialfällen zur mikropräparativen Trennung erfolgreich angewendet.

Sehr viel geringer ist die Auswahl der Methoden, die zur Reinigung von wasserunlöslichen Substanzen angewendet werden können. Trotz ihres großen apparativen Aufwandes ist die Gegenstromverteilung als Methode der Wahl anzusehen. Mit den in neuerer Zeit zugänglich gewordenen, in organischen Lösungsmitteln unlöslichen Dextran- und Polyacrylamidgelen ist aber auch eine Gelfiltration mit wasserunlöslichen Produkten möglich geworden.

Analytik : Aminosäurederivate und einfache Peptide aus nur wenigen Aminosäuren können durch die üblichen organischen Analysenmethoden identifiziert werden. Bei höheren Peptiden wird jedoch ihre Aussagekraft zunehmend geringer. Es hat sich daher eine spezielle Peptidanalytik entwickelt, die sich auf die Verfahren bezieht, die zur Struktur- und Sequenzanalyse auf dem Peptid- und Proteingebiet verwendet werden. Die Einheitlichkeit eines Produktes wird meistens chromatographisch oder elektrophoretisch bewiesen. Speziell bei Verbindungen, die keine freien funktionellen Gruppen haben (geschützte Peptide), ist die Dünnschichtchromatographie ein unentbehrliches Hilfsmittel. Allerdings reichen diese Methoden nicht in jedem Falle aus, da es speziell bei längeren Sequenzen keineswegs auszuschließen ist, daß Verunreinigungen die gleiche Wanderungsgeschwindigkeit zeigen wie das gewünschte Produkt. Gelegentlich werden auch Gegenstromverteilung oder Säulenchromatographie zum Beweis der Einheitlichkeit herangezogen.

An die Stelle der Elementaranalyse tritt bei höheren Peptiden die Aminosäureanalyse, die die Anwesenheit aller Aminosäuren und das Verhältnis der Aminosäuren untereinander angibt. Die zur chromatographischen Bestimmung erforderliche saure Totalhydrolyse führt zu einer mehr oder weniger starken Zersetzung einiger Aminosäuren (Serin, Threonin, Tyrosin, Cystein, Arginin); Trypthophan wird vollständig zerstört. Als Ausweg kann die alkalische Totalhydrolyse angewendet werden.

Eine unerläßliche Forderung in der Peptidanalytik ist der Nachweis der sterischen Reinheit, d. h. der Beweis, daß im Verlauf der Synthese keine Racemisierung eingetreten ist. Für diesen Zweck hat sich der Abbau mit proteolytischen Enzymen bewährt (*231*). Die wichtigsten Enzyme für diesen Zweck sind Trypsin (Spaltung nach basischen Aminosäuren) und Chymotrypsin (Spaltung bevorzugt, aber nicht ausschließlich nach aromatischen Aminosäuren). Ein schrittweiser Abbau vom Aminoende ist mit Aminopeptidasen, ein Abbau vom Carboxylende mit Carboxypeptidasen möglich (Carboxypeptidase A: Abspaltung von neutralen Aminosäuren, Carboxypeptidase B: Abspaltung nur basischer Aminosäuren). Eine weitere enzymatische Methode ist der Abbau nach Totalhydrolyse mit geeigneten L-Aminosäurebzw. D-Aminosäureoxydasen.

III. Synthese von Peptidwirkstoffen

A. Peptidhormone der Hypophyse

1. Adrenocorticotropes Hormon

Seit der ersten Synthese der N-terminalen ACTH-Eicosapeptidsequenz durch Boissonnas et al. (*46*) im Jahre 1956 sind etwa 40 weitere Teilsequenzen einschließlich ihrer Analoga, sowie die natürlichen Hormone von Schwein und Mensch mit 39 Aminosäureresten synthetisiert worden.

Literaturverzeichnis: SS. 107—119

Die Teilsequenzen umfassen den N-terminalen Bereich 1—16 bis 1—28. Ein Vergleich dieser Synthesen aus einem Zeitraum von 12 Jahren läßt den Fortschritt auf dem Gebiet der präparativen Peptidchemie erkennen. In Anbetracht der günstigen Verteilung der racemisierungsfreien Aminosäuren Glycin und Prolin über das gesamte Molekül (Gly^{10}, Gly^{14}, Pro^{19}, Pro^{24}, Gly^{27}) haben alle Autoren die Fragmentkondensation mit mehr oder weniger gleichen Schnittstellen gewählt. Wesentliche Unterschiede bestehen dagegen in der Wahl der Schutzgruppenkombinationen.

Synthese des α^{1-16}-ACTH von HOFMANN *et al.* *(82)*: Ausgehend von den Erfahrungen der α-MSH-Synthese und unter Verwendung von dort benutzten Teilsequenzen haben HOFMANN und Mitarbeiter das α^{1-16}-ACTH auf dem in *Abb. 16* angegebenen Wege synthetisiert. Die ε-Aminogruppen der Lysinreste wurden durch die Formylgruppe blockiert, das Arginin zunächst als Nitroverbindung, im weiteren Verlauf der Synthese in der protonisierten Form eingesetzt. Die C-terminale Carboxylgruppe und die γ-Carboxylgruppe des Glu^5-Restes wurden durch die Amidgruppe blockiert. Der im Verlauf der Synthese erforderliche Schutz der α-Aminogruppen erfolgt durch den Benzyloxycarbonylrest, der jeweils durch katalytische Hydrierung abgespalten wurde; das N-terminale Pentapeptid E 1—5 wurde in ungewöhnlicher Weise als N$^\alpha$-Acetylderivat verwendet. Die Abspaltung aller Schutzgruppen nach beendetem Aufbau der Peptidkette wurde mit 0,5 n HCl in Gegenwart von Thioglycolsäure im Verlauf von 80 Min. bei 100° durchgeführt. Unter diesen Bedingungen fand eine optimale Entfernung der Schutzgruppen, aber immer noch eine deutliche Zerstörung der Peptidkette statt. Bei einer späteren Synthese des (Val-NH$_2$)^{20}-α^{1-20}-ACTH ließ sich auf diesem Wege infolge einer sterischen Hinderung eine Hydrolyse der C-terminalen Amidbindung nicht mehr erreichen *(80)*.

Synthese des α^{1-19}-ACTH von LI *et al.* *(106)*: Unter Verwendung fast der gleichen Schnittstellen, aber mit einer anderen Schutzgruppenkombination haben LI und Mitarbeiter das α^{1-19}-ACTH synthetisiert *(Abb. 17,* S. 77). Die ε-Aminogruppen der Lysinreste und die Guanidogruppen der Argininreste wurden durch den Tosylrest, der Imidazolring des Histidins durch den Benzylrest geschützt. Zum Schutz der C-terminalen Carboxylgruppe diente der Methylester, von der Nonapeptid-Stufe D 11—19 an wurde die Carboxylgruppe nur durch Salzbildung blockiert. In ähnlicher Weise wurde die γ-Carboxylgruppe des Glutaminsäurerestes zunächst als Benzylester, bei der Kupplung des Tetrapeptids F 1—4 mit dem Hexadecapeptid F 5—19 aber als Salz eingesetzt. Die Verwendung der Tosylreste und des N^{Im}-Benzylrestes erforderte eine Schutzgruppenabspaltung mit Natrium in flüssigem Ammoniak und damit eine verlustreiche Reinigung.

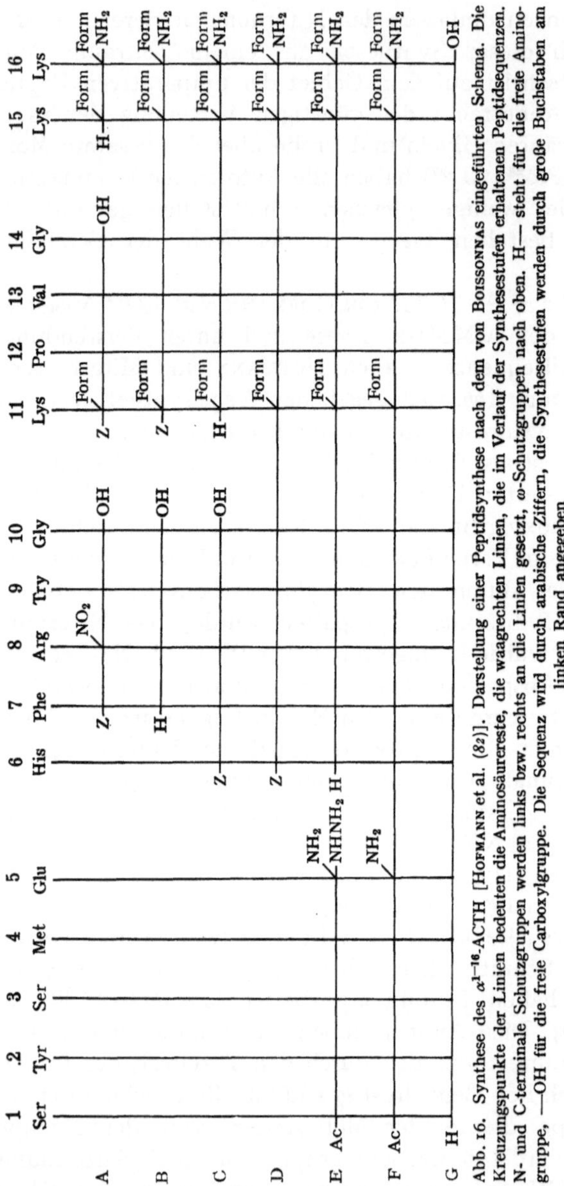

Abb. 16. Synthese des α^{1-16}-ACTH [Hofmann et al. (82)]. Darstellung einer Peptidsynthese nach dem von Boissonnas eingeführten Schema. Die Kreuzungspunkte der Linien bedeuten die Aminosäurereste, die waagrechten Linien, die im Verlauf der Synthesestufen erhaltenen Peptidsequenzen. N- und C-terminale Schutzgruppen werden links bzw. rechts an die Linien gesetzt, ω-Schutzgruppen nach oben. H— steht für die freie Aminogruppe, —OH für die freie Carboxylgruppe. Die Sequenz wird durch arabische Ziffern, die Synthesestufen werden durch große Buchstaben am linken Rand angegeben

Synthese des α^{1-24}-ACTH von Schwyzer *et al.* (*173*): Im Vergleich zu den vorangegangenen Synthesen bedeutete die Verwendung der leicht acidolytisch spaltbaren tert.-Butyloxycarbonylgruppe als ε-Aminoschutz und des ebenso leicht hydrolysierbaren tert.-Butylesters in der Synthese des α^{1-24}-ACTH durch Schwyzer und Mitarbeiter eine entscheidende

Literaturverzeichnis: SS. 107—119

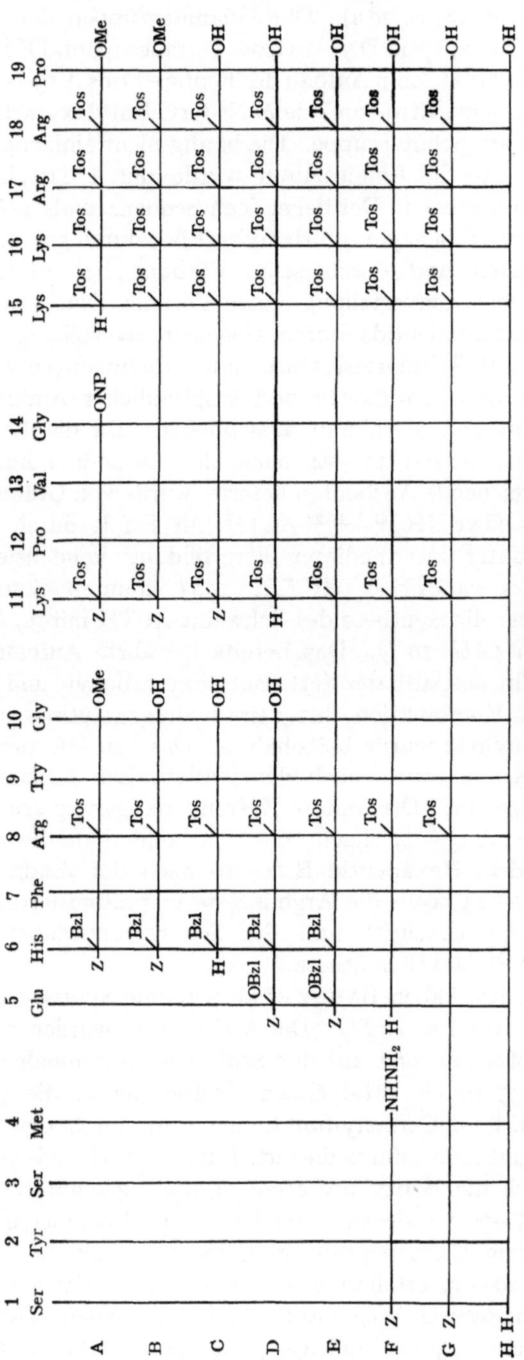

Abb. 17. Synthese des α^{1-19}-ACTH [Li et al. (106)]

Verbesserung *(Abb. 18*, S. 78 a). Die Guanidogruppen der Argininreste wurden bis zum Hexapeptid D 5—10 bzw. Tetradecapeptid E 11—24 durch den Nitrorest geschützt. Zum Aufbau des Pentapeptids A 15—19 diente der Tritylrest neben dem Nitro- und dem N^ϵ-tert.-Butyloxycarbonylrest als selektiv entfernbare Schutzgruppe. Die häufig nicht einfache Einführung dieser Schutzgruppe in Aminosäuren wurde durch Tritylierung einer Dipeptideinheit umgangen. Der bei einigen Sequenzen als α-Aminoschutz dienende p-Phenylazobenzyloxycarbonylrest bot durch günstige Kristallisationseigenschaften und durch seine Färbung bei Säulenchromatographie und Gegenstromverteilung einige Vorteile. Nach Reinigung des geschützten Tetracosapeptids durch Gegenstromverteilung wurden die Schutzgruppen mit Trifluoressigsäure unter Bedingungen entfernt, die eine Zerstörung der Peptidkette und empfindlicher Aminosäuren wie z. B. Tryptophan und Methionin ausschließen. Da die Kupplung der Teilsequenzen 1—10 und 11—24 nach der Carbodiimidmethode mitunter nicht ausreichende Ausbeuten lieferte, wurde von Geiger et al. *(61)* am Beispiel des (Tyr-NH$_2$)^{23}-α^{1-23}-ACTH mit Erfolg durch Zugabe von p-Nitrophenol unter intermediärer Esterbildung kondensiert.

Synthesen des natürlichen ACTH: 1963 konnten Schwyzer und Sieber *(177)* über die Synthese des Schweine-ACTH mit 39 Aminosäureresten berichten *(Abb. 19 I)*. Das bereits bewährte Aufbauprinzip mit acidolytisch leicht abspaltbaren tert.-Butyloxycarbonyl- und tert.-Butylestergruppen in Kombination mit dem hydrogenolytisch entfernbaren Benzyloxycarbonylrest wurde beibehalten. Das neu erforderliche Tetradecapeptid A 25—39 wurde nach der schrittweisen Methode (p-Nitrophenylester) aufgebaut. Die weitere Kettenverlängerung erfolgte mit der Octapeptidsäure A 17—24 nach der Carbodiimidmethode und anschließend mit dem Hexapeptid B 11—16 nach der Azidmethode. Ab Octapeptid B 17—24 lagen die Argininreste in protonisierter Form vor. Der letzte Kupplungsschritt und die Schutzgruppenabspaltung entsprachen der α^{1-24}-ACTH-Synthese.

Vier Jahre später haben Bajusz et al. *(17)* eine Synthese des Human-ACTH beschrieben *(Abb. 19 II)*. Die Argininreste wurden als N^G-Nitroderivate eingesetzt und erst auf der Stufe des C-terminalen Pentacosapeptids D 15—39 durch katalytische Hydrierung in die protonisierte Form übergeführt. ω-Carboxylfunktionen sind durch den tert.-Butylester, ω-Aminofunktionen durch die tert.-Butyloxycarbonyl- und α-Aminofunktionen durch die Benzyloxycarbonylgruppe geschützt. Das Pentacosapeptid D 15—39 wurde aus dem Pentapeptid-pentachlorphenylester C 15—19 und dem Eicosapeptidester C 20—39 erhalten. Die Synthese der Sequenz C 20—39 erfolgte über die isolierten Hydroxysuccinimidester. Auf einem zweiten Weg wurde der Pentacosapeptidester D 15—39 aus den Teilstücken 15—24 und 25—39 aufgebaut. Die Umsetzung des

E. Schröder und K. Lübke: Synthese von Peptiden und Peptidwirkstoffen

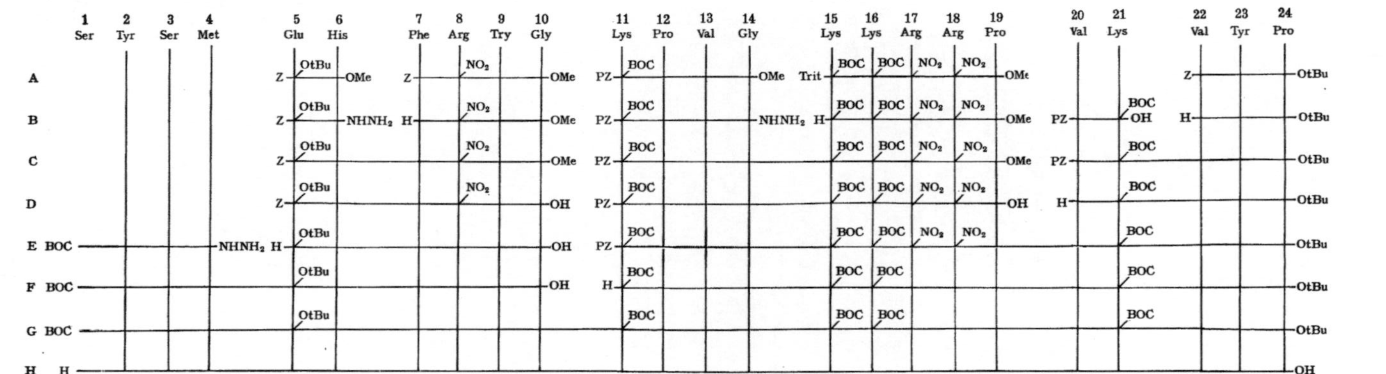

Abb. 18. Synthese des α^{1-24}-ACTH [Schwyzer und Kappeler (*173*)]

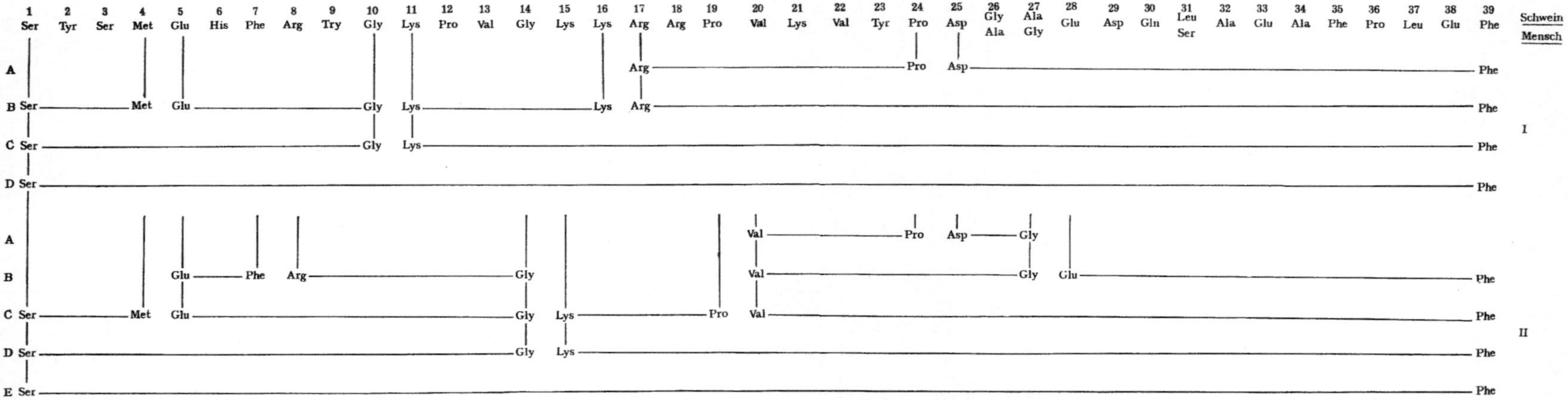

Abb. 19. I Synthese des αp-ACTH [Schwyzer und Sieber (*177*)], II Synthese des αh-ACTH [Bajusz et al. (*17*)]

N-terminal geschützten Tetradecapeptids E 1—14 mit dem Pentacosapeptidester E 15—39 erfolgte unter intermediärer Bildung eines
aktivierten Esters mit Dicyclohexylcarbodiimid/Pentachlorphenol.

Beziehungen zwischen Struktur und Aktivität (vgl. *78, 163*): Das kleinste noch
deutlich wirksame Fragment ist das N-terminale Tridecapeptidamid (0,1 IE/mg;
nat. ACTH 115 IE/mg). Eine Verlängerung der Peptidkette zum Hexadecapeptid-
amid verursacht keine Aktivitätssteigerung, erst die weitere Addition eines Arginin-
restes führt zu einem Anstieg auf 6 IE/mg. Mit der weiteren Verlängerung der Peptidkette steigt die Aktivität und bei 23 bis 24 Aminosäuren wird die volle Wirksamkeit
(auf Gewichtsbasis) erhalten. C-terminale Säureamide sind deutlich stärker wirksam
als die entsprechenden freien Säuren (*16, 139*). Die speziesbedingten Strukturunterschiede befinden sich in der für die biologische Aktivität nicht erforderlichen
Sequenz 27—33. Die positive Ladung des Gesamtmoleküls, die durch die basischen
Aminosäuren der Positionen 15—18 verursacht wird, ist offenbar ein bedeutender
Faktor für die Aktivität des Hormons. Aus den wenigen, bisher durch Aminosäureaustausch entstandenen Analoga von Teilsequenzen läßt sich erkennen, daß
z. B. ein Ersatz der Argininreste in Position 17 und 18 durch Lysin (*70*) oder durch
Ornithin (*204*) ohne Verlust der Wirksamkeit möglich ist. Auch ein Austausch des
Ser^1-Restes durch Glycin oder ein völliger Fortfall des Ser^1-Restes führt zu aktiven
Derivaten. Ähnlich kann auch der Tyr^2-Rest durch Phenylalanin bzw. der Ser^3-Rest
durch Alanin ersetzt werden. Bei den in den Positionen 1, 2 und 3 modifizierten
Verbindungen liegt die Aktivität im Bereich von 50—100 IE/mg (*62*). Nicht essen-
tiell ist auch der Methioninrest in Position 4, wie ein Austausch gegen Aminobutter-
säure im Fall eines Eicosapeptidamids mit 48 IE/mg zeigt (*79*).

Eine über die Aktivität des synthetischen α^{1-24}-ACTH (106 IE/mg) und des
natürlichen α^{1-39}-ACTH (115 IE/mg) hinausgehende Wirkung konnte am Beispiel
des $D\text{-}Ser^1\text{-}Nle^4\text{-}(Val\text{-}NH_2)^{25}\text{-}\alpha^{1-25}$-ACTH mit 625 IE/mg von GUTTMANN et al. (*71,*
vgl. *55*) und des D-Ser^1- bzw. D-Ala^1-α^{1-24}-ACTH mit 500—1000 IE/mg von KAP-
PELER et al. (*91*) demonstriert werden.

2. *Melanocytenstimulierende Hormone*

Parallel mit Arbeiten auf dem ACTH-Gebiet wurden von verschiedenen Arbeitsgruppen Synthesen des α- und β-MSH bzw. von Teil-
sequenzen dieser Wirkstoffe ausgearbeitet. Infolge der strukturellen
Übereinstimmung des α-MSH mit den ersten 13 Aminosäuren des ACTH
sind verschiedene Sequenzbereiche sowohl für Synthesen des α-MSH,
als auch für ACTH-Synthesen verwendet worden. Ähnlich wie beim
ACTH hat die Entwicklung neuer Schutzgruppen, Kupplungs- und
Reinigungsmethoden im Vergleich zu den ersten 1958—1960 publizierten
Synthesen, erst rationelle und zu Produkten mit voller biologischer
Aktivität führende Synthesen ermöglicht.

α-MSH: Bei der von BOISSONNAS et al. (*43, 67a*) beschriebenen
Synthese war die γ-Carboxylgruppe des Glu^5-Restes durch den Benzyl-
ester, die ε-Aminogruppe des Lys^{11}-Restes durch die Benzyloxycarbonyl-
gruppe geschützt *(Abb. 20 I)*. Als selektiv entfernbare Blockierung der
α-Aminogruppe diente der Tritylrest. Die Schutzgruppenabspaltung am
Endprodukt mußte mit HBr/Eisessig durchgeführt werden, da eine kata-

lytische Hydrierung durch die Anwesenheit von Methionin entfiel. Zur Vermeidung von Nebenreaktionen am Methioninrest wurde bei der Reaktion mit HBr/Eisessig in Gegenwart von Methyläthylsulfid gearbeitet. Bei den verwendeten Schnittstellen kann eine Racemisierung des Glu[5]- bzw. Try[9]-Restes nicht ausgeschlossen werden. Von Hofmann et al. (81) wurde als N[ε]-Schutz des Lys[11]-Restes die Formylgruppe und in Position 5 der Glutaminrest an Stelle eines Glutaminsäurederivates ver-

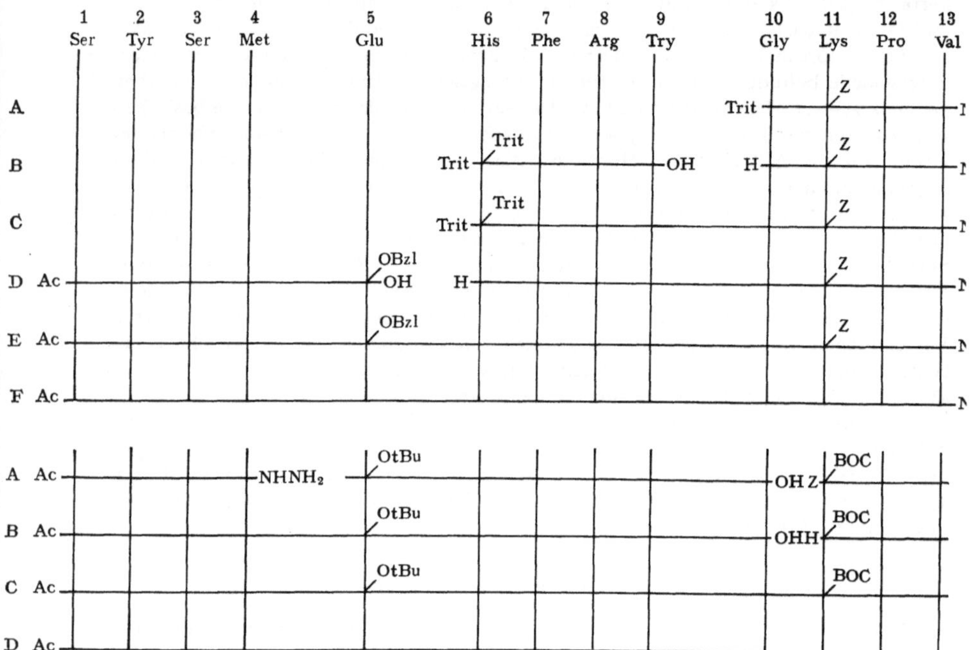

Abb. 20. I Synthese des α-MSH [Boissonnas et al. (43, 67a)], II Synthese des α-MSH [Schwyzer et al. (170)]

wendet. Eine selektive Entfernung der Formyl- und Amidgruppe neben dem N-terminalen Acetylrest ist hierbei nicht durchführbar. In der von Schwyzer et al. (170) beschriebenen Synthese wurden die ω-Funktionen durch die tert.-Butyloxycarbonylgruppe bzw. durch den tert.-Butylester blockiert (Abb. 20 II). Eine Schutzgruppenabspaltung mit Trifluoressigsäure war ohne Komplikationen möglich.

Unter Verwendung der gleichen Schnittstelle, jedoch mit N-terminalem Butyloxycarbonylrest und N[ε]-Phthalylrest in Position 11 konnte nach selektiver Abspaltung der tert.-Butyloxycarbonylgruppe die Möglichkeit geschaffen werden, rationell andere N-terminale Acylreste einzuführen. Die Entfernung der N[ε]-Pht-Gruppe gelang durch Hydrazin in schwach saurer Lösung (170).

Literaturverzeichnis: SS. 107—119

β-MSH: In der ersten von Schwyzer et al. *(174)* publizierten Synthese wurde an Stelle des Asparaginsäurerestes in Position 1 und an Stelle des Glutaminsäurerestes in Position 8 die entsprechenden ω-Säureamide Asparagin und Glutamin eingesetzt *(Abb. 21 I, S. 82)*. Diese Vereinfachung war erforderlich, da befriedigende Schutzgruppen für die ω-Carboxylfunktionen noch nicht bekannt waren. Der C-terminale Asparaginsäurerest wurde als Dimethylester verwendet und machte damit eine alkalische Verseifung nach beendetem Kettenaufbau notwendig. Die ε-Aminogruppen der Lysinreste waren durch den Tosylrest geschützt und erforderten eine Abspaltung mit Natrium in flüssigem Ammoniak. Auf diesem Wege konnte nur ein Produkt mit 1/100 der biologischen Aktivität des natürlichen Hormons isoliert werden *(169)*.

In einer 4 Jahre später von der gleichen Arbeitsgruppe publizierten Synthese *(172)* wurde das gleiche Aufbauschema *(Abb. 21 II)* verwendet. Der Schutz der ω-Funktionen erfolgte durch leicht acidolytisch spaltbare Reste, das Arginin wurde in der protonisierten Form eingesetzt. Das freie Peptid, das als wesentliche Verunreinigung das Methioninsulfoxydderivat enthielt, zeigte nach Gegenstromverteilung und Chromatographie an CM-Sephadex die volle biologische Aktivität.

Beziehungen zwischen Struktur und Aktivität (vgl. *78, 102, 163*): Die kürzeste aktive Peptidsequenz aus der α-MSH-Kette ist das Pentapeptid H-His-Phe-Arg-Try-Gly-OH mit 1,5—3·10⁴ E/g (α-MSH: 2·10¹⁰ E/g, Froschhauttest in vitro), das Bestandteil des α-MSH, β-MSH und ACTH ist. Eine N-terminale Verlängerung zur Sequenz 1—10 vergrößert die Aktivität auf 2,9·10⁶ E/g. Das N-terminale Octapeptid 1—8 ist inaktiv, die C-terminale Octapeptidsequenz zeigt 8·10⁶ E/g. Eine Entfernung des N-Acetylrestes von α-MSH vermindert die Aktivität um eine Zehnerpotenz. Das aktive Pentapeptid, das für die Melanocytenexpansion möglicherweise das aktive Zentrum darstellt, wurde von verschiedenen Autoren durch Aminosäureaustausch mit L- und D-Aminosäuren modifiziert. Im Falle des all-D-Pentapeptids konnte eine Hemmung des all-L-Pentapeptids in vitro und in vivo beobachtet werden *(222)*. Ein retro-enantio-Pentapeptid H-D-Try-D-Arg-D-Phe-D-His-D-Glu-OH ist inaktiv *(52)*.

3. Oxytocin und Vasopressin

Oxytocin: Durch Variation der Schnittstellen, Aminoschutzgruppen und Kupplungsmethoden konnten für das Oxytocin recht ideale Synthesebedingungen gefunden werden. Im Fall der S-Schutzgruppe wurde jedoch nach der vor fast 30 Jahren von Wood und du Vigneaud *(218)* eingeführten S-Benzylgruppe keine durchschlagende Verbesserung erzielt. Eine Übersicht über die von den einzelnen Autoren benutzten Syntheseprinzipien enthält die *Abb. 22.* In der historischen Synthese von du Vigneaud et al. *(58)* wurde in einer Fragmentkondensation [2 + + (3 + 4)]* zur Kupplung bevorzugt mit der Säurechlorid- und Tetra-

---

* Das heißt, zunächst wurden ein Tri- und ein Tetrapeptid kombiniert und das entstandene Heptapeptid mit einem Dipeptid umgesetzt.

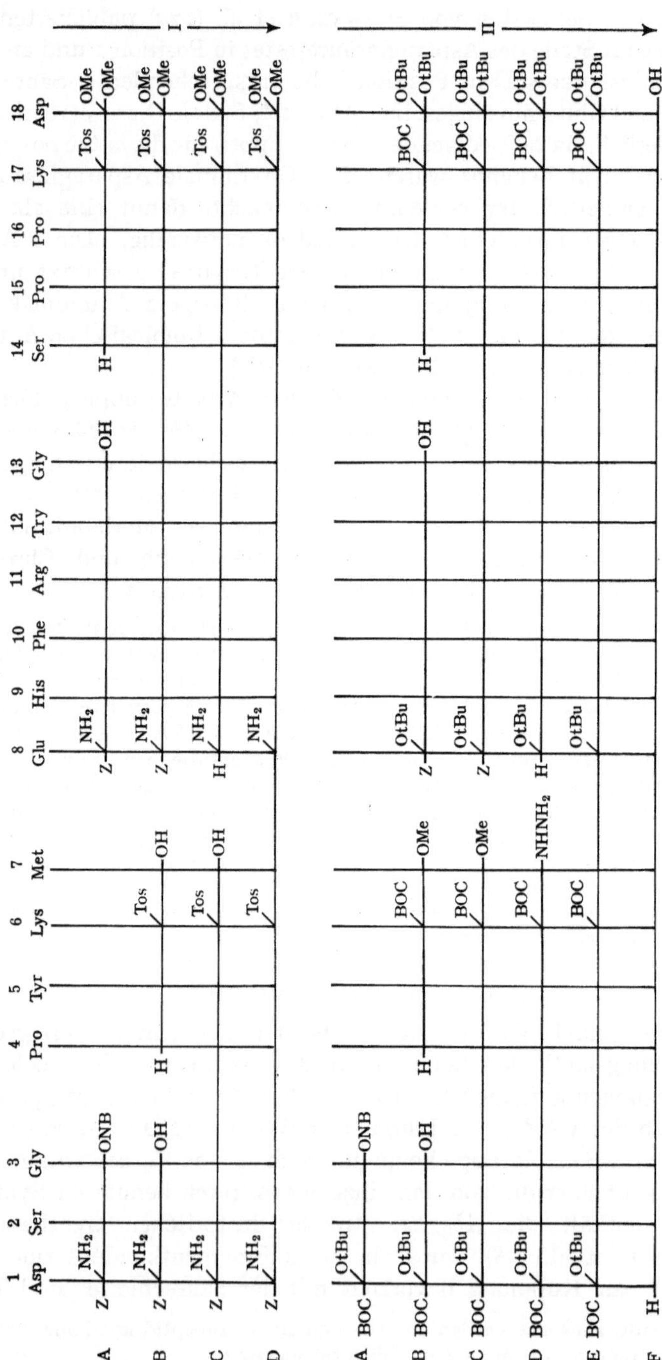

Abb. 21. I Synthese des β-MSH [Schwyzer et al. (*174*)], II Synthese des β-MSH (Schwyzer et al. (*172*))

äthylpyrophosphitmethode gearbeitet *(Abb. 22 I)*. Zum intermediären Schutz der α-Aminogruppe diente der Tosylrest, der nur unter gleichzeitiger Entfernung des S-Benzylrestes mit Natrium in flüssigem Am-

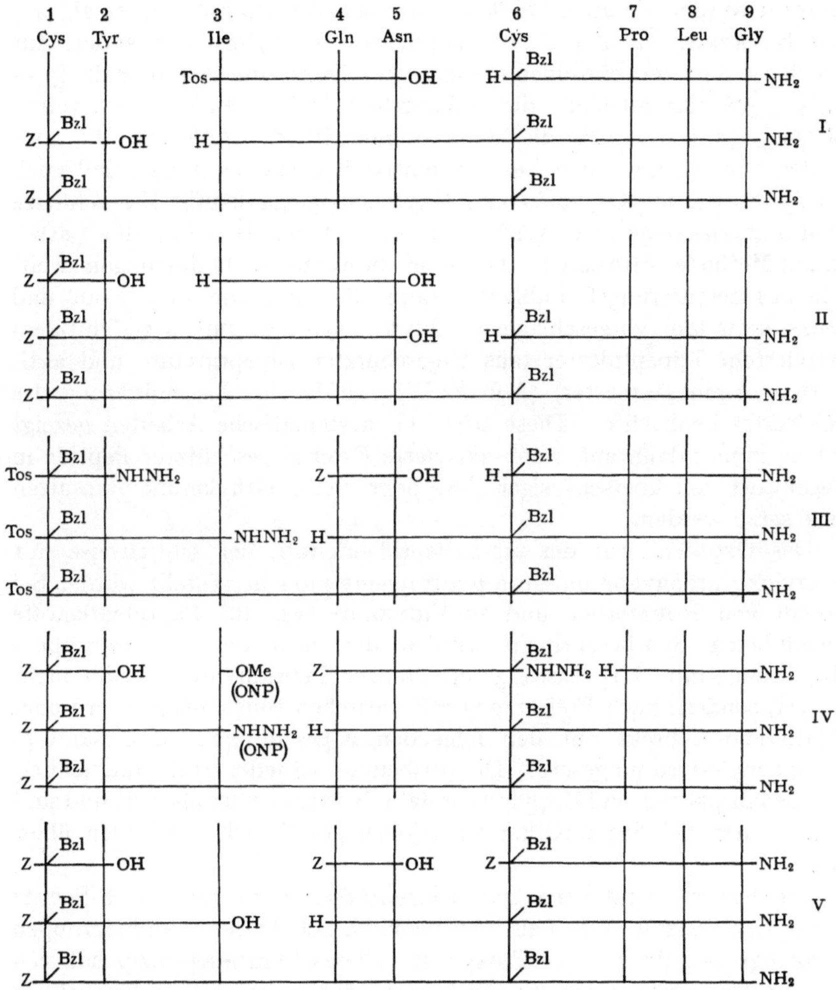

Abb. 22. Synthesen des Oxytocins. I Du VIGNEAUD et al. *(58)*, II BODANSZKY und DU VIGNEAUD *(33)*, III RUDINGER et al. *(144)*, IV BOISSONNAS et al. *(45)*, V BEYERMAN et al. *(31)*

moniak abgespalten werden kann und damit eine Re-benzylierung erforderte. Eine von BODANSZKY und DU VIGNEAUD *(33)* publizierte Modifizierung vermeidet durch Verlegung der Schnittstellen [(2 + 3) + 4] Nachteile der ersten Synthese *(Abb. 22 II)*. Zur Vereinigung der Fragmente zum voll geschützten Nonapeptid wurden die Carbodiimid-, Anhydrid- und o-Phenylenchlorphosphitmethoden verwendet. Bei einer weiteren Oxytocinsynthese benutzten Rudinger et al. (*144*) den Tosylrest bei dem Bauprinzip 3 + (2 + 4) nur für das voll geschützte Nonapeptid *(Abb. 22 III)*. Für den intermediären Schutz der α-Aminogruppen wurde, wie auch bei den Synthesen von Boissonnas et al. (*45*) und Beyerman et al. (*31*), der Benzyloxycarbonylrest verwendet. Zur Kondensation der einzelnen Fragmente diente im ersten Fall [3 + + (3 + 3)] ausschließlich die Azidmethode *(Abb. 22 IV)*, im zweiten Beispiel [3 + (2 + 4)] die Carbodiimidmethode *(Abb. 22 V)*. Das Fehlen von racemisierungsfreien Schnittstellen hat zur Folge, daß auch bei Synthesen von Oxytocin und Oxytocinanaloga häufig N-geschützte Peptidsäuren eingesetzt werden. Bevorzugt wurde dann die Carbodiimid-Methode verwendet, die unter bestimmten Bedingungen weitgehend racemisierungsfrei ablaufen kann. Bei einer von Jaquenoud und Boissonnas (*89*) vorgeschlagenen Oxytocinsynthese mit N-geschütztem aktiviertem Tripeptidester (aus N-geschützter Dipeptidsäure und aktiviertem Aminosäureester) (Abb. 22 IV), wurde eine Racemisierung des Ile^3-Restes beobachtet. Diese tritt, wie systematische Arbeiten gezeigt haben, immer dann auf, wenn aktivierte Ester N-geschützter Peptide in Gegenwart von überschüssiger Base oder von Triäthylammoniumsalzen umgesetzt werden.

Das Oxytocin war der erste Peptidwirkstoff, der schrittweise mit Benzyloxycarbonyl-aminosäure-p-nitrophenylester hergestellt wurde. Bei diesem von Bodanszky und du Vigneaud (*34*) für Peptidwirkstoffe eingeführten Syntheseprinzip wurden die nach der Schutzgruppenabspaltung mit HBr/Eisessig erhaltenen Peptidhydrobromide nicht isoliert, sondern nach Freisetzung mit basischen Ionenaustauschern oder Triäthylamin direkt mit den folgenden N-geschützten Aminosäure-p-nitrophenylestern umgesetzt. Die Ausbeuten bei jeder Stufe, die Schutzgruppenabspaltung und Kupplung umfaßt, betrugen mehr als 90% und sind somit denen bei Fragmentkondensationen gewöhnlich erhaltenen überlegen.

Die bisher diskutierten Oxytocinsynthesen enthielten als S-Schutz immer den Benzylrest und als intermediären Schutz der α-Aminogruppen bevorzugt den Benzyloxycarbonylrest. Dieses Syntheseprinzip hat sich für alle Oxytocinanaloga bewährt und wird heute fast ausschließlich verwendet; lediglich der Benzyloxycarbonylrest wird manchmal durch leichter abspaltbare Schutzgruppen wie tert.-Butyloxycarbonyl- oder p-Methoxybenzyloxycarbonyl ersetzt (*99*).

Abweichend vom Typ S-Benzyl/Benzyloxycarbonyl haben Velluz et al. (*206*) 1956 das Oxytocin ausschließlich mit Hilfe des S- und N-Tritylrestes nach einer Fragmentkondensation [3 + (2 + 4)] hergestellt, wobei von der selektiven N-Detritylierung durch 5 nHCl in Aceton Gebrauch gemacht wurde. Als Kondensationsmittel diente N,N'-Dicyclohexylcarbodiimid. Ungewöhnlich bei der Synthese war die erst auf der Stufe des vollgeschützten Nonapeptids durchgeführte Amidierung des C-terminalen Methylesters sowie der ω-Methylester des Glu^4- und Asp^5-Restes. GUTTMANN (67) beschrieb eine Oxytocinsynthese mit dem S-Äthylcarbamoyl-Schutz in Kombination mit der Benzyloxycarbonylgruppe nach dem schrittweisen Aufbau. Bei dem S-Äthylcarbamoyl-Rest kann die Abspaltung mit Natrium in flüssigem Ammoniak vermieden und mit flüssigem Ammoniak in Methanol (10 : 1) bei 25° erreicht werden. SAKAKIBARA und NOBUHARA (148) führten 1965 den S-p-Methoxybenzylrest ein, der ebenfalls mit Natrium in flüssigem Ammoniak oder auch mit siedender Trifluoressigsäure entfernt werden kann. Der eigentliche Vorteil dieses S-Schutzes wurde erst durch SAKAKIBARA et al. (149) offenbar, als es gelang, im Fall einer Arg^8-Vasopressin-Synthese den S-p-Methoxybenzylrest gemeinsam mit der als α-Aminoschutz verwendeten tert.-Amyloxycarbonylgruppe und der zum Schutz der Guanidogruppe verwendeten Nitrogruppe unter milden Bedingungen mit flüssigem Fluorwasserstoff in Gegenwart von Anisol bei 0° abzuspalten. Von PHOTAKI (132) wurde die S-Benzoylgruppe in Kombination mit dem Benzyloxycarbonyl-, tert.-Butyloxycarbonyl- und o-Nitrophenylsulfenylrest zur Oxytocin-Synthese eingesetzt. Die vorhandene Selektivität dieser Gruppen ermöglichte die Synthese von N-geschütztem (Z, BOC oder NPS) Oxytocein und Ringschluß vor Abspaltung der Aminoschutzgruppe.

Vasopressine: Synthesen der Vasopressine unterscheiden sich von denen des Oxytocins durch zusätzliche Schutzgruppen, die die basischen Aminosäuren Lysin bzw. Arginin in Position 8 erfordern. Im Fall von Lys^8-Vasopressin bzw. Lys^8-Vasopressin-Analoga wurde mit N$^\varepsilon$-Tosyl-Lysin gearbeitet, und zwar von MEIENHOFER und DU VIGNEAUD (114) nach der Fragmentkondensation [(2 + 3) + 4] mit N-terminalem Tosylschutz, von BOISSONNAS und HUGUENIN (47) über eine Fragmentkondensation [3 + (3 + 3)] unter ausschließlicher Verwendung des Benzyloxycarbonylschutzes und schließlich von BODANSZKY et al. (35) durch eine schrittweise Synthese. Die ersten von DU VIGNEAUD et al. (57, 94) 1954—1958 publizierten Synthesen des Arg^8-Vasopressins wurden mit protonisiertem Arginin durchgeführt. Wesentliche Verbesserungen, besonders in bezug auf die Endreinigung, boten die von HUGUENIN und BOISSONNAS (85) 1962 und von STUDER (195) 1963 beschriebenen Synthesen, bei denen in Fragmentkondensationen das N^G-Tosyl-Arginin verwendet wurde. Als Beispiel für eine andere Blockierung der ω-Aminofunktion soll eine Orn^8-Vasopressin-Synthese von BODANSZKY et al. (37) dienen. Ornithin wurde als N$^\delta$-Phthalylderivat eingesetzt und die Schutzgruppe nach Beendigung der Synthese durch Hydrazinolyse abgespalten. Anschließende Guanidierung lieferte dann Arg^8-Vasopressin.

Mit den Synthesen auf dem Oxytocin-/Vasopressingebiet ist das Problem der Dehydrierung der freien Thiolgruppen zum cyclischen Disulfid verbunden. Die Dehydrierung, die durch Luft oder andere Oxydationsmittel erreicht wird, kann zu verschiedenen Produkten führen. Neben der richtigen monomeren Struktur sind dimere Verbindungen mit parallelen oder antiparallelen Peptidketten sowie polymere Produkte zu erwarten. In einer systematischen Arbeit haben Heaton et al. (76) die Bildung der möglichen Produkte in Abhängigkeit von dem Abstand zwischen beiden Cysteinresten untersucht. Steht nur eine Aminosäure zwischen den beiden Cysteinresten, so wird die dimere antiparallele Form in 80%iger Ausbeute gebildet, bei 2 oder 3 Aminosäuren zwischen den Cysteinresten entsteht neben dieser dimeren Form auch die monomere Form. Bei 4 Aminosäuren zwischen den Cysteinresten, d. h. bei Strukturen, wie sie auch beim Oxytocin und Vasopressin vorliegen, wird mit 90% Ausbeute nur die monomere Form gebildet. Die Tatsache, daß der 20gliedrige cyclische Disulfidring bevorzugt gebildet wird, hat die Synthesen auf dem Oxytocin- und Vasopressingebiet sehr erleichtert.

Beziehungen zwischen Struktur und Aktivität (vgl. *78, 103, 143*): Gegenstand vieler Untersuchungen war immer wieder die Frage nach der Bedeutung des 20gliedrigen Disulfidringes für die biologische Aktivität. Ringvergrößerungen durch Einbau einer weiteren Aminosäure bzw. durch eine CH_2-Gruppe (β-Ala[4]-Oxytocin, γ-Mercaptobutyryl[1]-Oxytocin) führten immer zu einem Verlust der Aktivität. Ein cyclisches Analogon des Oxytocins, in dem ein S-Atom des Disulfidringes durch eine CH_2-Gruppe [Desamino-Carba[1]-Oxytocin (*145*)] ersetzt ist, zeigt dagegen noch eine bedeutende Wirkung. Analoga, bei denen ein oder beide S-Atome durch Selen ausgetauscht sind [6-Hemi-L-selenocystin-oxytocin, 1-Desamino-6-hemi-L-selenocystin-oxytocin, 1-Desamino-1,6-L-diselenocystin-oxytocin (*212, 213*)] stellen sogar hochaktive Verbindungen dar. Erstaunlicherweise erwies sich auch das Oxytocein [reduzierter Disulfidring (*224*)] als wirksam, nicht dagegen die entsprechende S-Methylverbindung und z. B. ein Ala[1]-Ala[6]-Arg[8]-Vasopressin, bei dem die SH-Gruppen also durch H-Atome ersetzt waren. Systematische Veränderungen der funktionellen Gruppen des Oxytocins ergaben, daß die phenolische Hydroxylgruppe in Position 2 nicht essentiell ist, jedoch einen wesentlichen Beitrag für eine volle Wirksamkeit liefert. Essentiell sind die Carboxamidgruppe in Position 5 und die C-terminale Amidgruppe, nicht dagegen die Carboxamidgruppe in Position 4. Als Derivate mit einer verstärkten Wirkung sind Desaminooxytocin (*84*) und die Desaminovasopressine (*86*) bekannt geworden, die bis zu 200% Aktivität besitzen. Aminoacylierte Derivate des Oxytocins und Vasopressins zeigen bei geringerer Aktivität einen prolongierten antidiuretischen und pressorischen Effekt (*92, 83*). Oxytocin-Inhibitor-Eigenschaften wurden bei Derivaten nachgewiesen, die an Stelle des Tyr[2]-Restes p-Methyl-, p-Äthylphenylalanin oder O-Äthyltyrosin enthielten (*230*).

Während das Oxytocin nur sehr geringe Vasopressin-Aktivitäten und die Vasopressine nur geringe Oxytocin-Aktivitäten haben, besitzen die beiden möglichen Hybridformen etwa gleich starke oxytocische und vasopressorische Eigenschaften. Die absolute Höhe dieser Aktivitäten liegt im Falle des Oxypressins (Phe[3]-Oxytocin, d. h. Vasopressinring und Oxytocinseitenkette) bei etwa 5—10% und im Falle des Arg[8]-Vasotocins (Ile[8]-Vasopressin bzw. Arg[8]-Oxytocin, d. h. Oxytocinring und Vasopressinseitenkette) bei etwa 50%, jeweils bezogen auf die originalen Wirkstoffe Oxytocin und Arg[8]-Vasopressin.

Für eine therapeutische Anwendung ist es von großer Bedeutung, daß es in einigen Fällen gelungen ist, die unterschiedlichen biologischen Wirkungen zu differenzieren. So besitzt das Ala[4]-Oxytocin (*68*) im Vergleich zum Oxytocin etwa 50% der Wirkung auf die Milchdrüse, aber weniger als 10% der uteronischen Wirksamkeit. Als Beispiele aus der Vasopressinreihe sind in *Tabelle 1* Analoga aufgeführt, die hohe antidiuretische Aktivitäten neben stark reduzierter vasopressorischer Wirkung entfalten.

Tabelle 1. Vasopressinanaloga mit differenzierter antidiuretischer und vasopressorischer Aktivität

	Antidiuretische Aktivität IE/mg	Vasopressorische Aktivität IE/mg	Antidiurese Blutdruck
Arg^8-Vasopressin*	400	400	1
Phe^2-Arg^8-Vasopressin*	350	122	2,9
Desamino-Arg^8-Vasopressin*	1300	370	3,5
Desamino-Phe^2-Arg^8-Vasopressin*......	800	30	27
D-Dab^8-Vasopressin**	120	4	30
Desamino-D-Dab^8-Vasopressin**	360	2	180

* Stürmer et al. (*201*).   ** Zaoral et al. (*227*).

B. Peptidhormone des Intestinaltraktes

1. Gastrin

Die von Anderson et al. (*8*) durchgeführte Synthese des Gastrins I (Schwein) stellt eine Kombination von schrittweiser Synthese (Sequenz 14—17 und 6—13) und Fragmentkondensation dar *(Abb. 23)*. Als Schutzgruppenkombination wird N-Benzyloxycarbonyl/tert.-Butylester verwendet. Bemerkenswert ist die selektive alkalische Verseifung des Tridecapeptidmethylesters (G 1—13) neben fünf ω-tert.-Butylester-gruppen. Verschiedene Nachteile, wie Isocyanatbildung des N-termi-nalen Pentapeptidazids F 1—5 infolge zu langsamer Reaktion mit der Sequenz F 6—13, schwierige Entfernung der nicht acylierten Anteile von F 6—13 und dadurch Polymerenbildung bei der nachfolgenden Kupplung führten zur Entwicklung des in *Abb. 24* angeführten Syntheseschemas, in dem die Endkupplung auf die Positionen 5/6 verlegt wurde (*9*). Nach Kupplung zum geschützten Dodecapeptidamid D 6—17 wurden sämtliche Schutzgruppen mit Trifluoressigsäure entfernt. Die nachfolgende Reak-tion mit dem Pentapeptid E 1—5 im Überschuß unter Salzblockierung der 5γ-Carboxylgruppen an Stelle der γ-tert.-Butylester und dadurch verminderte sterische Hinderung verlief besser als nach dem ersten Syntheseweg. Einmalige Chromatographie an Aminoäthylcellulose lieferte reines Gastrin I. Die Sequenz F 6—13 konnte nach dem gleichen Schema durch Kupplung mit Pyroglu-Gly-Pro-Try-Leu-azid auch zur Synthese des Human-Gastrin I verwendet werden (*18*). Eine zweite Synthese des Human-Gastrin I wurde von Morley (*119*) entwickelt. Das C-terminale Tetrapeptidamid wurde mit Hilfe der N-*o*-Nitrophenyl-sulfenylgruppe aufgebaut und war so mit dem β-tert.-Butylester-Schutz am Asp^{16}-Rest zugänglich *(Abb. 25)*. Die selektive Abspaltung der

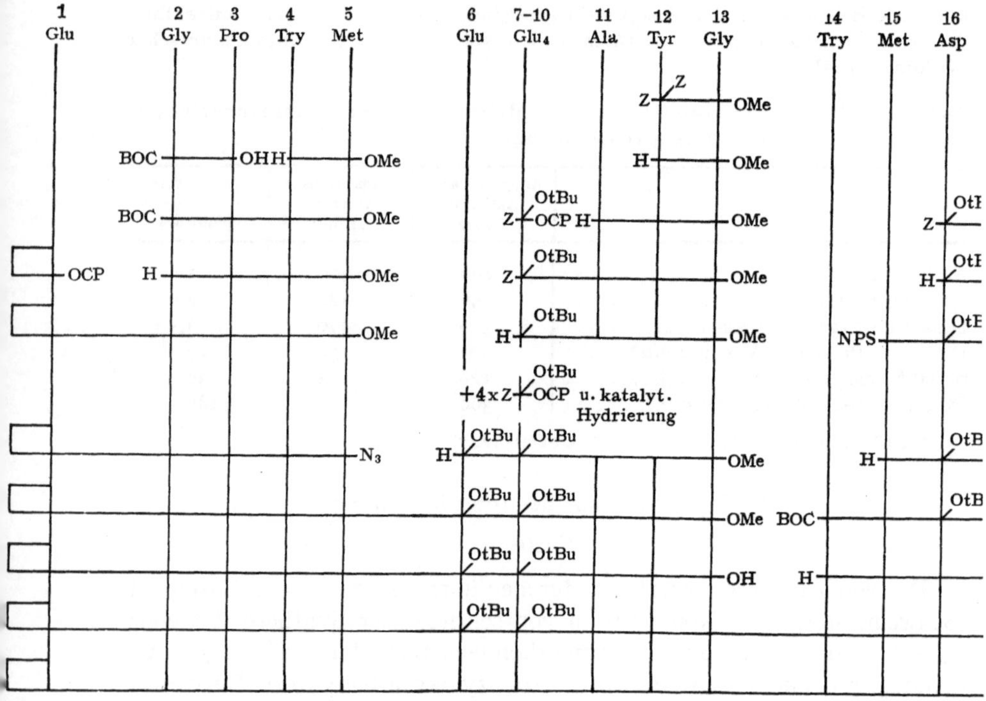

Abb. 23. Synthese des Gastrins I (Schwein) [Anderson et al. (*8*)]

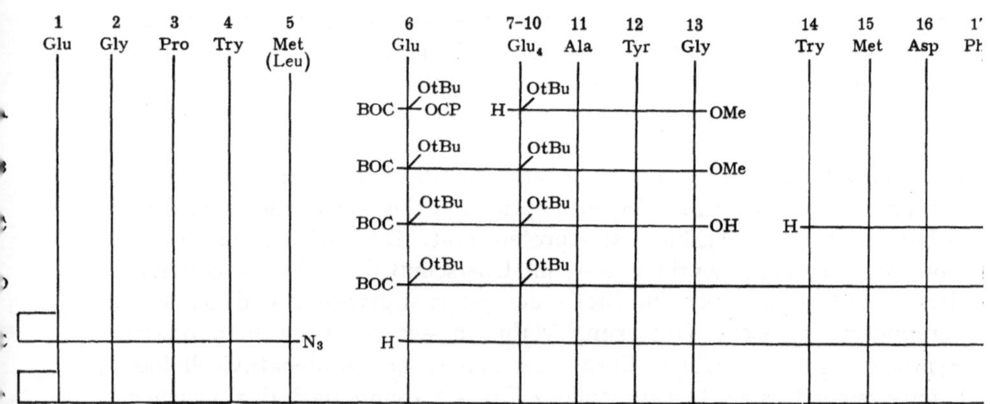

Abb. 24. Synthese des Gastrins I (Schwein, Mensch) [Anderson et al. (*9*)]

o-Nitrophenylsulfenylgruppe am G 14—17 gelang mit 4-Chlor-2-nitro-
thiophenol in Pyridin. Die Kupplung von H 14—17 mit H 1—13 bzw.
F 3—13 (als BOC-Derivat) wurde nach der Anhydridmethode unter Verwendung von Pivaloylchlorid durchgeführt. Zur Kupplung wurden auch

Literaturverzeichnis: SS. 107—119

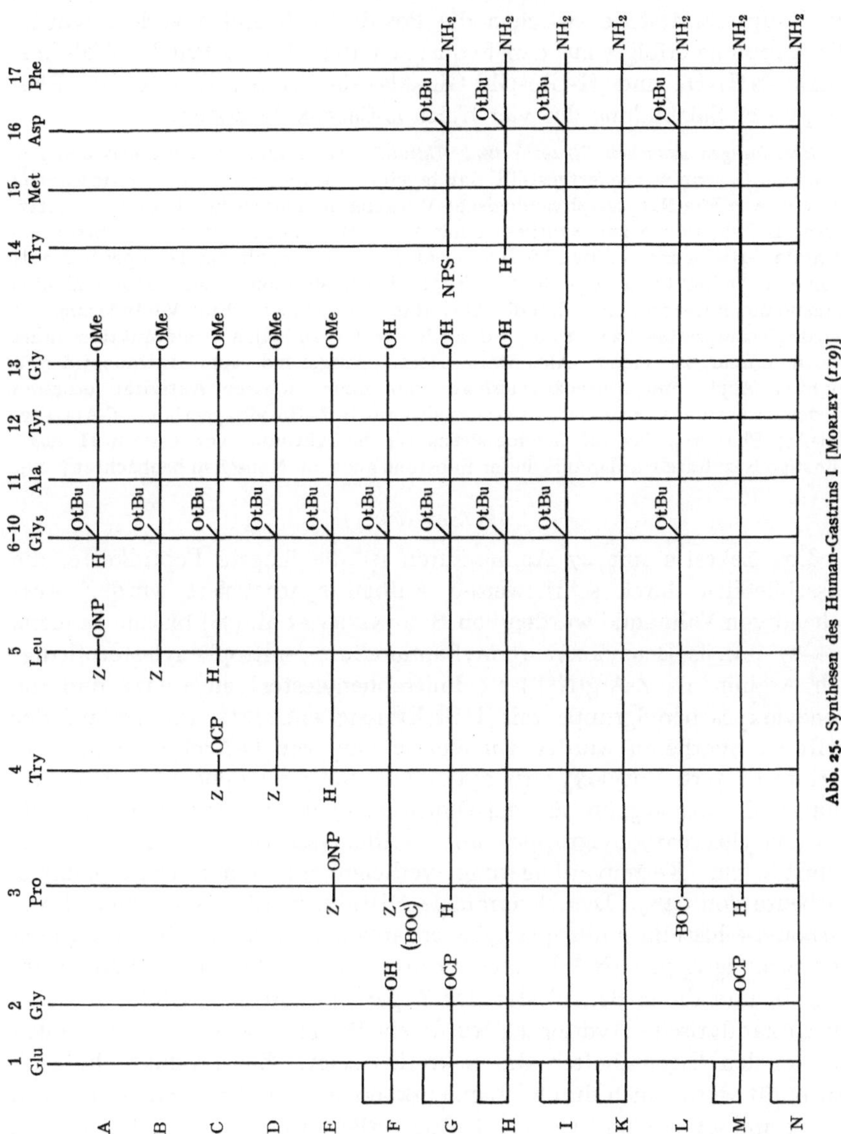

Abb. 25. Synthesen des Human-Gastrins I [MORLEY (*119*)]

die 2,4,5-Trichlorphenylester der Sequenz F 1—13 und H 3—13 verwendet. Die Umwandlung des Gastrins I (Schwein) in Gastrin II (Phenolsulfatester in Position 12) gelang durch Reaktion mit einem Pyridin-Schwefeltrioxyd-Komplex bei pH 10 (*64*).

Ein anderer Aufbau als für das Schweine- und Human-Gastrin I wurde von BEACHAM et al. (*19*) für das Schaf-Gastrin I gewählt, indem die Hauptschnittstelle zwischen die Positionen 6 und 7 verlegt wurde. Die Kupplung erfolgte aus den Sequenzen Pyroglu-Gly-Pro-Try-Val-Glu-(OtBu)-NHNH$_2$ und H-Glu-Glu-Glu-Ala-Ala-Tyr-Gly-Try-Met-Asp-Phe-NH$_2$ unter Salzbildung der vier freien ω-Carboxylgruppen.

Beziehungen zwischen Struktur und Aktivität: Im Laufe der synthetischen Arbeiten am Gastrin wurde festgestellt, daß bereits die C-terminale Tetrapeptidsequenz Try-Met-Asp-Phe-NH$_2$ die physiologische Wirkung des natürlichen Hormons besitzt. Dieser Befund führte zur Synthese einer Vielzahl von Analoga und zu Struktur-Aktivitäts-Beziehungen durch Morley et al. (*120*). Innerhalb des Tetrapeptidamids kann ein Aminosäureaustausch des Tryptophans, Methionins oder Phenylalanins vorgenommen werden, ohne daß die Aktivität verloren geht. Eine Veränderung des Asparaginsäurerestes bzw. zum Teil auch der C-terminalen Amidfunktion führt jedoch immer zu einem Aktivitätsverlust. Ausgehend vom Tetrapeptidamid konnten Acyl- und Aminoacylderivate mit einer höheren Aktivität erhalten werden. Als wirksamstes Derivat erwies sich das tert.-Butyloxycarbonyl-β-Ala-Try-Met-Asp-Phe-NH$_2$, das auf Gewichtsbasis 1/4 der Aktivität von Gastrin II zeigt. Günstige Resultate wurden mit dieser Substanz auch am Menschen beobachtet (*112*).

2. Sekretin

Das Sekretin mit 27 Aminosäuren ist die längste Peptidkette, die ausschließlich durch schrittweisen Aufbau synthetisiert wurde. Ausgehend von Valinamid wurden von Bodanszky et al. (*38*) bis zur Sequenz 17—27 jeweils Benzyloxycarbonyl-aminosäure-p-nitrophenylester [lediglich Arginin als Z-Arg(NO$_2$)-2,4-dinitrophenylester] eingesetzt und die Benzyloxycarbonylgruppe mit HBr/Eisessig entfernt. Im Verlauf der weiteren Synthesen wurden zur Vermeidung von O-Acetylierungen der Serinreste tert.-Butyloxycarbonyl-aminosäure-p-nitrophenylester (Threonin und Nitroarginin als 2,4-Dinitrophenylester) verwendet und die tert.-Butyloxycarbonylgruppe mit Trifluoressigsäure entfernt. Alle Schritte zur Kettenverlängerung verliefen mit einer Durchschnittsausbeute von 94%. Das N-terminale Histidin wurde als Bis-benzyloxycarbonyl-L-histidin-p-nitrophenylester ankondensiert. Nach Entfernung der Schutzgruppen (N^{Im}-Benzyloxycarbonyl am Histidin, N^{G}-Nitro am Arginin und die ω-Benzylester der Asparaginsäure und Glutaminsäure) durch katalytische Hydrierung wurde ein Produkt isoliert, das die vollen hormonalen Eigenschaften des Sekretins zeigt. Ein Produkt ähnlicher Aktivität wurde auch durch Fragmentkondensation der Partialsequenzen 9—13 und 14—27 und weitere Kettenverlängerung mit den Teilstücken 5—8 und schließlich 1—4 erhalten. Jedoch waren die Ausbeuten vergleichsweise schlechter.

C. Gewebshormone und verwandte Verbindungen

1. Angiotensine

Als Beispiel für die von der Arbeitsgruppe Schwyzer (*171*) auf dem Angiotensingebiet durchgeführten Synthesen soll das nach der Frag-

Literaturverzeichnis: SS. 107—119

mentkondensation erhaltene Val⁵-Angiotensin I dienen *(Abb. 26)*. Bei
der Herstellung des freien Peptids wurde zur Vermeidung von Nebenreaktionen an der Carbobenzoxygruppe bei einer alkalischen Verseifung
zuerst katalytisch hydriert und durch nachfolgende saure Verseifung der
C-terminale Methylester entfernt. Im Laufe dieser Reaktion wird gleichzeitig die β-Amidgruppe des Asp¹-Restes partiell entfernt, so daß nach
Gegenstromverteilung Val⁵-Angiotensin II I 1—10 und dessen Asp¹-
β-amid H 1—10 isoliert werden konnten.

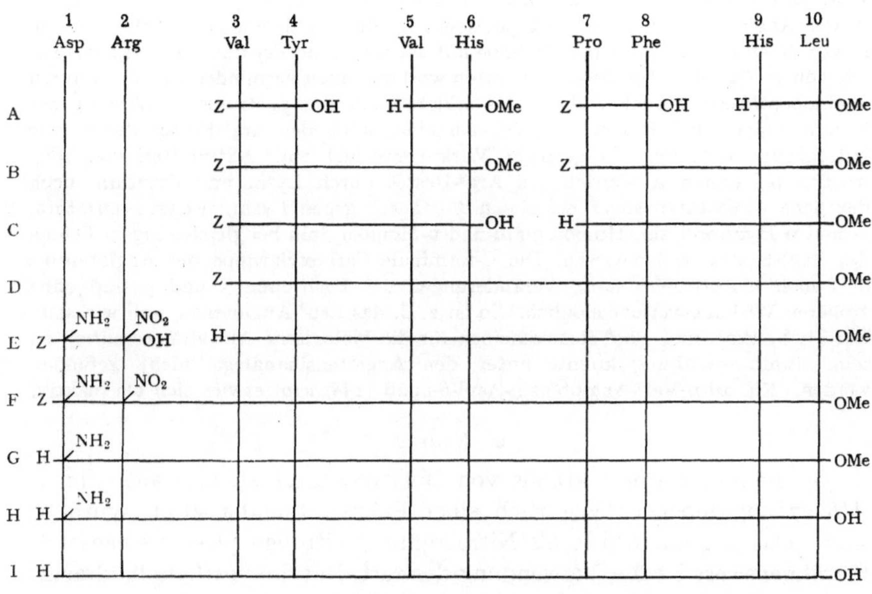

Abb. 26. Synthese des Val⁵-Angiotensins I und Synthese des Val⁵-Angiotensins I-Asp¹-β-amid
[SCHWYZER et al. (*171*)]

Nach dem heutigen Stand der Technik bestehen bei dem benutzten
Syntheseweg an allen Kupplungsstellen Gefahren der Racemisierung, die
in einer späteren Publikation von RINIKER und SCHWYZER (*141*) auch
für das Tyrosin am Asn¹-Val⁵-Angiotensin II und von SCHWARZ und
BUMPUS (*168*) bei einer analogen Synthese des Ile⁵-Angiotensin II für den
Arg²-Rest nachgewiesen werden konnte. Eine Ausschaltung von Race-
misierungen bot später ein von ARAKAWA und BUMPUS (*12*) publizierter
Syntheseweg, der zusätzlich durch Verwendung des p-Nitrobenzylesters
am Phe⁸-Rest und des Asp¹-β-benzylesters durch katalytische Hydrierung
direkt zum Ile⁵-Angiotensin II führte. Die Zwischenfragmente wurden
dabei durch Ankondensation jeweils nur einer Aminosäure bzw. beim
Dipeptid Z-Val-Tyr-NHNH₂, unter Verwendung der Azid-Methode er-

halten. Die Schutzgruppenkombination Benzyloxycarbonyl/p-Nitro-
benzylester bzw. Benzyloxycarbonyl/tert.-Butylester diente in der Folgezeit für die Darstellung zahlreicher Angiotensin-Analoga.

Beziehungen zwischen Struktur und Aktivität (vgl. *163*): Etwa 100 Analoga des
Ile⁵- und Val⁵-Angiotensin II sind in den letzten 10 Jahren synthetisiert worden und
ermöglichten folgende Strukturaktivitäts-Beziehung abzuleiten: Die β-Carboxyl-
gruppe des Asp¹-Restes kann als Amidgruppe vorliegen, ohne daß ein Wirkungsverlust eintritt. Ein Austausch des Asp¹-Restes gegen Glutaminsäure, Glutamin
oder Pyroglutaminsäure führt zu voll aktiven Derivaten; ein Austausch gegen Glycin
oder sogar ein Fehlen des Asp¹-Restes liefert Verbindungen mit 30—50% der Akti-
vität des natürlichen Wirkstoffes. Wirkungssteigerungen wurden mit D-Asp¹,
D-Asn¹-Angiotensin und mit Octapeptiden erzielt, bei denen der Asp¹- bzw. ein
D-Asp¹-Rest über die β-Carboxylgruppe mit der restlichen Peptidkette verknüpft ist.
Die höhere Aktivität bei diesen Derivaten wird auf einen verminderten Abbau durch
Aminopeptidasen zurückgeführt. Das erklärt auch eine protrahierte Wirkung der
Verbindungen. Bei Blockierung der Guanidogruppe des Arg²-Restes durch eine
Nitrogruppe tritt nur ein geringer Wirkungsverlust ein. Aktivitäten von 20%
wurden bei einem Austausch des Arg²-Restes durch Lysin und Ornithin, noch
deutliche Aktivitäten sogar bei einem Austausch gegen Lysin, D-Lysin, Ornithin,
L- und D-Hydroxylysin, Homoarginin und D-Homoarginin bei gleichzeitigem Fehlen
des Asp¹-Restes nachgewiesen. Die C-terminale Carboxylgruppe des Angiotensins
darf nicht substituiert sein. Veränderungen der Positionen 3 und 5 sind ohne
größeren Wirkungsverlust möglich. So ist z. B. das Leu³-Angiotensin voll wirksam.
Der His⁶-, Pro⁷- und Phe⁸-Rest scheinen für die biologische Aktivität essentiell zu
sein. Inhibitorwirkung konnte unter den Angiotensinanaloga nicht gefunden
werden. Ein all-D-Val⁵-Angiotensin-Asp¹-β-amid (*158, 210*) erwies sich als inaktiv.

2. Kinine

Bradykinin wurde erstmals von Boissonnas et al. (*44*) auf dem in
Abb. 27 angegebenen Weg nach einer Fragmentkondensation syntheti-
siert. Das Arginin wurde als Nitroarginin-p-nitrobenzylester eingesetzt.
Erwähnenswert ist die Verwendung der tert.-Butyloxycarbonylhydrazid-
gruppe, die gewählt wurde, da eine Hydrazinolyse von Methyl- bzw.
Äthylestern bei Anwesenheit des Nitroargininrestes nicht gelingt. Von
Nicolaides und De Wald (*123*) wurde für die Synthese des Brady-
kinins, ausgehend vom Nitroargininmethylester unter Verwendung von
Benzyloxycarbonyl-aminosäure-p-nitrophenylestern, die Methode der
schrittweisen Synthese verwendet. Das N-terminale Arginin wurde als
Z-Arg(Z)₂-OH eingesetzt. Bei der Decarboxylierung des Z-Phe-Ser-Pro-
Phe-Arg(NO₂)-OMe mit HBr/Eisessig wurde eine O-Acylierung des
Serinrestes beobachtet. Bei der alkalischen Verseifung des voll ge-
schützten Nonapeptids wurde der O-Acetylrest zusammen mit dem
C-terminalen Methylester unter gleichzeitiger Abspaltung *eines* ω-Benzyl-
oxycarbonylrestes am N-terminalen Argininrest entfernt. Eine Kombination schrittweiser Kettenaufbau/Fragmentkondensation unter Ver-
wendung der mit Trifluoressigsäure ebenfalls leicht entfernbaren N-tert.-
Amyloxycarbonyl-Schutzgruppe und des C-terminalen Benzylesters

wurde von SAKAKIBARA und INUKAI (*147*) beschrieben. Die für die Kupplung benötigten aktivierten Ester wurden nach einer neuen Methode intermediär durch Umesterung der N-tert.-Amyloxycarbonyl-aminosäuren mit Trifluoressigsäure-*p*-nitrophenylester, -2,4,5-trichlorphenylester bzw. -N-hydroxysuccinimidester hergestellt. Nach dieser Methode wurden jedoch nur N-geschützte Aminosäuren bzw. Peptide mit C-terminalem Glycin gekuppelt. Nach den Angaben der Autoren liefert die beschriebene Synthese ohne Reinigung Bradykinin mit voller Aktivität.

Für die Synthese des Kallidins wurde von PLESS et al. (*135*) auf das voll geschützte Bradykinin zurückgegriffen (vgl. Abb. 27; E 1—9) und nach Decarbobenzoxylierung mit HBr/Eisessig mit Z-Lys(Z)-ONP das

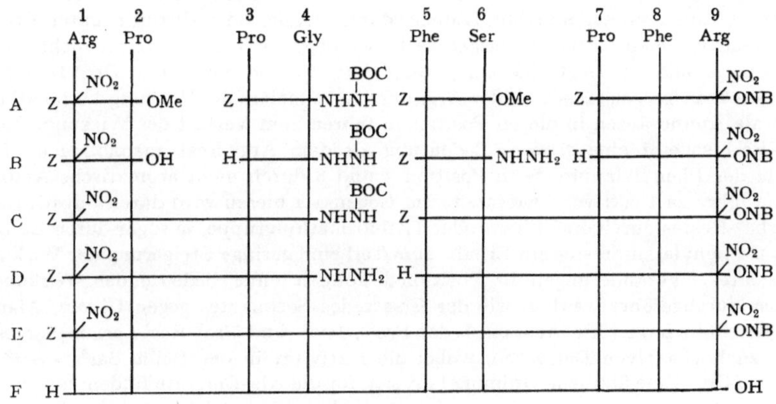

Abb. 27. Synthese des Bradykinins [BOISSONNAS et al. (*44*)]

geschützte Decapeptid aufgebaut, aus dem durch katalytische Hydrierung das freie Decapeptid erhalten wurde. In einer zweiten Synthese wurde das gleiche Bauprinzip gewählt, nur wurde an den Argininresten tosyliertes Bradykinin und entsprechend N^ε-tosyliertes Lysin eingesetzt. Die Abspaltung der Schutzgruppen ließ sich mit Natrium in flüssigem Ammoniak durchführen.

Die Synthese des Methionyl-lysyl-bradykinins gelang (*156*) durch Umsetzung von BOC-L-Met-L-Lys(BOC)-azid mit Bradykinin-Dihydrochlorid. In diesem Fall wurde die bei längeren argininhaltigen Peptiden häufig verwendete Technik der Protonisierung der Guanidinogruppen mit Erfolg verwendet.

Das erst kürzlich in seiner Struktur aufgeklärte Phyllokinin wurde von BERNARDI et al. (*23*) durch Fragmentkondensation hergestellt *(Abb. 28)*. Erwähnenswert bei dieser Synthese ist die Kombination der tert.-Butyloxycarbonylgruppe als Schutz der α-Aminogruppen mit der Benzyloxycarbonylgruppe als ω-Schutz des Arginins. Serin wurde vom Beginn der Synthese als O-Acetylderivat eingesetzt. Die Einführung der O-Sulfat-

Gruppe am C-terminalen Tyrosin gelang auf der Stufe der vollgeschützten Verbindung K 1—11 durch mehrstündige Behandlung mit 10 Äquivalenten Pyridin-Schwefeltrioxyd-Komplex in Pyridin. Katalytische Hydrierung und selektive alkalische Verseifung des O-Acetylrestes am Ser^6-Rest lieferte schließlich das freie Phyllokinin, das durch Gegenstromverteilung gereinigt werden konnte.

Beziehungen zwischen Struktur und Aktivität (vgl. *159, 163*): Die biologischen Eigenschaften der Kinine waren der Anlaß zur Synthese von etwa 150 Analoga des Bradykinins, Kallidins und Met-Lys-Bradykinis. Ziel dieser Arbeiten war in erster Linie eine Verlängerung der blutdrucksenkenden Wirkung unter Ausschaltung von Nebenwirkungen und die Auffindung von Bradykininantagonisten. Beides ist nicht erreicht worden. Die pharmakologischen Screeningteste erbrachten jedoch grundlegende Kenntnisse zur Struktur-Aktivität-Beziehung der Kinine:

Die terminalen Argininreste können durch Lysin oder Citrullin unter Erhalt einer relativ hohen Aktivität ausgetauscht werden. Daß die Basizität nicht allein ausschlaggebend ist, zeigt eine nur geringe Aktivität der Orn^1- und Orn^9-Derivate. Der gleichzeitige Austausch beider Argininreste reduziert die Wirkung beträchtlich. Neutrale Aminosäuren in diesen Positionen führen zum Verlust der Wirkung. Dem Arg^1-Rest scheint eine größere Bedeutung als dem Arg^9-Rest zuzukommen. Ein Ersatz der Phenylalaninreste in Position 5 und 8 durch nicht aromatische Aminosäuren führt zu inaktiven Derivaten. Im Gegensatz hierzu wird durch Substitution des Phe^8-Restes durch eine Fluor- oder Trifluormethylgruppe, ja sogar durch Einbau eines D-Phenylalaninrestes ein Erhalt, zum Teil eine geringe Steigerung der Wirkung beobachtet. Veränderungen in Position 6 können ohne bedeutenden Wirkungsverlust durchgeführt werden, wie der Ersatz des Serinrestes gegen Glycin, Alanin und Threonin zeigt. Ein Austausch der Pro^2-, Pro^3- und Pro^7-Reste gegen Sarcosin führt zu hochaktiven Derivaten, wobei die Aktivität in der Reihe $Sar^3 > Sar^2 > > Sar^7$ fällt. Eine ähnliche Ordnung läßt sich für die Ala-Derivate finden, nur liegen diese in ihrer absoluten Wirkung niedriger. Die höhere Wirkung der Sarcosinderivate könnte durch ihre Iminostruktur und der dadurch mit dem Prolin vorhandenen Ähnlichkeit zuzuschreiben sein. Ein Austausch des Gly^4-Restes durch Alanin, Prolin oder Serin liefert geringer wirksame Derivate. Der gleichzeitige Austausch von zwei oder mehr Aminosäuren führt zu praktisch inaktiven Peptiden. Eine Ausnahme macht das Ala^3-Gly^6-Bradykinin mit zirka 1/6 Bradykininaktivität, besonders aber das kürzlich beschriebene Depsipeptid Analogon Gly^6-$PheLac^8$-Bradykinin, bei dem am Kaninchenblutdruck bei doppelter Aktivität ein 2—3mal länger dauernder Depressoreffekt beobachtet werden konnte (*140*). Dieser Befund wird auf eine durch den Austausch Amid- gegen Esterbindung erklärbare 20—35% geringere Abbaurate des Peptids gegenüber Kaninchenserum-Kininase zurückgeführt. Verkürzung des Bradykininmoleküls am N- oder C-Terminus führt zu einem Aktivitätsverlust.

Analoga des Kallidins und Met-Lys-Bradykinins entsprechen in ihrer Wirksamkeit etwa den entsprechenden Bradykininanaloga. Eine 2—10mal höhere Aktivität konnte bei einigen Analoga des Met-Lys-Bradykinins gefunden werden, bei denen die N-terminale Aminosäure ausgetauscht wurde, z. B. Lys-Lys-Bradykinin (*157*). Einige Analoga des Bradykinins und Kallidins zeigen mitunter ein hohes Maß der Differenzierung in bezug auf ihre Wirkung an glattmuskulären Organen und am Blutdruck. Bei einigen von Stewart und Woolley (*194*) beschriebenen Bradykininanaloga konnten Andeutungen einer Anti-Bradykinin-Aktivität am Rattenuterus festgestellt werden. Retrobradykinin, all-D-Bradykinin und all-D-retro-Bradykinin erwiesen sich als inaktiv und zeigten auch keine Inhibitorwirkung (*207*).

Literaturverzeichnis: SS. 107—119

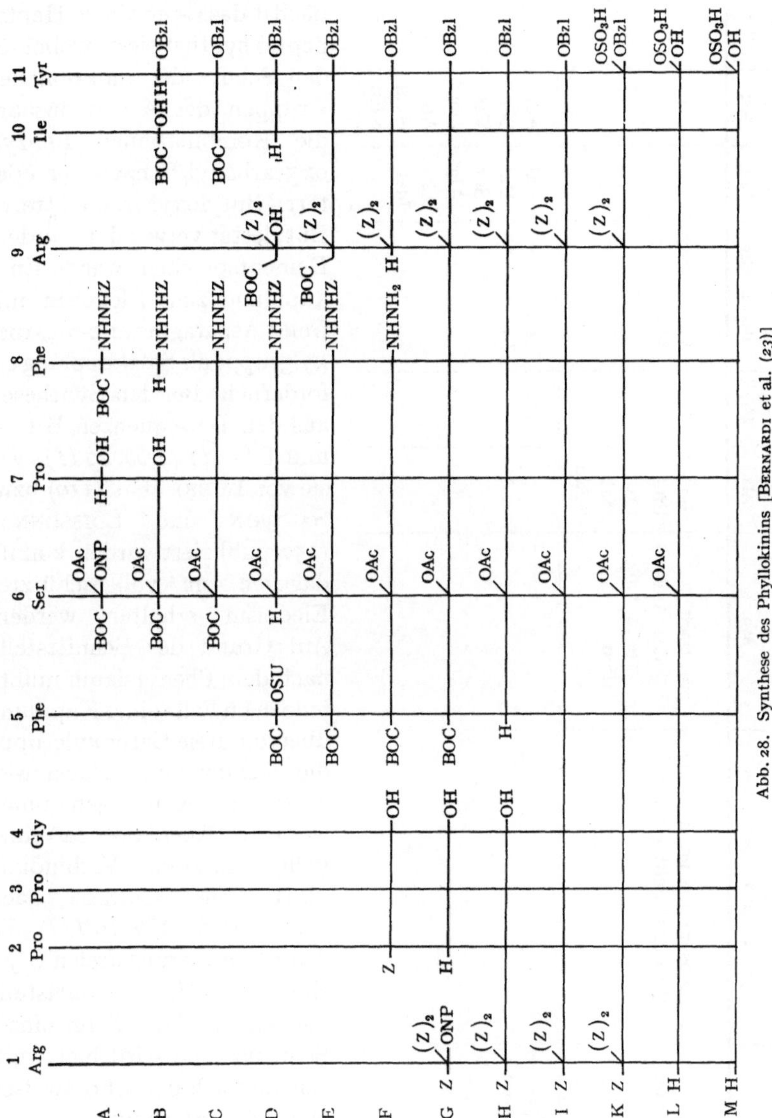

Abb. 28. Synthese des Phyllokinins [BERNARDI et al. (23)]

3. Eledoisin und Physalaemin

Eledoisin: Alle bisher bekannt gewordenen Synthesen des Eledoisins und seiner Analoga sind nach dem Prinzip der Fragmentkondensation durchgeführt worden. Entsprechend ihren Hauptschnittstellen sind drei prinzipielle Wege zu unterscheiden. Für die Synthesen nach *Abb. 29 I* wurde von SANDRIN und BOISSONNAS (*151*) bzw. LÜBKE et al. (*110*) zu-

     E. Schröder und K. Lübke:

Tabelle 2. Struktur-Wirksamkeits-Beziehungen am Eledoisin

	1 Pyroglu	2 Pro	3 Ser	4 Lys	5 Asp	6 Ala	7 Phe	8 Ile	9 Gly	10 Leu	11 Met	NH$_2$
Kein oder nur geringer Verlust der hypotensiven Aktivität	Aminosäuren sind für eine Wirkung nicht erforderlich				praktisch jeder Austausch möglich	Ser Phe Pro Met β-Ala	—	Val Phe	Pro Sar	—	Eti	
Starker oder vollständiger Verlust der hypotensiven Aktivität							Tyr	Leu Pro	—	Val Ile Ser Ala Asp Met	Ala Ser Val Leu Abu Nval Nleu	

nächst das C-terminale Heptapeptid synthetisiert, wobei für den Schutz der funktionellen Gruppen der Asparaginsäure die Kombinationen Benzyloxycarbonyl/Benzylester oder tert.-Butyloxycarbonyl/tert.-Butylester verwendet wurden. Dementsprechend war die Endkupplung zum Eledoisin mit freier Asparaginsäure-β-carboxylgruppe als Salzkupplung erforderlich. Bei den Synthesen aus den Teilsequenzen B 1—7 und B 8—11 *(Abb. 29 II)*, wie sie von Lübke et al. *(110)* bzw. Sandrin und Boissonnas *(152)* publiziert wurden, konnte dagegen ein vollgeschütztes Eledoisin erhalten werden. Auf Grund der Schnittstelle nach dem Phenylalanin mußte jedoch im Falle einer Kupplung über die freie Carboxylgruppe die Gefahr einer Racemisierung in Kauf genommen werden. Ebenfalls zu einer vollgeschützten Verbindung führte die Synthese nach Bajusz *(14)* *(Abb. 29 III)*, die den racemisierungsfreien Gly^9-Rest als Hauptschnittstelle verwendet. Die N-terminale Sequenz 1—9 wird hier weitgehend nach der schrittweisen Methode aufgebaut.

Beziehungen zwischen Struktur und Aktivität (vgl. *163*): Beim Eledoisin wird bereits mit der C-terminalen Heptapeptidsequenz praktisch die volle Aktivität erreicht *(161)*. Bei der schrittweisen Verlängerung dieses Heptapeptids werden mit dem Nona- und Decapeptid Analoga erhalten, die fast doppelt so wirksam sind wie das Eledoisin selbst (*24*). Darüber hinaus sind ungewöhnlich viele Analoga mit ausgetauschten Aminosäureresten bekannt geworden, die gleich oder sogar stärker wirksam sind als der Naturstoff. Die kürzeste noch wirksame Sequenz ist das C-

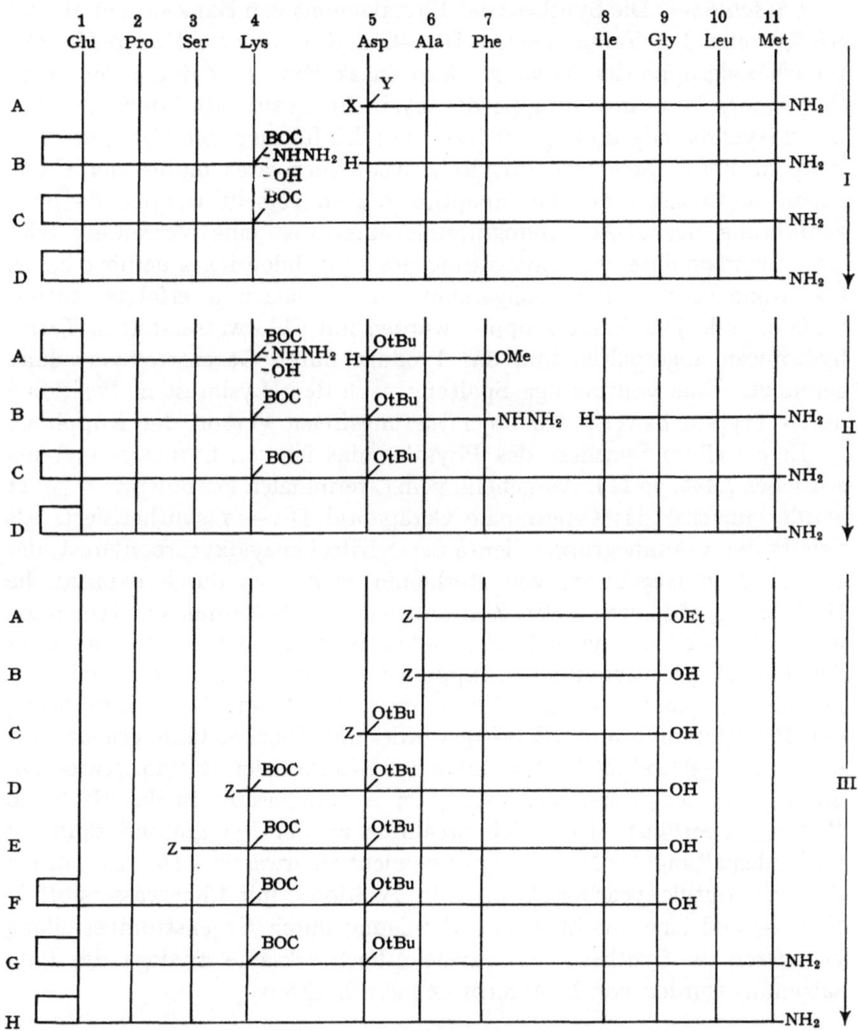

Abb. 29. Synthesen des Eledoisins. I und II [SANDRIN und BOISSONNAS (*151, 152*), LÜBKE et al. (*110*)], III [BAJUSZ (*14*)]

terminale Pentapeptid, und ein Austausch von Aminosäuren in diesem Bereich führt, von wenigen Ausnahmen abgesehen (z. B. Ethionin in Position 11), fast immer zu einer Reduktion der Aktivität. In *Tabelle 2* ist die Austauschbarkeit von Aminosäuren am Eledoisinmolekül schematisch zusammengestellt. Ungewöhnlich hochwirksame Verbindungen wurden auch durch Substitution funktioneller Gruppen erhalten. Verbindungen dieser Art sind die N^{α}-Acyl-Derivate der C-terminalen Hexapeptidsequenz H-Ala-Phe-Ile-Gly-Leu-Met-NH$_2$ (*111*) oder die N^{ε}-Acyl-Derivate der Sequenz H-Lys-Phe-Ile-Gly-Leu-Met-NH$_2$ (*22*, *25*). Ein all-D-Analogon der an sich voll wirksamen C-terminalen Gly^5-Val^8-Heptapeptidsequenz ist unwirksam (*164*).

Physalaemin: Die Synthese des Physalaemins von BERNARDI et al. (*26*) erfolgte aus den Teilsequenzen D 1—6 und D 7—11 *(Abb. 30 I)*. Die ω-Carboxylgruppe der Asparaginsäure in Position 3 ist durch den tert.-Butylester, die ε-Aminogruppe des Lysins in Position 6 durch die tert.-Butyloxycarbonylgruppe geschützt. Vor Einführung der Pyroglutaminsäure in Form ihres tert.-Butyloxycarbonylderivates mußte der C-terminale Methylester des Pentapeptids A 2—6 verseift werden, da nach Einführung des N-Acyl-pyroglutaminsäurerestes eine Verseifung oder eine Hydrazinolyse zur Aufspaltung des Pyrrolidonringes geführt hätte. Die Kondensation zum vollgeschützten Physalaemin erfolgte mittels Carbodiimid. Die Schutzgruppen wurden mit Chlorwasserstoff in Tetrahydrofuran abgespalten und das Produkt durch Gegenstromverteilung gereinigt. Eine vollständige Spaltung nach dem Lysinrest in Position 6 durch Trypsin beweist den racemisierungsfreien Verlauf der Kupplung.

Eine weitere Synthese des Physalaemins ist von CHILLEMI (*51*) beschrieben *(Abb. 30 II)*. Ausgehend vom C-terminalen Pentapeptid C 7—11 wurde zunächst das C-terminale Octapeptid D 4—11 synthetisiert. Als Schutz der α-Aminogruppe diente der p-Nitrobenzyloxycarbonylrest, der sich auch in Gegenwart von Methionin erfolgreich durch katalytische Hydrierung abspalten ließ. Zum Aufbau des N-terminalen Tripeptids mußte der zunächst synthetisierte Z-Ala-Asp(OtBu)-OMe vor Abspaltung der Benzyloxycarbonylschutzgruppe verseift werden, da der freie Dipeptidester quantitativ das Diketopiperazin bildete. Nach Einführung der Pyroglutaminsäure (B 1—3) wurde mit Diazomethan erneut verestert (C 1—3) und nach Abspaltung der N-terminalen Butyloxycarbonylgruppe und des β-tert.-Butylesters der Asparaginsäure in das Hydrazid (E 1—3) überführt. Eine Nebenreaktion an der Pyroglutaminsäure ist nach Abspaltung der N-Schutzgruppe nicht zu erwarten. Die Abspaltung der tert.-Butyloxycarbonylgruppe in Position 6 mit Chlorwasserstoff in Eisessig und eine anschließende Reinigung durch Gegenstromverteilung beendeten die Synthese. Verschiedene hochwirksame Analoga des Physalaemins wurden von BERNARDI (*22*) synthetisiert.

D. Peptidantibiotika

1. Gramicidin und verwandte Verbindungen

Als Beispiel für die Synthese eines cyclischen Peptidantibiotikums soll der Aufbau des Gramicidins S angeführt werden, wie er bereits 1957 von SCHWYZER und SIEBER (*175*) beschrieben wurde. Das Prinzip, die Cyclisierung eines linearen aktivierten Peptidesters, wird in gleicher Weise auch heute noch verwendet, lediglich durch die Entwicklung neuer Aminoschutzgruppen konnten Vereinfachungen erzielt werden. Wie *Abb. 31* zeigt, wurde nach der Fragmentkondensation zunächst der lineare Benzyloxycarbonyl-pentapeptidmethylester (I) hergestellt, der nach Decarbobenzoxylierung trityliert und verseift (IV) und mit dem Pentapeptidmethylester (II) in Gegenwart von 1-Cyclohexyl-3-(2-morpho-

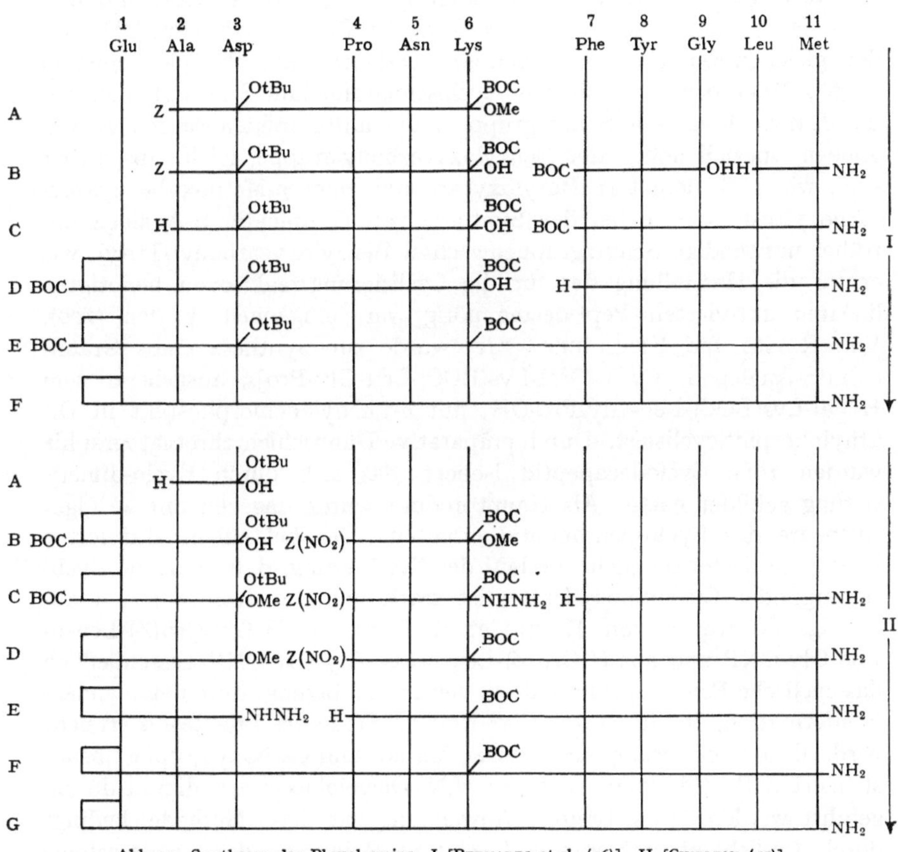

Abb. 30. Synthesen des Physalaemins. I [Bernardi et al. (26)], II [Chillemi (51)]

linyl)-4-äthyl-carbodiimid zum Trityl-decapeptidmethylester (VII) kondensiert wurde. Nach alkalischer Verseifung wurde mit Di-(*p*-nitrophenyl)-sulfit der Decapeptid-*p*-nitrophenylester (IX) hergestellt und nach Detritylierung mit Trifluoressigsäure (X) zur Vermeidung der Bildung linearer Polymere nach dem Verdünnungsprinzip in Dimethylformamid/Pyridin cyclisiert. Zur Entfernung linearer Produkte wurde mit basischen und sauren Ionenaustauschern behandelt; Gegenstromverteilung und Chromatographie an neutralem Al_2O_3 lieferten mit 28%iger

Ausbeute kristallisiertes Di-tosyl-gramicidin S (XI), von dem die Tosyl-
gruppen mit Natrium in flüssigem Ammoniak entfernt werden konnten.
Eine Vereinfachung der Synthese gelang durch Cyclisierung des aktivierten Pentapeptidesters (VI), bei der unter gleichen Bedingungen in
einer Verdopplungsreaktion Di-tosyl-gramicidin S gewonnen wurde
(*176*). Die Bildung des cyclischen Pentapeptids (Cyclo-semigramicidin S)
konnte nicht beobachtet werden. Erst später gelang es Waki und Izumiya (*211*) ausgehend von Z(OMe)-Val-Orn(Z)-Leu-D-Phe-Pro-OH, auf
dem gleichen Reaktionsweg neben 20% Di-benzyloxycarbonyl-gramicidin
S 16% Mono-benzyloxycarbonyl-cyclosemigramicidin S zu isolieren. Mit
der Entwicklung von Schutzgruppen, die unter milden sauren Bedingungen, speziell neben der Benzyloxycarbonylgruppe, leicht abspaltbar
sind, wie z. B. dem tert.-Butyloxycarbonyl- oder *p*-Methoxybenzyloxy-
carbonylrest, konnte bei der Synthese von Gramicidin S-Analoga der
früher notwendige Schutzgruppenwechsel Benzyloxycarbonyl/Trityl, wie
er für die Herstellung der für die Cyclisierungsreaktionen benötigten
linearen aktivierten Peptidester nötig war, umgangen werden (*100*).
Von Rothe und Eisenbeiss (*142*) wurde zur Synthese eines Grami-
cidin-S-Analogon, Cyclo-[Val-Lys(BOC)-Leu-Gly-Pro]$_2$, ausgehend vom
H-Val-Lys(BOC)-Leu-Gly-Pro-OH, mit *o*-Phenylenchlorphosphit in Di-
äthylphosphit cyclisiert. Durch präparative Dünnschichtchromatographie
wurden 10% Cyclodecapeptid isoliert, das sich durch Cyclo-dimeri-
sierung gebildet hatte. Als Hauptprodukt wurde dagegen mit 22%iger
Ausbeute das Cyclo-pentapeptid erhalten. Der bei einem aktivierten
Pentapeptidester mögliche Verlauf der Cyclisierung, d. h. einfache Cyclisierung oder Cyclodimerisierung ist stark von der Konformation abhängig. So wurde von Kondo et al. (*100*) aus H-Gly-Orn(Z)-Leu-D-
Phe-Gly-ONP und aus H-Orn(Z)-Leu-D-Phe-Gly-Gly-ONP ausschließlich
das cyclische Pentapeptid erhalten, vermutlich bedingt durch eine andere
Konformation, die durch den Gly^5-Rest an Stelle des Pro-Restes erreicht
wird. Eine Cyclisierung eines an der Amino- und Carboxylgruppe unsubstituierten Peptids kann auch mit N,N'-Dicyclohexylcarbodiimid durchgeführt werden. Eine breitere Anwendung hat diese Methode, bedingt
durch Löslichkeitsprobleme und schwere Reinigung des Cyclisierungs-
produktes nicht gefunden (vgl. jedoch Polymyxin, S. 102). Ähnlich selten
für die Cyclisierung von Naturstoffen wurde die Azid-, gemischte Anhydrid-,
Äthoxyacetylen- und N-Äthylen-5-phenylisoxazolium-3'-sulfonat-Methode
verwendet.

Beziehungen zwischen Struktur und Aktivität: Ein Austausch der Prolinreste im
Gramicidin S durch Sarcosylreste bewirkt keine Veränderung der Aktivität (*10*),
ein Austausch durch Glycylreste bewirkt sogar einen Anstieg der Aktivität
gegenüber *B. subtilis* (*11*). Cyclische Pentapeptide mit der Sequenz des Gramicidin S bzw. aktiver Analoga sind unwirksam. Ein Ersatz der Ornithylreste durch
Lysylreste ist nicht mit einem Aktivitätsverlust verbunden.

Literaturverzeichnis: SS. 107—119

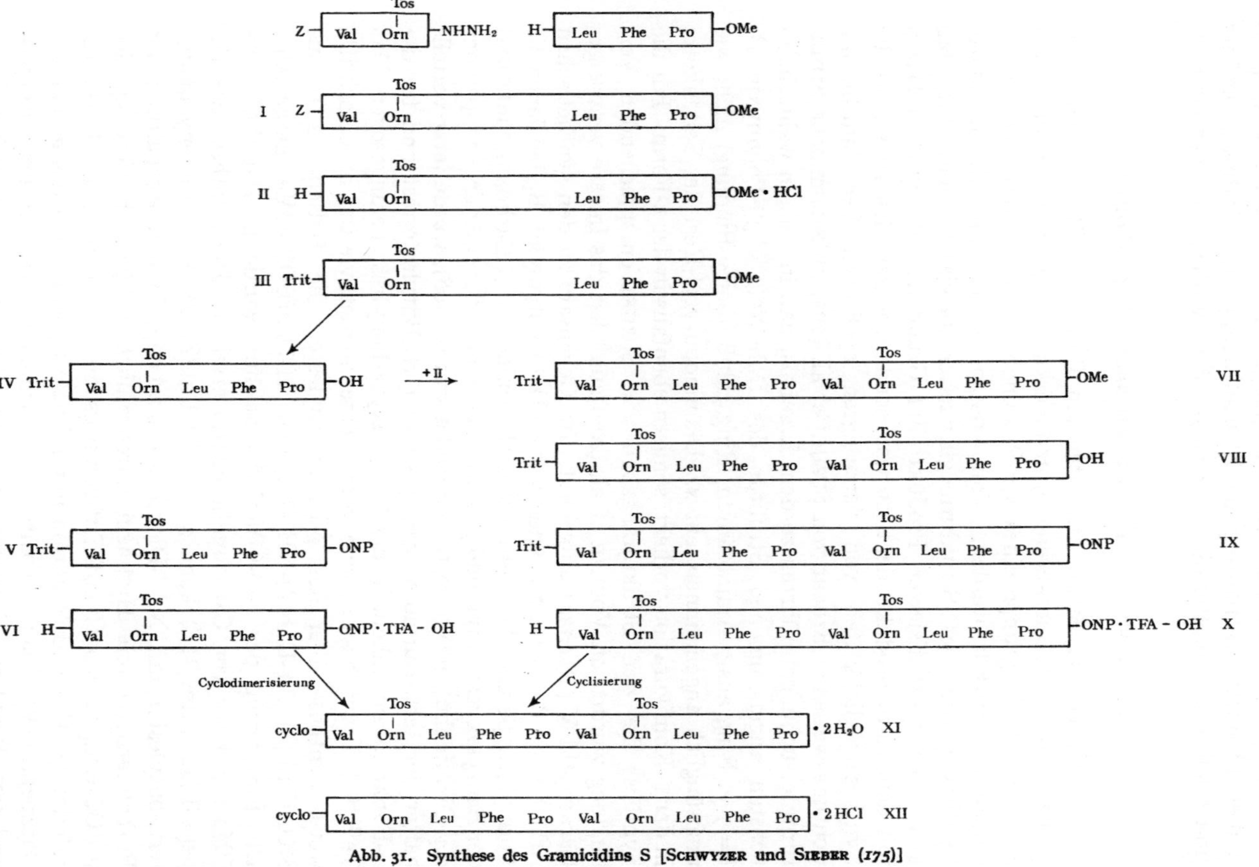

Abb. 31. Synthese des Gramicidins S [SCHWYZER und SIEBER (175)]

Kürzlich gelang Ohno et al. (*124*) aus Z(OMe)-Phe-D-Phe-Asn-Gln-Tyr-Val-Orn(Z)-Leu-D-Phe-Pro-ONP (hergestellt aus zwei Pentapeptideinheiten) durch Abspaltung der Z(OMe)-Gruppe mit Trifluoressigsäure und Cyclisierung in Pyridin bei 70—75° die Synthese des Tyrocidin A.

2. Polymyxin B_1 und verwandte Verbindungen

Auf Grund der Strukturarbeiten am Polymyxin B_1 konnten Ringgröße (7 oder 8 Aminosäurereste) und Verknüpfungsart (Ringschluß über α- oder γ-Aminogruppe eines α,γ-Diaminobutyrylrestes) zunächst nicht eindeutig ermittelt werden. Von Vogler et al. (*208*) durchgeführte Synthesen von vier Strukturmöglichkeiten bewiesen nicht nur, daß keines der synthetisierten Produkte auf Grund der biologischen Daten mit dem Naturprodukt identisch war, sondern waren gleichzeitig für die Synthesemöglichkeiten von verzweigten cyclischen Peptidantibiotika richtungsweisend. Erneut von Suzuki et al. (*202*) aufgenommene Strukturuntersuchungen führten zu dem Ergebnis, daß in der am wahrscheinlichsten angenommenen Struktur des Dab^1-N$^\alpha$(1',2',3')-Polymyxin B_1 (= 7 γ: Ring aus 7 Aminosäuren, Ringschluß über γ-Bindung) nicht wie ursprünglich angenommen, ein α,γ-Diaminobutyryl-Rest der Seitenkette in der D-Konfiguration vorliegt, sondern ebenfalls in der L-Form. Für die Synthese der den Strukturangaben von Suzuki entsprechenden Verbindung wurde von Vogler et al. (*209*) zunächst das lineare verzweigte Decapeptid V *(Abb. 32)* aufgebaut. Im Gegensatz zu den synthetischen Arbeiten am 8α-, 8γ-, 7α- und 7γ-(1'-D-Dab)-Polymyxin B_1, bei denen die Formylgruppe und der Methylester als Amino- bzw. Carboxylschutz und der Benzyloxycarbonyl- und Tosylrest als Schutz der ω-Aminogruppen der α,γ-Diaminobuttersäurereste verwendet wurde (*208*), wurde jetzt vorteilhafter der tert.-Butyloxycarbonyl- und tert.-Butylesterrest und für die ω-Funktion ausschließlich der Benzyloxycarbonylrest herangezogen. Das Teilstück I wurde ausgehend vom L-Threoninmethylester, durch schrittweise Synthese über den p-Nitrophenylester, die Teilsequenz II aus MOA-Dab(Z)-Thr-Dab(Z)-NHNH$_2$ und H-Dab(BOC)-OMe hergestellt. Die Einführung der (+)-6-Methyloctansäure wurde in H-Dab(Z)-Thr-OMe mit Hilfe von Carbonyldiimidazol erreicht. Die Synthese des Dipeptid-Derivates (IV) konnte aus Pht-Dab(Z)-OH und D-Phenylalanin-tert.-butylester mit N,N'-Dicyclohexylcarbodiimid und Abspaltung der Phthalylgruppe mit Hydrazin durchgeführt werden. Entfernung der γ-BOC-Gruppe von II mit Trifluoressigsäure, Umsetzung mit I, Hydrazidbildung (III) und Reaktion mit IV lieferte schließlich das geschützte Decapeptid (V). Nach Abspaltung des terminalen tert.-Butyloxycarbonyl- und tert.-Butylester-Schutzes wurde in Dimethylformamid/Dioxan mit der 40fachen Gewichtsmenge N,N'-Dicyclohexylcarbodiimid 3^1/2 Tage in

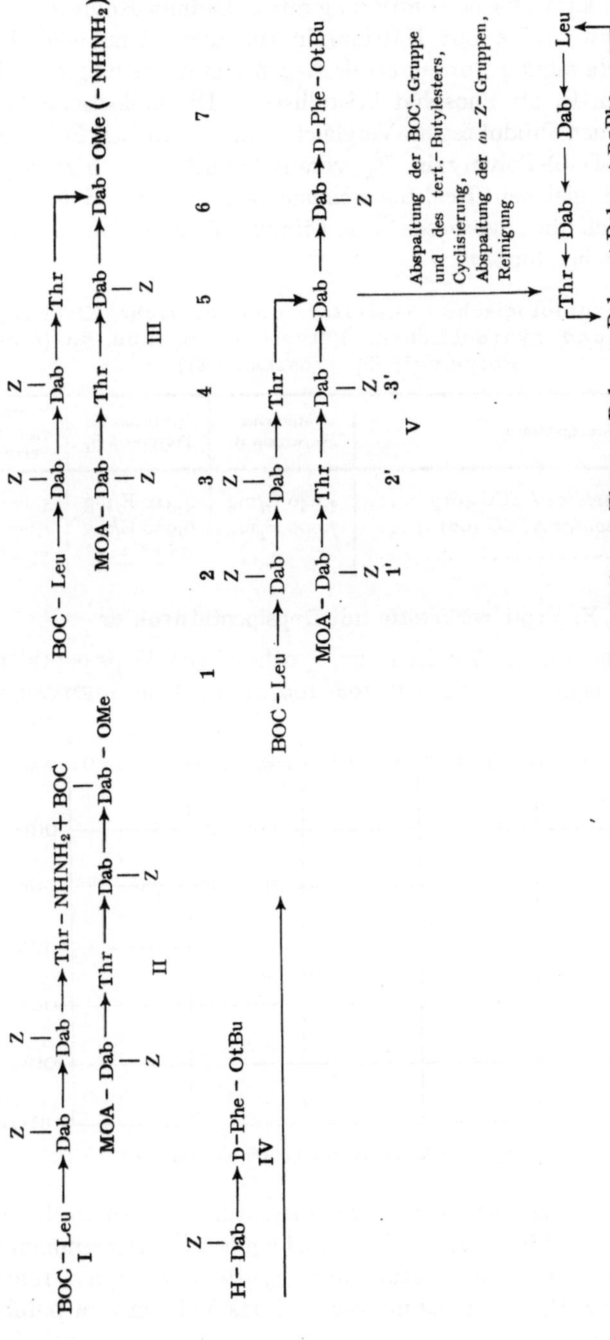

Abb. 32. Synthese des Polymyxins B₁ (all-L-Dab-7α-Verbindung) [VOGLER et al. (209)]

hoher Verdünnung cyclisiert. Eine Entfernung der ω-Benzyloxycarbonyl-gruppen durch katalytische Hydrierung mit Palladium-Kohle gelang nicht vollständig und mußte mit Natrium in flüssigem Ammoniak beendet werden. Die Reinigung wurde mittels Gegenstromverteilung erreicht und das Polymyxin B_1 als Phosphat kristallisiert. Die biologische Aktivität des synthetischen Produktes im Vergleich zum natürlichen Polymyxin B_1 und 7α-($1'$-D-Dab)-Polymyxin B_1 veranschaulicht die *Tabelle 3*.

Nach dem gleichen Reaktionsschema wurden von Studer et al. (*196, 197*) auch die Synthesen von Circulin A und Colistin A (Polymyxin E) durchgeführt.

Tabelle 3. Mikrobiologische Aktivität und optische Drehung von natürlichem und synthetischem Polymyxin B_1 und 7α-(1-D-Dab)-Polymyxin B_1 [Vogler (*208*)]

Mikroorganismen	Natürliches Polymyxin B_1	Synthetisches Polymyxin B_1	Synthetisches 7α-(1-D-Dab)-Polymyxin B_1
Brucella bronchiseptica ATCC 4617	4,940 E/mg	4,310 E/mg	5,260 E/mg
Klebsiella pneumoniae ATCC 100131	7,000 E/mg	6,900 E/mg	1,600 E/mg
$[\alpha]_D^{25}$ in H_2O	$72° \pm 2°$	$67,5° \pm 4°$	$53,2° \pm 2°$

E. Peptidwirkstoffe mit Depsipeptidstruktur

Nach methodischen Arbeiten zur Synthese von Depsipeptiden bzw. Peptoliden im engeren Sinne (vgl. *108, 109, 160*) gelang Shemyakin et al.

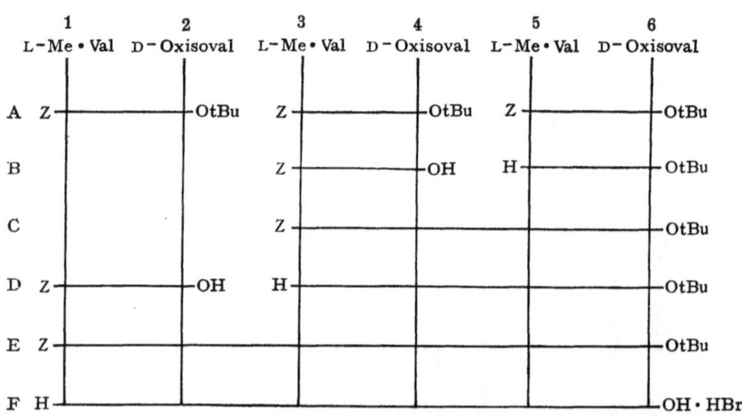

Abb. 33. Synthese des Enniatins B [Plattner et al. (*133*)]

(*184*) die Herstellung cyclischer, in der Literatur als Naturstoffsequenzen beschriebener Verbindungen. Eine fehlende Übereinstimmung der physikalischen Daten von Natur- und Syntheseprodukten führte zur Korrektur der z. B. für die Enniatine und das Valinomycin publizierten

Literaturverzeichnis: SS. 107—119

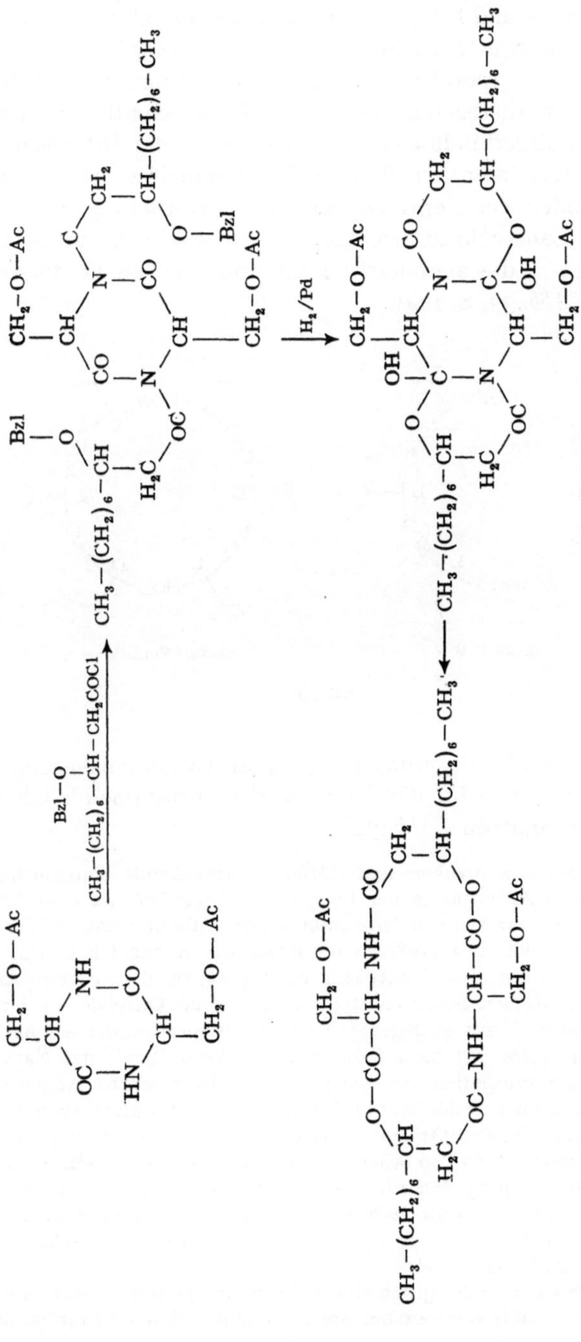

Abb. 34. Synthese des O,O′-Diacetylserratamolids [Shemyakin et al. (*182*)]

Strukturen. Quitt et al. (*138*) und Plattner et al. (*133*) konnten schließlich die Enniatine A und B [vgl. auch Shemyakin et al. (*188*)], Shemyakin
et al. das Valinomycin (*181*), Shemyakin et al. (*185*) bzw. Ovchinnikov
et al. (*126, 127*) die Sporidesmolide in Übereinstimmung mit der natürlichen Sequenz synthetisieren. Die Mehrzahl der Synthesen beginnt mit
der Bildung der Esterbindungen, bevorzugt nach der Benzolsulfochlorid-
methode. Die Kondensation dieser Peptolideinheiten erfolgt dann nach
üblichen Methoden der Peptidchemie. Die Cyclisierung wird in jedem
Falle über das Säurechlorid erreicht. Als Beispiel soll die Synthese des
Enniatin B dienen, das symmetrisch aus drei gleichen Peptolideinheiten
aufgebaut ist (*Abb. 33*, S. 104).

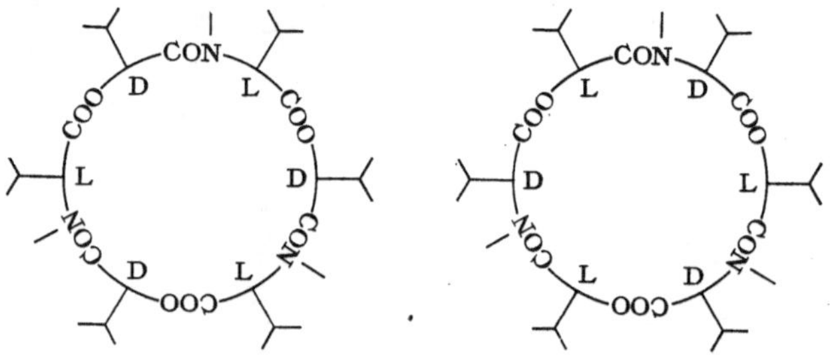

Abb. 35

Eine interessante Umlagerungsreaktion, die Cyclolumlagerung, konnten
Shemyakin et al. (*182*) für die Synthese des Serratamolids (als O,O′-Di-
acetylderivat) ausnutzen *(Abb. 34)*.

Beziehungen zwischen Struktur und Aktivität: Eingehende Untersuchungen zur
Frage Struktur und Aktivität in der Peptolidreihe wurden von dem Arbeitskreis
Shemyakin (*186, 191*) für Analoga der Enniatine, des Valinomycins und der Sporides-
molide und von Studer et al. (*198*) für die Enniatine durchgeführt. Das Struktur-
Aktivitäts-Problem wurde von Shemyakin et al. (*183*) für die Enniatin Antibiotika
auch vom topochemischen Standpunkt diskutiert. Der Antipode des Enniatin B,
das enantio-Enniatin B, mit entgegengesetzter Konfiguration der Aminosäure- und
Hydroxysäurereste *(Abb. 35)* besitzt die gleiche Aktivität wie das Naturprodukt.
In Analogie zu den synthetisierten Antipoden des Bradykinins, Angiotensins und
Oxytocins war auch im Fall des enantio-Enniatin, unter Voraussetzung eines stereo-
selektiven Receptors, Inaktivität zu erwarten. Dreht man jedoch eine der in Abb. 35
dargestellten Formeln um 60°, so fallen alle gleichen Asymmetriezentren zusammen,
während eine Esterbindung den Platz einer Amidbindung (und umgekehrt) ein-
nimmt. Relativ zum Receptor haben sich beim Enniatin B/enantio-Enniatin B
also nur Ester- und Amidbindung verschoben, ein Strukturunterschied, der keine
Veränderung der Aktivität bewirkt.
Ein Austausch einer Amid- gegen eine Esterbindung unter nahezu vollem Erhalt
der biologischen Aktivität konnte schon am Beispiel des PheLac[8]-Bradykinins gezeigt

werden (*190*). Die um ∼ 2 Zehnerpotenzen geringere biologische Aktivität eines Glyc[4]-Bradykinins beweist jedoch, daß Amid- und Esterbindung nicht in jedem Fall austauschbar sind.

Daß die am Enniatin B/enantio-Enniatin B gefundenen Ergebnisse auch für nicht-symmetrische Cyclopeptide zutreffen können, ergab das retro-enantio-Gly[5]-Gly[10]-Gramicidin, das die gleiche antibakterielle Aktivität aufweist wie Gly[5]-Gly[10]-Gramicidin S (*187*).

Literaturverzeichnis

1. ACHER, R., J. CHAUVET, M. T. CHAUVET et D. CREPY: Phylogénie des peptides hormonaux neurohypophysaires. Bull. soc. chim. biol. (Paris) **47**, 2279 (1965).

2. ALBERTSON, N. F.: Synthesis of Peptides with Mixed Anhydrides. Organ. React. **12**, 157 (1962).

3. ANASTASI, A., G. BERTACCINI and V. ERSPAMER: Pharmacological Data on Phyllokinin (Bradykinyl-Isoleucyl-Tyrosine O-Sulphate) and Bradykinyl-Isoleucyl-Tyrosine. Brit. J. Pharmacol. **27**, 479 (1966).

4. ANDERSON, G. W. and F. M. CALLAHAN: t-Butyl Esters of Amino Acids and Peptides and their Use in Peptide Synthesis. J. Amer. Chem. Soc. **82**, 3359 (1960).

5. ANDERSON, G. W., F. M. CALLAHAN and J. E. ZIMMERMAN: Synthesis of N-Hydroxysuccinimide Esters of Acyl Peptides by the Mixed Anhydride Method. J. Amer. Chem. Soc. **89**, 178 (1967).

6. ANDERSON, G. W. and A. C. McGREGOR: t-Butyloxycarbonylamino Acids and Their Use in Peptide Synthesis. J. Amer. Chem. Soc. **79**, 6180 (1957).

7. ANDERSON, G. W., J. E. ZIMMERMAN and F. M. CALLAHAN: The Use of Esters of N-Hydroxysuccinimide in Peptide Synthesis. J. Amer. Chem. Soc. **86**, 1839 (1964).

8. ANDERSON, J. C., M. A. BARTON, R. A. GREGORY, P. M. HARDY, G. W. KENNER, J. K. MacLEOD, J. PRESTON, R. C. SHEPPARD and J. S. MORLEY: Synthesis of Gastrin. Nature **204**, 933 (1964).

9. ANDERSON, J. C., G. W. KENNER, J. K. MacLEOD und R. C. SHEPPARD: Peptides. XXII. Syntheses of Porcine Gastrin I. Tetrahedron Suppl. 8, Part I, 39 (1966).

10. AOYAGI, H. and N. IZUMIYA: Studies of Peptide Antibiotics. V. Syntheses of Cyclic Penta- and Decapeptides with the L-Valyl-L-ornithyl-L-leucyl-D-phenylalanylsarcosyl Sequence. Bull. Chem. Soc. Japan **39**, 1747 (1966).

11. AOYAGI, H., T. KATO, M. OHNO, M. KONDO, M. WAKI, S. MAKISUMI and N. IZUMIYA: Studies on Peptide Antibiotics. III. Cyclo-(L-valyl-L-ornithyl-L-leucyl-D-phenylalanylglycyl)₂. Bull. Chem. Soc. Japan **38**, 2139 (1965).

12. ARAKAWA, K. and F. M. BUMPUS: An Improved Synthesis of Isoleucine[5] Angiotensin Octapeptide. J. Amer. Chem. Soc. **83**, 728 (1961).

13. ARAKAWA, K., M. NAKATANI and M. NAKAMURA: Purification of Human Angiotensin. Nature **214**, 278 (1967).

14. BAJUSZ, S.: A Novel Synthesis of Eledoisin. Acta Chim. Hungar. **42**, 383 (1964).

15. BAJUSZ, S. und K. MEDZIHRADSZKY: Erfahrungen über das Verhalten geschützter Peptide bei der Reaktion mit Natrium in flüssigem Ammoniak. In: G. T. Young (ed.), Peptides. Proc. 5th Europ. Sympos. Oxford 1962, p. 49. Oxford, London, New York, Paris: Pergamon Press. 1963.

16. — — Synthesis and Biological Activity of Corticotropin Fragments. Siehe Lit. *32*, p. 209.

17. BAJUSZ, S., K. MEDZIHRADSZKY, Z. PAULAY und ZS. LÁNG: Totalsynthese des menschlichen Corticotropins (αₕ-ACTH). Acta Chim. Hungar. **52**, 335 (1967).

18. Beacham, J., P. H. Bentley, R. A. Gregory, G. W. Kenner, J. K. MacLeod and R. C. Sheppard: Human Gastrin: Isolation, Structure and Synthesis: Synthesis of Human Gastrin I. Nature **209**, 585 (1966).

19. Beacham, J., P. H. Bentley, G. W. Kenner, J. J. Mendive and R. C. Sheppard: Gastrins from Some Mammalian Species. Siehe Lit. *32*, p. 235.

20. Berger, A., J. Noguchi and E. Katchalski: Poly-L-cysteine. J. Amer. Chem. Soc. **78**, 4483 (1956).

21. Bergmann, M. und L. Zervas: Über ein allgemeines Verfahren der Peptidsynthese. Ber. dtsch. chem. Ges. **65**, 1192 (1932).

22. Bernardi, L.: Synthetic Peptides Related to Physalaemin and Eledoisin. In: E. G. Erdös, N. Back and F. Sicuteri (ed.), Hypotensive Peptides. Proc. Internat. Sympos., Florence 1965, p. 86. New York: Springer, 1966.

23. Bernardi, L., G. Bosisio, R. de Castiglione and O. Goffredo: Synthesis of Phyllokinin, a Natural Bradykinin Analogue. Experientia **22**, 425 (1966).

24. Bernardi, L., G. Bosisio, F. Chillemi, G. de Caro, R. de Castiglione, V. Erspamer, A. Glaesser and O. Goffredo: Synthetic Peptides Related to Eledoisin. Experientia **20**, 306 (1964).

25. — — — — — — — — — Synthetic Peptides Related to Eledoisin. Experientia **21**, 695 (1965).

26. Bernardi, L., G. Bosisio, O. Goffredo and R. de Castiglione: Synthesis of Physalaemin. Experientia **20**, 490 (1964).

27. Bertaccini, G., J. M. Cei and V. Erspamer: Occurrence of Physalaemin in Extracts of the Skin of Physalaemus Fuscumaculatus and its Pharmacological Actions on Extravascular Smooth Muscle. Brit. J. Pharm. Chemotherap. **25**, 363 (1965).

28. Bethge, H. und F. A. Gries: Untersuchungen über die adrenocorticotrope Wirkung von β^{1-24}-Corticotropin beim Menschen. Arch. exp. Pathol. Pharmakol. **254**, 425 (1966).

29. Beyerman, H. C. and J. S. Bontekoe: The t-Butoxy Group, a Novel Hydroxyl-Protecting Group for Use in Peptide Synthesis with Hydroxy-Amino Acids. Rec. trav. chim. Pays-Bas **81**, 691 (1962).

30. — — Synthesis of the C-Terminal Tetrapeptide Sequence Derived from Glucagon. Rec. trav. chim. Pays-Bas **81**, 699 (1962).

31. Beyerman, H. C., J. S. Bontekoe and A. C. Koch: A Synthesis of Oxytocin. Rec. trav. chim. Pays-Bas **78**, 935 (1959).

32. Beyerman, H. C., A. van de Linde and W. Maassen van den Brink (ed.): Peptides. Proc. 8th Europ. Peptide Sympos. Noordwijk, Netherl. 1966. Amsterdam: North-Holland Publ. Co. 1967.

33. Bodanszky, M. and V. du Vigneaud: An Improved Synthesis of Oxytocin. J. Amer. Chem. Soc. **81**, 2504 (1959).

34. — — A Method of Synthesis of Long Peptide Chains Using a Synthesis of Oxytocin as an Example. J. Amer. Chem. Soc. **81**, 5688 (1959).

35. Bodanszky, M., J. Meienhofer and V. du Vigneaud: Synthesis of Lysine-vasopressin by the Nitrophenyl Ester Method. J. Amer. Chem. Soc. **82**, 3195 (1960).

36. Bodanszky, M. and M. A. Ondetti: Peptide Synthesis. New York: J. Wiley and Sons. 1966.

37. Bodanszky, M., M. A. Ondetti, C. A. Birkhimer and P. L. Thomas: Synthesis of Arginine-Containing Peptides through their Ornithine Analogs. Synthesis of Arginine Vasopressin, Arginine Vasotocin, and L-Histidyl-L-phenylalanyl-L-arginyl-L-tryptophylglycine. J. Amer. Chem. Soc. **86**, 4452 (1964).

38. Bodanszky, M., M. A. Ondetti, S. D. Levine, V. L. Narayanan, M. von Saltza, J. T. Sheehan, N. J. Williams and E. F. Sabo: Synthesis of a Heptacosapeptide Amide with the Hormonal Activity of Secretin. Chem. and Ind. **1966,** 1757.

39. Bodanszky, M. and D. Perlman: Are Peptide Antibiotics Small Proteins? Nature **204,** 840 (1964).

40. Bodanszky, M. and J. T. Sheehan: Active Esters and Resins in Peptide Synthesis. Chem. and Ind. **1964,** 1423.

41. Bodanszky, M., M. Szelke, E. Tömörkény and E. Weisz: Peptide Synthesis by Aminolysis of Active Esters. Chem. and Ind. **1955,** 1517.

42. Boissonnas, R. A.: Selectively Removable Amino Protective Groups Used in the Synthesis of Peptides. Adv. Organ. Chem. **3,** 159 (1963).

43. Boissonnas, R. A., St. Guttmann, R. L. Huguenin, P.-A. Jaquenoud et Ed. Sandrin: Synthèse de la L-histidyl-L-phénylalanyl-L-arginyl-L-tryptophanyl-glycyl-ε-CBO-L-lysyl-L-prolyl-L-valylamide. Helv. Chim. Acta **41,** 1867 (1958).

44. Boissonnas, R. A., St. Guttmann et P.-A. Jaquenoud: Synthèse de la L-arginyl-L-prolyl-L-prolyl-glycyl-L-phénylalanyl-L-séryl-L-prolyl-L-phénylalanyl-L-arginine, un nonapeptide présentant les propriétés de la bradykinine. Helv. Chim. Acta **43,** 1349 (1960).

45. Boissonnas, R. A., St. Guttmann, P.-A. Jaquenoud et J.-P. Waller: Une nouvelle synthèse de l'oxytocine. Helv. Chim. Acta **38,** 1491 (1955).

46. Boissonnas, R. A., St. Guttmann, J.-P. Waller and P.-A. Jaquenoud: Synthesis of a Polypeptide with ACTH-like Structure. Experientia **12,** 446 (1956).

47. Boissonnas, R. A. et R. L. Huguenin: Synthèse de la Lys^8-oxytocine (lysine-vasotocine) et nouvelle synthèse de la lysine-vasopressine. Helv. Chim. Acta **43,** 182 (1960).

48. Bruckner, V. and K. Medzihradszky (ed.): Proc. 7th Europ. Peptide Sympos., Budapest 1964; Acta Chim. Hungar. **44** (1965).

49. Callahan, F. M., G. W. Anderson, R. Paul and J. E. Zimmerman: The Tertiary Butyl Group as a Blocking Agent for Hydroxyl, Sulfhydryl and Amido Functions in Peptide Synthesis. J. Amer. Chem. Soc. **85,** 201 (1963).

50. Carpenter, F. H. and D. T. Gish: The Application of *p*-Nitrobenzyl Chloroformate to Peptide Synthesis. J. Amer. Chem. Soc. **74,** 3818 (1952).

51. Chillemi, F.: Sintesi della fisalemina, un endecapeptide biologicamente attivo del „Physalaemus fuscumaculatus". Gazz. chim. ital. **95,** 402 (1965).

52. Chung, D. and C. H. Li: The Synthesis of D-Tryptophanyl-D-arginyl-D-phenylalanyl-D-histidyl-D-glutamic Acid and its Effect on Melanotropic Activity. Biochim. Biophys. Acta **136,** 570 (1967).

53. Consden, R., A. H. Gordon, A. J. P. Martin and R. L. M. Synge: Gramicidin S: the Sequence of the Amino-acid Residues. Biochem. J. **41,** 596 (1947).

54. Determann, H.: Gelchromatographie. Gelfiltration, Gelpermeation, Molekülsiebe. Berlin, Heidelberg, New York: Springer-Verl. 1967.

55. Doepfner, W.: Biological Characterization of a New and Highly Potent Synthetic Analogue of Corticotrophin. Experientia **22,** 527 (1966).

56. Du Vigneaud, V. and O. K. Behrens: A Method for Protecting the Imidazole Ring of Histidine During Certain Reactions and its Application to the Preparation of L-Amino-N-Methylhistidine. J. Biol. Chem. **117,** 27 (1937).

57. Du Vigneaud, V., D. T. Gish, P. G. Katsoyannis and G. P. Hess: Synthesis of the Pressor-Antidiuretic Hormone, Arginine-Vasopressin. J. Amer. Chem. Soc. **80,** 3355 (1958).

58. Du Vigneaud, V., C. Ressler, J. M. Swan, C. W. Roberts, P. G. Katsoyannis and S. Gordon: The Synthesis of an Octapeptide Amide with the Hormonal Activity of Oxytocin. J. Amer. Chem. Soc. **75**, 4879 (1953).

59. Fontana, A., F. Marchiori, L. Moroder and E. Scoffone: New Removal Conditions of Sulfenyl Groups in Peptide Synthesis. Tetrahedron Letters **1966**, 2985.

60. Fridkin, M., A. Patchornik and E. Katchalski: A Synthesis of Cyclic Peptides Utilizing High Molecular Weight Carriers. J. Amer. Chem. Soc. **87**, 4646 (1965).

61. Geiger, R., K. Sturm und W. Siedel: Synthese eines biologisch aktiven Tricosapeptidamids mit der Aminosäuresequenz 1—23 des Corticotropins (ACTH). Chem. Ber. **97**, 1207 (1964).

62. Geiger, R., K. Sturm, G. Vogel und W. Siedel: Synthetische Analoga des Corticotropins. Zur Bedeutung der aminoterminalen Sequenz Ser-Tyr-Ser für die adrenocorticotrope Wirkung. Z. Naturforsch. **19b**, 858 (1964).

63. Greenstein, J. P. and M. Winitz: Chemistry of the Amino Acids. New York-London: J. Wiley and Sons. 1961.

64. Gregory, H., P. M. Hardy, D. S. Jones, G. W. Kenner and R. C. Sheppard: The Antral Hormone Gastrin. Nature **204**, 931 (1964).

65. Gross, F.: Angiotensin. Arch. exp. Pathol. Pharmakol. **245**, 196 (1963).

66. Guttmann, St.: On the Use of the Tosyl Group for the Protection of Basic Amino-Acids. In: G. T. Young (ed.), Peptides. Proc. 5th Europ. Sympos. Oxford 1962, p. 41. Oxford-London-New York-Paris: Pergamon Press. 1963.

67. — Synthèse du glutathion et de l'oxytocine à l'aide d'un nouveau groupe protecteur de la fonction thiol. Helv. Chim. Acta **49**, 83 (1966).

67a. Guttmann, St. et R. A. Boissonnas: Synthèse de l'α-mélanotropine (α-MSH) de Porc. Helv. Chim. Acta **42**, 1257 (1959).

68. — — Synthèse de la Sér⁴-oxytocine, de l'Ala⁴-oxytocine, de la Sér⁵-oxytocine et de l'Ala⁵-oxytocine. Helv. Chim. Acta **46**, 1626 (1963).

69. Guttmann, St. and J. Pless: On the Protection of the Guanidino Group of Arginine. Acta Chim. Hungar. **44**, 23 (1965).

70. Guttmann, St., J. Pless and R. A. Boissonnas: Synthesis of Peptides Related to ACTH. Acta Chim. Hungar. **44**, 141 (1965).

71. — — — Synthesis of a Highly Active Peptide Related to Corticotropin. Siehe Lit. *32*, p. 221.

72. Habermann, E.: Strukturaufklärung kininliefernder Peptide aus Rinderserum-Kininogen. Arch. exp. Pathol. Pharmakol. **253**, 474 (1966).

73. Habermann, E. und J. Jentsch: Über die Struktur des toxischen Bienengiftpeptids Melittin und deren Beziehung zur pharmakologischen Wirkung. Arch. exp. Pathol. Pharmakol. **253**, 40 (1966).

74. Handford, B. O., J. H. Jones, G. T. Young and T. F. N. Johnson: Amino Acids and Peptides. XXIV. The Use of Esters of 1-Hydroxypiperidine and of Other NN-Dialkylhydroxylamines in Peptide Synthesis and as Selective Acylating Agents. J. Chem. Soc. (London) **1965**, 6814.

75. Haux, P., H. Sawerthal und E. Habermann: Sequenzanalyse des Bienengift-Neurotoxin (Apamin) aus seinen tryptischen und chymotryptischen Spaltstücken. Z. physiol. Chem. **348**, 737 (1967).

76. Heaton, G. S., H. N. Rydon and J. A. Schofield: Polypeptides. III. The Oxidation of Some Peptides of Cysteine and Glycine. J. Chem. Soc. (London) **1956**, 3157.

77. Heller, H.: Neurohypophyseal Hormones. In: U. S. v. Euler and H. Helle (ed.): Comparative Endocrinology, vol. 1, p. 25. New York-London: Academic Press. 1963.

78. HOFMANN, K. and P. G. KATSOYANNIS: Synthesis and Function of Peptides of Biological Interest. In: H. Neurath (ed.), The Proteins: Composition, Structure and Function, vol. 1, p. 53. New York-London: Academic Press. 1963.

79. HOFMANN, K., R. D. WELLS, H. YAJIMA and J. ROSENTHALER: Studies on Polypeptides. XXVII. Elimination of the Methionine Residue as an Essential Functional Unit for in vivo Adrenocorticotropic Activity. J. Amer. Chem. Soc. **85**, 1546 (1963).

80. HOFMANN, K., H. YAJIMA, T.-Y. LIU, N. YANAIHARA, C. YANAIHARA and J. L. HUMES: Studies on Polypeptides. XXV. The Adrenocorticotropic Potency of an Eicosapeptide Amide Corresponding to the N-Terminal Portion of the ACTH Molecule; Contribution to the Relation between Peptide Chain-Length and Biological Activity. J. Amer. Chem. Soc. **84**, 4481 (1962).

81. HOFMANN, K., H. YAJIMA and E. T. SCHWARTZ: Studies on Polypeptides. XVII. The Synthesis of Three Acyltridecapeptide Amides Possessing a High Level of Melanocyte-expanding Activity in vitro. J. Amer. Chem. Soc. **82**, 3732 (1960).

82. HOFMANN, K., N. YANAIHARA, S. LANDE and H. YAJIMA: Studies on Polypeptides. XXIII. Synthesis and Biological Activity of a Hexadecapeptide Corresponding to the N-Terminal Sequence of the Corticotropins. J. Amer. Chem. Soc. **84**, 4470 (1962).

83. HONZL, J. and J. RUDINGER: Amino-Acids and Peptides. XXXIII. Nitrosyl Chloride and Butyl Nitrite as Reagents in Peptide Synthesis by the Azide Method; Suppression of Amide Formation. Collect. Czech. Chem. Comm. **26**, 2333 (1961).

84. HOPE, D. B., V. V. S. MURTI and V. DU VIGNEAUD: A Highly Potent Analogue of Oxytocin, Desamino-oxytocin. J. Biol. Chem. **237**, 1563 (1962).

85. HUGUENIN, R. L. et R. A. BOISSONNAS: Synthèses de la Phé²-arginine-vasopressine et de la Phé²-arginine-vasotocine et nouvelles synthèses de l'arginine-vasopressine et de l'arginine-vasotocine. Helv. Chim. Acta **45**, 1629 (1962).

86. HUGUENIN, R. L., E. STÜRMER, R. A. BOISSONNAS and B. BERDE: Desamino-arginine-vasopressin, an Analogue of Arginine-vasopressin with High Anti-diuretic Activity. Experientia **21**, 68 (1965).

87. JAKUBKE, H.-D.: Die Verwendung aktivierter Ester zur Peptidsynthese. Z. Chem. **6**, 52 (1966).

88. JAKUBKE, H.-D. und A. VOIGT: Über aktivierte Ester. VII. Untersuchungen über die peptidchemische Verwendbarkeit von Acylaminosäure-chinolyl-(8)-estern. Chem. Ber. **99**, 2419 (1966).

89. JAQUENOUD, P.-A. et R. A. BOISSONNAS: Synthèse de la Dé-Pro⁷-oxytocine, de la Dé-Leu⁸-oxytocine et de la Dé-Gly⁹-oxytocine. Helv. Chim. Acta **45**, 1462 (1962).

90. JORPES, E.: Sekretin. Nordisk Med. **76**, 965 (1966).

91. KAPPELER, H., B. RINIKER, W. RITTEL, P. DESAULLES, R. MAIER, B. SCHÄR and M. STAEHELIN: Synthesis and Biological Activity of Peptides Related to ACTH. Siehe Lit. *32*, p. 214.

92. KASAFÍREK, E., K. JOŠT, J. RUDINGER and F. ŠORM: Amino Acids and Peptides. LIV. Synthesis of Further Extended-chain Analogues of Oxytocin. Collect. Czech. Chem. Comm. **30**, 2600 (1965).

93. KASAFÍREK, E., B. RÁBEK, J. RUDINGER and F. ŠORM: Amino Acids and Peptides. LXVI. Synthesis of Ten Extended-chain Analogues of Lysine Vasopressin. Collect. Czech. Chem. Comm. **31**, 4581 (1966).

94. Katsoyannis, P. G., D. T. Gish and V. du Vigneaud: Synthetic Studies on Arginine-vasopressin: Condensation of S-Benzyl-N-carbobenzoxy-L-cysteinyl-L-tyrosyl-L-phenylalanyl-L-glutaminyl-L-asparagine and its O-Tosyl Derivative with S-Benzyl-L-cysteinyl-L-prolyl-L-arginylglycinamide. J. Amer. Chem. Soc. 79, 4516 (1957).

95. Kessler, W. und B. Iselin: Selektive Spaltung substituierter Phenylsulfenyl-Schutzgruppen bei Peptidsynthesen. Helv. Chim. Acta 49, 1330 (1966).

96. Khosla, M. C., R. R. Smeby and F. M. Bumpus: Solid-phase Peptide Synthesis of [L-Alanin³-L-isoleucine⁵]-angiotensin. II. Biochemistry 6, 754 (1967).

97. Kiryushkin, A. A., Yu. A. Ovchinnikov, I. V. Kozhevnikova and M. M. Shemyakin: Synthesis of Peptides on a Polymeric Support in Solution. Siehe Lit. 32, p. 100.

98. Kisfaludy, L. und S. Dualszky: p-Chlorcarbobenzoxyaminosäuren und -Peptide, I. Acta Chim. Hungar. 24, 301 (1960).

99. Klieger, E. und E. Schröder: Synthese von Ser³-Ileu⁸-Oxytocin und Des-amino-Ser⁴-Ileu⁸-Oxytocin (Desamino-Isotocin). Tetrahedron Letters 1965, 2067.

100. Kondo, M., H. Aoyagi, T. Kato and N. Izumiya: Studies of Peptide Antibiotics. VI. Syntheses of Cyclic Penta- and Decapeptides with a Glycyl-L-ornithyl-L-leucyl-D-phenylalanylglycyl Sequence. Bull. Chem. Soc. Japan 39, 2234 (1966).

101. Kovacs, J., L. Kisfaludy and M. Q. Ceprini: On the Optical Purity of Peptide Active Esters Prepared by N,N'-Dicyclohexylcarbodiimide and "Complexes" of N,N'-Dicyclohexylcarbodiimide-Pentachlorophenol and N,N'-Dicyclohexyl-carbodiimide-Pentafluorophenol. J. Amer. Chem. Soc. 89, 183 (1967).

102. Lande, S. and A. B. Lerner: The Biochemistry of Melanotropic Agents. Pharmacol. Rev. 19, 1 (1967).

103. Law, H. D.: Polypeptides of Medicinal Interest. In: G. P. Ellis and G. B. West (ed.): Progress in Medicinal Chemistry, vol. 4, p. 86. London: Butterworths. 1965.

104. Lenard, J. and A. B. Robinson: Use of Hydrogen Fluoride in Merrifield Solid-Phase Peptide Synthesis. J. Amer. Chem. Soc. 89, 181 (1967).

105. Li, C. H., L. Barnafi, M. Chrétien and D. Chung: Isolation and Amino-acid Sequence of β-LPH from Sheep Pituitary Glands. Nature 208, 1093 (1965).

106. Li, C. H., J. Meienhofer, E. Schnabel, D. Chung, T.-B. Lo and J. Rama-chandran: Synthesis of a Biologically Active Nonadecapeptide Corresponding to the First Nineteen Amino Acid Residues of Adrenocorticotropins. J. Amer. Chem. Soc. 83, 4449 (1961).

107. Loffet, A.: Synthesis of a Pentapeptide by the Merrifield Method. Experientia 23, 406 (1967).

108. Losse, G. und G. Bachmann: Chemie der Depsipeptide. I. Z. Chem. 4, 204 (1964).

109. — — Chemie der Depsipeptide. II. Z. Chem. 4, 241 (1964).

110. Lübke, K., E. Schröder, R. Schmiechen und H. Gibian: Über Peptid-synthesen, XXII. Synthese von Eledoisin und Glu⁵-Eledoisin. Liebigs Ann. Chem. 679, 195 (1964).

111. Lübke, K., G. Zöllner and E. Schröder: The Influence of Substitution or Omission of an Amino Group on the Hypotensive Activity of the C-Terminal Sequences of Eledoisin. In: E. G. Erdös, N. Back and F. Sicuteri (ed.): Hypotensive Peptides. Proc. Internat. Sympos. Florence 1965, p. 45. New York: Springer. 1966.

112. MAKHLOUF, G. M., J. P. A. McMANUS and W. I. CARD: Action of the Penta-peptide (ICI 50123) on Gastric Secretion in Man. Gastroenterology **51**, 455 (1966).

113. MARSHALL, G. R. and R. B. MERRIFIELD: Synthesis of Angiotensins by the Solid-Phase Method. Biochemistry **4**, 2394 (1965).

114. MEIENHOFER, J. and V. DU VIGNEAUD: Preparation of Lysine-vasopressin through a Cyrstalline Protected Nonapeptide Intermediate and Purification of the Hormone by Chromatography. J. Amer. Chem. Soc. **82**, 2279 (1960).

115. MERRIFIELD, R. B.: Solid Phase Peptide Synthesis. I. The Synthesis of a Tetra-peptide. J. Amer. Chem. Soc. **85**, 2149 (1963).

116. — Solid-Phase Peptide Synthesis. III. An Improved Synthesis of Bradykinin. Biochemistry **3**, 1385 (1964).

117. — Solid Phase Peptide Synthesis. IV. The Synthesis of Methionyl-lysyl-bradykinin. J. Organ. Chem. (USA) **29**, 3100 (1964).

118. MERRIFIELD, R. B. and A. MARGLIN: Progress in Solid Phase Peptide Syn-thesis. The Synthesis of Bovine Insulin. Siehe Lit. *32*, p. 85.

119. MORLEY, J. S.: Syntheses of Human Gastrin (I) and the Biological Properties of Analogues. Siehe Lit. *32*, p. 226.

120. MORLEY, J. S., H. J. TRACY and R. A. GREGORY: Structure-Function Rela-tionships in the Active C-Terminal Tetrapeptide Sequence of Gastrin. Nature **207**, 1356 (1965).

121. NAJJAR, V. A. and R. B. MERRIFIELD: Solid Phase Peptide Synthesis. VI. The Use of the *o*-Nitrophenylsulfenyl Group in the Synthesis of the Octadeca-peptide Bradykininylbradykinin. Biochemistry **5**, 3765 (1966).

122. NESVADBA, H., H. BACHMAYER und H. MICHL: Über die Synthese eines im Gift von *Bombina variegata* vorkommenden Hexapeptid-diamids. Monatsh. Chem. **96**, 1125 (1965).

123. NICOLAIDES, E. D. and H. A. DE WALD: Studies on the Synthesis of Poly-peptides. Bradykinin. J. Organ. Chem. (USA) **26**, 3872 (1961).

124. OHNO, M., T. KATO, S. MAKISUMI and N. IZUMIYA: Studies of Peptide Anti-biotics. IV. The Synthesis of Tyrocidine A. Bull. Chem. Soc. Japan **39**, 1738 (1966).

125. OKAWA, K.: Studies on Serine Peptides. I. Optical Resolution of O-Benzyl-DL-serine and Synthesis of DL-Serine Peptides. Bull. Chem. Soc. Japan **29**, 486 (1956).

126. OVCHINNIKOV, YU. A., A. A. KIRYUSHKIN and M. M. SHEMYAKIN: Total Syn-thesis of Sporidesmolide III. Tetrahedron Letters **1965**, 1111.

127. — — — Complete Synthesis of Sporidesmolides III and IV. Zhurn. Obshchei Khimii (USSR) **36**, 620 (1966).

128. PATCHORNIK, A., A. BERGER and E. KATCHALSKI: Carbobenzoxy Derivatives of Histidine, Imidazole and Benzimidazole. J. Amer. Chem. Soc. **79**, 6416 (1957).

129. PATCHORNIK, A., M. FRIDKIN and E. KATCHALSKI: Synthesis of Linear and Cyclic Peptides with the Aid of Insoluble Active Esters of Amino Acids and Peptides. Siehe Lit. *32*, p. 91.

130. PAULAY, Z. and S. BAJUSZ: A Novel Protection for the Guanidino Group of Arginine. Acta Chim. Hungar. **43**, 147 (1965).

131. PHILLIPS, J. G. and D. BELLAMY: Adrenocortical Hormones. In: U. S. v. Euler and H. Heller (ed.), Comparative Endocrinology, vol. I, p. 208. New York-London: Academic Press. 1963.

132. PHOTAKI, I.: A New Synthesis of Oxytocin Using S-Acyl Cysteines as Inter-mediates. J. Amer. Chem. Soc. **88**, 2292 (1966).

133. Plattner, Pl. A., K. Vogler, R. O. Studer, P. Quitt und W. Keller-Schierlein: Synthesen in der Depsipeptidreihe, 1. Mitt. Synthese von Enniatin B. Helv. Chim. Acta **46**, 927 (1963).

134. Pless, J. und R. A. Boissonnas: Über die Geschwindigkeit der Aminolyse von verschiedenen neuen, aktivierten, N-geschützten α-Aminosäure-phenylestern, insbesondere 2,4,5-Trichlorphenylestern. Helv. Chim. Acta **46**, 1609 (1963).

135. Pless, J., E. Stürmer, St. Guttmann und R. A. Boissonnas: Kallidin, Synthese und Eigenschaften. Helv. Chim. Acta **45**, 394 (1962).

136. Poduška, K., V. Gut, H. Zimmermannová, J. Rudinger and F. Šorm: Synthesis of Peptides Related to Trypsin Sequences. Acta Chim. Hungar. **44**, 165 (1965).

137. Poduška, K., H. Maassen van den Brink-Zimmermannová, J. Rudinger and F. Šorm: Some Contributions to the Chemistry of the o-Nitrophenylsulphenyl-Protecting Group. Siehe Lit. *32*, p. 38.

138. Quitt, P., R. O. Studer und K. Vogler: Synthesen in der Depsipeptid-Reihe. 2. Mitt. Synthese von Enniatin A. Helv. Chim. Acta **46**, 1715 (1963).

139. Ramachandran, J., D. Chung and C. H. Li: Adrenocorticotropins. XXXIV. Aspects of Structure-Activity Relationships of the ACTH Molecule. Synthesis of a Heptadecapeptide Amide, an Octadecapeptide Amide, and a Nonadecapeptide Amide Possessing High Biological Activities. J. Amer. Chem. Soc. **87**, 2696 (1965).

140. Ravdel, G. A., M. P. Filatova, L. A. Shchukina, T. S. Paskhina, M. S. Surovikina, S. S. Trapeznikova and T. P. Egorova: 6-Glycine-8-phenyllactic Acid Bradykinin. Its Synthesis, Biological Activity, and Splitting by Kininase (Carboxypeptidase N). J. Med. Chem. **10**, 242 (1967).

141. Riniker, B. und R. Schwyzer: Die sterische Einheitlichkeit des synthetischen Val^5-Hypertensin II-Asp^1-β-amids. Helv. Chim. Acta **44**, 658 (1961).

142. Rothe, M. und F. Eisenbeiss: Cyclische Peptide. XIV. Synthese eines Gramicidin S-Analogen, Cyclo-(Val-Lys(BOC)-Leu-Gly-Pro)$_2$ Z. Naturforsch. **21 b**, 814 (1966).

143. Rudinger, J. (ed.): Oxytocin, Vasopressin and their Structural Analogues. Proc. 2nd Internat. Pharmacol. Meeting, Prague 1963. New York: Macmillan (Pergamon), and Praha: Czechosl. Med. Press. 1964.

144. Rudinger, J., J. Honzl and M. Zaoral: Synthetic Studies in the Oxytocin Field. III. An Alternative Synthesis of Oxytocin. Collect. Czech. Chem. Comm. **21**, 202 (1956).

145. Rudinger, J. and K. Jošt: A Biologically Active Analogue of Oxytocin not Containing a Disulfide Group. Experientia **20**, 570 (1964).

146. Ruttenberg, M. A., T. P. King and L. C. Craig: The Chemistry of Tyrocidine. VI. The Amino Acid Sequence of Tyrocidine C. Biochemistry **4**, 11 (1965).

147. Sakakibara, S. and N. Inukai: The Trifluoroacetate Method of Peptide Synthesis. II. An Improved Synthesis of Bradykinin. Bull. Chem. Soc. Japan **39**, 1567 (1966).

148. Sakakibara, S., Y. Nobuhara, Y. Shimonishi and R. Kiyoi: A Synthesis of Oxytocin. Bull. Chem. Soc. Japan **38**, 120 (1965).

149. Sakakibara, S., Y. Shimonishi, M. Okada and Y. Kishida: Removal of Protective Groups by Anhydrous Hydrogen Fluoride. Siehe Lit. *32*, p. 44.

150. Sakakibara, S., M. Shin, M. Fujino, Y. Shimonishi, S. Inoue and N. Inukai: t-Amyloxycarbonyl as a New Protecting Group in Peptide Synthesis. I. The Synthesis and Properties of N-t-Amyloxycarbonylamino Acids and Related Compounds. Bull. Chem. Soc. Japan **38**, 1522 (1965).

151. SANDRIN, ED. and R. A. BOISSONNAS: Synthesis of Eledoisin. Experientia **18**, 59 (1962).

152. — — Synthèse d'analogues structuraux de l'élédoïsine. Helv. Chim. Acta **47**, 1294 (1964).

153. SARGES, R. and B. WITKOP: Gramicidin A. V. The Structure of Valine- and Isoleucine-gramicidin A. J. Amer. Chem. Soc. **87**, 2011 (1965).

154. — — Gramicidin. VII. The Structure of Valine- and Isoleucine-gramicidin B. J. Amer. Chem. Soc. **87**, 2027 (1965).

155. SCHNABEL, E.: Verbesserte Synthese von tert.-Butyloxycarbonyl aminosäuren durch pH-Stat-Reaktion. Liebigs Ann. Chem. **702**, 188 (1967).

156. SCHRÖDER, E.: Über Peptidsynthesen. Synthese von Methionyl-Lysyl-Bradykinin, einem Kinin aus Rinderblut. Experientia **20**, 39 (1964).

157. — Synthese und biologische Aktivität bradykininwirksamer Undeca-, Dodeca- und Tridecapeptide. Experientia **21**, 271 (1965).

158. — Über Peptidsynthesen, XL. 5. Mitt. über Angiotensin-Analoga Synthese des *all*-D-Val⁵-Angiotensin II-Asp¹-β amids. Liebigs Ann. Chem. **692**, 241 (1966).

159. SCHRÖDER, E. and R. HEMPEL: Bradykinin, Kallidin, and their Synthetic Analogues. Experientia **20**, 529 (1964).

160. SCHRÖDER, E. und K. LÜBKE: Peptolide. Experientia **19**, 57 (1963).

161. — — Über Peptidsynthesen. C-terminale Teilsequenzen des Eledoisins und eledoisinanaloger Verbindungen. Experientia **20**, 19 (1964).

162. — — The Peptides. Methods of Peptide Synthesis, vol. 1. New York-London: Academic Press. 1965.

163. — — The Peptides. Synthesis, Occurrence, and Action of Biologically Active Peptides, vol. 2. New York-London: Academic Press. 1966.

164. SCHRÖDER, E., K. LÜBKE und R. HEMPEL: D-Aminosäurehaltige Analoga des Eledoisins. Experientia **21**, 70 (1965).

165. SCHRÖDER, E. und S. MATTHES: Reinigung synthetischer Peptide mit der trägerfreien präparativen Durchfluß-Elektrophorese. J. Chromatogr. **17**, 189 (1965).

166. SCHWARTZ, I. L. and L. M. LIVINGSTON: Cellular and Molecular Aspects of the Antidiuretic Action of Vasopressins and Related Peptides. Vitamins and Horm. **22**, 261 (1964).

167. SCHWARZ, H. and K. ARAKAWA: The Use of *p*-Nitrobenzyl Esters in Peptide Synthesis. J. Amer. Chem. Soc. **81**, 5691 (1959).

168. SCHWARZ, H. and F. M. BUMPUS: Synthesis of an Optically Pure Tetrapeptide Contained in Angiotensin. J. Amer. Chem. Soc. **81**, 890 (1959).

169. SCHWYZER, R.: Recent Synthetic Developments in the Field of Polypeptide Hormones. In: H. Peeters (ed.), Protides of the Biological Fluids. Proc. 9th Colloq., Bruges 1961, p. 27. Amsterdam-New York: Elsevier. 1962.

170. SCHWYZER, R., A. COSTOPANAGIOTIS und P. SIEBER: Zwei Synthesen des α-Melanotropins (α-MSH) mit Hilfe leicht entfernbarer Schutzgruppen. Helv. Chim. Acta **46**, 870 (1963).

171. SCHWYZER, R., B. ISELIN, H. KAPPELER, B. RINIKER, W. RITTEL und H. ZUBER: Synthese hochwirksamer Dekapeptide mit der Aminosäuresequenz des Val⁵-Hypertensins I (L-Asparagyl-L-arginyl-L-valyl-L-tyrosyl-L-valyl-L-histidyl-L-prolyl-L-phenylalanyl-L-histidyl-L-leucin und L-Asparaginyl-L-arginyl-L- valyl-L-tyrosyl-L-valyl-L-histidyl-L-prolyl-L-phenylalanyl-L-histidyl-L-leucin). Helv. Chim. Acta **41**, 1273 (1958).

172. — — — — — — Die Synthese des β-Melanotropins (β-MSH) mit der Aminosäuresequenz des bovinen Hormons. Helv. Chim. Acta **46**, 1975 (1963).

173. Schwyzer, R. und H. Kappeler: Synthese eines Tetracosapeptides mit hoher corticotroper Wirksamkeit: β^{1-24}-Corticotropin. Helv. Chim. Acta **46**, 1550 (1963).

174. Schwyzer, R., H. Kappeler, B. Iselin, W. Rittel und H. Zuber: Synthese und biologische Aktivität von geschützten Polypeptidsequenzen des β-Melanophoren-stimulierenden Hormons (β-MSH) des Rindes. Helv. Chim. Acta **42**, 1702 (1959).

175. Schwyzer, R. und P. Sieber: Die Synthese von Gramicidin S. Helv. Chim. Acta **40**, 624 (1957).

176. — — Verdoppelungsreaktionen beim Ringschluß von Peptiden. I. Synthese von Gramicidin S und von bis-homo-Gramicidin S aus den Pentapeptid-Einheiten. 7. Mitt. über homodet cyclische Polypeptide. Helv. Chim. Acta **41**, 2186 (1958).

177. — — Total Synthesis of Adrenocorticotropic Hormone. Nature **199**, 172 (1963).

178. Schwyzer, R., P. Sieber und H. Kappeler: Zur Synthese von N-t-Butyloxycarbonyl-aminosäuren. Helv. Chim. Acta **42**, 2622 (1959).

179. Sewing, K.-Fr.: Die Bedeutung von Gastrin in der Magensaftsekretion. Dtsch. med. Wschr. **91**, 1506 (1966).

180. Sheehan, J. C. and G. P. Hess: A New Method of Forming Peptide Bonds. J. Amer. Chem. Soc. **77**, 1067 (1955).

181. Shemyakin, M. M., N. A. Aldanova, E. I. Vinogradova and M. Yu. Feigina: The Structure and Total Synthesis of Valinomycin. Tetrahedron Letters **1963**, 1921.

182. Shemyakin, M. M., Yu. A. Ovchinnikov, V. K. Antonov, A. A. Kiryushkin, V. T. Ivanov, V. I. Shchelokov and A. M. Shkrob: Total Synthesis of Serratamolide. I. Synthesis of O,O'-Diacetylserratamolide. Tetrahedron Letters **1964**, 47.

183. Shemyakin, M. M., Yu. A. Ovchinnikov, V. T. Ivanov and A. V. Evstratov: Topochemical Approach in Studies of the Structure-Activity Relation: Enantio-enniatin B. Nature **213**, 412 (1967).

184. Shemyakin, M. M., Yu. A. Ovchinnikov, V. T. Ivanov and A. A. Kiryushkin: The Structure of Enniatins and Related Antibiotics. Tetrahedron **19**, 581 (1963).

185. Shemyakin, M. M., Yu. A. Ovchinnikov, V. T. Ivanov, A. A. Kiryushkin and K. Kh. Khalilulina: Depsipeptides. XLII. Structure and Complete Synthesis of Sporidesmolides I and II. Zhurn. Obshchei Khimii (USSR) **35**, 1399 (1965).

186. Shemyakin, M. M., Yu. A. Ovchinnikov, V. T. Ivanov, A. A. Kiryushkin, G. L. Zhdanov and I. D. Ryabova: The Structure-Antimicrobial Relation of Depsipeptides. Experientia **19**, 566 (1963).

187. Shemyakin, M. M., Yu. A. Ovchinnikov, V. T. Ivanov and I. D. Ryabova: Topochemical Approach to the Structure-Activity Relation, Retroenantio-Gly^{5,10}-Gramicidin S. Experientia **23**, 326 (1967).

188. Shemyakin, M. M., Yu. A. Ovchinnikov, A. A. Kiryushkin and V. T. Ivanov: Studies in Depsipeptide Chemistry, 25. The Structure and Total Synthesis of Enniatins A and B. Izvest. Akad. Nauk. SSSR, Khim. Ser. **1965**, 1623.

189. Shemyakin, M. M., Yu. A. Ovchinnikov, A. A. Kiryushkin and I. V. Kozhevnikova: Synthesis of Peptides in Solution on a Polymeric Support. I. Synthesis of Glycylglycyl-L-leucylglycine. Tetrahedron Letters **1965**, 2323.

190. SHEMYAKIN, M. M., L. A. SHCHUKINA, E. I. VINOGRADOVA, G. A. RAVDEL and YU. A. OVCHINNIKOV: Mutual Replaceability of Amide and Ester Groups in Biologically Active Peptides and Depsipeptides. Experientia **22**, 535 (1966).

191. SHEMYAKIN, M. M., E. I. VINOGRADOVA, M. YU. FEIGINA, N. A. ALDANOVA, N. F. LOGINOVA, I. D. RYABOVA and I. A. PAVLENKO: The Structure-Antimicrobial Relation for Valinomycin Depsipeptides. Experientia **21**, 548 (1965).

192. SHIPOLINI, R., A. F. BRADBURY, G. L. CALLEWAERT and C. A. VERNON: The Structure of Apamin. Chem. Commun. **1967**, 679.

193. SKEGGS, L. T., Jr., K. E. LENTZ, J. R. KAHN and N. P. SHUMWAY: The Synthesis of a Tetradecapeptide Renin Substrate. J. Exp. Medicine **108**, 283 (1958).

194. STEWART, J. M. and D. W. WOOLLEY: Threonine Analogs of Bradykinin Designed as Antimetabolites. Biochemistry **3**, 700 (1964).

195. STUDER, R. O.: Vergleich von synthetischem und natürlichem Arginin-vasopressin. Helv. Chim. Acta **46**, 421 (1963).

196. STUDER, R. O., W. LERGIER, P. LANZ, E. BÖHNI und K. VOGLER: Synthesen in der Polymyxinreihe, 10. Mitt. Die Synthese von Colistin A (Polymyxin E_1). Helv. Chim. Acta **48**, 1371 (1965).

197. STUDER, R. O., W. LERGIER und K. VOGLER: Synthesen in der Polymyxinreihe, 11. Mitt. Die Synthese von Circulin A. Helv. Chim. Acta **49**, 974 (1966).

198. STUDER, R. O., P. QUITT, E. BÖHNI und K. VOGLER: Synthesen in der Depsipeptid-Reihe, 4. Mitt. Struktur und mikrobiologische Aktivität in der Enniantin-Reihe. Monatsh. Chem. **96**, 461 (1965).

199. STÜRMER, E.: Plasmakinine. Schweiz. med. Wschr. **96**, 1667 (1966).

200. STÜRMER E. und A. FANCHAMPS: Eledoisin-Chemie, Pharmakologie, klinisch-experimentelle und vorläufige therapeutische Erfahrungen. Dtsch. med. Wschr. **90**, 1012 (1965).

201. STÜRMER, E., R. L. HUGUENIN, R. A. BOISSONNAS and B. BERDE: Deamino[1]-phenylalanine[2]-arginine[8]-vasopressin: A Peptide with Highly Selective Antidiuretic Activity. Experientia **21**, 583 (1965).

202. SUZUKI, T., K. HAYASHI, K. FUJIKAWA and K. TSUKAMOTO: Contribution to the Elucidation of the Structure of Polymyxin B_1. J. Biochemistry (Tokyo) **54**, 555 (1963).

203. TASCHNER, E., C. WASIELEWSKI und J. F. BIERNAT: Neue Veresterungsmethoden in der Peptidchemie, IV. Darstellung von tert.-Butylestern N-acylierter Aminosäuren mit Hilfe von tert.-Butylacetat. Liebigs Ann. Chem. **646**, 119 (1961).

204. TESSER, G. I. und R. SCHWYZER: Synthese des 17,18-Diornithin-β-corticotropin-(1—24)-tetracosapeptides, eines biologisch aktiven Analogons des adrenocorticotropen Hormons. Helv. Chim. Acta **49**, 1013 (1966).

205. TSOU, C.-L., Y.-C. DU and G.-J. XÜ: The Reduction of Insulin and its Benzyl Derivatives by Sodium in Liquid Ammonia and the Regeneration of Activity from the Reduced Products. Sci. Sinica (Peking) **10**, 332 (1961).

206. VELLUZ, L., G. AMIARD, J. BARTOS, B. GOFFINET et R. HEYMÈS: Accès à l'ocytocine de synthèse, à l'aide d'intermédiaires S,N-tritylés. Bull. soc. chim. France **1956**, 1464.

207. VOGLER, K. and P. LANZ: Synthesis of All-D-Isomers in the Bradykinin Series. In: E. G. ERDÖS, N. BACK and F. SICUTERI (ed.), Hypotensive Peptides. Proc. Internat. Sympos. Florence 1965, p. 14. New York: Springer. 1966.

208. VOGLER, K. and R. O. STUDER: The Chemistry of the Polymyxin Antibiotics. Experientia **22**, 345 (1966).

209. Vogler, K., R. O. Studer, P. Lanz, W. Lergier und E. Böhni: Synthesen in der Polymyxin-Reihe, 9. Mitt. Synthese von Polymyxin B_1. Helv. Chim. Acta **48**, 1161 (1965).

210. Vogler, K., R. O. Studer, W. Lergier und P. Lanz: Synthese von All-D-Val[5]-Angiotensin II-Asp[1]-β-Amid. Helv. Chim. Acta **48**, 1407 (1965).

211. Waki, M. and N. Izumiya: Cyclosemigramicidin S. J. Amer. Chem. Soc. **89**, 1278 (1967).

212. Walter, R. and V. du Vigneaud: 6-Hemi-L-selenocystine-oxytocin and 1-Deamino-6-hemi-L-selenocystine-oxytocin, Highly Potent Isologs of Oxytocin and 1-Deamino-oxytocin. J. Amer. Chem. Soc. **87**, 4192 (1965).

213. — — 1-Deamino-1,6-L-selenocystine-oxytocin, a Highly Potent Isolog of 1-Deamino-oxytocin. J. Amer. Chem. Soc. **88**, 1331 (1966).

214. Weygand, F., D. Hoffmann und E. Wünsch: Peptidsynthesen mit Dicyclohexylcarbodiimid unter Zusatz von N-Hydroxysuccinimid. Z. Naturforsch. **21 b**, 426 (1966).

215. Weygand, F. und K. Hunger: Acylierung von Aminosäuren mit p-Methoxybenzyloxycarbonyl-azid. Chem. Ber. **95**, 1 (1962).

216. Weygand, F. und U. Ragnarsson: Peptidsynthesen mit Dicyclohexylcarbodiimid unter Zusatz von N-Hydroxysuccinimid. II. Peptidsynthesen an einem festen Träger. Z. Naturforsch. **21 b**, 1141 (1966).

217. Wiegershausen, B. und I. Paegelow: Die Rolle der Kinine im pathologischen Geschehen. Deutsche Gesundheitsw. **21**, 2092 (1966).

218. Wood, J. L. and V. du Vigneaud: Racemization of Benzyl-l-Cysteine, with a New Method of Preparing d-Cystine. J. Biol. Chem. **130**, 109 (1939).

219. Wünsch, E., A. Fontana und F. Drees: Zur Entacylierung von N-(2-Nitrophenyl)-sulfenyl peptiden bei Anwesenheit von Tryptophan in der Peptidsequenz. Z. Naturforsch. **22 b**, 607 (1967).

220. Wünsch, E., G. Fries und A. Zwick: Beiträge zur Peptidsynthese. V. Darstellung und peptidsynthetische Verwendung von O-Benzyl-L-tyrosin. Chem. Ber. **91**, 542 (1958).

221. Wünsch, E. und G. Fürst: Racematspaltung von O-Benzyl-DL-serin. Z. physiol. Chem. **329**, 109 (1962).

222. Yajima, H. and K. Kubo: Studies on Peptides. II. Synthesis and Physiological Properties of D-Histidyl-D-phenylalanyl-D-arginyl-D-tryptophylglycine, an Optical Antipode of an Active Fragment of α-Melanocyte-Stimulating Hormone. J. Amer. Chem. Soc. **87**, 2039 (1965).

223. Yamashiro, D.: Partition Chromatography of Oxytocin on "Sephadex". Nature **201**, 76 (1964).

224. Yamashiro, D., D. Gillessen and V. du Vigneaud: Oxytoceine and Deamino-oxytoceine. Biochemistry **5**, 3711 (1966).

225. Young, J. D., E. Benjamini, J. M. Stewart and C. Y. Leung: Immunochemical Studies on Tobacco Mosaic Virus Protein. V. The Solid-Phase Synthesis of Peptides of an Antigenically Active Decapeptide of Tobacco Mosaic Virus Protein and the Reaction of these Peptides with Antibodies to the Whole Protein. Biochemistry **6**, 1455 (1967).

226. Zahn, H., T. Okuda and Y. Shimonishi: Merrifield Synthesis of Human Insulin Chains and Their Alteration During the Sodium Treatment. Siehe Lit. *32*, p. 108.

227. Zaoral, M., J. Kolc and F. Šorm: Amino Acids and Peptides. LXXI. Synthesis of 1-Deamino-8-D-γ-Aminobutyrine-Vasopressin, 1-Deamino-8-D-Lysine-Vasopressin, and 1-Deamino-8-D-Arginine-Vasopressin. Collect. Czech. Chem. Comm. **32**, 1250 (1967).

228. ZERVAS, L. (ed.): Peptides. Proc. 6th Europ. Sympos., Athens 1963. Oxford, London, Edinburgh, New York, Paris, Frankfurt: Pergamon Press. 1965.

229. ZERVAS, L., D. BOROVAS and E. GAZIS: New Methods in Peptide Synthesis. I. Tritylsulfenyl and o-Nitrophenylsulfenyl Groups as N-Protecting Groups. J. Amer. Chem. Soc. **85**, 3660 (1963).

230. ZHUZE, A. L., K. JOŠT, E. KASAFÍREK and J. RUDINGER: Amino Acids and Peptides. XLV. Analogues of Oxytocin with O-Ethyltyrosine, *p*-Methylphenylalanine, and *p*-Ethylphenylalanine Replacing Tyrosine. Collect. Czech. Chem. Comm. **29**, 2648 (1964).

231. ZUBER, H.: Die Verwendung enzymatischer Reaktionen für die Reinheitsprüfung und Strukturaufklärung von Peptiden und Proteinen. Chimia **14**, 405 (1960).

(Eingelaufen am 25. Oktober 1967)

Insulin
Structure, Synthesis and Biosynthesis of the Hormone
By **Anthony C. Trakatellis** and **Gerald P. Schwartz**,
Upton, New York

With 18 Figures

Contents

Synthesis of the A 10–21 dodecapeptide 133. — Synthesis of the
A 5–9 Pentapeptide Azide 135. — Synthesis of the A 1–4 Tetrapeptide Azide 135. — Synthesis of the Sheep Insulin A Chain 135

Synthesis of the B 10–30 Heneicosapeptide 139. — Final Steps
in the Syntheses of the Sheep (Bovine) B Chain 142

The authors wish to express their gratitude to Dr. P. G. Katsoyannis for many helpful discussions and to Drs. E. Popenoe and M. Montjar for the critical reading of the manuscript. They also wish to thank Miss Roberta Klimaski for her assistance in preparing the bibliography. — This work was supported by the United States Atomic Energy Commission.

Abbreviations

$A\text{-}SSO_3^-$: S-sulfonated derivative of the A chain.

$B\text{-}SSO_3^-$: S-sulfonated derivative of the B chain.

Bz: Benzyl.

ONBz: p-Nitrobenzyl ester.

OBz: Benzyl ester.

Z: Carbobenzoxy.

OMe: Methyl ester.

OPNP: p-Nitrophenyl ester.

OBu: t-Butyl ester.

Ac: Acetyl.

Tos: p-Toluenesulfonyl.

I. Introduction

Insulin, a protein hormone indispensable for regulation of blood sugar, is produced by the β-cells of the islets of pancreas. The islet formations were described in 1867 by Paul Langerhans; and in 1889 Mering and Minkowski (*94*) observed that removal of the pancreas caused diabetes in experimental animals. This observation signaled the beginning of the fascinating history of insulin, marked through the years by intensive investigations and brilliant achievements. Laguesse (*79*) advanced the speculation that the islets of Langerhans were organs of internal secretion. Opie (*100, 101*) observed that the pathological pancreatic changes associated with diabetes were localized in the islets. Lane (*80*) was the first to differentiate the α- and β-cells of the islets and Homans (*41*) found that the β-cells are chiefly affected in diabetes. The climax of these investigations was the isolation of insulin in 1921 by Banting and Best (*5, 6*) and the crystallization of this hormone by Abel *(1a)* in 1926.

Sanger and co-workers (see below) elucidated the primary structure of this hormone after carrying out a brillant experimental program during the decade 1945–1955. Thus, insulin was the first protein of completely known covalent structure; and the way was opened for the next dramatic step, which was the first chemical synthesis of a protein, namely of sheep insulin. This monumental task was achieved by Katsoyannis and co-workers (*49–60, 63, 64, 66, 69*) with the synthesis of the S-sulfonated derivatives of the A and B chains of sheep insulin and the combination of the sulfhydryl forms of these synthetic chains to yield insulin. Soon thereafter, Zahn and co-workers (*93*) reported the synthesis in the fully protected form of the A and B chains of the same protein; and two years later the synthesis of the corresponding chains of bovine insulin was reported by Wang, Niu et al. (*76, 99, 146*). Finally, the synthesis of human insulin, the first human protein ever synthesized, was reported in 1966–1967 by Katsoyannis and co-workers (*51, 53, 59, 61, 62, 69*).

Continuous research through the years has enriched our knowledge of the covalent structure of insulins from various species, the steric structure of the hormone, the relation of structure and function, the mechanisms of its physiological action and its immunological behavior. The international literature on insulin has grown so rapidly in recent years that this review will be limited to coverage of the following topics relating to this hormone: structure, relationship of structure and function, chemical synthesis and biosynthesis by the β-cells of the islets of Langerhans.

II. Primary Structure of Insulins

A combination of original techniques enabled Sanger and his co-workers to elucidate the complete covalent structure of insulin. Although these pioneering techniques have been modified and improved, and even though additional techniques are now available, the approach utilized by Sanger still serves as a model and guide for contemporary investigations on the primary structure of proteins.

The identification of the N-terminal residues of insulin was accomplished with the dinitrophenyl method (DNP) introduced in 1945 by Sanger (*108*). This method (*12, 31, 108, 135*) is a general technique for labelling free amino-groups in peptides and proteins, and it has been widely applied in protein structural analysis. When the DNP-method was applied to bovine insulin, three DNP-derivatives were isolated: ε-DNP-lysine, DNP-phenylalanine and DNP-glycine. The lysine residues were obviously bound within the polypeptide chains, while phenylalanine and glycine were N-terminal residues of the insulin molecule (*111*). On the basis of a molecular weight of 12 000, assumed for insulin at that time (*112*), it was deduced from the quantitative results that the hormone contained four polypeptide chains with two phenylalanine N-terminal residues and two glycine N-terminal residues. However, when the molecular weight of insulin was accurately determined to be 6000 (*37*) it became apparent that this protein must contain two polypeptide chains. Since it seemed probable that these two chains were joined together by disulfide bridges, performic acid oxidation was used (*109*) to split the insulin molecule according to the reaction:

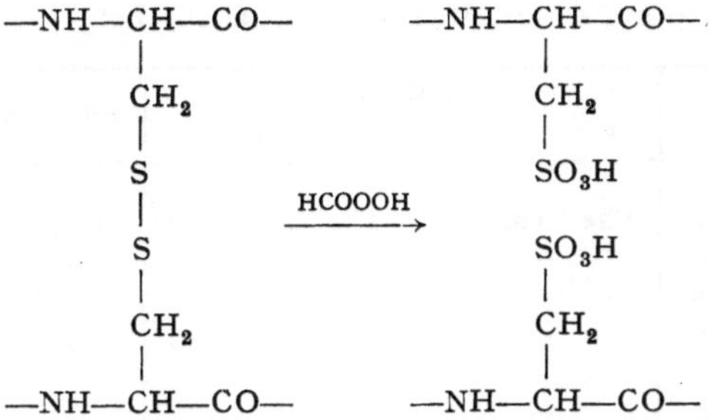

$$
\begin{array}{ccc}
\text{—NH—CH—CO—} & & \text{—NH—CH—CO—} \\
| & & | \\
CH_2 & & CH_2 \\
| & \xrightarrow{\ \text{HCOOOH}\ } & | \\
S & & SO_3H \\
| & & \\
S & & SO_3H \\
| & & | \\
CH_2 & & CH_2 \\
| & & | \\
\text{—NH—CH—CO—} & & \text{—NH—CH—CO—} \\
(1) & & (2)
\end{array}
$$

The cystine residues of insulin (1) upon oxidation with performic acid were converted to cysteic acid residues (2). From the oxidized insulin mixture, two fractions were isolated by precipitation methods, fraction A (A chain) with N-terminus glycine and fraction B (B chain) with N-terminus phenylalanine (*110*).

Separation and identification of the various peptides produced with partial acid or enzymatic hydrolysis led to the elucidation of the entire amino acid sequences of the A and B chains of insulin (*113–116*). Finally, the complete structure of the hormone *(Fig. 1)* was established with the determination of the arrangement of the disulfide bonds within the molecule of insulin (*107*). Essentially similar methods were applied in the elucidation of the primary structure of various insulins.

As may be seen from *Table 1*, several mammalian species have insulins with identical A and/or B chains. According to SMITH (*125*) the fact that insulin sequences from such diverse species (pig, dog, sperm whale, fin whale) are identical, suggests that this is the original sequence of mammalian insulin. All other sequences can be derived from this one with only a few mutations. Unlike the other mammals so far examined, the rat produces two insulins (Table 1, p. 124).

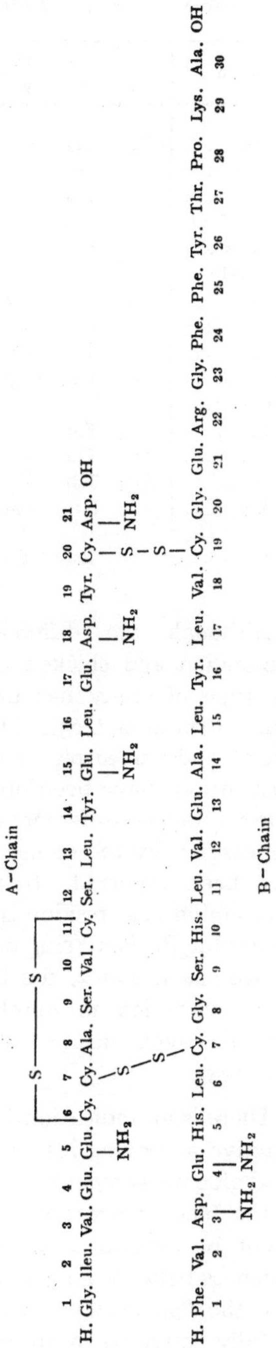

Fig. 1. Structure of bovine insulin

Table 1. Amino Acid Sequences of Mammalian Insulins*

Insulin	A Chain				B Chain						References
	4	8	9	10	1	2	3	27	29	30	
Bovine	Glu	Ala	Ser	Val	Phe	Val	Asn	Thr	Lys	Ala	(107)
Porcine	—	Thr	—	Ileu	—	—	—	—	—	—	(11)
Dog	—	Thr	—	Ileu	—	—	—	—	—	—	(125)
Sperm Whale	—	Thr	—	Ileu	—	—	—	—	—	—	(47)
Fin Whale ..	—	Thr	—	Ileu	—	—	—	—	—	—	(34)
Sei Whale ..	—	—	—	Thr	—	—	—	—	—	—	(47)
Sheep......	—	—	Gly	Val	—	—	—	—	—	—	(11)
Goat	—	—	Gly	Val	—	—	—	—	—	—	(11)
Elephant ...	—	—	Gly	Val	—	—	—	—	—	Thr	(125)
Horse	—	Thr	Gly	Val							(39)
Human	—	Thr	—	Ileu	—	—	—	—	—	Thr	(97)
Rabbit	—	Thr	—	Ileu	—	—	—	—	—	Ser	(125)
Rat 1	Asp	Thr	—	Ileu	—	—	Lys	—	—	Ser	(125)
Rat 2	Asp	Thr	—	Ileu	—	—	Lys	—	Met	Ser	(125)
Chicken	—	His	Asn	Thr	Ala	Ala	—	Ser	—	—	(125)

* Sequences not listed are identical.

Although the differences in sequences originally observed in mammalian and chicken insulins were relatively few, the extension of this type of study has unexpectedly revealed a surprising variety of insulin structures (125). Thus, numerous differences were found between porcine and guinea pig or fish insulins *(Table 2)*. At present amino acid substitutions have been observed in 29 of the 51 positions in the insulin molecule but no differences in the specific activity of the various insulins as measured by the mouse convulsion method or the blood glucose assay have been reported. In toadfish and anglerfish insulins (Table 2) an extra amino acid residue is found at the N-terminus of the B chain, and the residue B_{30} is missing in all fish insulins examined so far. In addition, in toadfish insulin 2, the B_{29} residue is also missing. There is only one known variation in number of residues of the A chain of insulins, that of coypu insulin which contains an extra C-terminal aspartic acid (125).

There is an indication that guinea pig insulin may be only one fourth as active as bovine insulin when measured by the mouse convulsion or blood glucose assays. However, this insulin is fully active when measured by the blood glucose assay in the guinea pig (125). This observation cannot be explained on the basis of the numerous differences in the sequences between guinea pig insulin and the other mammalian hormones since the fish insulins, which also show a large number of differences, are fully active when measured by the mouse convulsion assay.

References, pp. 152—160

Table 2. Comparison of Amino Acid Sequence of Porcine, Guinea Pig and Fish Insulins

A Chain

	1	2	3	4	5	6	7	8	9	10	11	12	13	14	15	16	17	18	19	20	21	References
Porcine	Gly.	Ileu.	Val.	Glu.	Gln.	Cys.	Cys.	Thr.	Ser.	Ileu.	Cys.	Ser.	Leu.	Tyr.	Glu.	Leu.	Glu.	Asn.	Tyr.	Cys.	Asn.	*(11)*
Guinea Pig .	—	—	—	Asp.	—	—	—	—	Gly.	Thr.	—	Thr.	Arg.	His.	—	—	—	Ser.	—	—	—	*(125)*
Bonito	—	—	(His.	—	—	—	—	His.	Lys.	Pro.	—	Asx.	—	) Phe.	—	—	—	—	—	—	—	*(73, 74)*
Toad Fish 1 .	—	—	—	—	—	—	—	(His.	Arg.	Pro.)	—	Asp.	Ileu.	Phe.	Asp.	—	Gln.	Ser.	—	—	—	*(125)*
Toad Fish 2 .	—	—	—	—	—	—	—	(His.	Arg.	Pro.)	—	Asp.	Lys.	Phe.	Asp.	—	Gln.	Ser.	—	—	—	*(125)*

B Chain (B 1—18)

	1	2	3	4	5	6	7	8	9	10	11	12	13	14	15	16	17	18	References
Porcine	Phe.	Val.	Asn.	Gln.	His.	Leu.	Cys.	Gly.	Ser.	His.	Leu.	Val.	Glu.	Ala.	Leu.	Tyr.	Leu.	Val.	*(11)*
Guinea Pig .	—	—	Ser.	Arg.	—	—	—	—	—	Asn.	—	—	—	Thr.	—	—	Ser.	—	*(125)*
Bonito	Ala.	Ala.	—	(Pro.	—	—	)—	—	—	—	—	—	—	—	—	—	—	—	*(73, 74)*
Toad Fish 1 .	Met.(Ala.	Pro.	Pro.)	—	—	—	—	—	—	—	—	—	Asp.	—	—	—	—	—	*(125)*
Toad Fish 2 .	Met.(Ala.	Pro.	Pro.)	—	—	—	—	—	—	—	—	—	Asp.	—	—	—	—	—	*(125)*
Angler Fish .	Val.	Ala.	Pro.	(Ala.	—	)—	—	—	—	—	—	—	Asp.	—	—	—	—	—	

B Chain (B 19—30)

	19	20	21	22	23	24	25	26	27	28	29	30
Porcine	Cys.	Gly.	Glu.	Arg.	Gly.	Phe.	Phe.	Tyr.	Thr.	Pro.	Lys.	Ala.
Guinea Pig .	—	Glu.	Asp.	Asp.	—	—	—	—	Ileu.	—	—	Asp.
Bonito	—	(—	—	)—	—	—	—	—	Glu.	—	—	*
Toad Fish 1 .	—	—	Asp.	—	—	—	—	—	Asn.	—	—	*
Toad Fish 2 .	—	—	Asp.	—	—	—	—	—	Asn.	Ser.	*	*
Angler Fish .	—	—	Asp.	—	—	—	—	—	Asn.	Ser.	—	*

Sequences not listed are identical.

Amino acids in parentheses are not definitely established.

Asx = not known if Asn or Asp.

* = residue missing.

III. Splitting of Insulin and Isolation of the A and B Chains

A. Splitting of Insulin with Peroxyacids

Treatment of insulin with performic acid *(Fig. 2)* results in the splitting of the hormone (3) into its A (4) and B (5) chains by conversion of the cystine to cysteic acid residues (*109, 110*). The oxidized chains can be separated by precipitation methods (*110*), by countercurrent distribution (*103*) or by ion exchange chromatography (*96*). The oxidized chains can be further degraded by proteolytic enzymes such as trypsin, chymotrypsin, pepsin, papain, mould protease, subtilisin and rennin (*133*).

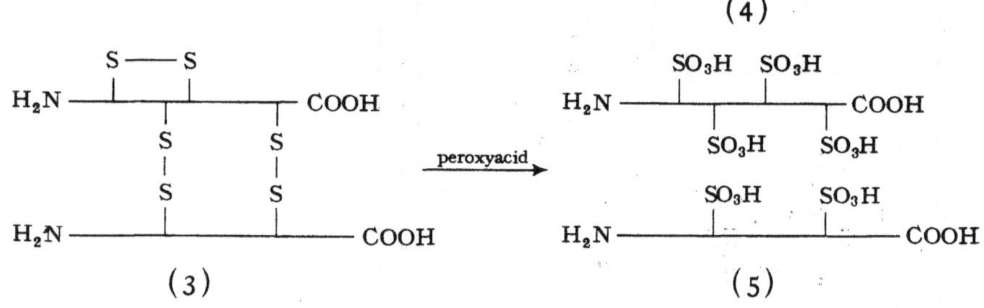

Fig. 2. Splitting of insulin with peroxyacids

B. Splitting of Insulin by Reduction of the SS Bonds

Reduction of SS groups in a protein is accomplished with the use of a thiol such as 2-mercaptoethanol, thioglycolic acid, 2-mercaptoethylamine, etc. Usually, the reaction is carried out in the presence of an agent such as urea, guanidine hydrochloride or detergent which cause the unfolding of the protein (*119*). Upon treatment of insulin with lithium thioglycolate at pH 5, only the intrachain SS bond, between residues A 6 and A 11, is reduced (*86*). If the reaction is carried out in the presence of 8 M urea, all three SS bonds of the hormone are reduced (*86*). Complete reduction of the three SS bonds of insulin can also be achieved by using high concentrations of 2-mercaptoethanol (*134*); and the hormone (3) is split to its A (6) and B (7) chains by conversion of the cystine to cysteine residues *(Fig. 3)*. The sulfhydryl groups of the reduced A and B chains are readily oxidizable. Alkylation of the sulfhydryl groups with either iodoacetic acid or its amide (*1, 23, 24, 45, 72, 86, 134, 149*) produces the carboxymethylated chains (8, 9). Aminoethylation with bromoethylamine (*87*) has also been used (*4*). The separation of the alkylated chains can be accomplished by gel filtration (*1, 23*).

C. Splitting of Insulin by Oxidative Sulfitolysis

The disulfide bonds of proteins can be cleaved upon reaction with sulfite (*2, 3, 102, 130*). This reaction with disulfides has been described by CLARKE (*21*) and by LUGG (*88*), and the equilibrium and kinetics of the reaction by CECIL and MCPHEE (*18, 89*) and by STRICKS and KOLTHOFF (*128, 129*). The reaction of sulfite with disulfides is illustrated by the following equation:

$$RSSR + SO_3^- = RS^- + RSSO_3^-$$

oxidizing agent

By the addition of a mild oxidizing agent such as Cu^{++} or sodium tetrathionate, the RS^- is oxidized back to RSSR, and eventually all the disulfide is converted to the S-sulfonated form. Treatment of insulin with sodium sulfite and with addition of Cu^{++} or sodium tetrathionate *(Fig. 4)*, in the presence of 8M urea or 8M guanidine hydrochloride (*3, 17, 26, 28, 67, 104*), results in the cleavage of the hormone (**3**) and the conversion of the A and B chains to the corresponding S-sulfonated derivatives (**10, 11**).

The separation and purification of the S-sulfonated derivatives of the insulin chains (A-SSO$^-$, B-SSO$_3^-$) is accomplished by ion exchange chromatography on Dowex (*26*) or by gel filtration (*138*). Separation also can be accomplished by three other methods, described by KATSOYANNIS and his co-workers (*67*), i. e., chromatography on carboxymethylcellulose or aminoethylcellulose and continuous flow electrophoresis. The three methods were used by these investigators for the large scale preparation of the S-sulfonated chains of bovine and porcine insulin, with a yield, based on the amount of insulin used, of 60–70% for the A-SSO$_3^-$ and 60–73% for the B-SSO$_3^-$. Identical materials were obtained by all three methods, as judged by several criteria. Thus, amino acid analysis of acid hydrolysates of the purified chains gave data in excellent

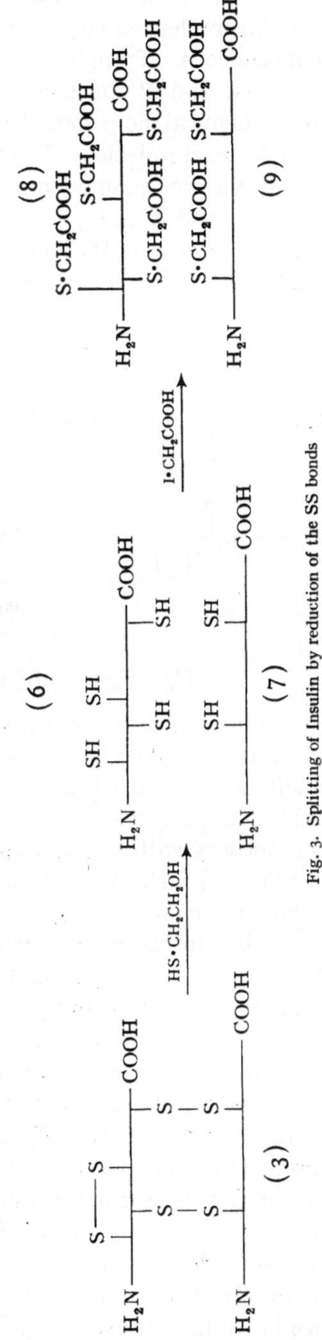

Fig. 3. Splitting of Insulin by reduction of the SS bonds

agreement with the theoretically expected values. On paper and thin layer electrophoresis at two pH values the isolated chains were homogeneous. Complete digestion of the $A-SSO_3^-$ by leucine amino peptidase and of the $B-SSO_3^-$ by amino peptidase M indicates that the stereochemical homogeneity of the chains is preserved during the preparation and isolation. The S-sulfonated derivatives of the insulin chains are stable compounds and are devoid of insulin activity (67). The $A-SSO_3^-$ and $B-SSO_3^-$ are readily converted to the corresponding sulfhydryl forms and therefore represent the most suitable form of the A and B chains for insulin resynthesis experiments.

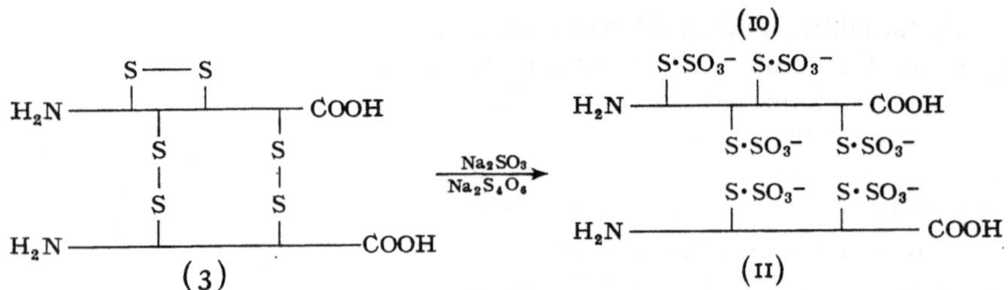

Fig. 4. Sulfitolysis of insulin

IV. Recombination of the Insulin Chains

Reduction of a mixture of $A-SSO_3^-$ and $B-SSO_3^-$ followed by air oxidation, enabled DIXON and WARDLAW (26) and DU et al. (28) to obtain small but definite insulin activity. The fact that the activity of the oxidation products was due to insulin was shown by neutralization of this activity with insulin antiserum (26) and by the isolation of crystalline insulin (28). The recombination reaction is illustrated in *Figure 5*. The S-sulfonated chains (10, 11) are converted with a thiol to the corresponding sulfhydryl forms (6, 7) which upon air oxidation combine to yield insulin. Although early reports by LANGDON (81), ZAHN (154), and VOLFIN et al. (143) had indicated that preparations of the individual A or B chains may possess insulin activity, subsequent experiments with well purified chains showed that none of the chain derivatives have a significant effect on the response of the assay systems to added insulin (104). Furthermore, it was shown that $A-SSO_3^-$ and $B-SSO_3^-$ and their reduced and oxidized products do not possess any insulin activity (26, 67). Consequently, the insulin activity of a recombination mixture of the A and B chains is a reliable measure of the insulin formed by interaction of these chains. Initially, the over-all yield of insulin formed upon recombination of the A and B chains was of the order of 1–2% (26, 28); later the yields of insulin were improved to 5–10% (105, 137, 150). More recently DU et al.

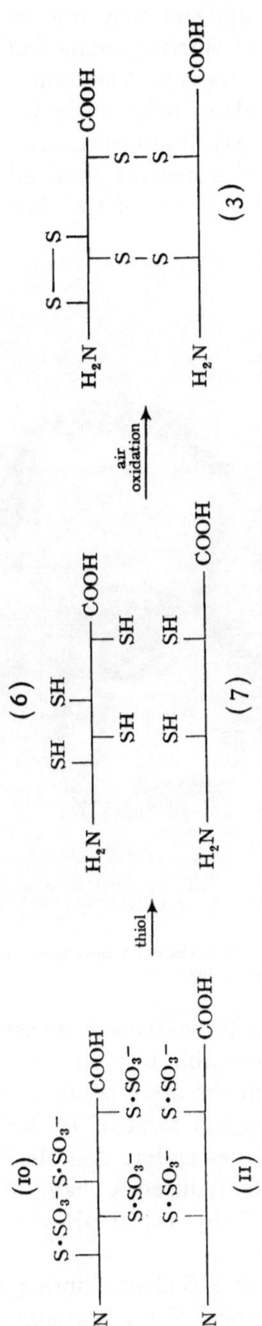

Fig. 5. Regeneration of insulin by interaction of the sulfhydryl forms of the A and B chains

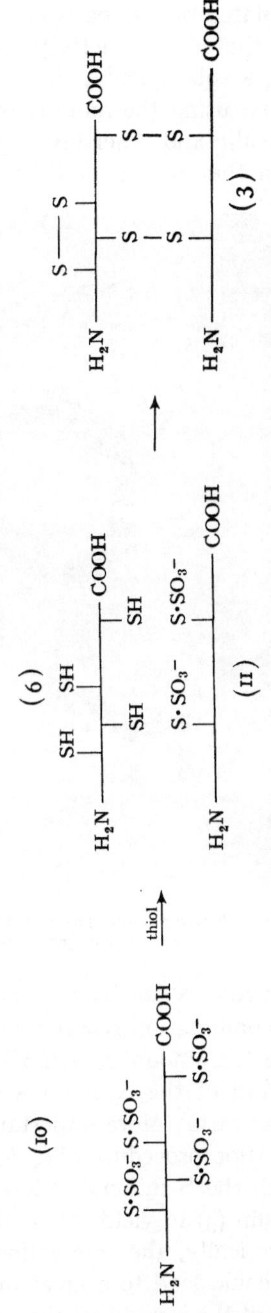

Fig. 6. Regeneration of insulin by combination of the sulfhydryl form of the A chain with the S-sulfonated derivative of the B chain

(*27*) claimed a yield of 50%; however, it appears that this yield was not calculated on the basis of the amounts of starting chains but on the basis of the specific activity of the final product. The same authors report in a subsequent publication (*99*) yields of only 10–12%. PRUITT et al. (*105*) using the method of Du et al. (*27*) obtained a yield of only 5% of insulin and other investigators using this method obtained a yield (based on the amount of B-SSO$_3^-$ used) of 12–16% of the theory (*59*).

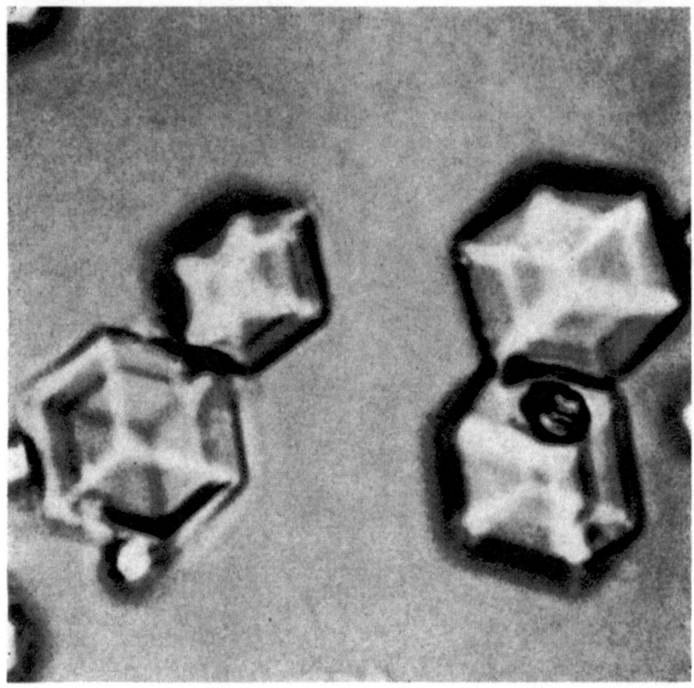

Fig. 7. Regenerated bovine insulin, produced by interaction of A and B chains. According to KATSOYANNIS et al. [From: Biochemistry **6**, 2642 (1967)]

KATSOYANNIS and co-workers (*59, 68*), after extensive investigation of the recombination reaction conditions, were able to obtain 50% yield (based on the amount of B-SSO$_3^-$ used) from the interaction of the sulfhydryl form of the A chain with the sulfhydryl form of B chain in a 5 : 1 molar ratio. More importantly, these investigators described a new recombination procedure *(Fig. 6)* in which the reduced A chain (**6**) reacts at 2° with the S-sulfonated form of the B chain (**11**) at pH 9.6–10.0 to form insulin (**3**) in yields of 60–80% (*59, 68*).

Theoretically, the interaction of the A and B chains during recombination could lead to a great number of isomers of the hormone and to polymers of the individual chains. Thus, the probability of the formation

References, pp. 152—160

of insulin by the simultaneous oxidation of its reduced chains was considered small (*71*). However, the fact that interaction of the two chains produces insulin in 60–80% yields (*59*, *68*), implies that the necessary information for the complementarity and covalent linking of the insulin chains to form the hormone is contained within the primary structure of the chains.

V. Isolation of the Regenerated Hormone from Recombination Mixtures

The isolation of the regenerated hormone from recombination mixtures can be accomplished through extraction with acidified alcohol (*27*, *28*), but this method is unsatisfactory because the over-all recovery of pure

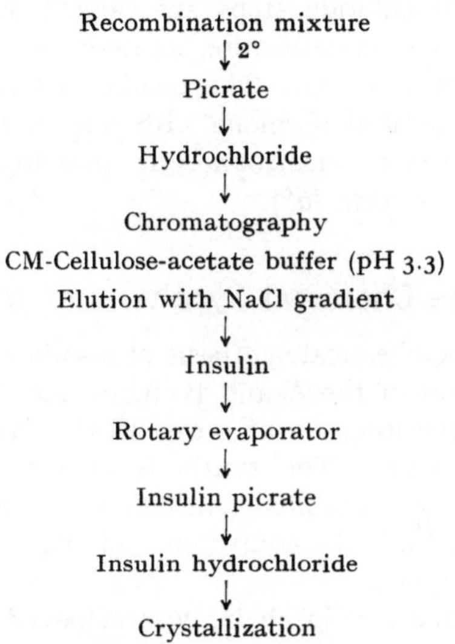

Fig. 8. Procedure for the isolation of insulin from recombination reaction mixtures (*68*)

insulin is very low. A satisfactory resolution of the components of the recombination mixture of the A and B chains of insulin and good recovery of the hormone was obtained by KATSOYANNIS and co-workers (*68*) with column chromatography on carboxymethylcellulose with urea-acetate buffer. Six components were identified by these investigators in recombination mixtures. Two of the components had an amino acid composition corresponding to the A chain and two other components had a composition corresponding to the B chain. These A and B chain components did not possess any insulin activity. The remaining two components were found to possess insulin activity and were tentatively designated as insulin I and insulin II. Insulin II was identical with the natural hormone with respect to amino acid composition, specific activity, electrophoretic and chromatographic mobility, infrared spectrum and crystalline form *(Fig. 7)*. Insulin I was often contaminated with varying amounts of B chain and therefore its potency ranged from 16 to 22 IU/mg. The relative amounts of insulins I and II are greatly dependent on the method used for recombining the A and B chains and also on the processing of the recombination mixture prior to chromatography. Insulins I and II were found to be interconvertible but the nature of the mechanism underlying this interconversion is not known (*68*). The over-all recovery of the regenerated insulin from the recombination mixtures was 50–65%.

KATSOYANNIS and his co-workers (*68*) also introduced an even simpler isolation procedure which is shown diagramatically in *Fig. 8* (p. 131). The picrate and hydrochloride steps are carried out as described by RANDALL (*106*) and the crystallization as described by RANDALL (*106*) or EPSTEIN and ANFINSEN (*29*). The insulin isolated by this method is identical with the natural hormone with respect to amino acid composition, specific activity, electrophoretic mobility, chromatographic mobility and crystalline form (*68*).

VI. The Chemical Synthesis of Insulin

The problem of the chemical synthesis of insulin consists of two basic parts: (a) the synthesis of the A and B chains and (b) the combination of these chains and the isolation of the synthetic insulin from the combination reaction mixture. The synthesis of the insulin chains also includes the purification of the final synthetic products and the analytical tests necessary to establish the equivalence of the synthetic chains with their natural counterparts.

The synthesis of the insulin chains was achieved by the application of the "fragment condensation approach" (*48*). Peptide subunits were synthesized stepwise using amino acids with their α-amino group protected with carbobenzoxy group and were condensed to give larger polypeptides. This process was repeated until the total synthesis of the A and B chain in the protected form was complete. The proper selection of peptide subunits and their condensation by procedures that do not cause racemization was of critical importance for the ultimate synthetic success. Condensation of fragments containing a C-terminal optically active amino acid residue was usually accomplished by the azide procedure, to avoid the risk of racemization, whereas condensation of peptide subunits with glycine as the C-terminus were coupled by other methods as well (see *40, 118*).

In addition to the carbobenzoxy group used for the protection of the α-amino function of the amino acids, the benzyl group was employed for protection of the sulfhydryl functions and the imidazole nitrogen of histidine, the p-toluenesulfonyl group for protection of the ε-amino group of lysine, and the guanidino group of arginine, and the benzyl, p-nitrobenzyl and t-butyl groups for masking carboxylic functions.

The activation of the N-protected amino acids was achieved by conversion to the corresponding p-nitrophenyl ester (*8*), and in some instances, by the carbodiimide (*120*) or the mixed anhydride (*139*) procedures.

Since S-sulfonated derivatives of the A and B chains were desirable, deblocking of the protected synthetic A and B chains was followed by oxidative sulfitolysis.

A. Synthesis of the Insulin Chains

1. Synthesis of the A Chain of Insulin

a. Sheep Insulin A Chain

The synthesis of the A chain of sheep insulin was accomplished by KAT-SOYANNIS and co-workers (*66*) by two synthetic routes. Three peptide fragments corresponding to residues A 1–4, A 5–9 and A 10–21 were synthesized and condensed by two different synthetic routes to give the protected A chain of sheep insulin. The first synthetic route includes the condensation of the dodecapeptide A 10–21 with the pentapeptide A 5–9 to give the heptadecapeptide A 5–21. Condensation of the latter derivative with the tetrapeptide A 1–4 affords the A chain in the protected form (*66*).

In the second synthetic route, peptides A 1–4 and A 5–9 were condensed to yield the A 1–9 nonapeptide fragment which was then coupled with the A 10–21 dodecapeptide derivative to yield the protected A-chain (*55, 60, 64, 66*). Although the final products by both synthetic routes are identical by many criteria employed, the first synthetic pathway affords better yields.

Synthesis of the A 10–21 Dodecapeptide. The synthesis of the A 10–21 dodecapeptide is summarized in *Fig. 9* (p. 134) (*63*). Condensation of dipeptides (**12**) and (**13**) by the mixed anhydride procedure affords the tetrapeptide N-carbobenzoxy-L-asparaginyl-L-tyrosyl-S-benzyl-L-cysteinyl-L-asparagine p-nitrobenzyl ester (**14**). Treatment of (**14**) with hydrogen bromide in acetic acid gives the amino-free tetrapeptide (**15**). Interaction of (**15**) with N-carbobenzoxy-γ-benzyl-L-glutamic p-nitrophenyl ester yields the protected A 17–21 pentapeptide (**16**) which is transformed into the partially protected derivative (**17**) by treatment with hydrogen bromide in acetic acid.

N-Carbobenzoxy-O-benzyl-L-tyrosyl-L-glutaminyl-L-leucine methyl ester (**19**), prepared by reaction of the dipeptide (**18**) with the p-nitrophenyl ester of N-carbobenzoxy-O-benzyl-L-tyrosine, is decarbobenzoxy-

lated and debenzylated on exposure to hydrogen bromide in acetic acid.
The resulting product (**20**) is allowed to react with N-carbobenzoxy-L-
leucine *p*-nitrophenyl ester to give the protected A13–16 tetrapeptide
ester (**21**). Treatment of (**21**) with hydrazine affords the hydrazide derivative (**22**) which is transformed into the azide (**23**) by reaction with
nitrous acid.

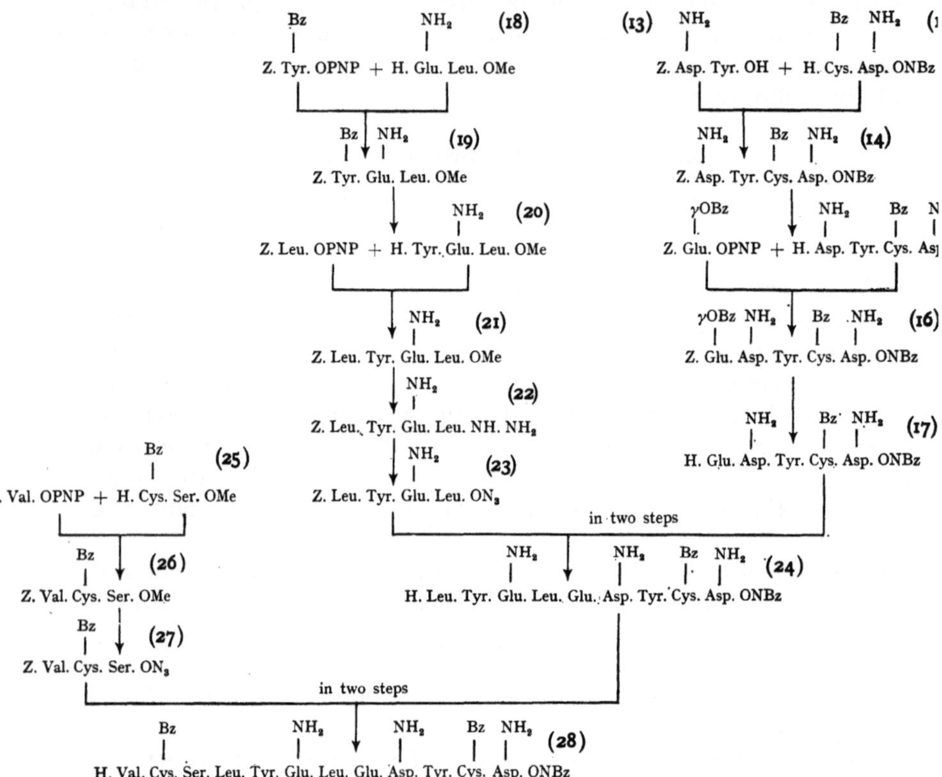

Fig. 9. Synthesis of the A 10—21 dodecapeptide (Katsoyannis et al. (*63*))

Interaction of the tetrapeptide azide A13–16 (**23**) with the A17–21
pentapeptide (**17**) gives the protected A13–21 C-terminal nonapeptide
which on exposure to hydrogen bromide in acetic acid is transformed into
the amino-free derivative (**24**).

Condensation of N-carbobenzoxy-L-valine *p*-nitrophenyl ester with
the dipeptide (**25**) gives N-carbobenzoxy-L-valyl-S-benzyl-L-cysteinyl-
L-serine methyl ester (**26**), which in turn is converted into the A10–12
tripeptide azide (**27**). Interaction of the latter compound (**27**) with the
A13–21 nonapeptide (**24**) produces the protected C-terminal dodeca-

peptide A 10–21, which on exposure to hydrogen bromide in acetic acid affords the amino-free derivative (**28**).

Synthesis of the A 5–9 Pentapeptide Azide (**29**). This derivative was obtained from the corresponding methyl ester which in turn was prepared by the stepwise *p*-nitrophenyl ester method in the usual way (*66*).

Synthesis of the A 1–4 Tetrapeptide Azide (**32**). This derivative was obtained from the ester via the hydrazide (*64*).

Synthesis of the Sheep Insulin A Chain. The final steps in the synthesis of the sheep insulin A chain are shown in *Figure 10.* Condensation of the A 5–9 pentapeptide azide (**29**) with the A 10–21 dodecapeptide (**28**) gives the protected A 5–21 heptadecapeptide (**30**). Exposure of this derivative to hydrogen bromide in trifluoracetic acid affords the corresponding amino-free derivative (**31**). Interaction of the latter product (**31**) with the A 1–4 tetrapeptide azide (**32**) yields the protected A chain derivative (**33**). Treatment of (**33**) with trifluoroacetic acid removes the t-butyl group and treatment with sodium and liquid ammonia removes the remaining protecting groups.

The resulting reduced product is treated with sodium sulfite and sodium tetrathionate (oxidative sulfitolysis) to give the S-sulfonated derivative of sheep insulin A chain (**34**).

The S-sulfonated derivative of the A chain is purified by gel filtration on Sephadex. Amino acid analysis of the material obtained by acid or enzymatic hydrolysis (leucine aminopeptidase) shows a composition in molar ratios in excellent agreement with the theoretically expected values for the A chain of sheep insulin. The synthetic A chain behaves as a single component and exhibits identical mobility with the natural counterpart on paper chromatography in two solvent systems and on thin layer electrophoresis. Also the synthetic A chain exhibits identical IR spectrum and optical rotation with the natural chain.

The synthesis of the sheep insulin A chain was achieved by ZAHN and co-workers (*93, 156*) by a similar synthetic route *(Fig. 11).* Condensation of N-carbobenzoxy-γ-t-butyl-L-glutamyl-L-asparaginyl-L-tyrosine azide with S-benzyl-L-cysteinyl-L-asparagine followed by decarbobenzoxylation yields the A 17–21 C-terminal pentapeptide (**35**) (*153*).

The A 10–16 heptapeptide azide (**36**) is obtained by the usual procedures from the corresponding ester. Condensation of (**35**) and (**36**) affords the protected A 10–21 C-terminal dodecapeptide which on exposure to hydrogen bromide in trifluoroacetic acid gives the amino-free derivative (**37**). Interaction of N-carbobenzoxy-L-glutamine *p*-nitrophenyl ester with the product obtained by decarbobenzoxylation and saponification of the tetrapeptide ester derivative (**38**) affords the protected pentapeptide A 5–9 which on exposure to hydrogen bromide in acetic acid gives the partially protected derivative (**39**). Coupling of

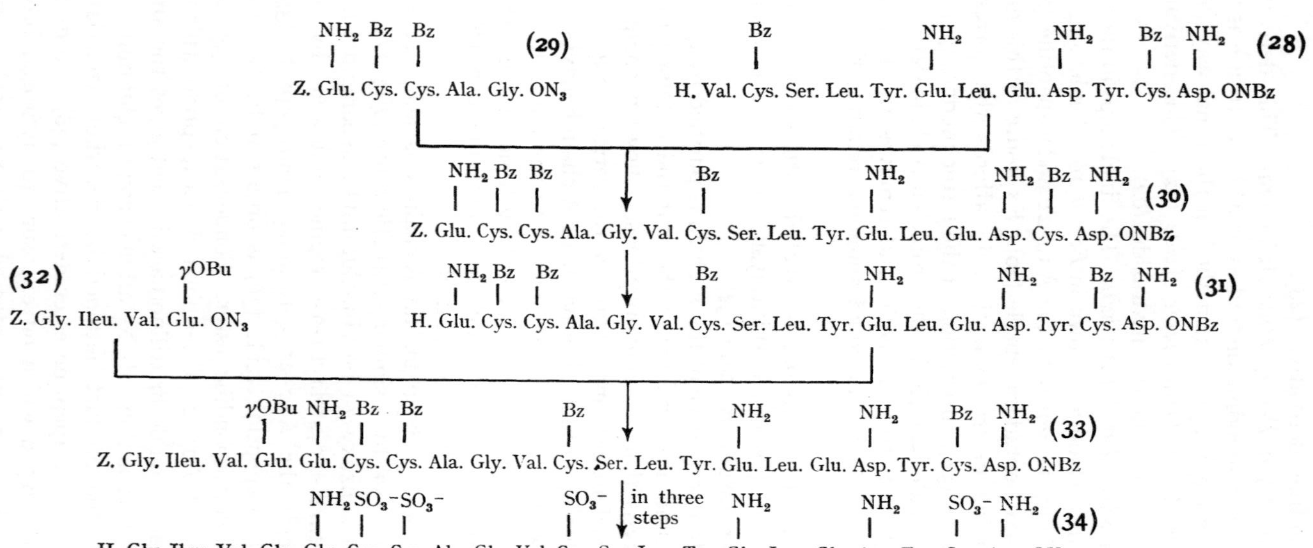

Fig. 10. Final steps in the synthesis of the S-sulfonated derivative of the A chain of sheep insulin (KATSOYANNIS et al. (53, 66))

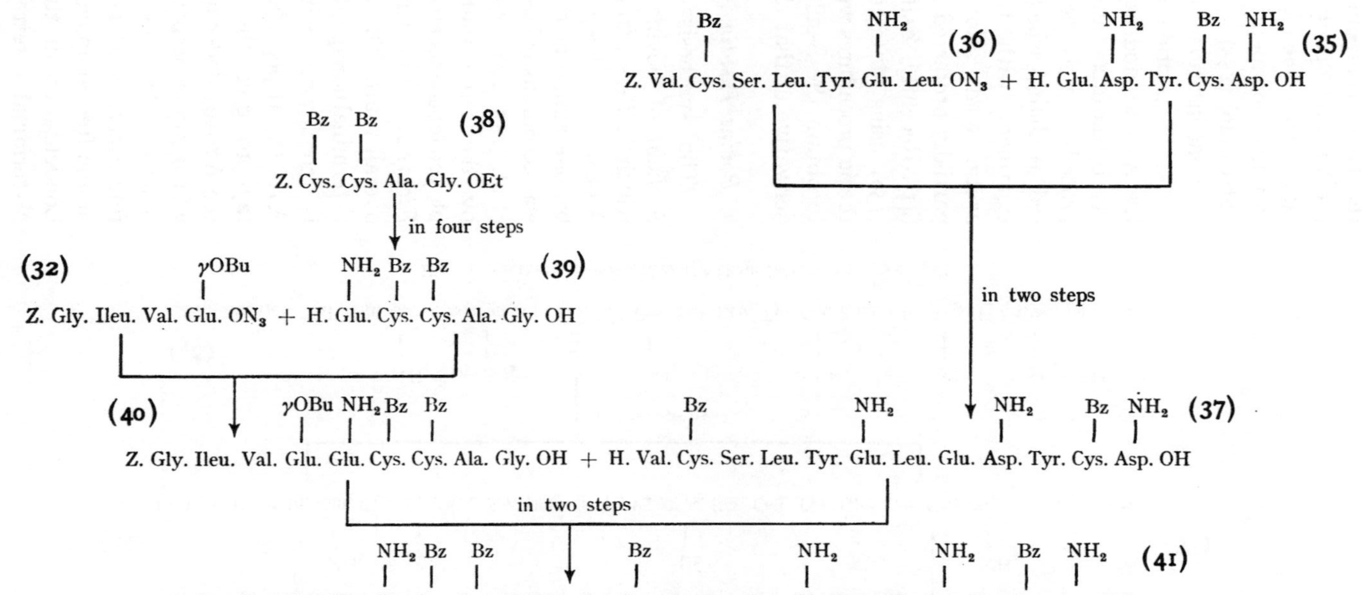

Fig. 11. Synthesis of the protected sheep insulin A chain (ZAHN et al. (93, 156))

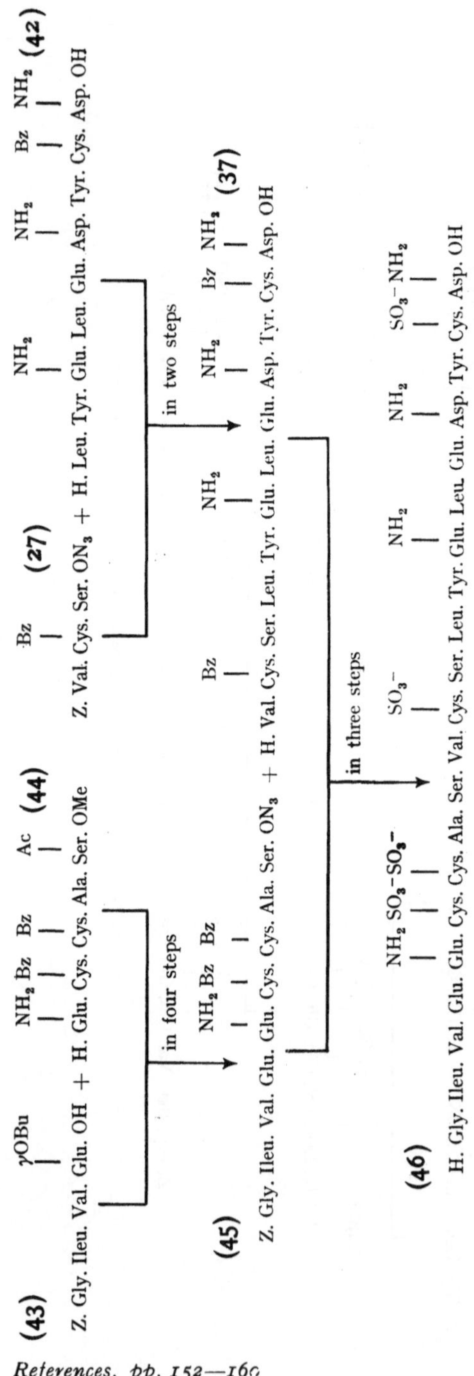

Fig. 12. Synthesis of bovine insulin A chain (Wang et al. (145, 146))

the latter compound with the A1–4 tetrapeptide azide (32) yields the A1–9 N-terminal nonapeptide derivative (40) (155).

Condensation of the A1–9 nonapeptide (40) with the A10–21 dodecapeptide (37) by the mixed anhydride procedure gives the protected A chain, which upon treatment with trifluoroacetic acid affords the partially protected A chain (41). Recently, Zahn et al. (158) using different synthetic procedures were able to obtain the S-sulfonated derivative of that chain.

b. *Bovine Insulin A Chain*

The synthesis of the A chain of bovine insulin which differs from the A chain of sheep insulin by one amino acid residue was accomplished by Wang et al. (145, 146) essentially by the same routes employed for the sheep A chain. *Figure 12* summarizes the overall route. The A13–21 C-terminal nonapeptide (42) is condensed with the A10–12 tripeptide azide (27) to give the protected A10–21 dodecapeptide, which upon exposure to hydrogen bromide in trifluoroacetic acid gives the amino-free component (37). Condensation of the A1–4 N-terminal tetrapeptide

(43) with the A5–9 pentapeptide (44) by the carbodiimide procedure yields the protected A1–9 nonapeptide ester which eventually is converted to the azide (45). Interaction of the A1–9 nonapeptide (45) with the A10–21 dodecapeptide (37) affords the protected bovine A chain derivative which is deblocked and sulfitolyzed to give the S-sulfonated A chain (46).

c. Human (Porcine) Insulin A Chain

The synthesis of the A chain of human insulin which is identical with the A chain of porcine insulin was accomplished by KATSOYANNIS and co-workers (62, 65) by essentially the same synthetic routes which they employed in the synthesis of the sheep A chain. As was the case in the synthesis of sheep A chain, the condensation of A1–4 tetrapeptide with the A5–21 heptadecapeptide affords better yields of synthetic product than the condensation of the A1–9 nonapeptide with the A10–21 dodecapeptide.

The synthesis of the A chain S-sulfonate by the A1–4 plus A5–21 condensation, is summarized in *Figure 13* (p. 140). The A10–12 tripeptide azide (47), is coupled with the A13–21 nonapeptide (24) to give the C-terminal A10–21 dodecapeptide of human A chain (70) which on exposure to hydrogen bromide in trifluoroacetic acid is transformed into the amino-free derivative (48). Reaction of A5–9 pentapeptide azide (49) (70) with the A10–21 dodecapeptide (48) results in the formation of the protected A5–21 heptadecapeptide. The A5–21 heptadecapeptide is decarbobenzoxy-lated with hydrogen bromide in trifluoroacetic acid to give the partially protected derivative (50). Reaction of (50) with the A1–4 N-terminal tetra-peptide azide (32) produces the protected A chain derivative. Treatment of the protected A chain with trifluoroacetic acid and sodium in liquid ammonia, followed by sulfitolysis, gives the S-sulfonated derivative of human A chain (51). This derivative is purified by chromatography on Sephadex.

2. Synthesis of the B Chain of Insulin

a. Sheep (Bovine) Insulin B Chain

The amino acid sequence of the B chain of sheep and bovine insulins are identical (Table 1, p. 124). This chain was synthesized by KATSOYANNIS and co-workers (53, 56) by the condensation of two peptide fragments, namely a nonapeptide (B1–9) with a heneicosapeptide (B10–30).

Synthesis of the B10–30 Heneicosapeptide. The synthesis of the B10–30 heneicosapeptide is summarized in *Figure 14* (p. 141). The B21–30 C-terminal decapeptide (52) is prepared by stepwise addition of appropriate carbobenzoxyamino acids (58) and is converted by hydrogenolysis

A. C. TRAKATELLIS and G. P. SCHWARTZ:

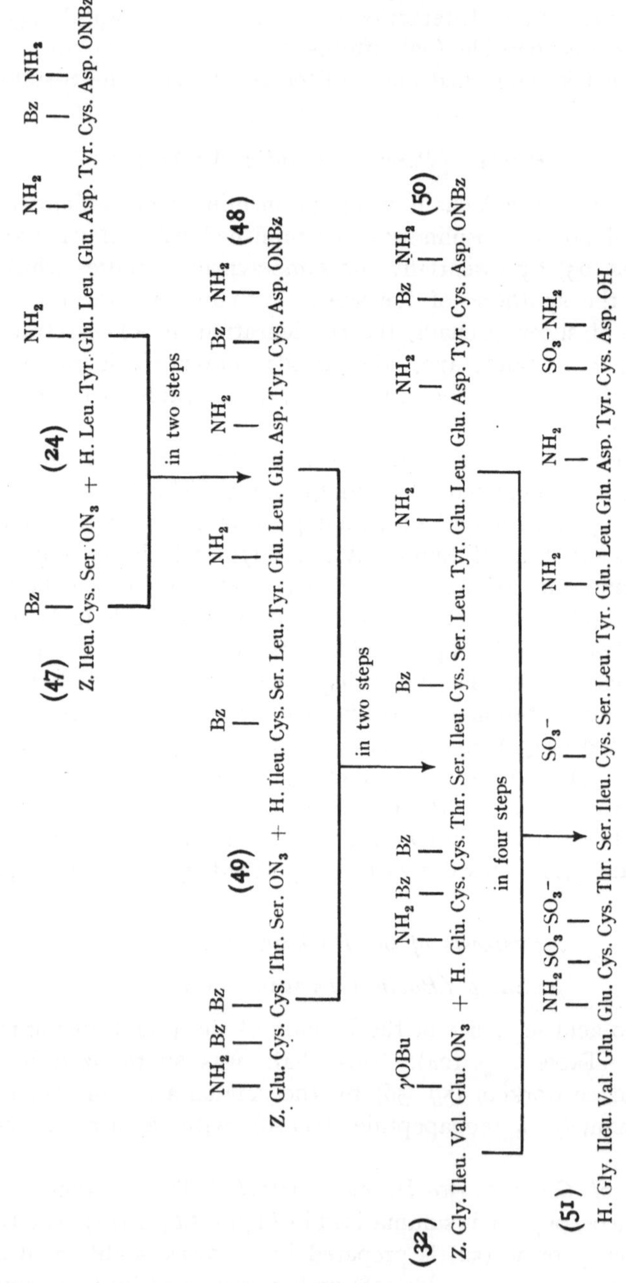

Fig. 13. Synthesis of human insulin A chain (KATSOYANNIS (65))

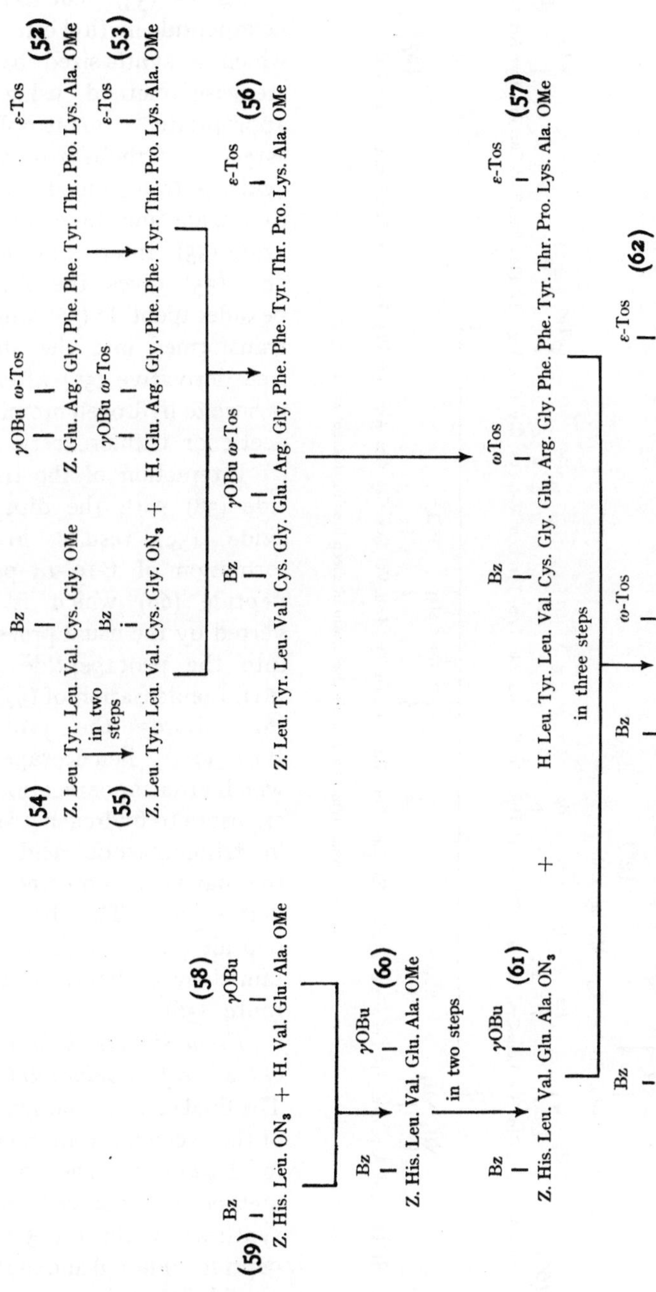

Fig. 14. Synthesis of the C-terminal heneicosapeptide of sheep insulin B chain (KATSOYANNIS (53))

to the partially protected derivative (53). The B15–20 hexapeptide methyl ester (54), which is synthesized by the stepwise method using the appropriate p-nitrophenyl esters of carbobenzoxyamino acids, is transformed into the hydrazide and then into the azide (55). Interaction of (53) and (55) gives the B15–30 hexadecapeptide (56) which is transformed into the amino-free derivative (57) after exposure to hydrogen bromide in acetic or trifluoroacetic acid.

Interaction of the tripeptide (58) with the dipeptide azide (59) results in the formation of B10–14 pentapeptide (60) which is converted by the usual procedure into the pentapeptide azide (61). Condensation of (57) and (61) affords the protected B10–30 heneicosapeptide which after saponification and exposure to hydrogen bromide in trifluoroacetic acid gives the partially protected derivative (62). This heneicosapeptide was originally obtained by a slightly different route (56).

Final Steps in the Synthesis of the Sheep (Bovine) B Chain. The final steps in the synthesis of the B chain are summarized in *Figure 15*. The stepwise method is employed for the synthesis of the B1–5 pentapeptide azide (63) and the tetrapeptide B6–9 derivatives (64).

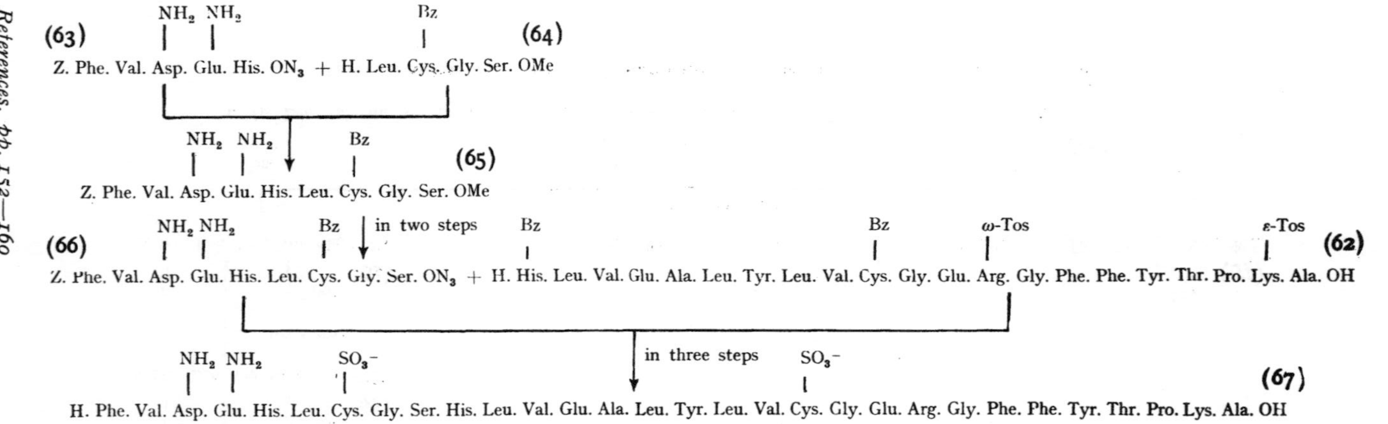

Fig. 15. Final steps in synthesis of B chain of sheep insulin (KATSOYANNIS (53))

Interaction of these compounds affords the protected N-terminal nona-peptide B1–9 (**65**) which is converted to the azide (**66**) via the hydraz.de. Condensation of the B1–9 azide (**66**) with the B10–20 peptide fragment yields the protected B chain. Deblocking by sodium in liquid ammonia and sulfitolysis of the resulting product affords the S-sulfonated B chain (**67**) which is purified by chromatography on a carboxymethylcellulose column.

MEIENHOFER et al. (*92, 93*) have also synthesized the sheep B chain by using a different approach (*Fig. 16*, p. 144) in which the fragments B1–8 and B9–30 were used in the final coupling step.

The heptapeptide (**68**) is condensed with tripeptide (**69**) by the mixed anhydride method to give the C-terminal decapeptide which after hydrogenation affords the amino-free derivative (**70**) (*117*).

Condensation of (**70**) with the hexapeptide (**71**) by the carbodiimide procedure yields the hexadecapeptide fragment which is converted to the partially protected derivative (**72**) on exposure to hydrogen bromide in trifluoroacetic acid (*91*). The hexapeptide azide (**73**) prepared from the corresponding methyl ester (*91*) in the usual way is condensed with (**72**) to give the B9–30 docosapeptide which after treatment with hydrogen bromide in trifluoroacetic acid gives (**74**). The B1–8 octapeptide (**75**) synthesized by the stepwise method (*159*) is coupled with (**74**) by the mixed anhydride procedure to yield the protected B chain derivative of sheep insulin (**76**).

NIU et al. (*98, 99*) have also synthesized the B chain by condensation of the B1–8 and B9–30 fragments. These fragments, however, were prepared by different routes and protecting groups than those used by MEIENHOFER et al. (*92, 93*). A summary of the synthesis is shown in *Figure 17*. Interaction of the B17–20 tetrapeptide *p*-nitrophenyl ester (**77**) with the C-terminal B21–30 decapeptide (**78**) followed by saponification and exposure to hydrogen bromide in trifluoroacetic acid affords the B17–30 tetradecapeptide (**79**) (*77*).

Interaction of (**79**) with the B9–16 octapeptide azide (**80**) gives the B9–30 docosapeptide which is transformed into the amino-free derivative (**81**) by exposure to hydrogen bromide in trifluoroacetic acid (*76*). Condensation of (**81**) with the B1–8 octapeptide azide (**82**) (*19*) affords the protected B chain derivative which after deblocking and sulfitolysis yields the B chain S-sulfonate (**83**).

b. Human Insulin B Chain

The synthesis of the B chain of human insulin which differs from the B chain of sheep insulin by one amino acid residue was accomplished by KATSOYANNIS and co-workers (*53, 61*) essentially by the same routes employed for the sheep B chain.

A. C. TRAKATELLIS and G. P. SCHWARTZ:

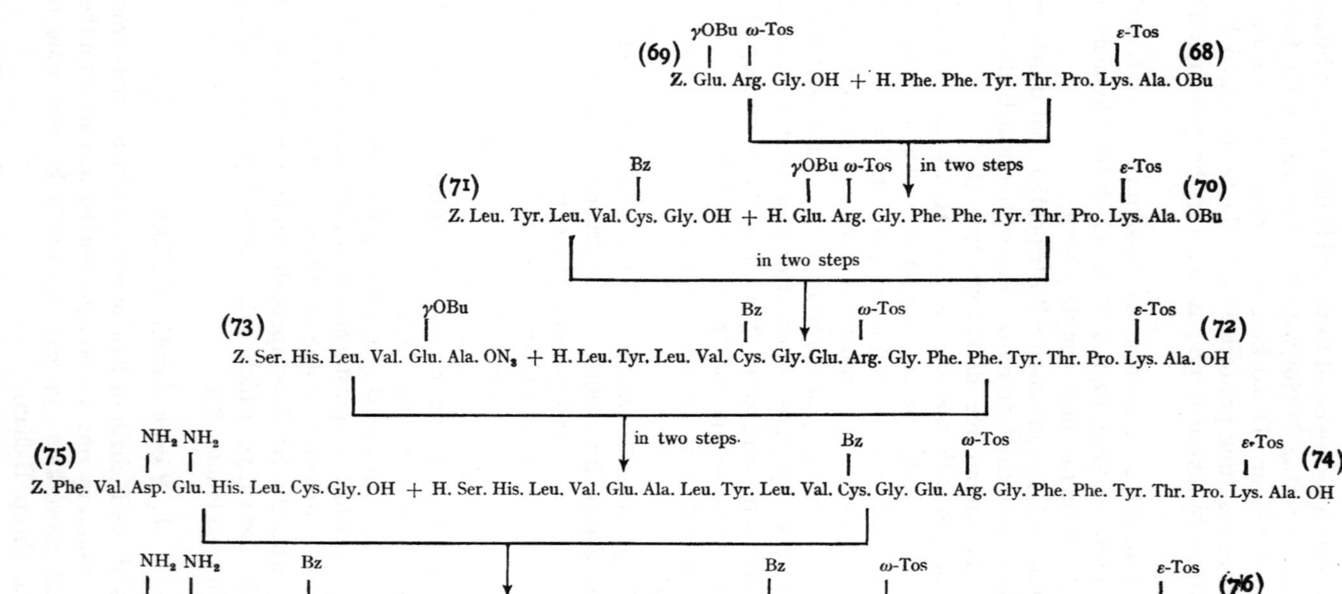

Fig. 16. Synthesis of the protected B chain of sheep insulin (NEIENHOFER et al. (92, 93))

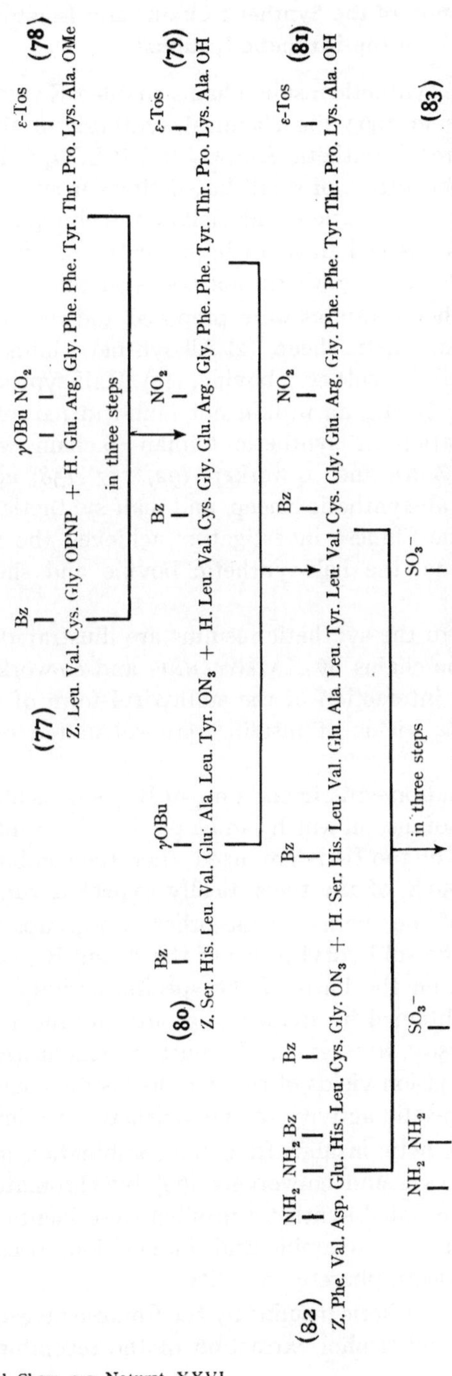

Fig. 17. Synthesis of the bovine insulin B chain (Niu et al. (98, 99))

B. Combination of the Synthetic Chains and Isolation
of the Synthetic Insulins

Combination of the synthetic insulin chains enabled Katsoyannis and co-workers to achieve in 1963 the chemical synthesis of the hormone. Reduction of a mixture of synthetic A-SSO$_3$ and B-SSO$_3$ followed by air oxidation led to the generation of small but definite insulin activity (*49, 60*). Since then, new combination and isolation techniques introduced by these investigators enabled them to improve the yields of synthetic insulins and to isolate these synthetic hormones in pure form (*59, 69*).

All in all, six synthetic insulins were prepared and isolated in highly purified form: (1) All-synthetic sheep, (2) All-synthetic human, (3) Half-synthetic sheep, (4) Half-synthetic bovine, (5) Half-synthetic porcine and (6) Half-synthetic B$_A$H$_B$, an insulin not found in nature which was produced by combination of synthetic human B chain with natural bovine A chain (*69*). Zahn and co-workers (*92, 157, 158*) accomplished the synthesis of the all-synthetic sheep and half-synthetic sheep and bovine insulins and the Chinese investigators achieved the synthesis of the all-synthetic bovine, the half-synthetic bovine and sheep insulins (*75, 98, 99, 146*).

All data pertinent to the synthetic insulins are illustrated in *Table 3*. The combination of the chains by Katsoyannis and co-workers (*59, 68, 69*) was performed by interaction of the sulfhydryl form of the A chain with B-SSO$_3^-$ and the yields of insulin were calculated on the basis of the B-SSO$_3^-$ used.

In a typical combination experiment 5 mg of B-SSO$_3^-$ is used and thus the maximum amount of insulin which can be produced is 8 mg or 200 IU. If only 4 mg of insulin or 100 IU is produced, then the combination yield of such a reaction is 50% of the theoretically expected value.

The combination of the chains by the other two groups was carried out by interaction of the sulfhydryl forms of the A and B chains and the yields were calculated on the basis of the specific activity of the final product. The yields obtained by these two groups are much lower than that obtained by Katsoyannis et al. It must be emphasized that the variation of the combination yields of the various insulin chains used has no reflection on the specific activity of the synthetic insulins produced.

Isolation of the synthetic insulins from the combination mixture was achieved by Katsoyannis and co-workers (*69*) by chromatography on CM-cellulose; and the isolated synthetic insulins were identical with the natural hormones with respect to amino acid composition, specific activity, chromatographic and electrophoretic mobility.

The isolation of the synthetic insulins by the Chinese investigators (*75*) was effected through acid-alcohol extraction of the recombination mix-

Table 3. Combination Yields, Isolation Recoveries and Specific Activities of Insulins Synthesized by Interaction of A and B Chains

Type of Chains Used for Combination	Insulin Produced	Combination Yield %	Over-All Recovery* %	Specific Activity***	References
Synthetic Sheep A + Natural Bovine B	Sheep (half-synthetic)	30–38	39	25**	Katsoyannis et al. (69)
Synthetic Sheep A + Natural Bovine B	Sheep (half-synthetic)	3.5	†	27**	Zahn et al. (158)
Synthetic Bovine A + Natural Bovine B	Bovine (half-synthetic)	5–9	5	23**	Wang et al. (146)
Natural Bovine A + Synthetic Sheep B	Bovine (half-synthetic)	17–19	37	22**	Katsoyannis et al. (69)
Natural Bovine A + Synthetic Sheep B	Bovine (half-synthetic)	0.65	†	†	Meienhofer et al. (92)
Natural Bovine A + Synthetic Bovine B	Bovine (half-synthetic)	5–10	5	23**	Niu et al. (98, 99)
Synthetic Sheep A + Synthetic Sheep B	Sheep (all-synthetic)	12–16	43	25**	Katsoyannis (69)
Synthetic Sheep A + Synthetic Sheep B	Sheep (all-synthetic)	0.2–1	†	†	Zahn et al. (157)
Synthetic Bovine A + Synthetic Bovine B	Bovine (all-synthetic)	1–2	5	23**	Kung et al. (75)
Synthetic Human A + Natural Bovine B	Porcine (half-synthetic)	30–42	52	22	Katsoyannis et al. (69)
Natural Porcine A + Natural Bovine B	Porcine	35–50	42	25	Katsoyannis et al. (69)
Natural Bovine A + Synthetic Human B	Insulin $B_A H_B$	18–23	35	22**	Katsoyannis et al. (69)
Synthetic Human A + Synthetic Human B	Human (all synthetic)	15–22	51	24	Katsoyannis et al. (69)

* The insulin activity present in the combination mixture taken as 100%.

** Obtained in crystalline form.

*** International units (IU) per mg. The international standard of insulin (identical with the USP standard) contains 24 units per mg.

† Not reported.

ture, but the over-all recovery of the synthetic hormones was very low and, specifically in the isolation of all-synthetic bovine insulin, it was of the order of a few hundredths of a milligram.

Because of the low recovery of the synthetic bovine insulin, important analytical data and especially amino acid composition were not obtained.

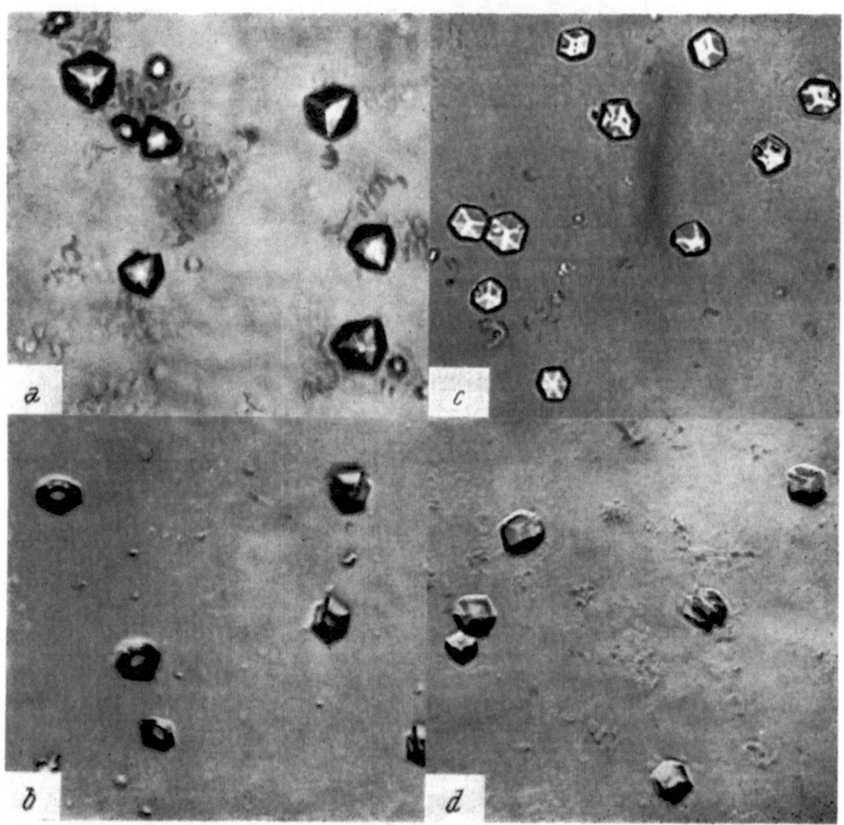

Fig. 18. Synthetic crystalline zinc insulins: (a) Half-synthetic sheep insulin (synthetic sheep A chain plus natural bovine B chain); (b) Half-synthetic bovine insulin (natural bovine A chain plus synthetic sheep B chain); (c) All-synthetic sheep insulin (synthetic sheep A chain plus synthetic sheep B chain); (d) Half-synthetic insulin $B_A H_B$ (natural bovine A chain plus synthetic human B chain). According to Katsoyannis et al. [From: Biochemistry 6, 2656 (1967)]

Zahn and co-workers (158) using the acid-alcohol extraction have isolated only the half-synthetic sheep insulin, but data for the half-synthetic bovine and all-synthetic sheep insulin have not been presented.

The all-synthetic sheep, the half-synthetic sheep and bovine and the half-synthetic insulin $B_A H_B$ were obtained in crystalline form *(Fig. 18)*.

References, pp. 152—160

The synthesis of insulin $B_A H_B$, which has not yet been found in nature and which is endowed with the full potency of the natural hormones, constitutes the first synthesis of an insulin analog which embodies within its structure the naturally occurring A and B chains (hybrid insulin). The synthesis of this protein indicates that it is possible to synthesize insulins which do not occur in nature by combining A and B chains of naturally occurring hormones. Whether combinations of all-natural chains can always be accomplished and whether the insulin thus produced would have the full potency of the natural hormone remains to be seen. Finally, the methods so far developed in the field of synthetic insulins open unlimited possibilities for the synthesis of insulin analogs and for the study of the relationship between chemical structure and hormonal activity of insulin.

VII. Relation of Structure of Insulin to Biological Activity

The field of hormone research has been handicapped through the years because the mechanism of action of hormones at the molecular level has not yet been elucidated. However, extensive studies have been carried out by various investigators to establish relationship between structure and hormonal activity. Several lines of experimentation were followed, such as the effects of chemical modification or enzymatic degradation of the hormone on its activity.

Selective chemical modification of functional groups in peptides and proteins is very difficult to accomplish and this handicaps studies on the relation of chemical changes to biological activity. Furthermore, if selective chemical modification of a group is attainable in some instances, the results are still very difficult to interpret in terms of the requirement of the group for hormonal activity. This is the case because such modification of a group may also bring about changes in the steric structure of the molecule. Therefore, only general observations can be made.

Modification of the amino groups of insulin by reaction with acetic anhydride, nitrous acid or formaldehyde usually does not cause significant change in the potency of the hormone (*30, 32, 84*). In contrast with these results, a progressive loss of activity was observed upon treatment of bovine insulin with fluorescein isothiocyanate (*10, 136*). This compound was found to react with the amino group of insulin in A 1, B 1 and B 29 to yield the mono-, di- or trisubstituted derivatives (*10*). The mono-derivative had the fluorescein group attached to the N-terminal phenylalanine (B 1) and possessed 40% of the insulin activity; the di-derivative (B 1, A 1) had only 4% of the hormone activity; and the tri-derivative (B 1, A 1 and B 29) had little activity, if any. A loss of activity was also obtained upon treatment of insulin with aromatic isocyanates (*32*).

It must be noted that carbamylation of insulin with cyanate does not produce any change in activity (*22*). All these results are difficult to interpret because different groups attached to the same groups of the insulin molecule produce different effects on its hormonal activity.

Sulfation of the aliphatic hydroxyl groups did not cause any significant loss of activity (*32*); whereas esterification of the carboxyl groups (*16, 33*), reduction of at least one disulfide bond (*141, 151*) or photooxidation of the histidine residues (*147, 148*) resulted in loss of hormonal potency. Similarly, loss of activity was observed after acylation or extensive iodination of the tyrosine residues (*32*) or extensive reaction of the histidine and tyrosine residues with diazobenzene sulfonic acid (*32*) or sulfation of these residues (*124*).

The amide bonds in insulin are sensitive to acid treatment which results in their removal. This is especially true for the very labile amide bond of the C-terminus asparagine A21 (*14, 15, 20, 122*). Therefore, it is not surprising that samples of commercial insulin contain various amounts of deamido-insulins, since the isolation of the hormone from pancreas is effected by acid extraction. The monodeamido-insulin which has been isolated from commercial samples by countercurrent distribution (*35, 36*) or partition chromatography (*20*) is fully active. Chromatography on CM-cellulose was also used recently for the isolation of the deamidated insulins (*25*).

Digestion of insulin with carboxypeptidase results in the removal of the C-terminal alanine (B30). The dealanine-insulin thus produced was isolated by countercurrent distribution and found to be fully active (*123*). Prolonged action of carboxypeptidase resulted in the removal of both C-terminals (A21 and B30). The dealanine-deasparagine-insulin was isolated by partition chromatography and possessed only 5% of the hormonal activity (*38, 121*). Upon digestion of insulin with trypsin the peptide bonds between arg. gly (B22–23) and lys. ala (B29–30) are cleaved. The resulting deoctapeptide-insulin was isolated by partition chromatography (*152*), and by chromatography on DEAE-Sephadex (*9*) or CM-cellulose (Trakatellis, unpublished data) and was found to be inactive.

Finally, an insulin missing the C-terminal tripeptide pro. lys. ala has been synthesized and isolated by Katsoyannis and co-workers (*54*) and was found to possess almost full biological activity. From degradative studies, it appears that the terminal A21 residue and the residues B22–27 may play an important role in the hormonal function of the insulin molecule. However, it must be emphasized that the elimination of these residues change the conformation of the hormone (*13*) and the loss of hormonal activity could be due to this change. Therefore, it is not yet clear, if the carboxyl terminal of the A chain and the B22–27 residues of the B chain are involved in what might be called an active site, or if these groups are needed to hold the molecule in a particular steric structure which is necessary for the expression of hormonal activity.

With the chemical synthesis of insulin, the "synthetic approach" to the relation of structure to function is now feasible and it is anticipated that it will be as illuminating for insulin as it has been for oxytocin and vaso-pressin.

VIII. Biosynthesis of Insulin

Strong evidence has accumulated in recent years in favor of the following scheme of protein biosynthesis. DNA directs the formation of complementary strands of messenger RNA which possess the necessary genetic information to determine the sequential arrangement of amino acids in proteins. The messenger RNA attaches itself to ribosomes to form a multiple ribosomal structure, the polysome. The amino acids in an activated state are transported by transfer RNA molecules to the polysomes where the genetic message is decoded during movement of messenger RNA relative to the ribosomes. Each ribosome constitutes a condensing site through which the messenger RNA passes with suc-cessive exposure of codons and at which a polypeptide chain is growing. The biosynthesis of insulin appears to follow the above general pattern (*82, 83, 85, 90, 131, 132, 140, 142*).

Biosynthetic studies on insulin with mammalian pancreases are handicapped because of the prescence of the enzymes of the exocrine pancreas and because the islets constitute only about 1% of the total organ. A more suitable system is obviously the intact islets which can be separated from the pancreas by the procedure of MOSKALEWSKI (*78, 95*). Tumors of the β-cells could also provide a useful system (*126, 127*) but their occurrence in man is infrequent. One of the most suitable systems is the islet tissue of certain teleost fishes (*42–44, 46, 82, 83*). In these fishes the islets form a separate organ, the Brockmann body, or principal islet (*7*).

Very little is known about the mechanisms involved in insulin bio-synthesis and its regulation.

Two fundamental possibilities have to be considered regarding the assembly of the insulin molecule: (a) The hormone is synthesized as an end-to-end protein precursor (proinsulin) followed by formation of the disulfide bonds and enzymatic cleavage at one or more points. (b) The cell synthesizes the A and B chains of the hormone, which then combine to form insulin. The high yields of insulin obtained by KATSOYANNIS and co-workers (*59, 68*) in the in vitro recombination of the A and B chains suggest that the necessary information for complementarity and co-valent linking of the chains is contained within their primary structure, and therefore, the synthesis of insulin in vivo by such a mechanism should not be considered unlikely. Evidence for this mechanism was presented by Humbel (*43, 44*) from studies with anglerfish islets. The anglerfish insulin contains three prolines. Assuming that this insulin is synthesized end-to-end, short-time incubation of the islets with labelled proline ought to produce a gradient of specific radioactivities in the prolines. The lowest specific activity would be associated with the proline residing in the part of the molecule where synthesis starts, and the highest specific activity would be associated with the proline found at the opposite end. Under the assumption that all three prolines are inserted in the growing protein chain by the same t-RNA species, the data obtained by Humbel are not compatible with a single-chain precursor of insulin. The search for a "proinsulin" in pancreatic extracts by Wang and Carpenter (*144*) was also unsuccessful. In contrast with the above results, evidence for the existence of a "proinsulin" has recently been presented by Steiner et al. (*126, 127*).

It appears that the final rejection of one of these two mechanisms must await more experimental data.

The complete elucidation of the mechanisms involved in insulin biosynthesis and its regulation might well be expected to illuminate the still unknown nature of diabetes mellitus.

References

1. Anfinsen, C. B. and E. Haber: Studies on the Reduction and Re-formation of Protein Disulfide Bonds. J. Biol. Chem. **236**, 1361 (1961).

1a. Abel, J. J.: Crystalline Insulin. Proc. Nat. Acad. Sci. (USA) **12**, 132 (1926).

2. Bailey, J. L.: The Preparation of Proteins and Peptides Containing S-Sulfonate Groups. Biochem. J. **67**, 21 P (1957).

3. Bailey, J. L. and R. D. Cole: Studies on the Reaction of Sulfite with Proteins. J. Biol. Chem. **234**, 1733 (1959).

4. Baldesten, A.: Separation of the Aminoethylated A and B Chains of Insulin. Acta Chem. Scand. **20**, 270 (1966).

5. Banting, F. G. and C. H. Best: Internal Secretion of Pancreas. J. Lab. Clin. Med. **7**, 251 (1922).

6. Banting, F. G., C. H. Best, J. B. Collip and J. J. MacLeod: The Preparation of Pancreatic Extracts Containing Insulin. Proc. Trans. Roy. Soc. Canada [3] **16**, 27 (1922).

7. Bergmann, W.: Innersekretorische Drüsen. I. Schilddrüse — Epithelkörperchen — Langerhanssche Inseln. In: W. v. Möllendorff (ed.), Handbuch der mikroskopischen Anatomie des Menschen, Bd. 6, 2. Teil, S. 263. Berlin: Springer-Verlag. 1939.

8. Bodanszky, M. and V. du Vigneaud: A Method of Synthesis of Long Peptide Chains Using a Synthesis of Oxytocin as an Example. J. Amer. Chem. Soc. **81**, 5688 (1959).

9. Bromer, W. W. and R. E. Chance: Preparation and Characterization of Desoctapeptide-insulin. Biochim. Biophys. Acta **133**, 219 (1967).

10. BROMER, W. W., S. K. SHEEHAN, A. W. BERNS and E. R. ARQUILLA: Preparation and Properties of Fluoresceinthiocarbamyl Insulins. Biochemistry **6**, 2378 (1967).

11. BROWN, H., F. SANGER and R. KITAI: The Structure of Pig and Sheep Insulins. Biochem. J. **60**, 556 (1955).

12. CANFIELD, R. E. and C. B. ANFINSEN: Concepts and Experimental Approaches in the Determination of the Primary Structure of Proteins. In: H. Neurath, The Proteins, Vol. I, p. 311. New York: Academic Press. 1963.

13. CARPENTER, F. H.: Relationship of Structure to Biological Activity of Insulin as Revealed by Degradative Studies. Amer. J. Med. **40**, 750 (1966).

14. CARPENTER, F. H. and A. CHRAMBACH: On the Amide Content of Insulin Fractions Isolation by Partition Column Chromatography and Countercurrent Distribution. J. Biol. Chem. **237**, 404 (1962).

15. CARPENTER, F. H. and S. L. HAYES: Electrophoresis on Cellulose Acetate of Insulin and Insulin Derivatives. Correlation with Behavior on Countercurrent Distribution and Partition Column Chromatography. Biochemistry **2**, 1272 (1963).

16. CARR, F. H., K. CULHANE, A. T. FULLER and S. W. F. UNDERHILL: A Reversible Inactivation of Insulin. Biochem. J. **23**, 1010 (1929).

17. CECIL, R. and U. E. LOENING: The Reaction of the Disulfide Bonds of Insulin with Sodium Sulfite. Biochem. J. **66**, 18P (1957).

18. CECIL, R. and J. R. McPHEE: A Kinetic Study of the Reactions on Some Disulfides with Sodium Sulfite. Biochem. J. **60**, 496 (1955).

19. CHEN, C. C., W. T. HUANG and C. I. NIU: Synthesis of the Peptide Fragment of the B-Chain of Insulin. VI. Synthesis of Derivatives of the N-Terminal Octapeptide. Sci. Sinica **13**, 1235 (1964).

20. CHRAMBACH, A. and F. H. CARPENTER: Partition Column Chromatography of Insulin. Production and Separation of Transformation Products. J. Biol. Chem. **235**, 3478 (1960).

21. CLARKE, H. T.: The Action of Sulfite upon Cystine. J. Biol. Chem. **97**, 235 (1932).

22. COLE, R. D.: On the Transformation of Insulin in Concentrated Solutions of Urea. J. Biol. Chem. **236**, 2670 (1961).

23. CRESTFIELD, A. M., S. MOORE and W. H. STEIN: The Preparation and Enzymatic Hydrolysis of Reduced and S-Carboxymethylated Proteins. J. Biol. Chem. **238**, 622 (1963).

24. CRESTFIELD, A. M., J. SKUPIN, S. MOORE and W. H. STEIN: Reduction of Disulfide Bonds in Proteins by Sodium Borohydride. Federat. Proc. (Amer. Soc. Exp. Biol.) **19**, 341 (1960).

25. DILLON, W. W. and R. G. ROMANS: Heterogeneity of Insulin. II. Chromatography of Insulin on Carboxymethyl Cellulose in Urea Containing Buffers. Canad. J. Biochem. **45**, 221 (1967).

26. DIXON, G. H. and A. C. WARDLAW: Regeneration of Insulin Activity from the Separated and Inactive A and B Chains. Nature **188**, 721 (1960).

27. DU, Y. C., R. Q. JIANG and C. L. TSOU: Conditions for Successful Resynthesis of Insulin from its Glycyl and Phenylalanyl Chains. Sci. Sinica **14**, 229 (1965).

28. DU, Y. C., Y. S. ZHANG, Z. X. LU and C. L. TSOU: Resynthesis of Insulin from its Glycyl and Phenylalanyl Chains. Sci. Sinica **10**, 84 (1961).

29. EPSTEIN, C. J. and C. B. ANFINSEN: The Use of Gel Filtration in the Isolation and Purification of Beef Insulin. Biochemistry **2**, 461 (1963).

30. EVANS, R. L. and H. A. SAROFF: A Physiologically Active Guanidinated Derivative of Insulin. J. Biol. Chem. **228**, 295 (1957).

31. Fraenkel-Conrat, H., J. I. Harris and A. L. Levy: Recent Developments in Techniques for Terminal and Sequence Studies in Peptides and Proteins. In: D. Glick (ed.), Methods of Biochemical Analysis, Vol. II, p. 359. New York: Interscience. 1955.

32. Fraenkel-Conrat, J. and H. Fraenkel-Conrat: The Essential Groups of Insulin. Biochim. Biophys. Acta. **5**, 89 (1950).

33. Glendening, M. B., D. M. Greenberg and H. Fraenkel-Conrat: Biologically Active Insulin Sulfate. J. Biol. Chem. **167**, 125 (1947).

34. Hama, H., K. Titani, S. Sakaki and K. Narita: The Amino Acid Sequence of Fin Whale Insulin. J. Biochem. (Tokyo) **56**, 285 (1964).

35. Harfenist, E. J. and L. C. Craig: Countercurrent Distribution of Insulin. J. Amer. Chem. Soc. **73**, 877 (1951).

36. — — Countercurrent Distribution Studies with Insulin. J. Amer. Chem. Soc. **74**, 3083 (1952).

37. — — The Molecular Weight of Insulin. J. Amer. Chem. Soc. **74**, 3087 (1952).

38. Harris, J. I. and C. H. Li: The Biological Activity of Enzymatic Digests of Insulin. J. Amer. Chem. Soc. **74**, 2945 (1952).

39. Harris, J. I., F. Sanger and M. A. Naughton: Species Differences in Insulin. Arch. Biochem. Biophys. **65**, 427 (1956).

40. Hofmann, K. and P. G. Katsoyannis: Synthesis and Function of Peptides of Biological Interest. In: H. Neurath (ed.), The Proteins, Vol. I, p. 53. New York: Academic Press. 1963.

41. Homans, J.: Degeneration of the Islands of Langerhans Associated with Experimental Diabetes in the Cat. J. Med. Res. **30**, 49 (1914).

42. Humbel, R. E.: Studies on Isolated Islets of Langerhans (Brockmann's Bodies) of Teleost Fishes. II. Evidence for Insulin Biosynthesis in vitro. Biochim. Biophys. Acta **74**, 96 (1963).

43. — Biosynthesis of the Two Chains of Insulin. Proc. Nat. Acad. Sci. (USA) **53**, 853 (1965).

44. — Biosynthesis of Insulin. Amer. J. Med. **40**, 672 (1966).

45. Humbel, R. E. and A. M. Crestfield: Isolation and Partial Structural Analysis of Insulin from the Separate Islet Tissue of *Lophius piscatorius*. Biochemistry **4**, 1044 (1965).

46. Humbel, R. E. and A. E. Renold: Studies on Isolated Islets of Langerhans (Brockmann's Bodies) of Teleost Fishes. I. Metabolic Activity in vitro. Biochim. Biophys. Acta **74**, 84 (1963).

47. Ishihara, Y., T. Saito, Y. Ito and M. Fujino: Structure of Sperm- and Sei-Whale Insulins and their Breakdown by Whale Pepsin. Nature **181**, 1468 (1958).

48. Katsoyannis, P. G.: Peptide Synthesis and Protein Structure. J. Polymer Sci. **49**, 51 (1961).

49. — Synthetic Studies on the A and B Chains of Insulin. Vox Sanguinis **9**, 227 (1964). Presented as the first Edwin J. Cohn Memorial Lecture of the 15th Annual Scientific Conference of Protein Foundation on November 25, 1963.

50. — The Synthesis of the A and B Chains of Insulin and their Combination to Generate Insulin Activity. Excerpta Med., Intern. Congr. Ser. **83**, 1216 (1964).

51. — The Chemical Synthesis of Human and Sheep Insulin. Amer. J. Med. **40**, 652 (1966).

52. — Synthesis of Insulin: Availability of A and B Chains Readily Leads to the Synthesis of this Protein. Science **154**, 1509 (1966).

53. — Synthetic Insulins. Recent Progr. Hormone Res. **23**, 505 (1967).

54. — Synthetic Insulins. 6th Congr. Intern. Diabetes Fed., Stockholm 1967. Excerpta Med., Intern. Congr. Ser. (in press).

55. KATSOYANNIS, P. G., K. FUKUDA and A. TOMETSKO: Insulin Peptides. VI. The Synthesis of a Partially Protected Nonapeptide Corresponding to the First Nine Amino Acid Residues of the A Chain of Insulin. J. Amer. Chem. Soc. 85, 1681 (1963).

56. KATSOYANNIS, P. G., K. FUKUDA, A. TOMETSKO, K. SUZUKI and M. TILAK: Insulin Peptides. X. The Synthesis of the B Chain of Insulin and its Combination with Natural or Synthetic A Chain to Generate Insulin Activity. J. Amer. Chem. Soc. 86, 930 (1964).

57. KATSOYANNIS, P. G., K. SUZUKI and A. TOMETSKO: Insulin Peptides. IV. The Synthesis of a Protected Decapeptide Containing the C-Terminal Sequence of the A Chain of Insulin. J. Amer. Chem. Soc. 85, 1139 (1963).

58. KATSOYANNIS, P. G. and M. TILAK: Insulin Peptides. VIII. A Synthetic Heptadecapeptide Derivative Corresponding to the C-Terminal Sequence of the B Chain of Insulin. J. Amer. Chem. Soc. 85, 4028 (1963).

59. KATSOYANNIS, P. G. and A. TOMETSKO: Insulin Synthesis by Recombination of A and B Chains: a Highly Efficient Method. Proc. Nat. Acad. Sci. (USA) 55, 1554 (1966).

60. KATSOYANNIS, P. G., A. TOMETSKO and K. FUKUDA: Insulin Peptides. IX. The Synthesis of the A Chain of Insulin and its Combination with Natural B Chain to Generate Insulin Activity. J. Amer. Chem. Soc. 85, 2863 (1963).

61. KATSOYANNIS, P. G., A. TOMETSKO, J. GINOS and M. TILAK: Insulin Peptides. XI. The Synthesis of the B Chain of Human Insulin and its Combination with the Natural A Chain of Bovine Insulin to Generate Insulin Activity. J. Amer. Chem. Soc. 88, 164 (1966).

62. KATSOYANNIS, P. G., A. TOMETSKO and C. ZALUT: Insulin Peptides. XII. Human Insulin Generation by Combination of Synthetic A and B Chains. J. Amer. Chem. Soc. 88, 166 (1966).

63. — — — Insulin Peptides. XIII. The Synthesis of the Dodecapeptide Derivative Containing the C-Terminal Sequence of the A Chain of Sheep Insulin. J. Amer. Chem. Soc. 88, 5618 (1966).

64. — — — Insulin Peptides. XIV. Synthetic Peptide Derivatives Related to the N-Terminal of the A Chain of Sheep Insulin (Positions 1–9). J. Amer. Chem. Soc. 88, 5622 (1966).

65. — — — Insulin Peptides. XVII. The Synthesis of the A Chain of Human (Porcine) Insulin and its Isolation as the S-Sulfonated Derivative. J. Amer. Chem. Soc. 89, 4505 (1967).

66. KATSOYANNIS, P. G., A. TOMETSKO, C. ZALUT and K. FUKUDA: Insulin Peptides. XV. The Synthesis of the A Chain of Sheep Insulin and its Combination with Synthetic or Natural B Chain to Produce Insulin. J. Amer. Chem. Soc. 88, 5625 (1966).

67. KATSOYANNIS, P. G., A. TOMETSKO, C. ZALUT, S. JOHNSON and A. C. TRAKATELLIS: Studies on the Synthesis of Insulin from Natural and Synthetic A and B Chains. I. Splitting of Insulin and Isolation of the S-Sulfonated Derivatives of the A and B Chains. Biochemistry 6, 2635 (1967).

68. KATSOYANNIS, P. G., A. C. TRAKATELLIS, S. JOHNSON, C. ZALUT and G. (P.) SCHWARTZ: Studies on the Synthesis of Insulin from Natural and Synthetic A and B Chains. II. Isolation of Insulin from Recombination Mixtures of Natural A and B Chains. Biochemistry 6, 2642 (1967).

69. KATSOYANNIS, P. G., A. C. TRAKATELLIS, C. ZALUT, S. JOHNSON, A. TOMETSKO, G. (P.) SCHWARTZ and J. GINOS: Studies on the Synthesis of Insulin from Natural and Synthetic A and B Chains. III. Synthetic Insulins. Biochemistry 6, 2656 (1967).

70. Katsoyannis, P. G., C. Zalut and A. M. Tometsko: Insulin Peptides. XVI. The Synthesis of a Nonapeptide and a Dodecapeptide Derivative Related to the A Chain of Human Insulin (Positions 1–9 and 10–21). J. Amer. Chem. Soc. 89, 4502 (1967).

71. Kauzmann, W.: Relative Probabilities of Isomers in Cystine-containing Randomly Coiled Polypeptides. In: R. Benesch et al. (ed.), Sulfur in Proteins, p. 93. Proc. Sympos. Falmouth, Mass., 1958. New York: Academic Press. 1959.

72. Kimmel, J. R. and A. J. Parcells: S-Carboxymethyl Papain. Federat. Proc. (Amer. Soc. Exp. Biol.) 19, 341 (1960).

73. Kotaki, A.: Studies on Insulin. III. On the Structure of the Alanyl Chain of Bonito Insulin. J. Biochem. (Tokyo) 51, 301 (1962).

74. — Studies on Insulin. V. On the Structure of the Glycyl Chain of Bonito Insulin II. J. Biochem. (Tokyo) 53, 61 (1963).

75. Kung, Y. T., Y. C. Du, W. T. Huang, C. C. Chen, L. T. Ke, S. C. Hu, R. Q. Jiang, S. Q. Chu, C. I. Niu, J. Z. Hsu, W. C. Chang, L. L. Chen, H. S. Li, Y. Wang, T. P. Loh, A. H. Chi, C. H. Li, P. T. Shi, Y. H. Yieh, K. L. Tang and H. Y. Hsing: Total Synthesis of Crystalline Insulin. Sci. Sinica 15, 544 (1966).

76. Kung, Y. T., L. T. Ke and C. I. Niu: Synthesis of the Peptide Fragments of the B Chain of Insulin. X. Synthesis of a Derivative of the C-Terminal Docosapeptide of the B Chain of Insulin. Sci. Sinica 15, 221 (1966).

77. Kung, Y. T., L. T. Ke, C. I. Niu and S. C. Hu: Synthesis of the Peptide Fragments of the B Chain of Insulin. VII. Synthesis of a Derivative of the C-Terminal Tetradecapeptide of the B Chain of Insulin. Sci. Sinica 13, 1245 (1964).

78. Lacy, P. E. and M. Kostianovsky: Method for the Isolation of Intact Islets of Langerhans from the Rat Pancreas. Diabetes 16, 35 (1967).

79. Laguesse, E.: Sur la formation des îlots de Langerhans dans le pancréas. C. R. séances soc. biol. 45, 819 (1893).

80. Lane, M. A.: The Cytological Characters of the Areas of Langerhans. Amer. J. Anat. 7, 409 (1907).

81. Langdon, R. G.: Biological Activities of the Phenylalanyl Chain of Insulin. J. Biol. Chem. 235, pc 15 (1960).

82. Lazarow, A.: Functional Characterization and Metabolic Pathways of the Pancreatic Islet Tissue. Recent Progr. Hormone Res. 19, 489 (1963).

83. Lazarow, A., G. E. Bauer and A. W. Lindall: In: S. E. Brolin (ed.), The Structure and Metabolism of the Pancreatic Islets: Protein Synthesis in Islet Tissue, p. 203. London: Pergamon Press. 1964.

84. Li, C. H.: Preparation and Properties of Dinitrophenyl-$NH_{(\varepsilon)}$-Insulin. Nature 178, 1402 (1956).

85. Light, A. and M. V. Simpson: Studies on the Biosynthesis of Insulin. I. The Paper Chromatographic Isolation of C^{14}-Labeled Insulin from Calf Pancreas Slices. Biochim. Biophys. Acta 20, 251 (1956).

86. Lindley, H.: The Reduction of the Disulfide Bonds of Insulin. J. Amer. Chem. Soc. 77, 4927 (1955).

87. — A New Synthetic Substrate for Trypsin and its Application to the Determination of the Amino Acid Sequence of Proteins. Nature 178, 647 (1956).

88. Lugg, J. W. H.: The Application of Phospho-18-tungstic Acid (Folin's Reagent) to the Colorimetric Determination of Cystine, Cysteine and Related Substances. I. The Reduction of Phospho-18-tungstic Acid by Various Substances. Biochem. J. 26, 2144 (1932).

89. McPhee, J. R.: Further Studies on the Reactions of Disulfides with Sodium Sulfite. Biochem. J. 64, 22 (1956).

90. MALLORY, A., G. H. SMITH and K. W. TAYLOR: The Incorporation of Tritium-Labelled Amino Acids into Insulins in Rat Pancreas in vitro. Biochem. J. **91**, 484 (1964).

91. MEIENHOFER, J.: Synthese der Insulinsequenz B 9—20. Z. Naturforsch. **19 b**, 114 (1964).

92. MEIENHOFER, J. und E. SCHNABEL: Eine Synthese der B-Kette des Insulins. Z. Naturforsch. **20 b**, 661 (1964).

93. MEIENHOFER, J., E. SCHNABEL, H. BREMER, O. BRINKHOFF, R. ZABEL, W. SROKA, H. KLOSTERMEYER, D. BRANDENBURG, T. OKUDA und H. ZAHN: Synthese der Insulinketten und ihre Kombination zu insulinaktiven Präparaten. Z. Naturforsch. **18 b**, 1120 (1963).

94. MERING, J. v. und O. MINKOWSKI: Diabetes mellitus nach Pankreasextirpation. Arch. exp. Pathol. Pharmakol. **26**, 371 (1889).

95. MOSKALEWSKI, S.: Isolation and Culture of the Islets of Langerhans of the Guinea Pig. Gen. Comp. Endocrin. **5**, 342 (1965).

96. MYCEK, M. J., D. D. CLARKE, A. NEIDLE and H. WAELSCH: Amine Incorporation into Insulin as Catalyzed by Transglutaminase. Arch. Biochem. Biophys. **84**, 528 (1959).

97. NICOL, D. S. H. W. and L. F. SMITH: Amino Acid Sequence of Human Insulin. Nature **187**, 483 (1960).

98. NIU, C. I., Y. T. KUNG, W. T. HUANG, L. T. KE, C. C. CHEN, Y. C. CHEN, Y. C. DU, R. Q. JIANG, C. L. TSOU, S. C. HU, S. Q. CHU and K. Z. WANG: Successful Synthesis of Crystalline Insulin from its Natural A Chain and the Synthetic B Chain. Sci. Sinica **14**, 1386 (1965).

99. — — — — — — — — — — — Synthesis of Crystalline Insulin from its Natural A Chain and the Synthetic B Chain. Sci. Sinica **15**, 231 (1966).

100. OPIE, E. L.: On the Relation of Chronic Interstitial Pancreatitis to the Islands of Langerhans and to Diabetes mellitus. Exper. Med. **5**, 397 (1901).

101. — Diabetes mellitus Associated with Hyaline Degeneration of the Islands of Langerhans of the Pancreas. Bull. Johns Hopkins Hosp. **12**, 263 (1901).

102. PECHÈRE, J. F., G. H. DIXON, R. H. MAYBURY and H. NEURATH: Cleavage of Disulfide Bonds in Trypsinogen and α-Chymotrypsinogen. J. Biol. Chem. **233**, 1364 (1958).

103. PIERCE, J. G.: Separation of the Glycyl from the Phenylalanyl Chain of Oxidized Insulin by Countercurrent Distribution. J. Amer. Chem. Soc. **77**, 184 (1955).

104. PRUITT, K. M., J. CANTRELL and B. R. BOSHELL: The Effect of Insulin Derivatives on the Insulin Response of Assays in vitro. Biochim. Biophys. Acta **115**, 329 (1966).

105. PRUITT, K. M., B. S. ROBISON and J. H. GIBBS: Study of Biological Activity Regenerated by the Oxidation of Fully Reduced Insulin. Biopolymers **4**, 351 (1966).

106. RANDALL, S. S.: The Small-scale Preparation of Crystalline Insulin. Biochim. Biophys. Acta **90**, 472 (1964).

107. RYLE, A. P., F. SANGER, L. F. SMITH and R. KITAI: The Disulfide Bonds of Insulin. Biochem. J. **60**, 541 (1955).

108. SANGER, F.: The Free Amino Groups of Insulin. Biochem. J. **39**, 507 (1945).

109. — Oxidation of Insulin by Performic Acid. Nature **160**, 295 (1947).

110. — Fractionation of Oxidized Insulin. Biochem. J. **44**, 126 (1949).

111. — The Terminal Peptides of Insulin. Biochem. J. **45**, 563 (1949).

112. — Chemistry of Insulin. Brit. Med. Bull. **16**, 183 (1960).

113. SANGER, F. and E. O. P. THOMPSON: The Amino Acid Sequence of the Glycyl Chain of Insulin. 1. The Identification of Lower Peptides from Partial Hydrolysates. Biochem. J. **53**, 353 (1953).

114. — — The Amino Acid Sequence of the Glycyl Chain of Insulin. 2. The Investigation of Peptides from Enzymic Hydrolysates. Biochem. J. **53**, 366 (1953).

115. SANGER, F. and H. TUPPY: The Amino Acid Sequence in the Phenylalanyl Chain of Insulin. 1. The Idendification of Lower Peptides from Partial Hydrolysates. Biochem. J. **49**, 463 (1951).

116. — — The Amino Acid Sequence in the Phenylalanyl Chain of Insulin. 2. The Investigation of Peptides from Enzymic Hydrolysates. Biochem. J. **49**, 481 (1951).

117. SCHNABEL, E.: Neusynthese der Insulinsequenz B 21—30. Z. Naturforsch. **19 b**, 120 (1964).

118. SCHRÖDER, E. and K. LÜBKE: The Peptides, Vol. I. New York: Academic Press. 1966.

119. SELA, M., F. H. WHITE, Jr. and C. B. ANFINSEN: The Reductive Cleavage of Disulfide Bonds and its Application to Problems of Protein Structure. Biochim. Biophys. Acta **31**, 417 (1959).

120. SHEEHAN, J. C. and G. P. HESS: A New Method of Forming Peptide Bonds. J. Amer. Chem. Soc. **77**, 1067 (1955).

121. SLOBIN, L. I. and F. H. CARPENTER: Action of Carboxypeptidase-A on Bovine Insulin. Preparation of Desalanine-Desasparagine-Insulin. Biochemistry **2**, 16 (1963).

122. — — The Labile Amide in Insulin. Preparation of Desalanine-Desamido-Insulin. Biochemistry **2**, 22 (1963).

123. — — Kinetic Studies on the Action of Carboxypeptidase-A on Bovine Insulin and Related Model Peptides. Biochemistry **5**, 499 (1966).

124. SLUYTERMAN, L. A. Ae. and J. M. KWESTROO-VAN DEN BOSCH: Sulphation of Insulin and Electrophoresis of the Products Obtained. Biochim. Biophys. Acta **38**, 102 (1960).

125. SMITH, L. F.: Species Variation in the Amino Acid Sequence of Insulin. Amer. J. Med. **40**, 662 (1966).

126. STEINER, D. F., D. CUNNINGHAM, L. SPIGELMAN and B. ATEN: Insulin Biosynthesis: Evidence for a Precursor. Science **157**, 697 (1967).

127. STEINER, D. F. and P. E. OYER: The Biosynthesis of Insulin and a Probable Precursor of Insulin by a Human Islet Cell Adenoma. Proc. Nat. Acad. Sci. (USA) **57**, 473 (1967).

128. STRICKS, W. and I. M. KOLTHOFF: Equilibrium Constants of the Reactions of Sulfite with Cystine and with Dithiodiglycolic Acid. J. Amer. Chem. Soc. **73**, 4569 (1951).

129. STRICKS, W., I. M. KOLTHOFF and R. C. KAPOOR: Equilibrium Constants of the Reaction Between Sulfite and Oxidized Glutathione. J. Amer. Chem. Soc. **77**, 2057 (1955).

130. SWAN, J. M.: Thiols, Disulphides and Thiosulphates: Some New Reactions and Possibilities in Peptide and Protein Chemistry. Nature **180**, 643 (1957).

131. TAYLOR, K. W. and D. G. PARRY: The Incorporation of Tritium-Labelled Amino Acids into Insulin in Ox Pancreas in vitro. Biochem. J. **89**, 94 P (1963).

132. TAYLOR, K. W., D. G. PARRY and G. H. SMITH: Biosynthetic Labelling of Mammalian Insulins in vitro. Nature **203**, 1144 (1964).

133. THOMPSON, E. O. P.: The Selective Degradation of Proteins. Adv. Organ. Chem. **1**, 149 (1960).

134. THOMPSON, E. O. P. and I. J. O'DONNELL: Quantitative Reduction of Disulfide Bonds in Proteins using High Concentrations of Mercaptoethanol. Biochim. Biophys. Acta **53**, 447 (1961).

135. THOMPSON, E. O. P. and A. R. THOMPSON: Paper Chromatography in the Study of the Structure of Peptides and Proteins. Fortschr. Chem. organ. Naturstoffe **12**, 270 (1955).

136. TIETZE, F., G. E. MORTIMORE and N. R. LOMAX: Preparation and Properties of Fluorescent Insulin Derivatives. Biochim. Biophys. Acta **59**, 336 (1962).

137. TSOU, C. L., Y. C. DU and G. J. XÜ: The Reduction of Insulin and its Benzyl Derivatives by Sodium in Liquid Ammonia and the Regeneration of Activity from the Reduced Products. Sci. Sinica **10**, 332 (1961).

138. VARANDANI, P. T.: A Convenient Preparation of Reduced and S-Sulfonated A and B Chains of Insulin. Biochim. Biophys. Acta **127**, 246 (1966).

139. VAUGHAN, J. R., Jr.: Acylalkyl Carbonates as Acylating Agents for the Synthesis of Peptides. J. Amer. Chem. Soc. **73**, 3547 (1951).

140. VAUGHAN, M. and C. B. ANFINSEN: Non-uniform Labeling of Insulin and Ribonuclease Synthesized in vitro. J. Biol. Chem. **211**, 367 (1954).

141. VIGNEAUD, V. DU, A. FITCH, E. PEKAREK and W. W. LOCKWOOD: The Inactivation of Crystalline Insulin by Cysteine and Glutathione. J. Biol. Chem. **94**, 233 (1931).

142. VOELKER, I., E. SCHÜMANN und C. V. HOLT: Biosynthese des Insulins. I. Mitt. Darstellung von biosynthetisch markiertem ^{35}S-Insulin. Biochem. Z. **335**, 382 (1962).

143. VOLFIN, P., A. M. CHAMBAUT, D. EBOUÉ-BONIS, H. CLAUSER, O. BRINKHOFF, H. BREMER, J. MEIENHOFER and H. ZAHN: Biological Activity of Natural and Synthetic Insulin A Chain Preparations on the Isolated Rat Diaphragm. Nature **203**, 408 (1964).

144. WANG, S. S. and F. H. CARPENTER: A Compositional Assay for Insulin Applied to a Search for Proinsulin. J. Biol. Chem. **240**, 1619 (1965).

145. WANG, Y., J. Z. HSU, W. C. CHANG, L. L. CHENG, C. Y. HSING, A. H. CHI, T. P. LOH, C. H. LI, P. T. SHI and Y. H. YIEH: A Preliminary Report on the Synthesis of the A Chain of Bovine Insulin. Sci. Sinica **13**, 2030 (1964).

146. WANG, Y., J. Z. HSU, W. C. CHANG, L. L. CHENG, H. S. LI, C. Y. HSING, P. T. SHI, T. P. LOH, A. H. CHI, C. H. LI, Y. H. YIEH and K. L. TANG: Partial Synthesis of Crystalline Bovine Insulin from Synthetic A Chain and Natural B Chain. Sci. Sinica **14**, 1887 (1965).

147. WEIL, L., T. S. SEIBLES and T. T. HERSKOVITS: Photooxidation of Bovine Insulin Sensitized by Methylene Blue. Arch. Biochem. Biophys. **111**, 308 (1965).

148. WEITZEL, G., W. SCHAEG, G. BODEN und B. WILLMS: Einfluß der Photooxidation auf Histidingehalt und Aktivität von Insulin. Liebigs Ann. Chem. **689**, 248 (1965).

149. WILSON, S., M. A. APRILE and L. SASAKI: Passive Cutaneous Anaphylaxis Induced in Guinea Pigs by Insulins and their Component Chains. Canad. J. Biochem. **44**, 989 (1966).

150. WILSON, S., G. H. DIXON and A. C. WARDLAW: Resynthesis of Cod Insulin from its Polypeptide Chains and the Preparation of Cod-Ox "Hybrid" Insulins. Biochim. Biophys. Acta **62**, 483 (1962).

151. WINTERSTEINER, O.: The Action of Sulfhydryl Compounds on Insulin. J. Biol. Chem. **102**, 473 (1933).

152. YOUNG, J. D. and F. H. CARPENTER: Isolation and Characterization of Products Formed by the Action of Trypsin on Insulin. J. Biol. Chem. **236**, 743 (1961).

153. Zabel, R. und H. Zahn: Synthese eines Pentapeptid-Derivates mit der C-terminalen Sequenz A 17—21 der Insulin-A-Kette. Z. Naturforsch. **20 b**, 650 (1965).

154. Zahn, H.: Discussion. In: B. S. Leibel and G. A. Wrenshall, On the Nature and Treatment of Diabetes, p. 87. New York: Excerpta Medica Found. 1965.

155. Zahn, H., H. Bremer, W. Sroka und J. Meienhofer: Synthese eines Nonapeptid-Derivates mit der N-terminalen Sequenz A 1—9 der A-Kette des Schafinsulins. Z. Naturforsch. **20 b**, 646 (1965).

156. Zahn, H., H. Bremer und R. Zabel: Synthese einer teilgeschützten A-Kette des Schafinsulins. Z. Naturforsch. **20 b**, 653 (1965).

157. Zahn, H., O. Brinkhoff, J. Meienhofer, E. F. Pfeiffer, H. Ditschuneit und Ch. Gloxhuber: Kombination synthetischer Insulinketten zu biologisch aktiven Präparaten. Z. Naturforsch. **20 b**, 666 (1965).

158. Zahn, H., W. Danho und B. Gutte: Eine neue Synthese der A-Kette des Schafinsulins und deren Vereinigung mit natürlicher B-Kette zu kristallinem vollaktivem Insulin. Z. Naturforsch. **21 b**, 763 (1966).

159. Zahn, H., J. Meienhofer und H. Klostermeyer: Eine Synthese der Insulinsequenz B 1—8 mit am Imidazol-Stickstoff ungeschütztem Histidin. Z. Naturforsch. **19 b**, 110 (1964).

(Received, November 22, 1967)

Makrotetrolide

Von **W. Keller-Schierlein** und **H. Gerlach**, Zürich

Mit 5 Abbildungen

Inhaltsübersicht

I. Vorkommen und Isolierung

Die Makrotetrolide Nonactin und seine Homologen bilden eine Gruppe von cyclischen Estern hoher biologischer Wirksamkeit, die bisher ausschließlich in Kulturen von Actinomycetenstämmen nachgewiesen wurden. Innerhalb der Actinomyceten, insbesondere der Gattung Streptomyces, scheinen diese Verbindungen ziemlich verbreitet zu sein. Unsere eigenen Untersuchungen haben zur Auffindung von über 20 Makrotetrolide produzierenden Stämmen geführt und in verschiedenen anderen Laboratorien sind Stoffwechselprodukte isoliert worden, die sich als Nonactin oder Gemische von Nonactin mit seinen Homologen erwiesen. So konnten die Antibiotica Werramycin (*30*), SQ 15 859 (*7*), N-329 A (*22*), Fluorin (*5*) und Lustericin (*28*) durch direkten Vergleich mit Nonactin oder Makrotetrolidgemischen identifiziert werden. Von anderen in der Literatur beschriebenen Verbindungen aus Actinomyceten, die für einen Vergleich nicht erhältlich waren, wie etwa dem Longisporin (*18*), kann eine Identität oder nahe Verwandtschaft mit Nonactin aus den Eigenschaften angenommen werden.

Die Makrotetrolide fallen oft in beträchtlichen Mengen an. Es wurden Stämme gefunden, die über 5 g Makrotetrolidgemisch pro Liter Kultur produzieren.

Die Makrotetrolide sind lipophile neutrale Verbindungen und finden sich nach der Filtration der Kulturbrühen im Filterrückstand angereichert. Sie lassen sich daraus leicht mit Aceton extrahieren. Aus den Filtraten, die geringere Mengen dieser Antibiotica enthalten, können sie durch Extraktion mit Äthylacetat oder Methylenchlorid gewonnen werden. Das Nonactin zeichnet sich durch große Kristallisationsfreudigkeit aus und kristallisiert oft schon aus den Rohextrakten in langen farblosen Nadeln aus (4). Schwierig ist dagegen die Reindarstellung der Homologen des Nonactins, die sich in den Mutterlaugen vorfinden. Beck und Mitarb. (1) verwendeten dafür die Säulenchromatographie an der 300fachen Menge Kieselgel, eine Methode, die sich praktisch nur für die Trennung kleinerer Substanzmengen eignet. Für die Isolierung höherer Homologer (Makrotetrolide B, C, D und G) benützten Keller-Schierlein und Mitarb. (9, 16) eine mehrfache Craig-Verteilung zwischen Petroläther und 85%igem wäßrigem Methanol, in Kombination mit der Chromatographie an Kieselgel.

Tabelle 1. Charakteristische Eigenschaften der Makrotetrolide

Makrotetrolid	Schmelzpunkt	Bruttoformel	$[\alpha]_D$ (Chlf.)	Rf*	M+**
Nonactin	148°	$C_{40}H_{64}O_{12}$	0°	0,62	736
Monactin	63—64°	$C_{41}H_{66}O_{12}$	+ 2°	0,48	750
Dinactin	66—67°	$C_{42}H_{68}O_{12}$	+ 2,5°	0,32	764
Trinactin	67—68°	$C_{43}H_{70}O_{12}$	+ 1,5°	0,15	778
Substanz G	Öl	$C_{43}H_{70}O_{12}$	—	0,29	778
Substanz D	63—65°	$C_{44}H_{72}O_{12}$	—	0,11	792
Substanz C	70—71°	$C_{45}H_{74}O_{12}$	—	0,13	806
Substanz B	56—58°	$C_{46}H_{76}O_{12}$	—	0,10	820
Peliomycin.......	160—164°	$C_{46}H_{76}O_{14}$	— 74,2°	—	—

* Dünnschichtchromatographie auf Kieselgel G (Merck); Fließmittel: Chloroform-Äthylacetat 1 : 2 (1, 9). Für die Auftrennung der höheren Homologen (B—D) eignet sich besser Äthylacetat allein als Fließmittel. Die Makrotetrolide geben intensiv fluoreszierende Flecke nach Besprühen mit konz. Schwefelsäure und Erhitzen auf 140°.

** Massenzahl des Molekülions im Massenspektrum (16).

Mit Hilfe dieser Methoden sind bisher 8 Makrotetrolide isoliert worden, von denen 7 in kristalliner Form erhalten wurden. Ihre Eigenschaften sind in der *Tabelle 1* zusammengestellt. Nach einer Privatmitteilung von D. Perlman ist ferner das Antibioticum Peliomycin (27) in die Verwandtschaft der Makrotetrolide zu zählen.

Literaturverzeichnis: SS. 187—189

Für die Identifizierung der einzelnen Makrotetrolide ist zunächst hauptsächlich auf die Dünnschichtchromatographie auf Kieselgel abgestellt worden (vgl. Tabelle 1). Die Trennung ist überraschend gut, sofern minimale Mengen an Substanz aufgetragen werden. Mit 1 μg und mehr pro Substanzfleck tritt dagegen keine Trennung mehr ein. Die Methode ist daher für die Einheitlichkeitsprüfung der Präparate nur sehr beschränkt anwendbar. Die IR-Absorptionsspektren eignen sich gut für die Charakterisierung der Gruppe als Ganzes; dagegen sind die Unterschiede zwischen den einzelnen Homologen so geringfügig, daß eine sichere Unterscheidung nicht möglich ist. Selbst die NMR-Spektren unterscheiden sich auf den ersten Blick kaum voneinander (1). Allerdings nimmt die Komplexität der CH_3-C-Region bei den Homologen etwas zu, und die relativen Intensitäten der einzelnen Signalgruppen verändern sich in Abhängigkeit von der Konstitution (vgl. Kap. III, S. 165). Für die Analyse von Substanzgemischen ist aber auch die NMR-Spektroskopie von geringem Wert.

Als zuverlässigste Methode für die Identifizierung der Komponenten und für die Reinheitsbestimmung hat sich die Massenspektrometrie bewährt. Da den Massenspektren zudem eine wesentliche Bedeutung für die Strukturaufklärung der höheren Homologen zukommt, wird ihnen ein separates Kapitel gewidmet (S. 177).

Hervorgehoben sei an dieser Stelle noch die optische Inaktivität des Nonactins und die sehr geringen spezifischen Drehungen seiner Homologen. Diese Eigenschaft ist für die Beurteilung der Stereochemie (Kap. IV, S. 169) von Bedeutung.

II. Die Konstitution des Nonactins

Die einfachste Verbindung der Reihe, das Nonactin (4) bildet farblose Kristallnadeln vom Schmelzp. 148°*. Im UV-Absorptionsspektrum ist oberhalb 210 mμ kein ausgeprägtes Maximum zu erkennen. Das IR-Spektrum zeigt ein scharfes Maximum bei 1726 cm^{-1}, das für Estercarbonyle charakteristisch ist. Hydroxylbanden sind nicht vorhanden, und nach Zerewitinow lassen sich keine aktiven Wasserstoffe nachweisen. Nonactin reagiert nicht mit 2,4-Dinitrophenylhydrazin.

Aus den Analysen ergibt sich die Formel $(C_{10}H_{16}O_3)_n$. Nachdem die zuerst ausgeführten ebullioskopischen Molekulargewichtsbestimmungen auf einen Wert von n = 3 hatten schließen lassen, konnte die Bruttoformel später auf Grund röntgenologischer Daten (6) auf $C_{40}H_{64}O_{12}$ (n = 4) korrigiert werden. Die exakte Molekulargewichtsbestimmung durch Massenspektrometrie (16) bestätigte diese Formel ($M^+ = 736$).

Die alkalische Hydrolyse von Nonactin lieferte nahezu 4 Mole einer flüssigen Carbonsäure $C_{10}H_{18}O_4$. Da diese Säure (Nonactinsäure) in alkalischer Lösung mit der 2-epi-Nonactinsäure im Gleichgewicht steht, ist die rohe Nonactinsäure ein Gemisch zweier diastereomerer Verbindungen. Die Trennung der beiden Methylester gelang durch Chromato-

---

* Geringe, durch Dünnschichtchromatographie kaum nachweisbare Beimengungen von Homologen können den Schmelzpunkt erheblich erniedrigen.

graphie an Kieselgel. Bei der alkalischen Verseifung der beiden sterisch einheitlichen Methylester entstand wieder das gleiche Gemisch der beiden epimeren Säuren.

Stereochemisch einheitlich verlief dagegen die reduktive Spaltung von Nonactin mit Lithiumaluminiumhydrid zum Diol, $C_{10}H_{20}O_3$, das einheitliche, gut kristallisierende Derivate lieferte. Durch Reduktion des reinen Nonactinsäuremethylesters zum C_{10}-Diol bzw. von 2-epi-Nonactinsäuremethylester zu einem isomeren Diol, $C_{10}H_{20}O_3$, konnte die Zuordnung der beiden Ester zur genuinen bzw. zur epi-Reihe vorgenommen werden. Das Problem der Konstitutionsaufklärung des Nonactins, $C_{40}H_{64}O_{12}$, war damit zunächst auf die Konstitutionsaufklärung eines C_{10}-Bausteins zurückgeführt.

Die Formel (5, S. 166) der Nonactinsäure beruht im wesentlichen auf folgenden Umwandlungs- und Abbaureaktionen (6): Die Nonactinsäure liefert bei der C-Methylbestimmung nach Kuhn und Roth annähernd 2 Mole Essigsäure. Als weiteres Produkt dieser Oxydation wurde Bernsteinsäure nachgewiesen. Die Nonactinsäure besitzt ein sekundäres Hydroxyl, denn ihr Methylester liefert bei der milden Oxydation mit Chrom(VI)-oxid den entsprechenden Ketosäure-methylester (11), der gemäß seinem NMR-Spektrum und der positiven Jodoformreaktion ein Methylketon ist. Bei der alkalischen Hydrolyse dieses Esters trat eine Fragmentierung ein: es ließen sich zirka 0,8 Mol Aceton als 2,4-Dinitrophenylhydrazon fassen. Nach diesen ersten Abbauversuchen mußte die Nonactinsäure folgende Atomgruppierungen enthalten: CH_3—CHOH— CH_2—, (C)—CH_2—CH_2—(C), CH_3—(C) und COOH. Das vierte Sauerstoffatom, das sich durch chemische Reaktionen und spektroskopische Befunde weder als Hydroxyl- noch als Carbonylsauerstoff zu erkennen gab, mußte als Äthersauerstoff vorliegen.

Das Kohlenstoffgerüst der Nonactinsäure konnte aus den Produkten einer Ätherspaltung mit 64%iger Bromwasserstoffsäure abgeleitet werden. Diese wurden mit Wasserstoff in Gegenwart von Palladiumkohle reduktiv entbromiert und die erhaltenen Säuren mit Diazomethan verestert. Durch Chromatographie ließen sich zwei Hauptprodukte in reiner Form isolieren. Das eine, $C_{11}H_{20}O_3$, erwies sich als 8-Desoxynonactinsäuremethylester (13, S. 166), der später auch auf übersichtlichere Weise bereitet wurde*. Wichtiger für die Konstitutionsaufklärung war der zweite Methylester, $C_{11}H_{22}O_3$ (14), ein aliphatischer Hydroxycarbonsäuremethylester. Die Hydroxylfunktion wurde über das p-Toluolsulfonat und das Bromid und anschließende reduktive Entbromierung entfernt. Es entstand dabei die α-Methylpelargonsäure (15), die als kristallines Anilid mit einem synthetischen Präparat verglichen wurde.

---

* Siehe auch den Nachtrag, S. 186.

Literaturverzeichnis: SS. 187—189

Durch Oxydation des Hydroxycarbonsäure-methylesters (**14**) mit Chrom(VI)-oxid in Aceton entstand der entsprechende Ketocarbonsäure-methylester (**16**), dessen Verseifung und Decarboxylierung das Nonanon-(3) (**17**) lieferte. Dieses wurde durch Vergleich mit einem synthetischen Präparat identifiziert. Dadurch ist die Lage des Hydroxyls im Methylester (**14**) festgelegt.

Diese Abbauresultate, zusammen mit den oben abgeleiteten Atomgruppierungen, erlauben den eindeutigen Schluß auf Formel (**5**, S. 166) für die Nonactinsäure. Die epi-Nonactinsäure muß sich von ihr durch verschiedenen räumlichen Bau am C-2 unterscheiden, da die beiden epimeren Methylester bei der milden Oxydation zwei verschiedene Ketoester der Konstitution (**11**) liefern.

Das Nonactin setzt sich aus vier Nonactinsäureeinheiten, unter Abzug von 4 Molen Wasser, derart zusammen, daß keine freien Carboxyl- und Hydroxylfunktionen übrigbleiben. Dies ist nur möglich durch Bildung eines cyclischen Esters mit vier gleichartigen Esterbindungen gemäß Formel (**1**).

Die Formel (**1**) des Nonactins ist im besten Einklang mit dem NMR-Spektrum (*1*).

Dieses zeigt 5 Gruppen von Signalen: 2 Dublette zu je 12 Protonen bei δ 1,02 und 1,19 ppm entsprechen den 8 CH_3—(CH)-Gruppen, von denen je 4 gleichartig sind. Ein stark strukturierter Signalhaufen zwischen δ 1,3 und 2,1 ppm entspricht den insgesamt 12 Methylengruppen. Die 4 Protonen in α-Stellung zu den Estercarbonylen geben ein Quintett mit Schwerpunkt bei δ 2,4 ppm. Eine Signalgruppe bei δ 3,8 ppm ist den 8 Protonen neben den Äthersauerstoffen zuzuordnen. Schließlich findet sich ein Quartett bei δ 4,9 ppm für die 4 Wasserstoffatome neben Estersauerstoffen. Die Integrale der einzelnen Signalgruppen entsprechen der Anzahl der jeweils zugeordneten Protonen bestens.

Die gute Übereinstimmung des NMR-Spektrums mit der auf chemischem Weg abgeleiteten Konstitutionsformel (**1**) des Nonactins darf als Beweis dafür genommen werden, daß im Verlauf der oben erwähnten Abbaureaktionen keine Umlagerung stattgefunden hat.

III. Die Konstitution der Nonactinhomologen

Während das Nonactin bei der reduktiven Spaltung mit Lithiumaluminiumhydrid 4 Mole eines einheitlichen Diols $C_{10}H_{20}O_3$ (**8**) liefert, erhält man aus den Makrotetroliden Monactin, Dinactin und Trinactin, daneben ein zweites Spaltprodukt, das bei der Dünnschichtchromatographie einen höheren R_f-Wert zeigt und dessen Analysen auf die Zusammensetzung $C_{11}H_{22}O_3$ stimmen (*1*). Die höheren Homologen G, D, C und B geben unter den gleichen Bedingungen noch ein drittes Diol, $C_{12}H_{24}O_3$ (*9*).

Entsprechend der früher beobachteten leichten Epimerisierung der Nonactinsäure in alkalischer Lösung gibt die Hydrolyse von Monactin, Dinactin und Trinactin (*1*) je ein Gemisch von 4 Säuren, das durch Verteilungschromatographie in Nonactinsäure und 2-epi-Nonactinsäure einerseits und in Homononactinsäure und 2-epi-Homononactinsäure anderseits getrennt wurde. Das Gemisch der beiden letzteren konnte kristallisiert werden und gab auf die Formel $C_{11}H_{20}O_4$ passende Analysen. Die Trennung der beiden epimeren C_{11}-Säuren erfolgte wiederum durch Chromatographie der Methylester.

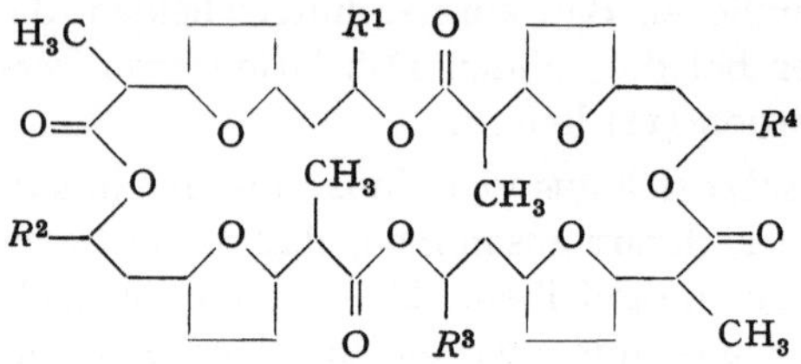

(1) $R^1 = R^2 = R^3 = R^4 = -CH_3$; Nonactin
(2) $R^1 = R^2 = R^3 = CH_3$, $R^4 = -C_2H_5$; Monactin
(3) $R^1 = R^3 = -CH_3$, $R^2 = R^4 = -C_2H_5$; Dinactin
(4) $R^1 = -CH_3$, $R^2 = R^3 = R^4 = -C_2H_5$; Trinactin

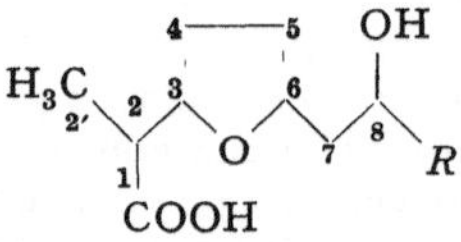

(5) $R = CH_3$; Nonactinsäure
(6) $R = C_2H_5$; Homononactinsäure
(7) $R = CH(CH_3)_2$; Bishomononactinsäure

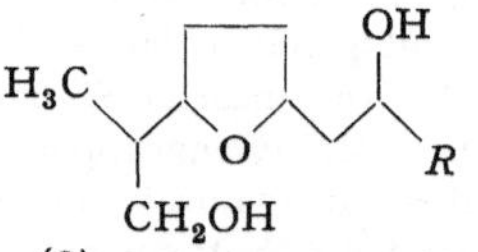

(8) $R = CH_3$; C_{10}-Diol
(9) $R = C_2H_5$; C_{11}-Diol
(10) $R = CH(CH_3)_2$; C_{12}-Diol

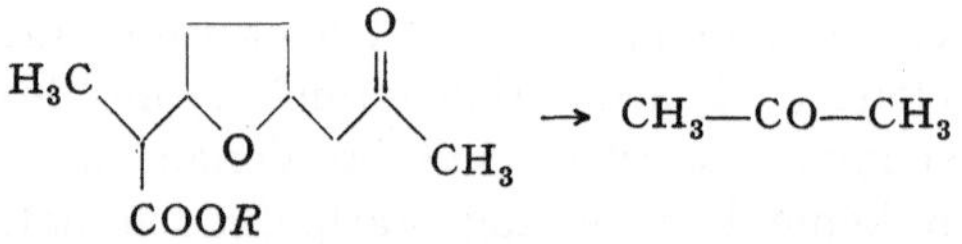

(11) $R = H$; 8-Dehydrononactinsäure (12) Aceton

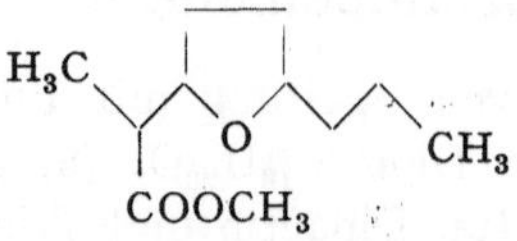

(13) 8-Desoxynonactinsäure-methylester

$CH_3—CH(COOCH_3)—CHOH—(CH_2)_5—CH_3$

(14) β-Hydroxy-α-methylpelargonsäure-methylester

$CH_3—CH(COOH)—(CH_2)_6—CH_3$

(15) α-Methylpelargonsäure

$CH_3—CH(COOCH_3)—CO—(CH_2)_5—CH_3$

(16) β-Keto-α-methylpelargonsäure-methylester

$CH_3—CH_2—CO—(CH_2)_5—CH_3$

(17) Nonanon-(3)

Literaturverzeichnis: SS. 187—189

(**18**) 8-Dehydrohomononactinsäure   (**19**) Äthyl-methylketon

CH_3—CO—CH_2—CH_3

(**20**) $R^1 = R^2 = H$
(**21**) $R^1 = H$, $R^2 = CH_3$
(**22**) $R^1 = R^2 = CH_3$

(**23**) C_9-Diol

(**24**) $R = H$; 8-Dehydro-bishomo-nonactinsäure
(**25**) $R = CH_3$

(**26**)   (**27**)

Ns = Nonactinsäurereste,   Hs = Homononactinsäurereste.

Die chemischen Umwandlungen der Homononactinsäure und des C_{11}-Diols verliefen analog denen der Nonactinsäure (5) und des C_{10}-Diols (8). Es sollen hier nur die Abweichungen erwähnt werden, die zur Aufstellung der Formeln (6) bzw. (9) für die homologen Spaltprodukte führten (*1*):

a) Bei der Oxydation des C_{11}-Diols nach Kuhn und Roth entsteht ein Gemisch von Essigsäure und Propionsäure, wie sich durch Dünnschichtchromatographie der *p*-Phenylphenacylester leicht feststellen ließ. Die Summe der flüchtigen Säuren betrug wieder nahezu 2 Mole.

b) Die Ketosäure (**18**), die durch milde Oxydation des C_{11}-Diols mit Chrom(VI)-oxid erhalten wurde, gab bei der alkalischen Fragmentierung Äthyl-methylketon (**19**).

c) Ein Vergleich der NMR-Spektren von Nonactinsäure-methylester und 2-epi-Nonactinsäure-methylester mit denen der beiden epimeren Homononactinsäure-methylester führte ebenfalls zur Ableitung der Formel (6) für die C_{11}-Säuren.

Oberhalb δ 2,2 ppm unterschieden sich die Spektren der entsprechenden C_{10}- und C_{11}-Hydroxysäure-methylester nicht signifikant. Der Signalhaufe der Methylenprotonen (δ zirka 1,4 bis 2,2 ppm) ist bei den Estern der C_{11}-Säuren etwas größer als bei den niedrigeren Homologen; die Integrale entsprechen 8 anstatt 6 Protonen. Signifikante Unterschiede findet man im Gebiet der Methylsignale (vgl. *Tabelle 2*). Die Dublette bei δ 1,08 und 1,18 ppm, die den Methylgruppen in α-Stellung zu den

Estercarbonylen zugeordnet werden müssen, sind in beiden homologen Reihen gleich. Hingegen fehlt in der Homo-Reihe das Dublett für die zweite CH_3—(CH)-Gruppe bei δ 1,12 ppm. An dessen Stelle tritt ein Triplett bei δ 0,91 ppm für eine CH_3—(CH_2)-Gruppe auf, die das Vorhandensein einer Äthyl- statt einer Methylgruppe am Carbinol-C-Atom beweist.

Tabelle 2. NMR-Spektren von Methylestern

Verbindung	d (J = 7)	d (J = 7)	t (J = 7)
Nonactinsäure-methylester	1,08 ppm	1,12 ppm	—
2-epi-Nonactinsäure-methylester	1,18 ppm	1,12 ppm	—
Homononactinsäure-methylester	1,08 ppm	—	0,91 ppm
2-epi-Homononactinsäure-methylester ..	1,18 ppm	—	0,92 ppm

Ein strenger Beweis für die angegebene Beziehung zwischen der Nonactinsäure und der Homononactinsäure und gleichzeitig ein Beweis für die übereinstimmende relative Konfiguration an drei der vier Chiralitätszentren der beiden Verbindungen, wurde durch eine direkte Verknüpfung über ein gemeinsames Abbauprodukt erbracht. Optisch aktive 8-Dehydrononactinsäure (**11**, $R = H$, S. 166) und 8-Dehydro-homononactinsäure (**18**), beide hergestellt durch Oxydation der entsprechenden sterisch einheitlichen Diole (**8**) bzw. (**9**) mit Chrom(VI)-oxid, wurden mit Trifluorperessigsäure umgesetzt. Es trat dabei zum Teil Spaltung der Bindung zwischen C-8 und C-9 ein, wobei erwartungsgemäß die gewünschte Reaktion bei der C_{11}-Ketosäure (**18**) glatter verlief als beim niedrigeren Homologen (**11**). Die rohen Oxydationsprodukte, die u. a. die Halbester (**20**) bzw. (**21**) enthielten, wurden direkt mit Lithiumaluminiumhydrid reduziert und die erhaltenen Alkoholgemische durch Verteilungschromatographie getrennt. Aus beiden Ausgangsmaterialien wurden chemisch identische Diole $C_9H_{18}O_3$ (**23**) isoliert und durch IR-Spektren und R_f-Werte miteinander identifiziert. (Über die entgegengesetzte Chiralität der beiden Produkte — für beide war Dinactin das ursprüngliche Ausgangsmaterial — vgl. Kap. IV, S. 169.)

Das C_{12}-Diol, das neben den oben beschriebenen Diolen (**8**) und (**9**) bei der reduktiven Spaltung der Makrotetrolide B, C, D und G (Tabelle 1, S. 162) erhalten wurde, besitzt, wie aus folgenden Untersuchungen hervorgeht (**9**), die Konstitution (**10**, S. 166):

a) Das NMR-Spektrum des C_{12}-Diols besitzt im 1-ppm-Gebiet Signale von 3 statt nur 2 C-Methylgruppen, die alle als Dublette mit J zirka 7 cps. auftreten.

b) Durch Oxydation des Diols mit Chrom(VI)-oxid gelangte man zur Ketosäure (**24**), die in Form ihres Methylesters chromatographisch rein erhalten wurde. Die Oxydation mit Trifluorperessigsäure, die in diesem Falle besonders glatt die Bindung zwischen C-8 und C-9 angriff, gab vorwiegend den Ester (**22**). Nach Reduktion mit Lithiumaluminium-hydrid wurde wiederum das Diol $C_9H_{18}O_3$ (**23**) erhalten, das nach IR-Spektrum und R_f-Wert mit dem entsprechenden Abbauprodukt aus C_{10}-Diol und C_{11}-Diol identisch war.

Die quantitative Zusammensetzung der einzelnen Makrotetrolide folgt aus den nachgewiesenen Bausteinen und den durch die Massenspektren erhärteten Bruttoformeln und ergibt das in der *Tabelle 3* dargestellte Bild.

Tabelle 3. Bausteine der Makrotetrolide

Verbindung	Bruttoformel	Bausteine		
		Ns*	Hs*	Bs*
Nonactin	$C_{40}H_{64}O_{12}$	4	—	—
Monactin	$C_{41}H_{66}O_{12}$	3	1	—
Dinactin	$C_{42}H_{68}O_{12}$	2	2	—
Trinactin	$C_{43}H_{70}O_{12}$	1	3	—
Substanz G	$C_{43}H_{70}O_{12}$	2	1	1
Substanz D	$C_{44}H_{72}O_{12}$	1	2	1
Substanz C	$C_{45}H_{74}O_{12}$	1	1	2
Substanz B	$C_{46}H_{76}O_{12}$	—	2	2

* Ns = Nonactinsäurereste, Hs = Homononactinsäurereste, Bs = Bishomo-nonactinsäurereste.

Die Anzahl der einzelnen Bausteine wird bestätigt durch die Ausbeute an den einzelnen Diolen bei der reduktiven Spaltung reiner Makrotetrolide. Eine weitere Bestätigung ergibt sich aus den bei der Massenspektrometrie entstehenden Fragmenten (vgl. Kap. V, S. 177). Wir können demnach für das Monactin die Konstitutionsformel (**2**, S. 166), für das Trinactin die Formel (**4**) schreiben. Für das Dinactin sind zunächst zwei Anordnungen möglich, die in abgekürzter Schreibweise durch die Formeln (**26**) und (**27**) dargestellt werden. Erst auf Grund des Massenspektrums (Kap. V) konnte eindeutig zugunsten der Formel (**26**) = (**3**) entschieden werden. Auch die höheren Homologen B, C, D und G sind analog der Formel (**1**) aufgebaut, wobei in allen Fällen mehrere verschiedene Anordnungen der Bausteine möglich sind, zwischen denen bisher nicht entschieden werden konnte (vgl. Kap. IV und V).

IV. Stereochemie der Makrotetrolide

Ungeachtet der Anwesenheit von 16 asymmetrischen C-Atomen im Molekül des Nonactins ist dieses optisch inaktiv, und seine Spalt-produkte (Nonactinsäure, Nonactindiol usw.) sind Racemate. Nonactin selber ist demnach entweder ein racemisches Gemisch oder wahrschein-licher eine meso-Verbindung. Die Homologen des Nonactins zeigen sehr geringe optische Drehungen, dagegen wurden bei deren Spaltung optisch aktive Abbauprodukte erhalten. Diese für Naturstoffe ungewöhnliche Stereochemie ist Gegenstand des vorliegenden Abschnittes.

Die Strukturaufklärung der Makrotetrolide läßt sich in zwei Teilaufgaben trennen. Einmal muß die absolute und relative Konfiguration der als Bausteine dienenden Hydroxysäuren, die je vier asymmetrische C-Atome enthalten, aufgeklärt werden. Sodann ist die Anordnung der zum Teil enantiomeren Bausteine im Tetralactonring zu bestimmen.

1. Konfiguration der Bausteine

Im Verlaufe der Konstitutionsaufklärung wurden die drei homologen 8-Dehydrosäuren (11, S. 166), (18) und (24) in bis auf das Vorzeichen der optischen Drehung identische C_9-Diole (23) übergeführt (vgl. Kap. III, S. 165). Durch diese Verknüpfung wird gleichzeitig die analoge relative Konfiguration der Kohlenstoffatome 2, 3 und 6 in der Nonactinsäure, Homononactinsäure und Bishomononactinsäure bewiesen. Aus 8-Dehydrononactinsäure (11) und 8-Dehydrohomononactinsäure (18) mit entgegengesetztem Vorzeichen der optischen Drehung, die beim Abbau des Dinactins erhalten werden, gewinnt man enantiomere C_9-Diole (23), d. h. die beiden Säuren gehören verschiedenen sterischen Reihen an. Beim Übergang der optisch aktiven C_{10}- und C_{11}-Diole (8) und (9) in die entsprechenden Verbindungen mit 8-epi-Konfiguration beobachtet man eine gleichsinnige Verschiebung der optischen Drehung. Dieser Übergang wird experimentell erreicht durch Oxydation der Diole (8) und (9) mit Chrom(VI)-oxid zu den Ketosäuren (11) und (18), gefolgt von einer Reduktion der Methylester dieser Säuren mit Lithiumaluminiumhydrid. Diese Beobachtungen zeigen, daß Nonactinsäure und Homononactinsäure mit gleichem Vorzeichen der optischen Drehung eine analoge Konfiguration an den asymmetrischen Kohlenstoffatomen 2, 3, 6 und 8 besitzen.

Zur Bestimmung der relativen Konfiguration an den Kohlenstoffatomen 3 und 6 (*11*) wurde in der rechtsdrehenden Ketosäure (*18*) die Carboxylgruppe in eine Methylketongruppierung übergeführt (ausgehend vom Säurechlorid durch Umsetzung mit Diazomethan zum Chlormethylketon und anschließende Reduktion mit Zink und Eisessig) und das entstandene Diketon (28) durch eine zweifache Baeyer-Villiger-Spaltung zum Diol (29) abgebaut. Durch Umwandlung in das Bis-*p*-toluolsulfonat und das Dibromid, gefolgt von einer katalytischen Hydrierung mit Palladiumkohle in Kaliumhydroxid-Lösung, wurden die beiden Hydroxylfunktionen des Diols (29) entfernt. Das so entstandene 2,5-Diäthyltetrahydrofuran war optisch inaktiv und muß deshalb *cis*-Konfiguration besitzen. Daraus läßt sich ableiten, daß auch in der Ketosäure (18) und damit in der Nonactinsäure und in der Homononactinsäure die Substituenten am Tetrahydrofuranring in *cis*-Stellung zueinander stehen. Diese Zuordnung ließ sich durch die Überführung des racemischen C_{10}-Diols (8) in das 5-Isopropyl-2-hydroxymethyl-tetrahydrofuran (31) mit bekannter *cis*-Konfiguration bestätigen.

Nonactinsäure und 2-epi-Nonactinsäure unterscheiden sich, ebenso wie die entsprechenden an C-2 epimeren 8-Desoxynonactinsäuren (13) in ihrer Acidität (*11*). Die Säuren mit genuiner Konfiguration an C-2 besitzen um 0,2 bis 0,3 Einheiten höhere pK*$_{MCS}$-Werte *(Tabelle 4)*. Man kann den beobachteten Unterschied in der Acidität der Säuren der Ausbildung von intramolekularen Wasserstoffbrücken zwischen Carboxylgruppe und Äthersauerstoff zuschreiben. Bei den entsprechenden, an C-2 epimeren 8-Desoxyalkoholen (32) läßt sich die intramolekulare Wasserstoffbrücke durch Aufnahme der IR-Spektren bei verschiedener

Tabelle 4. pK-Werte von epimeren Nonactinsäurederivaten und Lage des Dubletts der Methylgruppe an C-2 in den NMR-Spektren der Methylester

Verbindung	pK*$_{MCS}$	(d, J = 7 cps)
Nonactinsäure	7,21	1,08 ppm
2-epi-Nonactinsäure	7,00	1,18 ppm
8-Desoxynonactinsäure.	7,27	1,02 ppm
2-epi-8-Desoxynonactinsäure	6,96	1,12 ppm

Verdünnung in Tetrachlorkohlenstoff direkt nachweisen. Der Alkohol mit natürlicher Konfiguration an C-2 zeigt nur die Absorption einer „gebundenen" Hydroxylgruppe bei 3540 cm^{-1}; der 2-epi-Alkohol zeigt dagegen auch bei großer Verdünnung zwei Banden bei 3635 und 3520 cm^{-1}, von denen diejenige mit der höheren Wellenzahl charakteristisch ist für eine „freie" Hydroxylgruppe.

Je zwei typische gestaffelte Konformationen der beiden relativen Konfigurationsmöglichkeiten an C-2 und C-3, A und B, sind in *Abb. 1* als Newman-Projektionen dargestellt. Auf Grund von bewährten Über-

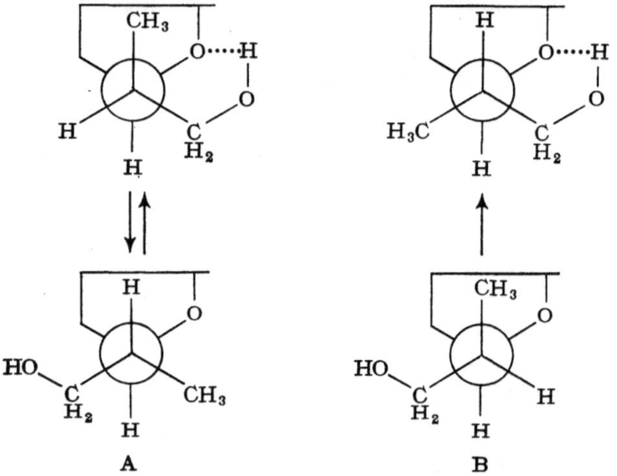

A   B

Abb. 1. Bildung von intramolekularen Wasserstoffbrücken bei den an C-2 epimeren 8-Desoxy-alkoholen. Gestaffelte Konformationen in Newman-Projektion

legungen kann man dem Alkohol mit genuiner Konfiguration an C-2 die relative Konfiguration B zuordnen. Hier ist die zur Ausbildung der Wasserstoffbrücke notwendige synclinale Konformation diejenige mit der stabilsten Anordnung der großen Substituenten. Bei der zweiten möglichen relativen Konfiguration A sind im Gegensatz dazu bei beiden, zur Ausbildung intramolekularer Wasserstoffbrücken befähigten Konformationen die großen Substituenten vicinal kumuliert, was zur Bevorzugung der dritten, antiperiplanaren Konformation ohne intramolekulare Wasserstoffbrücken führt. Analoge Überlegungen gelten auch für die an C-2 epimeren Säurepaare. Die relative Konfiguration an C-2 und C-3 der genuinen Säuren muß der relativen Konfigurationsmöglichkeit B entsprechen, da die Ausbildung intramolekularer Wasserstoffbrücken eine schwächere Säure mit einem höheren pK$^{*}_{MCS}$-Wert ergibt.

Die NMR-Spektren der Methylester der beiden erwähnten Säurepaare weisen ebenfalls auf eine Bevorzugung der antiperiplanaren Konformation in der 2-epi-Reihe A hin. Die Methylgruppe an C-2 der 2-epi-Methylester befindet sich in synclinaler Lage zum Äthersauerstoff des Tetrahydrofuranringes und damit in seinem „deshielding"-Bereich. Wie aus Tabelle 4 ersichtlich ist, liegt das Dublett dieser Methylgruppe in den NMR-Spektren der Methylester mit 2-epi-Konfiguration A um 0,1 ppm bei niedrigeren Feldstärken als das entsprechende Dublett der Methylester mit der genuinen Konfiguration B.

Auch im IR-Spektrum der an C-8 epimeren 1-Desoxy-alkohole (33) treten in verdünnter Tetrachlorkohlenstoff-Lösung OH-Banden verschiedener Lage und Intensität auf (*11*). Diesmal ist beim 8-epi-Alkohol auch bei großer Verdünnung keine Absorption einer freien Hydroxylgruppe nachzuweisen, während der Alkohol mit natürlicher Konfiguration am C-8 bei großer Verdünnung zwei Banden bei 3525 und 3620 cm^{-1} aufweist. Diese Beobachtung führte in Verbindung mit einer Konformationsanalyse zur Zuordnung der in Formel (35) eingezeichneten relativen Konfiguration an C-6 und C-8 der Nonactinsäure. Durch die Bestimmung der absoluten Konfiguration an C-6 und C-8 (vgl. unten) wird diese Zuordnung bestätigt.

$(30,\ \text{S. } 171) \rightarrow$

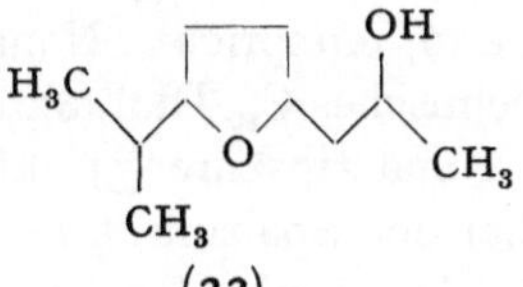

(33)

$(31) \longrightarrow$

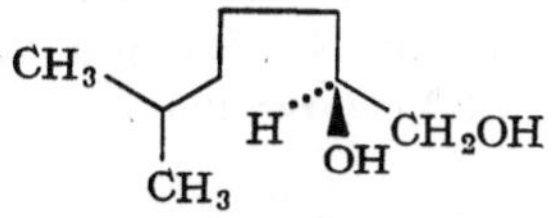

(34)

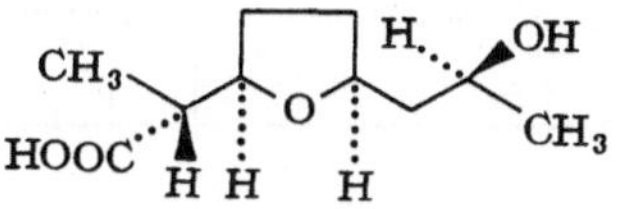

(35):
(-)-Nonactinsäure.

Ausgehend vom rechtsdrehenden C_{10}-Diol (8, S. 166) aus Dinactin wurde rechtsdrehendes 5-Isopropyl-2-hydroxymethyl-tetrahydrofuran (31) hergestellt. Dieses wurde durch Ätherspaltung mit Acetylbromid-Zinkchlorid und anschließende reduktive Enthalogenierung in das linksdrehende 6-Methyl-1,2-dihydroxyheptan (34) übergeführt, dem in Analogie zum linksdrehenden 1,2-Dihydroxybutan und 1,2-Dihydroxyhexan die S-Konfiguration zugeteilt wurde (*11*).

Zur Bestimmung der Chiralität des asymmetrischen Kohlenstoffatoms 8 der Nonactinsäure wurde das rechtsdrehende 1-Desoxydiol (33) mit der genuinen Konfiguration an C-8 nach Horeau (15) mit α-Phenylbuttersäure-anhydrid umgesetzt und dabei eine bevorzugte Veresterung mit der rechtsdrehenden α-Phenylbuttersäure festgestellt (11). Diese Beobachtung ist nach den bisherigen Erfahrungen nur mit einer S-Konfiguration an C-8 im rechtsdrehenden 1-Desoxydiol (33) vereinbar. Für die linksdrehende Nonactinsäure folgt daraus die in Formel (35, S. 173) angegebene (2R, 3R, 6S, 8S)-Konfiguration, während die rechtsdrehende Homononactinsäure aus Dinactin die (2S, 3S, 6R, 8R)-Konfiguration besitzen muß.

2. Konfiguration von Nonactin und seinen Homologen

Die Bestimmung des Vorzeichens der optischen Drehung der aus den einzelnen Makrotetroliden durch Reduktion mit Lithiumaluminiumhydrid erhaltenen Diole (8), (9) und (10, S. 166) gab Auskunft über die Chiralität der Bausteine. Dabei war zu beachten, daß z. B. dem (+)-C_{10}-Diol (8) die (—)-Nonactinsäure (5) und dem (—)-C_{11}-Diol (9) die (+)-Homononactinsäure (6) entspricht. Nonactin selbst liefert bei der reduktiven Spaltung racemisches C_{10}-Diol (8), d. h. es ist aus je zwei der beiden Enantiomeren der Nonactinsäure (5) aufgebaut (6). Die Homologen des Nonactins geben bei der Spaltung teils optisch reine, teils partiell racemische Diole mit $^1/_3$ der maximalen Drehung. In *Tabelle 5* ist angegeben, welche Enantiomere der drei homologen Hydroxysäuren (5), (6) und (7) (S. 166) (abgekürzt als Ns, Hs und Bs) die vier Bausteine der betreffenden Makrotetrolide darstellen.

Tabelle 5. Richtung der optischen Drehungen der Hydroxysäuren aus Makrotetroliden

Verbindung	Nonactinsäure		Homononactinsäure		Bishomononactinsäure	
	(—)	(+)	(—)	(+)	(—)	(+)
Nonactin	2	2	—	—	—	—
Monactin........	2	1	—	1	—	—
Dinactin	2	—	—	2	—	—
Trinactin........	1	—	1	2	—	—
Substanz G......	2	—	—	1	—	1
Substanz D......	1	—	—	2	1	—
Substanz C	1	—	—	1	1	1
Substanz B			nicht	gemessen		

Alle aufgefundenen Kombinationen der vier Bausteine setzen sich aus je zwei (—)- und (+)-Hydroxysäuren zusammen. Die Beteiligung von zwei linkshändigen und zwei rechtshändigen Hydroxysäureresten scheint

Literaturverzeichnis: SS. 187—189

eines der Aufbauprinzipien der Makrotetrolide darzustellen. Die Kenntnis von Zahl, Art und Chiralität der Bausteine genügt jedoch nicht zur Festlegung der Konfiguration (und zum Teil der Konstitution) der Makrotetrolide. Zusätzlich muß man noch die Anordnung dieser Bausteine kennen. Zur Illustration der großen Zahl der Möglichkeiten sei im Folgenden gezeigt, auf wieviele Arten die Makrotetrolide, deren vier Bausteine bekannt sind, zu Tetralactonen aufgebaut werden können.

Für Nonactin (1, S. 166) sind die vier Strukturmöglichkeiten (36) A—D denkbar. Für Dinactin ergeben sich zwei mögliche chirale Strukturen (37) A und B. Für Monactin lassen sich die drei chiralen Strukturen (38) A—C, für Trinactin die drei chiralen Strukturen (39) A—C, für Makrotetrolid G die drei chiralen Strukturen (40) A—C und für Makrotetrolid D die drei chiralen Strukturen (41) A—C formulieren. Noch vielfältiger liegen die Verhältnisse beim Makrotetrolid C, für das die sechs chiralen Strukturmöglichkeiten (42) A—F bestehen. Die Konfigurationen der Bausteine von Makrotetrolid B sind nicht bekannt. Für diese Verbindung sind zwei chirale Strukturen möglich, sofern die beiden Homononactinsäurereste und die beiden Bishomononactinsäurereste je die gleiche absolute Konfiguration besitzen (vgl. das analog gebaute Dinactin 37). Wenn je beide Enantiomeren der zwei homologen Säuren als Bausteine dienen, sind sechs Strukturen für Makrotetrolid B möglich, die sich in zwei Enantiomerenpaare und zwei meso-Formen aufteilen lassen.

$$
\begin{array}{cc}
\begin{array}{cc}
(+)\mathrm{Ns} \!-\! (+)\mathrm{Ns} & (-)\mathrm{Ns} \!-\! (-)\mathrm{Ns} \\
| \qquad\quad | & | \qquad\quad | \\
(+)\mathrm{Ns} \!-\! (+)\mathrm{Ns} & (-)\mathrm{Ns} \!-\! (-)\mathrm{Ns}
\end{array}
&
\begin{array}{cc}
(+)\mathrm{Ns} \!-\! (+)\mathrm{Ns} & (-)\mathrm{Ns} \!-\! (-)\mathrm{Ns} \\
| \qquad\quad | & | \qquad\quad | \\
(+)\mathrm{Ns} \!-\! (-)\mathrm{Ns} & (-)\mathrm{Ns} \!-\! (+)\mathrm{Ns}
\end{array}
\\[2mm]
\text{A (Racemat)} & \text{B (Racemat)}
\end{array}
$$

$$
\begin{array}{cc}
\begin{array}{c}
(+)\mathrm{Ns} \!-\! (+)\mathrm{Ns} \\
| \qquad\quad | \\
(-)\mathrm{Ns} \!-\! (-)\mathrm{Ns}
\end{array}
&
\begin{array}{c}
(+)\mathrm{Ns} \!-\! (-)\mathrm{Ns} \\
| \qquad\quad | \\
(-)\mathrm{Ns} \!-\! (+)\mathrm{Ns}
\end{array}
\end{array}
$$

C (meso-Form mit   D (meso-Form mit vier-
Symmetriezentrum C_i)   zähliger Drehspiegelachse S_4)

(36) Stereochemische Formulierungsmöglichkeiten für Nonactin

$$
\begin{array}{cc}
\begin{array}{c}
(-)\mathrm{Ns} \!-\! (-)\mathrm{Ns} \\
| \qquad\quad | \\
(+)\mathrm{Hs} \!-\! (+)\mathrm{Hs}
\end{array}
&
\begin{array}{c}
(-)\mathrm{Ns} \!-\! (+)\mathrm{Hs} \\
| \qquad\quad | \\
(+)\mathrm{Hs} \!-\! (-)\mathrm{Ns}
\end{array}
\\[2mm]
\text{A} & \text{B}
\end{array}
$$

(37) Strukturelle Formulierungsmöglichkeiten für Dinactin

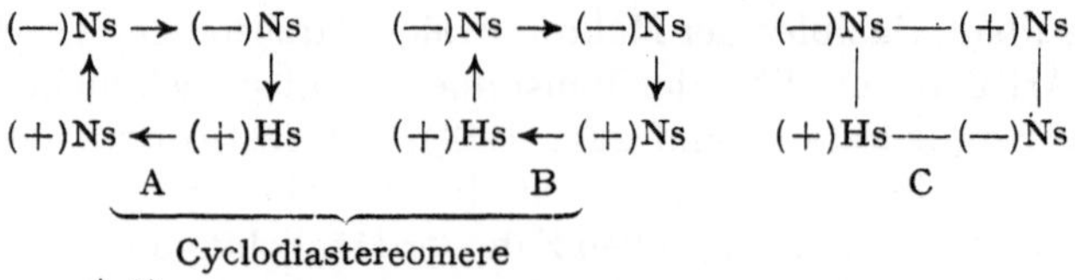

$$
\begin{array}{ccc}
(-)\text{Ns} \rightarrow (-)\text{Ns} & (-)\text{Ns} \rightarrow (-)\text{Ns} & (-)\text{Ns} \!-\!\!-\! (+)\text{Ns} \\
\uparrow \qquad \downarrow & \uparrow \qquad \downarrow & | \qquad\quad | \\
(+)\text{Ns} \leftarrow (+)\text{Hs} & (+)\text{Hs} \leftarrow (+)\text{Ns} & (+)\text{Hs} \!-\!\!-\! (-)\text{Ns} \\
\text{A} & \text{B} & \text{C}
\end{array}
$$

Cyclodiastereomere

(38) Stereochemische Formulierungsmöglichkeiten für Monactin*

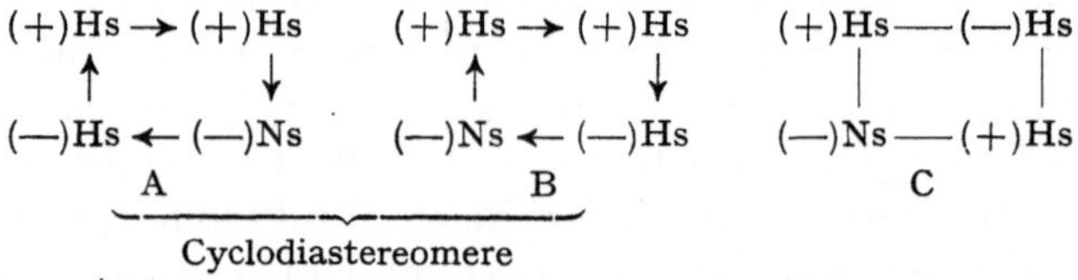

$$
\begin{array}{ccc}
(+)\text{Hs} \rightarrow (+)\text{Hs} & (+)\text{Hs} \rightarrow (+)\text{Hs} & (+)\text{Hs} \!-\!\!-\! (-)\text{Hs} \\
\uparrow \qquad \downarrow & \uparrow \qquad \downarrow & | \qquad\quad | \\
(-)\text{Hs} \leftarrow (-)\text{Ns} & (-)\text{Ns} \leftarrow (-)\text{Hs} & (-)\text{Ns} \!-\!\!-\! (+)\text{Hs} \\
\text{A} & \text{B} & \text{C}
\end{array}
$$

Cyclodiastereomere

(39) Stereochemische Formulierungsmöglichkeiten für Trinactin

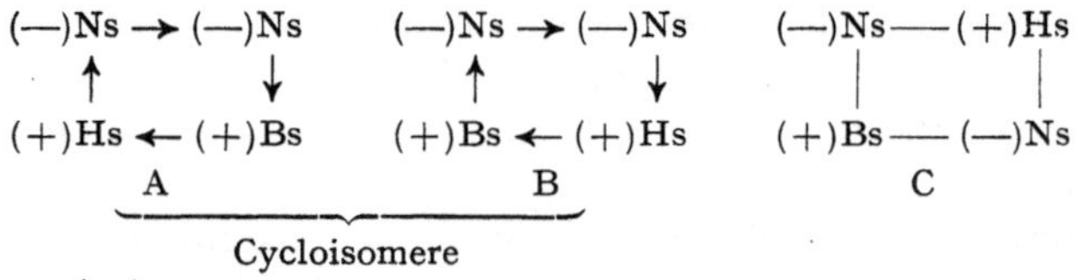

$$
\begin{array}{ccc}
(-)\text{Ns} \rightarrow (-)\text{Ns} & (-)\text{Ns} \rightarrow (-)\text{Ns} & (-)\text{Ns} \!-\!\!-\! (+)\text{Hs} \\
\uparrow \qquad \downarrow & \uparrow \qquad \downarrow & | \qquad\quad | \\
(+)\text{Hs} \leftarrow (+)\text{Bs} & (+)\text{Bs} \leftarrow (+)\text{Hs} & (+)\text{Bs} \!-\!\!-\! (-)\text{Ns} \\
\text{A} & \text{B} & \text{C}
\end{array}
$$

Cycloisomere

(40) Strukturelle Formulierungsmöglichkeiten für Makrotetrolid G

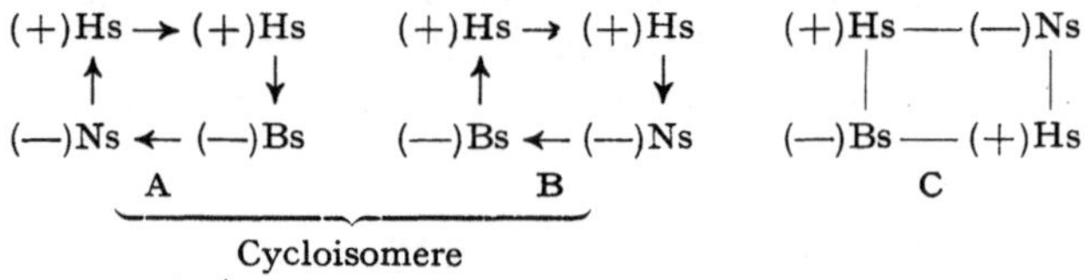

$$
\begin{array}{ccc}
(+)\text{Hs} \rightarrow (+)\text{Hs} & (+)\text{Hs} \rightarrow (+)\text{Hs} & (+)\text{Hs} \!-\!\!-\! (-)\text{Ns} \\
\uparrow \qquad \downarrow & \uparrow \qquad \downarrow & | \qquad\quad | \\
(-)\text{Ns} \leftarrow (-)\text{Bs} & (-)\text{Bs} \leftarrow (-)\text{Ns} & (-)\text{Bs} \!-\!\!-\! (+)\text{Hs} \\
\text{A} & \text{B} & \text{C}
\end{array}
$$

Cycloisomere

(41) Formulierungsmöglichkeiten für Makrotetrolid D

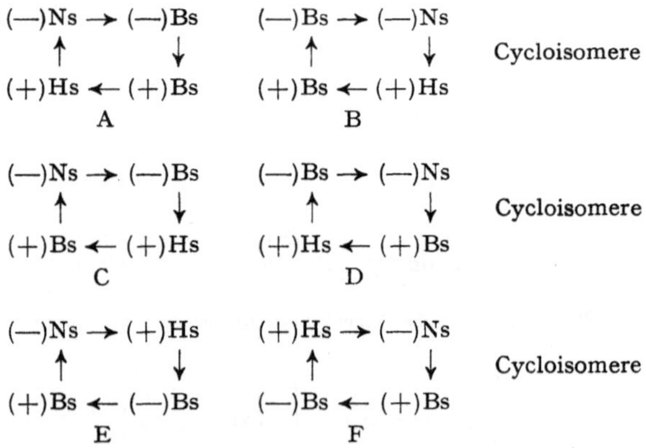

$$
\begin{array}{cc}
(-)\text{Ns} \rightarrow (-)\text{Bs} & (-)\text{Bs} \rightarrow (-)\text{Ns} \\
\uparrow \qquad \downarrow & \uparrow \qquad \downarrow \\
(+)\text{Hs} \leftarrow (+)\text{Bs} & (+)\text{Bs} \leftarrow (+)\text{Hs} \\
\text{A} & \text{B}
\end{array}
\qquad \text{Cycloisomere}
$$

$$
\begin{array}{cc}
(-)\text{Ns} \rightarrow (-)\text{Bs} & (-)\text{Bs} \rightarrow (-)\text{Ns} \\
\uparrow \qquad \downarrow & \uparrow \qquad \downarrow \\
(+)\text{Bs} \leftarrow (+)\text{Hs} & (+)\text{Hs} \leftarrow (+)\text{Bs} \\
\text{C} & \text{D}
\end{array}
\qquad \text{Cycloisomere}
$$

$$
\begin{array}{cc}
(-)\text{Ns} \rightarrow (+)\text{Hs} & (+)\text{Hs} \rightarrow (-)\text{Ns} \\
\uparrow \qquad \downarrow & \uparrow \qquad \downarrow \\
(+)\text{Bs} \leftarrow (-)\text{Bs} & (-)\text{Bs} \leftarrow (+)\text{Bs} \\
\text{E} & \text{F}
\end{array}
\qquad \text{Cycloisomere}
$$

(42) Die sechs Strukturmöglichkeiten für Makrotetrolid C

---

* Die Pfeile geben die Richtung der Esterbindungen —CO → O— an, entsprechend der in der Peptidchemie gebräuchlichen Bezeichnung der Richtung der Amidbindungen.

Literaturverzeichnis: SS. 187—189

Isomerenpaare, die das gleiche Verteilungsmuster aber verschiedene Anordnung (= Ringrichtung) der Bausteine aufweisen, sind Beispiele von Cyclodiastereomeren und Cycloisomeren, wie sie von PRELOG und Mitarb. (*8, 10, 25*) diskutiert wurden. Es soll hier besonders auf diese Isomeriefälle hingewiesen werden, da sie die Strukturaufklärung wesentlich erschweren.

In den Fällen, in denen sich die Bausteine der Makrotetrolide in ihrer Konstitution unterscheiden, lassen sich gewisse Informationen über ihre Anordnung durch eine Analyse der Massenspektren gewinnen. Eine eindeutige Strukturzuordnung mit Hilfe dieser Methode ließ sich nur für das Dinactin vornehmen. Bei anderen Verbindungen der Reihe konnten einzelne Möglichkeiten ausgeschlossen werden. Einige Einzelheiten betreffend die Anordnung der Bausteine werden im Kap. V näher diskutiert.

V. Massenspektren der Makrotetrolide

Sowohl für die Identifizierung und Reinheitsprüfung der Makrotetrolide wie auch für die Abklärung gewisser struktureller Einzelheiten hat sich die Massenspektrometrie als besonders wertvoll erwiesen (*16*). Das Massenspektrum des Nonactins *(Abb. 2)* ist, abgesehen vom Bereich kleiner Massenzahlen, auffallend einfach. Im wesentlichen sind es vier Peaks, die besonders herausgehoben sind. Das Ion der Masse 736 ist das Molekülion und damit direkt mit der Bruttoformel des Antibioticums verknüpft. Das Signal ist genügend stark, daß noch 5—10% von homologen Beimengen erkannt werden können, eine Menge, die sich durch Dünnschichtchromatographie kaum mehr erfassen läßt.

Die Signale der Massenzahlen 553, 369, und 185 fallen durch eine regelmäßige Beziehung zum Molekulargewicht auf. Es ist nämlich $553 = {}^3/_4\,M + 1$, $369 = {}^1/_2\,M + 1$ und $185 = {}^1/_4\,M + 1$. Daraus drängt sich die Vermutung auf, daß diese Fragmente durch Abspaltung von

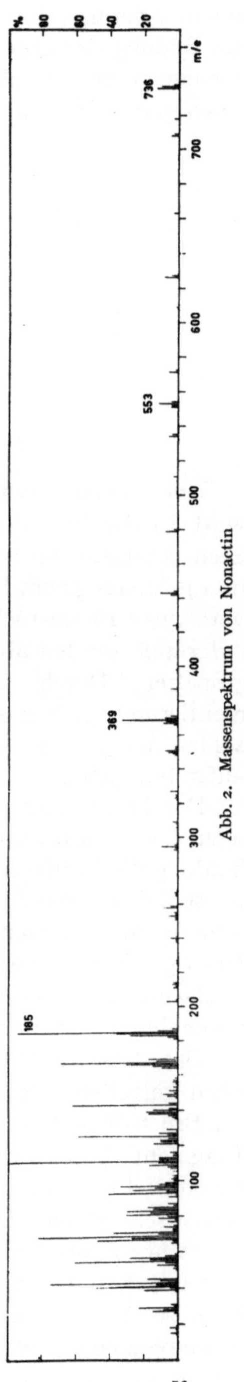

Abb. 2. Massenspektrum von Nonactin

einem, zwei und drei Bausteinen aus dem Gesamtmolekül entstehen. Für die Bildung der drei Bruchstücke ist ein gemeinsamer Mechanismus anzunehmen und da ihre Massenzahlen ungerade sind, muß es sich um einen gemischten Mechanismus handeln.

Ein Fragmentierungsmechanismus, der bei Estern häufig angetroffen wird, ist die McLafferty-Umlagerung, wie sie in der Formel (43) bei der Stelle a angedeutet ist. Sie resultiert in einer Spaltung der C—C-Bindung in α,β-Stellung zum Estercarbonyl und in der Verschiebung eines Protons vom einen Bruchstück auf das andere. Bindungen in α,β zu einer Carbonylgruppe werden aber auch bei einfachen Fragmentierungen bevorzugt gespalten. Durch eine McLafferty-Umlagerung bei a und eine Fragmentierung bei b, c oder d müssen nun aber Fragmente mit den Massenzahlen m/e 553, 369 und 185 entstehen, wie sie im Spektrum bevorzugt auftreten (Abb. 2).

Die Frage nach der Ursache für den gemischten Mechanismus ist nicht leicht zu beantworten. Wir nehmen an, daß in der cyclischen Verbindung die Konformation an der Stelle a für eine Umlagerung ungünstig ist, so daß als erster Schritt eine Fragmentierung eintritt. Nach der Ringöffnung bei b, c oder d kann sich dagegen bei a eine für die Umlagerung günstige Konformation ausbilden. Durch eine doppelte McLafferty-Umlagerung bei ae, af oder ag sind dagegen die im Spektrum deutlich erkennbaren Fragmente mit m/e 626, 442 und 258 zu erklären.

Das Massenspektrum des Monactins (*16*) bestätigt aufs beste die oben getroffenen Zuordnungen. Neben dem Peak bei m/e 185 für ein C_{10}-Bruchstück finden wir neu einen solchen bei m/e 199 für das C_{11}-Fragment. Das Verhältnis der beiden Signale entspricht nicht genau dem theoretischen von 3 : 1, da die Fragmentierung offenbar nicht rein statistisch erfolgt. Die Abweichung ist aber relativ gering, so daß eine ungefähre Abschätzung des Verhältnisses der am Aufbau eines Makrotetrolids beteiligten Bausteine aus dem Massenspektrum erlaubt ist. Eine Bestätigung für die Anwesenheit eines C_{11}-Bausteins neben 3 Nonactinsäureeinheiten ergibt sich aus dem Gebiet um $^3/_4$ M, wo das Signal

M-183 (m/e 567) von einem kleineren mit M-197 (m/e 553) begleitet ist. Im Gebiet um $^1/_2$ M finden wir zwei fast gleich starke Signale bei m/e 369 und 383, entsprechend einem C_{20}- und einem C_{21}-Bruchstück, wie sie aus dem Zerfall von Monactin (38) in zwei ungefähre Hälften zu erwarten sind. Ein Entscheid zwischen den drei möglichen Anordnungen (38) A, B und C (S. 176) ist auf Grund des Massenspektrums nicht möglich.

(44)

Beim Dinactin (44) wird die aus dem chemischen Abbau abgeleitete Zusammensetzung aus je zwei Nonactinsäure- und Homononactinsäureresten durch zwei fast gleich hohe Signale bei m/e 185 und 199 bzw. m/e 567 und 581 bestätigt. Ein Entscheid zwischen den beiden möglichen Anordnungen der Bausteine (37) A und B sollte sich aus der Region um $^1/_2$ M treffen lassen. Bei der Anordnung (37) B sind nämlich bei der Spaltung in zwei ungefähre Hälften nur Bruchstücke mit 21 C-Atomen (m/e 383) zu erwarten, während die Anordnung (37) A zu Fragmenten mit m/e 369, 383 und 397 im Verhältnis 1 : 2 : 1 führen sollte. Das Massenspektrum des reinsten verfügbaren Dinactins zeigt in diesem Gebiet praktisch nur das Signal mit m/e 383, was als Beweis für die Anordnung (37) B = (3, S. 166) gelten muß. Da vom chemischen Abbau her bekannt ist (vgl. Kap. IV, S. 174), daß beide Nonactinsäureeinheiten des Dinactins die (2R, 3R, 6S, 8S)- und beide Homononactinsäureeinheiten die (2S, 3S, 6R, 8R)-Konfiguration besitzen, ist das Dinactin damit das erste und bisher einzige Makrotetrolid, für das die vollständige Konstitution und Konfiguration gemäß der Formel (44) experimentell gesichert ist. Aus biogenetischen Gründen ist anzunehmen, daß die alternierende Anordnung der Bausteine verschiedener Chiralität (45) für alle Verbindungen der Reihe gleichermaßen gilt.

$$\begin{array}{ccc} (+) & \!\!\!\!-\!\!\!\!- & (-) \\ | & & | \\ (-) & \!\!\!\!-\!\!\!\!- & (+) \end{array}$$

(45)

Es gibt Anzeichen dafür (9), daß in den Makrotetrolidgemischen geringere Mengen von Isomeren des Dinactins vorkommen, die einer anderen Anordnung der Bausteine entsprechen. Geringfügigen Beimengungen von solchen Isomeren sind die sehr schwachen Signale bei m/e 369 und 397 im Spektrum des Dinactins zuzuschreiben. Ihre Intensitäten machen weniger als 5% von I_{383} aus.

Interessant ist die Gegenüberstellung der Massenspektren der Isomeren Trinactin und Substanz G, die beide ein Molekülion mit $M^+ = 778$ geben. In der Region um $^1/_4$ M zeigen sich wesentliche Unterschiede, indem Trinactin nur die beiden Signale m/e 185 und 199 zeigt (Aufbau aus einem Nonactinsäure- und drei Homononactinsäureresten), während das Spektrum der Substanz G zudem einen starken Peak bei m/e 213 für ein C_{12}-Fragment aufweist *(Tabelle 6)*.

Tabelle 6. Massenspektren der Makrotetrolide

Verbindung	Intensitäten in % des Basispeaks										
	$^1/_4$ M-Region			$^1/_2$ M-Region					$^3/_4$ M-Region		
	m/e										
	185	199	213	369	383	397	411	425	M-183	M-197	M-211
Nonactin	95	—	—	33	—	—	—	—	12	—	—
Monactin	91	27	—	16	12	—	—	—	1,7	0,6	—
Dinactin.........	90	99	—	4,5	53	2	—	—	2,5	2,6	—
Trinactin	29	75	4	—	10	6	—	—	0,9	2,6	—
Substanz G	87	68	37	6	21	16	2,6	—	2,6	1,8	2,0
Substanz D	42	97	39	—	10	4,2	3,8	—	0,8	2,0	1,5
Substanz C	38	65	99	—	11	12	6,5	2,0	0,8	1,0	2,0
Substanz B	—	30	28	—	—	3	13	1,5	—	1,2	2,0

Zwischen den drei Strukturmöglichkeiten (39) A, B und C (S. 176) für das Trinactin läßt sich auf Grund des Massenspektrums nicht unterscheiden, da alle drei Strukturen im Gebiet um $^1/_2$ M das gleiche Fragmentierungsmuster erwarten lassen, nämlich Signale für ein C_{21}- und ein C_{22}-Bruchstück bei m/e 383 und 397 mit etwa gleicher Intensität. Diese Signale treten im Spektrum (Tabelle 6) tatsächlich deutlich hervor.

Von den drei für das Makrotetrolid G möglichen Strukturformeln (40) A, B und C besitzen die zwei Cycloisomeren (A und B) das gleiche Verteilungsmuster der Bausteine und sind durch Massenspektrometrie daher nicht unterscheidbar. Sie lassen im $^1/_2$ M-Gebiet die vier Peaks bei m/e 369, 383, 397 und 411 mit etwa gleichen Intensitäten erwarten. Dagegen läßt die Struktur (40) C eine Fragmentierung erwarten, die in diesem Spektralgebiet nur zu Fragmenten mit 21 und 22 C-Atomen (m/e 383 und 397) führt. Im Massenspektrum der Substanz G (Tabelle 6) überwiegen diese beiden letzteren Signale sehr stark, so daß der Hauptkomponente des Präparates die Struktur (40) C zugeordnet werden muß. Peaks geringer Intensität bei m/e 369 und 411 zeigen aber, daß die Verbindung noch Beimengungen enthält, denen die Konstitutionen (40) A und B zukommen. In Übereinstimmung damit sind die Diole, die bei der reduktiven Spaltung der Substanz G erhalten wurden, optisch nicht ganz rein. Dies ist zu erwarten, wenn ein Gemisch vorliegt, dessen alle Kom-

Literaturverzeichnis: SS. 187—189

ponenten der allgemeinen stereochemischen Grundsequenz (45) gehorchen, die aber in konstitutioneller Hinsicht verschiedene Verteilungsmuster der Bausteine besitzen.

. Auf eine nähere Diskussion der Massenspektren der Verbindungen B, C und D kann hier verzichtet werden, da sie keine neuen Anhaltspunkte zur generellen Deutung dieser Spektren liefern. Es zeigte sich (*16*), daß auch diese Präparate zwar frei von Homologen aber nicht ganz einheitlich in bezug auf die Bausteinsequenz waren. Eine eindeutige Strukturzuordnung war in keinem Falle möglich. Eine Zusammenstellung der wichtigsten experimentellen Daten ist in der Tabelle 6 wiedergegeben. Abbildungen finden sich in der Literatur (*16*).

VI. Biologische Wirkung und elektrochemisches Verhalten

Die ersten Angaben über die antibiotische Wirksamkeit von Nonactin waren recht widersprüchlich. So konnten CORBAZ et al. (*4*) für reines Nonactin keine antibakterielle Wirkung finden, obwohl die Rohkulturen eine deutliche Aktivität gegen Mycobakterien gezeigt hatten. NISHIMURA und Mitarb. (*22*) fanden eine hohe Aktivität gegen den Pilz *Piricularia oryzae*, während eine Reihe von Bakterien unempfindlich zu sein scheinen. Später wiesen dagegen WALLHÄUSSER und Mitarb. (*30*) eine hohe Wirksamkeit gegen Gram-positive und säurefeste Bakterien für das Nonactin und seine Homologen nach (vgl. auch *17*). BENNETT und Mitarb. (*2*) machten es wahrscheinlich, daß die ursprünglich negativen Resultate darauf zurückzuführen waren, daß das Nonactin wegen seiner Unlöslichkeit in Wasser und wasserhaltigen Lösungen gar nicht an den Wirkungsort (siehe unten) gelangte. Nachdem man gelernt hatte, diesem Umstand durch veränderte Testanordnungen Rechnung zu tragen, konnten reproduzierbare, hohe Aktivitäten gegen eine Vielzahl von Mikroorganismen nachgewiesen werden. MEYERS und Mitarb. (*19*) geben die in der *Tabelle 7* verzeichneten minimalen Wirkungskonzentrationen (MIC) gegen *Staphylococcus aureus* und *Mycobacterium bovis* an. Es fällt dabei besonders die Wirkungszunahme vom Nonactin zu seinen Homologen auf, ein Effekt, der auch in anderen Laboratorien beobachtet werden konnte. ZÄHNER anderseits macht auf die besonders hohe Wirksamkeit gegen zwei Pilzarten — *Botrytis cinerea* und *Paecilomyces varioti* — aufmerksam (vgl. *9*).

Die Makrotetrolide sind demnach Antibiotica von recht breitem Wirkungsspektrum, indem hohe Aktivitäten gegen Gram-positive und säurefeste Bakterien sowie Pilze beobachtet wurden.

Die Wirkung der Makrotetrolide beschränkt sich aber nicht auf Mikroorganismen. Eine hohe cytostatische Wirksamkeit wurde sowohl an Earles L-Zellkulturen (s. Tabelle 7) wie auch an Ehrlichs Asciteszellen (*2*) nachgewiesen. Die Toxicität am lebenden Tier hängt sehr stark von der Applikationsart ab (*2*). Eine mit Sesamöl verdünnte Chloroform-lösung von Nonactin erwies sich bei subcutaner Verabreichung an der Maus als ziemlich giftig (LD_{50} = 45 mg/kg). Toxische Erscheinungen am Menschen können schon beim Arbeiten im Laboratorium durch Kontakt mit Makrotetrolidstaub auftreten; Hautausschläge und Augenentzündungen sind die wesentlichen Folgen.

Tabelle 7. Wirkung von Makrotetroliden gegen Bakterien und Zellkulturen

Verbindung	MIC in μg/ml		
	Staph. aureus	Mycobact. bovis	Earles L-Zellen
Nonactin	0,95	0,96	0,060
Monactin	0,08	0,12	0,010
Dinactin..........	0,05	0,04	0,004
Trinactin	0,04	0,04	0,005

Eingehende Untersuchungen über die Wirkungsweise verdanken wir Lardy und seiner Arbeitsgruppe (*12—14*), die als Testobjekte Mitochondrien aus Rattenleber benützten. Die Makrotetrolide sind danach in erster Linie starke Inhibitoren der oxydativen Phosphorylierung, der Bildung des energiereichen Adenosintriphosphats (ATP) aus Adenosindiphosphat (ADP) und anorganischem Phosphat, wobei die für diese Reaktion aufzuwendende Energie durch die Dehydrierung der reduzierten Form von Nicotinsäureamid-dinucleotid (NADH) geliefert wird:

$$NADH + H^+ + {}^{1}/_{2} O_2 + 3\,ADP + 3\,H_3PO_4 \rightarrow NAD^+ + 3\,ATP + 4\,H_2O$$

Hand in Hand mit dieser Hemmung geht eine Stimulierung der Zellatmung und eine vermehrte Hydrolyse von ATP in den Mitochondrien. Die letztere führen Lardy und Mitarb. auf eine induzierte Aktivierung von ATPase zurück.

Eine wesentliche Beobachtung bei diesen Versuchen war diejenige einer starken Abhängigkeit der Wirkung von den in der umgebenden Lösung anwesenden anorganischen Kationen. Eine hohe Wirkung (zirka 10^{-10} M Lösung von Monactin, Dinactin und Trinactin bzw. 10^{-9} M Lösung von Nonactin) wurde in Gegenwart von K^+- und Rb^+-Ionen beobachtet. Um etwa 2 Zehnerpotenzen höher war die minimale wirksame Konzentration an Makrotetroliden, wenn nur Na^+-Ionen vorhanden waren. Bei Abwesenheit von Na^+, K^+- und Rb^+, aber in Gegenwart von Li^+- oder Ammoniumionen war nur eine sehr geringe Aktivität feststellbar. *Abb. 3* gibt eine solche Ionenabhängigkeit der Monactinwirkung wieder.

Die ausgedehnten Versuche von Lardy und Mitarb. (*12—14*) führten zum Schluß, daß die primäre Wirkung der Makrotetrolide darin besteht, daß die Durchlässigkeit der Mitochondrienmembranen für verschiedene Metallionen wesentlich beeinflußt wird. Die zuerst beobachteten Wirkungen — Entkoppelung der oxydativen Phosphorylierung, vermehrte Hydrolyse von ATP, Steigerung der Zellatmung — sind dann eine Folge der Akkumulierung von Alkaliionen in den Mitochondrien. Eine Bestätigung fand diese Vermutung durch direkte Messung der Ionen-Transport-Aktivität von Monactin (*12*).

Mit den energetischen Beziehungen dieser Aktivität setzten sich PRESSMAN und Mitarb. (*3*) auseinander, die eine analoge Wirkung für das Cyclodepsipeptid Valinomycin gefunden hatten. Die Akkumulierung von Kaliumionen in den Mitochondrien benötigt Energie. Diese wird geliefert durch die Hydrolyse von ATP zu ADP und anorganischem Phosphat. Die von LARDY beobachtete „Induzierung der ATP-Hydrolyse" ist demnach aus energetischen Gründen notwendigerweise mit dem Ionentransport verknüpft. Die Energie, die durch die Hydrolyse eines ATP-Moleküls geliefert wird, ermöglicht das „Einpumpen" von etwa 3,6 Kaliumionen. Mit dem erhöhten Eindringen von Alkaliionen ist anderseits ein „Ausstoßen" von Wasserstoffionen verbunden (*12, 26*). Aus den energetischen Zusammenhängen ergibt sich, daß der Kaliumionentransport durch die Mitochondrienmembran nicht eine bloße Diffusion sein kann. Es handelt sich um einen aktiven, energieverbrauchen-

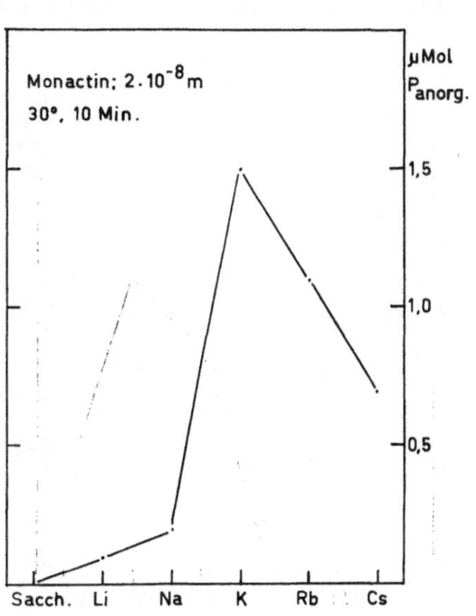

Abb. 3. ATP-Hydrolyse in Mitochondrien in Gegenwart von Monactin und verschiedenen Alkalimetallionen (Sacch. = Saccharose allein; Alkali-Konzentration je 60 mMol/l). $P_{anorg.}$ = durch Hydrolyse von ATP gebildetes anorganisches Phosphat). Nach LARDY und Mitarb. (*13, 14*)

den Prozeß. Die durch Makrotetrolide veränderte Membran kann bildhaft als eine „Ionenpumpe" bezeichnet werden (*26*). Die Akkumulierung von Kaliumionen in den Mitochondrien ist mit einem Anschwellen dieser Zellbestandteile verbunden, die etwa 5% des Mitochondrienvolumens ausmachen kann (*14, 26*).

Modellmäßig kann man sich vorstellen, daß sich die Makrotetrolidmoleküle an die Membran anlagern (oder sich in diese einlagern), derart, daß eine Art „Ionensieb" entsteht. Die Sieb-Vorstellung ist aber recht oberflächlich, was einerseits daraus hervorgeht, daß die relativ großen Ionen K⁺ und Rb⁺ leicht, die viel kleineren Ionen Na⁺ und Li⁺ viel schwerer durch die Membran transportiert werden. Anderseits kann ein bloßes Sieb keinen energieverbrauchenden Prozeß bewirken. Es muß

demnach eine direkte Wechselwirkung zwischen den aktiven (mit Makrotetrolidmolekülen besetzten) Stellen der Membran und den Alkaliionen
bestehen. Es kann daher angenommen werden, daß die beobachtete
Ionenselektivität nicht eine Eigenschaft der Mitochondrienmembran,
sondern der Makrotetrolide selber ist.

MUELLER und RUDIN (*21*) ist es gelungen, Membrane aus ganz anderem
Material (Lipide aus Rinderhirn, Herzlipide, Lecithin, Sphingomyelin)
mit Makrotetroliden zu imprägnieren. Diese Membrane zeigten eine

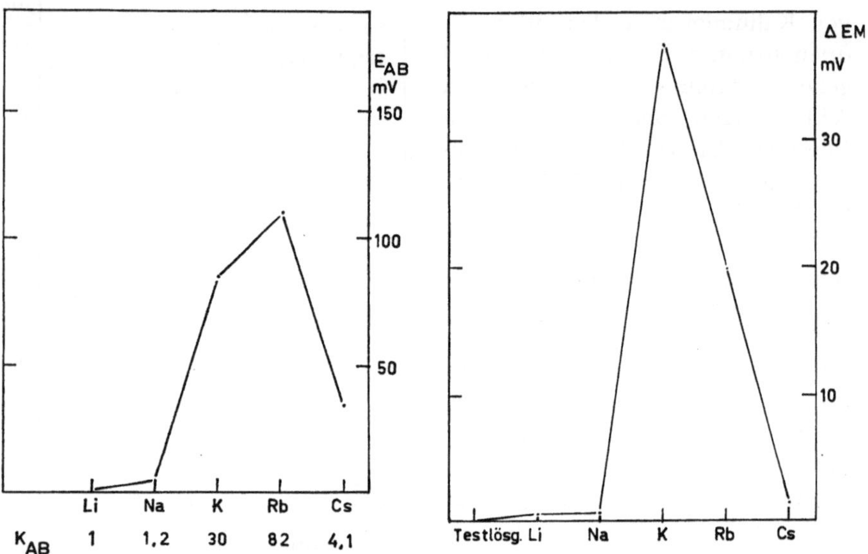

Abb. 4. Biionische Potentiale (E_{AB}) an Lipid-
membranen mit Dinactin. K_{AB} = Selektivitäts-
koeffizient. Nach MUELLER und RUDIN (*21*). [Aus:
Biochem. Biophys. Res. Comm. **26**, 398 (1967)]

Abb. 5. EMK-Messung an Makrotetrolidmembranen
auf Glasfritten. Anordnung der Meßzellen und Zusammensetzung der Testlösung, siehe (*29*)

ähnliche Ionenspezifizität wie die entsprechend behandelten Mitochondrienmembranen. Während LARDY und Mitarb. (*12—14*) wie auch
PRESSMAN und Mitarb. (*3, 20, 26*) vorwiegend das biochemische Verhalten
zur Beurteilung des Ionentransportes benützten, konnten MUELLER und
RUDIN durch direkte elektrochemische Messungen eine Ionendiskrimi-
nierung bestimmen. So bildete sich zwischen einer 0,05 N NaCl-Lösung
und einer 0,05 N KCl-Lösung, die durch eine aktive Membran getrennt
waren, ein Potential von 150 mV aus (negativ auf der K⁺-Seite). Die
Ionenspezifizität geht aus *Abb. 4* hervor. Die Art des Trägermaterials
war dabei praktisch ohne Einfluß auf die Meßwerte.

Einen Schritt weiter gingen STEFANAC und SIMON (*29*). Durch
Imprägnieren von inertem Trägermaterial (Filterpapier, Glasfritten etc.)

Literaturverzeichnis: SS. 187—189

mit übersättigten Makrotetrolidlösungen und Verdunsten des Lösungsmittels konnten Membrane hergestellt werden, die aus reinen Makrotetroliden bestanden. Diese wurden in einer elektrochemischen Meßkette eingesetzt und durch Bestimmen der EMK konnte eine ganz analoge Ionenabhängigkeit festgestellt werden, wie sie LARDY an biologischem Material gefunden hatte *(Abb. 5)*.

Die gute Übereinstimmung der Kurven in Abb. 3 und 5 läßt den Schluß zu, daß die biologische Wirkung der Makrotetrolide eine direkte Folge ihres elektrochemischen Verhaltens ist. Es stellt sich nun die Frage nach der Art der Wechselwirkung zwischen den Makrotetrolidmolekülen und den Alkaliionen. Eine naheliegende Vermutung besteht darin, daß die Makrotetrolide Komplexbildner für die einwertigen Metallionen mit unterschiedlichen Komplexstabilitäten sind. Der Beweis für die Richtigkeit dieser Annahme wurde von SIMON und Mitarb. (*24*) erbracht. Einerseits wurden beständige kristalline Komplexe des Nonactins und Monactins mit den Rhodaniden des Kaliums und Ammoniums hergestellt und analysiert* *(Tabelle 8)*, anderseits ergab die Messung der Stabilitätskonstanten einiger ausgewählter Komplexe einen direkten Zusammenhang mit dem elektrochemischen und biologischen Verhalten *(Tabelle 9)*.

Tabelle 8. **Eigenschaften der kristallisierten Alkalirhodanidkomplexe von Makrotetroliden**

Komplex	Schmelzpunkt	Bruttoformel*	IR-Spektra in Chloroform	
			ν (CO)	ν (CNS)
Nonactin-kalium-rhodanid	244—246°	$C_{41}H_{64}O_{12}KNS$	1710 cm^{-1}	2055 cm^{-1}
Nonactin-ammonium-rhodanid	198—202°	$C_{41}H_{68}O_{12}N_2S$	1719 cm^{-1}	2055 cm^{-1}
Monactin-kalium-rhodanid	240—242°	$C_{42}H_{66}O_{12}KNS$	1710 cm^{-1}	2055 cm^{-1}
Monactin-ammonium-rhodanid	200—205°	$C_{42}H_{70}O_{12}N_2S$	1719 cm^{-1}	2055 cm^{-1}

* In Übereinstimmung mit den Mikroanalysen.

Tabelle 9. **Komplexbildungskonstanten in Methanol bei 30°**

Ligand	K in kg · Mol^{-1} für die Kationen		$\dfrac{K_{K^+}}{K_{Na^+}}$
	Na$^+$	K$^+$	
Nonactin	$(1,3 \pm 0,2) \cdot 10^2$	$(5,0 \pm 0,7) \cdot 10^3$	38 ± 5
Monactin	$(1,1 \pm 0,1) \cdot 10^3$	$(2,5 \pm 1,0) \cdot 10^5$	230 ± 70

* Über die Bedeutung der Röntgen-Strukturanalyse des Nonactin-Kalium-rhodanid-Komplexes siehe Nachtrag auf S. 186.

Es erwies sich nämlich, daß die Komplexe des Kaliums um ein vielfaches stabiler sind als diejenigen des Natriums und auch die Unterschiede im biologischen Verhalten von Nonactin und Monactin (S. 182) finden ihre Parallele in den Komplexbildungskonstanten, die für Monactin um eine bis zwei Größenordnungen höher gefunden wurden als für Nonactin.

Die erwähnten Komplexe sind in aprotischen Lösungen stabil. Beim Auflösen in Wasser werden sie hydrolytisch gespalten, wobei das ursprüngliche Makrotetrolid auskristallisiert. Die Makrotetrolide scheinen auch ein beträchtliches Komplexierungsvermögen für Wasserstoffionen zu besitzen, worauf die Isolierung eines kristallisierten Nonactin-perchlorats durch Dutcher (7) hinweist.

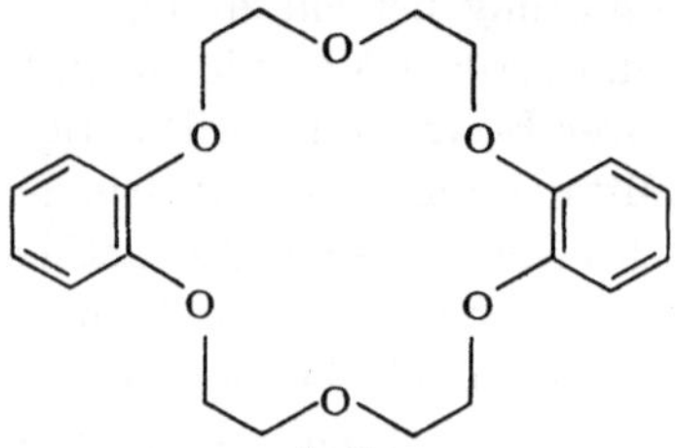

Es sei noch darauf hingewiesen, daß einige weitere Antibiotica bekannt sind, die sich biologisch den Makrotetroliden ähnlich verhalten. Zum Teil sind es ebenfalls makrocyclische Verbindungen, nämlich die Cyclodepsipeptide Valinomycin (*12, 20, 21, 26*) und die Enniatine (*21*). Überraschenderweise gehören dazu aber auch einige offenkettige Polypeptide, nämlich die Gramicidine A—D (*12, 26*). Es wird angenommen, daß sich diese Verbindungen in einer ringähnlichen Konformation an die Mitochondrienmembran anlagern. Das Komplexbildungsvermögen dieser Verbindungen ist noch nicht untersucht worden. Hingegen sind kürzlich einige synthetische Verbindungen hergestellt worden, die mit Alkaliionen und mit anderen Metallionen kristalline Komplexe geben (*23*). Interessanterweise handelt es sich auch hierbei um makrocyclische Polyäther vom Typus der Verbindung (46). Die biologischen Eigenschaften der letzteren und ihrer Derivate sind noch nicht bekannt.

Nachtrag

Seit der Fertigstellung des Manuskripts sind zwei Abhandlungen erschienen, auf deren kurze Erwähnung hier nicht verzichtet werden darf. Eine teilweise stereospezifisch verlaufende Synthese des 8-Desoxynonactinsäure-methylesters (*13*, S. 166) bestätigt die relative *cis*-Stellung der beiden Substituenten am Tetrahydrofuranring (vgl. Kap. IV, S. 170). Der entscheidende Schritt der Synthese (*8a*) bestand in der katalytischen Hydrierung eines 2,5-disubstituierten Furans mittels eines Rhodiumkatalysators, die erfahrungsgemäß ausschließlich zu *cis*-disubstituierten Produkten führt.

Von weitertragender Bedeutung ist die Röntgen-Strukturanalyse des in Kap. VI (S. 185) erwähnten Nonactin-Kaliumrhodanid-Komplexes [Kilbourn et al. (*16a*)]. Es ergab sich daraus:

a. Die aus dem chemischen und physikalischen Verhalten von Nonactinsäure und ihren Derivaten abgeleitete relative Konfiguration (35, S. 173) wird in allen Einzelheiten bestätigt.

b. Die bisher nur für das Dinactin experimentell belegte alternierende Anordnung der (+)- und (—)-Bausteine (allgemeine Formel 45, S. 179) gilt auch für das Nonactin. Für das komplexchemische Verhalten scheint diesem Aufbauschema eine entscheidende Bedeutung zuzukommen. Die früher geäußerte Vermutung (S. 179), daß diese Anordnung allen Makrotetroliden eigen ist, gewinnt damit an Wahrscheinlichkeit.

c. Das Kaliumion koordiniert acht Sauerstoffatome der Ligandmoleküle, die die Ecken eines nahezu regelmäßigen Hexaeders besetzen. Es sind dies die vier Äthersauerstoffe der Tetrahydrofuranringe und die vier Carbonylsauerstoffe der Estergruppen.

d. Um die unter *c* erwähnte Komplexierungsart zu ermöglichen, ist das Nonactinmolekül derart gefaltet, daß das Komplexion eine nahezu kugelige Gestalt annimmt. Die Konformation des macrocyclischen Ringes erinnert dabei an eine Figur, wie sie z. B. durch die Naht eines Tennisballs beschrieben wird. Im Gegensatz dazu liegt das freie Nonactin im Kristallgitter in einer weitgehend flachen Konformation vor, wie sich aus den Abmessungen der Elementarzelle ergibt (*6 a*).

Literaturverzeichnis

1. BECK, J., H. GERLACH, V. PRELOG und W. VOSER: Stoffwechselprodukte von Actinomyceten, 35. Mitt. Über die Konstitution der Makrotetrolide Monactin, Dinactin und Trinactin. Helv. Chim. Acta **45**, 620 (1962).

2. BENNETT, R. E., S. A. BRINDLE, N. A. GIUFFRE, P. W. JACKSON, J. KOWALD, F. E. PANSY, D. PERLMAN and W. H. TREJO: Production of a Cytotoxic Agent, SQ 15.859, by *Streptomyces chrysomallus*. Antimicrobial Agents Chemother. **1961**, 169.

3. COCKRELL, R. S., E. J. HARRIS and B. C. PRESSMAN: Energetics of Potassium Transport in Mitochondria Induced by Valinomycin. Biochemistry **5**, 2326 (1966).

4. CORBAZ, R., L. ETTLINGER, E. GÄUMANN, W. KELLER-SCHIERLEIN, F. KRADOLFER, L. NEIPP, V. PRELOG und H. ZÄHNER: Stoffwechselprodukte von Actinomyceten, 3. Mitt. Nonactin. Helv. Chim. Acta **38**, 1445 (1955).

5. DENISOVA, S. I., G. A. OVCHINNIKOVA und G. P. MENSHIKOV: Über das Kohlenstoffskelett einer Hydroxysäure, Hydrolyseprodukt des Antibioticums Fluorin. Zhurn. Obshchei Khim. **33**, 2058 (1963).

6. DOMINGUEZ, J., J. D. DUNITZ, H. GERLACH und V. PRELOG: Stoffwechselprodukte von Actinomyceten, 32. Mitt. Über die Konstitution von Nonactin. Helv. Chim. Acta **45**, 129 (1962).

6a. DUNITZ, J. D.: The Interpretation of Pseudo-Orthorhombic Diffraction Patterns. Acta Crystallogr. **17**, 1299 (1964), sowie unveröffentlichte neue Ergebnisse von DUNITZ und Mitarb.

7. DUTCHER, J. D.: Isolation and Characterization of a Cytotoxic Agent, SQ 15.859, from *Streptomyces chrysomallus*. Antimicrobial Agents Chemother. **1961**, 173.

8. Gerlach, H., G. Haas und V. Prelog: Über zwei cycloisomere Triglycyl-tri-L-alanyle. Helv. Chim. Acta **49**, 603 (1966).

8a. Gerlach, H. und E. Huber: Stoffwechselprodukte von Mikroorganismen, 60. Mitt. Synthese der 8-Desoxy-nonactinsäure. Helv. Chim. Acta **50**, 2087 (1967).

9. Gerlach, H., R. Hütter, W. Keller-Schierlein, J. Seibl und H. Zähner: Stoffwechselprodukte von Mikroorganismen, 58. Mitt. Neue Makrotetrolide aus Actinomyceten. Helv. Chim. Acta. **50**, 1782 (1967).

10. Gerlach, H., Yu. A. Ovchinnikov und V. Prelog: Über cycloenantiomere cyclo-Hexaalanyle und ein cycloenantiomeres cyclo-Diglycyl-tetraalanyl. Helv. Chim. Acta **47**, 2294 (1964).

11. Gerlach, H. und V. Prelog: Über die Konfiguration der Nonactinsäure. Liebigs Ann. Chem. **669**, 121 (1963).

12. Graven, S. N., H. A. Lardy and S. Estrada: Antibiotics as Tools for Metabolic Studies. VIII. Effect of Nonactin Homologs on Alkali Metal Cation Transport and Rate of Respiration in Mitochondria. Biochemistry **6**, 365 (1967).

13. Graven, S. N., H. A. Lardy, D. Johnson and A. Rutter: Antibiotics as Tools for Metabolic Studies. V. Effect of Nonactin, Monactin, Dinactin, and Trinactin on Oxidative Phosphorylation and Adenosin Triphosphatase Induction. Biochemistry **5**, 1729 (1966).

14. Graven, S. N., H. A. Lardy and A. Rutter: Antibiotics as Tools for Metabolic Studies. VI. Damped Oscillatory Swelling of Mitochondria Induced by Nonactin, Monactin, Dinactin, and Trinactin. Biochemistry **5**, 1735 (1966).

15. Horeau, A.: Principe et applications d'une nouvelle méthode de détermination des configurations dite „par dédoublement partiel". Tetrahedron Letters **1961**, 506.

16. Keller-Schierlein, W., H. Gerlach and J. Seibl: Identification Problems in the Macrotetrolide Series. Antimicrobial Agents Chemother. **1966**, 644.

16a. Kilbourn, B. T., J. D. Dunitz, L. A. R. Pioda and W. Simon: Structure of the K^+ Complex with Nonactin, a Macrotetrolide Antibiotic Possessing Highly Specific K^+ Transport Properties. J. Mol. Biol. **30**, 559 (1967).

17. Köhler, W., H. Thrum und R. Schlegel: Antibakterielles Wirkungsspektrum von Nonactin unter besonderer Berücksichtigung von Corynebacteriaceae. Zbl. Bakteriol. Parasitenk. **194**, 457 (1964).

18. Menshikov, G. P. und M. M. Rubinstein: Isolierung des neuen Antibioticums Longisporin und seine chemische Charakterisierung. Zhurn. Obshchei Khim. **26**, 2035 (1956).

19. Meyers, E., F. E. Pansy, D. Perlman, D. A. Smith and F. L. Weisenborn: The in vitro Activity of Nonactin and its Homologs: Monactin, Dinactin, and Trinactin. J. Antibiotics (Japan) A **18**, 128 (1965).

20. Moore, C. and B. C. Pressman: Mechanism of Action of Valinomycin on Mitochondria. Biochem. Biophys. Res. Comm. **15**, 562 (1964).

21. Mueller, P. and D. O. Rudin: Development of K^+-Na^+ Discrimination in Experimental Bimolecular Lipide Membranes by Macrocyclic Antibiotics. Biochem. Biophys. Res. Comm. **26**, 398 (1967).

22. Nishimura, H., M. Mayama, T. Kimura, A. Kimura, Y. Kawamura, K. Tawara, Y. Tanaka, S. Okamoto and H. Kyotani: Two Antibiotics Identical with Nonactin and Valinomycin Obtained from a *Streptomyces tsusimaensis* n. sp. J. Antibiotics (Japan) A **17**, 11 (1964).

23. Pedersen, C. J.: Cyclic Polyethers and their Complexes with Metal Salts. J. Amer. Chem. Soc. **89**, 2495 (1967).

24. Pioda, L. A. R., H. A. Wachter, R. E. Dohner und W. Simon: Komplexe von Nonactin und Monactin mit Natrium-, Kalium- und Ammoniumionen. Helv. Chim. Acta **50**, 1373 (1967).

25. PRELOG, V. und H. GERLACH: Cycloenantiomerie und Cyclodiastereomerie. Helv. Chim. Acta 47, 2288 (1964).
26. PRESSMAN, B. C.: Induced Active Transport of Ions in Mitochondria. Proc. Nat. Acad. Sci. (USA) 53, 1076 (1965).
27. SCHMITZ, H., S. B. DEAK, K. E. CROOK, Jr. and I. R. HOOPER: Peliomycin, a New Cytotoxic Agent. I. Production, Isolation, and Characterization. Antimicrobial Agents Chemother. 1963, 89.
28. SHIBATA, M., K. NAKAZAWA, M. INOUE, J. TERUMICHI and A. MIYAKE: A New Antibiotic, Lustericin, Produced by a Streptomyces No. 10.400. Annu. Rep. Takeda Res. Labs. 17, 19 (1958).
29. STEFANAC, Z. and W. SIMON: Ion Specific Electrochemical Behaviour of Macrotetrolides in Membranes. Microchem. J. 12, 125 (1967).
30. WALLHÄUSSER, K. H., G. HUBER, G. NESEMANN, P. PRÄVE und K. ZEPF: Die Antibiotica FH 3582 A und B und ihre Identität mit Nonactin und seinen Homologen. Arzneimittelforsch. 14, 356 (1964).

(Eingelaufen am 1. Oktober 1967)

Limonoid Bitter Principles

By **DAVID L. DREYER,** Pasadena, California

Contents

I. Introduction

The limonoid bitter principles are a class of C_{26} degraded triterpenes believed to arise as oxidation products of tetracyclic triterpenes. A second group of related bitter principles, the simaroubolides, occur as further breakdown products in this series (*30, 55, 97*). Limonoids are a stereochemically homogeneous group and are all biogenetically derivable from a tirucallol type tetracyclic triterpene.

The limonoids and simaroubolides occur in three very closely related plant families, the Rutaceae, Meliaceae and Simaroubaceae. The Rutaceae is a plant family with extensive and diversified synthetic ability. Many of its metabolites are highly oxidized and modified relative to the normal groups of natural products. Striking in this regard is the relatively frequent occurrence of epoxy groups in its metabolites. Limonoids occurring in the Meliaceae show, in general, much greater structural variation and complexity than those in the Rutaceae. They vary from relatively simple compounds to highly modified C_{26}-terpenoids which have undergone extensive structural alterations.

The distinctive structural features of limonoids arise from (a) the catabolism of a normal steroid side-chain and formation of a β-substituted furan ring with loss of four carbon atoms, and (b) the frequent expansion of a carbocyclic ring into a lactone or ring-opened derivative. The latter process is formally analogous to a Baeyer-Villiger reaction or photochemical ring opening of an alicyclic ketone, as in andirobin and nimbin (see Sections 9 and 10, pp. 216, 219). This ring opening can be further modified by reclosure to give limonoids of the highest structural complexity yet found. Such is the case in methyl angolensate or mexicanolide and swietenine (see Section 11, p. 223).

The numbering system of limonoids (p. 193) is the same as for their biogenetic precursors, the tetracyclic triterpenes, and is retained in the highly rearranged limonoids of the mexicanolide type (see Section 11, p. 223).

There has been a rapid expansion in studies of limonoid bitter principles in the past six years. This review is largely concerned with a survey of the main chemical features and biogenetic interrelationships of these bitter principles.

II. Structure Determination and Chemistry of Limonoids

1. Limonin (1)

Limonin, the major limonoid in *Citrus* seeds, has been known for over a hundred years (*21*) and, as a reasonably accessible material, is the most extensively studied member of the limonoids. Its chemical structure determination was reported by Arigoni et al. (*15*) in 1960 and confirmed by X-ray crystallographic studies on the iodoacetate of epilimonol (*16, 17*). Interest in limonin has been maintained because of its possible commercial production as a by-product in the citrus processing industry and because it is responsible for an undesirable bitterness in certain processed citrus products (*67, 83*).

The presence of limonin is responsible for bitterness in some processed citrus products, especially orange juices from California Navel oranges and Valencia oranges grown in Australia. Limonin also contributes to the bitterness of products made from grapefruit, although the flavanone naringin is the major bitter material in grapefruit (*92*). Limonin does not cause bitterness in citrus fruit or in freshly prepared juice. The bitterness developes only after the juice has been heated or allowed to stand for 12–24 hours. In order to explain this delayed bittering process, it has been proposed that limonin exists in the fresh fruit as a nonbitter precursor. Kefford (*83*) in Australia has proposed, however, that the delayed bittering period is the length of time required for the limonin to diffuse from the tissue into the juice. Limonin must be in solution in order to taste bitter.

Maier and Beverly (*91*) have recently shown that the delayed bittering is actually due to the presence of a form of limonin in which one of the lactone groups is opened. The ring-opened limonic acid is easily studied by paper electrophoresis in buffered solution and detected by Ehrlich's reagent. This form is stable at the nearly neutral pH of the tissue. When membrane walls are broken during juicing, the ring-opened form is closed to limonin by the acidic juice (pH ~ 3.5).

The structure determination of many limonoids subsequently isolated has followed the same pattern as that used for limonin. The proposed biogenetic origin of limonin from a tetracyclic triterpene, suggested by Arigoni et al. (*15*) has received substantial circumstantial support from the structures of limonoids subsequently isolated. The X-ray crystallographic structure determination shows that limonin can be formulated

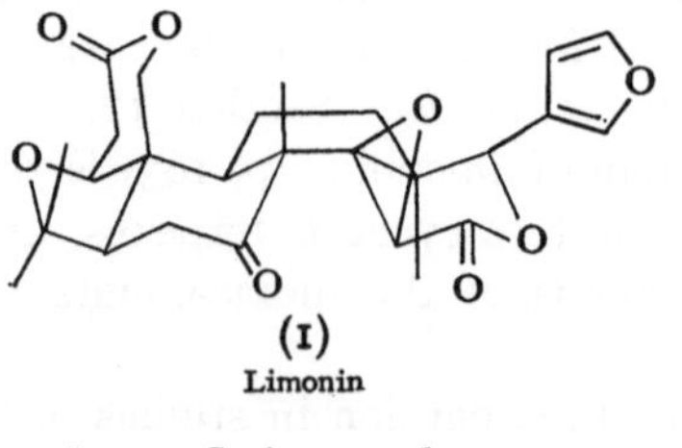

(1)

Limonin

as (1). Limonin has a boat *C*-ring and two severe 1,3-diaxial methyl-methylene interactions. The X-ray structure determination shows that the *A*-ring is in a half-boat conformation (*16, 17*).

Investigations on limonin have revealed a number of rearrangements and chemical conversions which have been previously summarized (*52, 113*). The chemistry of limonin is dominated by its base-catalyzed conversions. Studies of strongly acid-catalyzed reactions have been less successful due to the presence of the acid-sensitive furan ring. The conversions of limonin (**1**) to limonin diosphenol (**2**) (*20*) and of limonol (**3**) to merolimonol (**4**) (*53, 95*) *(Chart 1)* have had the greatest impact on further work in this field.

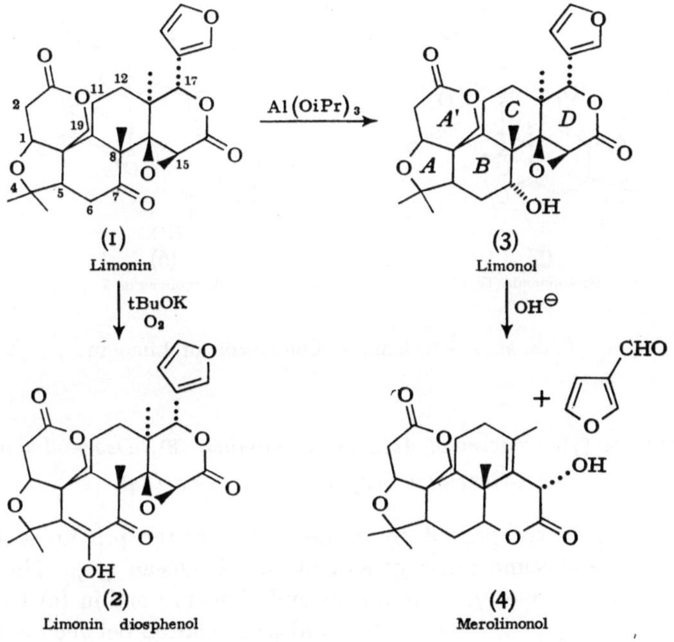

(**1**)
Limonin

(**3**)
Limonol

Al(OiPr)₃

tBuOK
O₂

OH⊖

CHO

(**2**)
Limonin diosphenol

(**4**)
Merolimonol

Chart 1. Some Important Reactions of Limonin

Recent photochemical studies (*62*) on limonin have shown that it can be converted into two new photoproducts *(Chart 2)*. The major product, photolimonin II (**5**), which shows properties very similar to those of limonin, appears to be a C-8 stereoisomer. The second product, photolimonin I (**6**), is a *B*-ring-opened aldehyde. The existence of both of these photoproducts might be predicted from classical photochemical considerations on saturated cyclic ketones. Photolimonin I is of special interest because of its structural similarity to andirobin and methyl angolensate (see Section 9, p. 216) both of which are naturally occurring *B*-ring *seco*-limonoids. In theory, a similar fragmentation could account for the formation of andirobin from a gedunin derivative (Section 9).

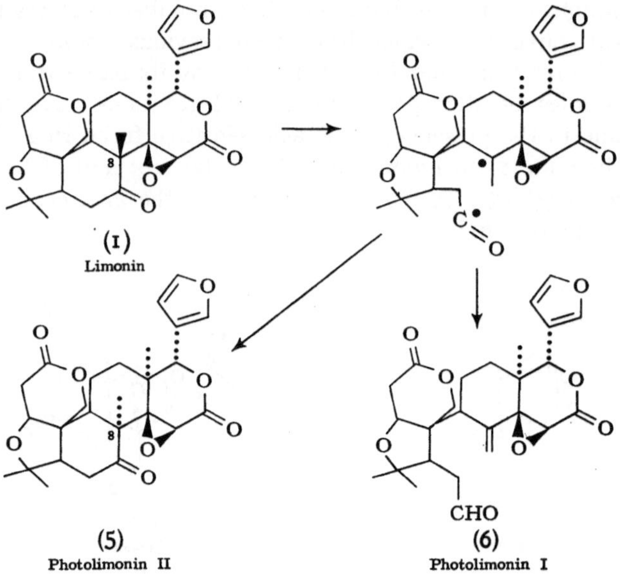

Chart 2. Photochemical Conversion of Limonin

2. Obacunone (7), Obacunoic Acid (10), Nomilin (8), Deacetylnomilin (9), 7α-Obacunol (12) and Veprisone (14)

Limonin is accompanied by three minor bitter principles in most *Citrus* seeds and some other genera of the Rutaceae (*58*). These three limonoids, obacunone (7), nomilin (8) and deacetylnomilin (9) have been interconverted (*57, 66, 68*) *(Chart 3)* and a correlation between obacunone and limonin has been established through a lengthy series of transformations (*86*).

The correlation between obacunone and limonin, which involves, in part, the limonol (3) to merolimonol (4) conversion, established the identity of all the stereochemical centers with the exception of C-17. The stereochemistry at C-17 is judged to be the same in both systems by (a) common botanical origin; (b) similarity of the chemical shifts in the nuclear magnetic resonance spectra in both systems for H-17 (*56*) and (c) similarity of the ORD and CD curves in the short wavelength region of the two systems (*61*).

The stereochemistry at the I-position in nomilin (8) and deacetyl-nomilin (9) has not yet been rigorously established in spite of substantial efforts in this direction (*61*). Present circumstantial evidence (*57, 58*) suggests that the final stages in the formation of limonin are as follows: deacetylnomilin (**9**) → nomilin (**8**) → obacunone (**7**) → obacunoic acid (**10**) → isoobacunoic acid (**11**) → limonin (**1**).

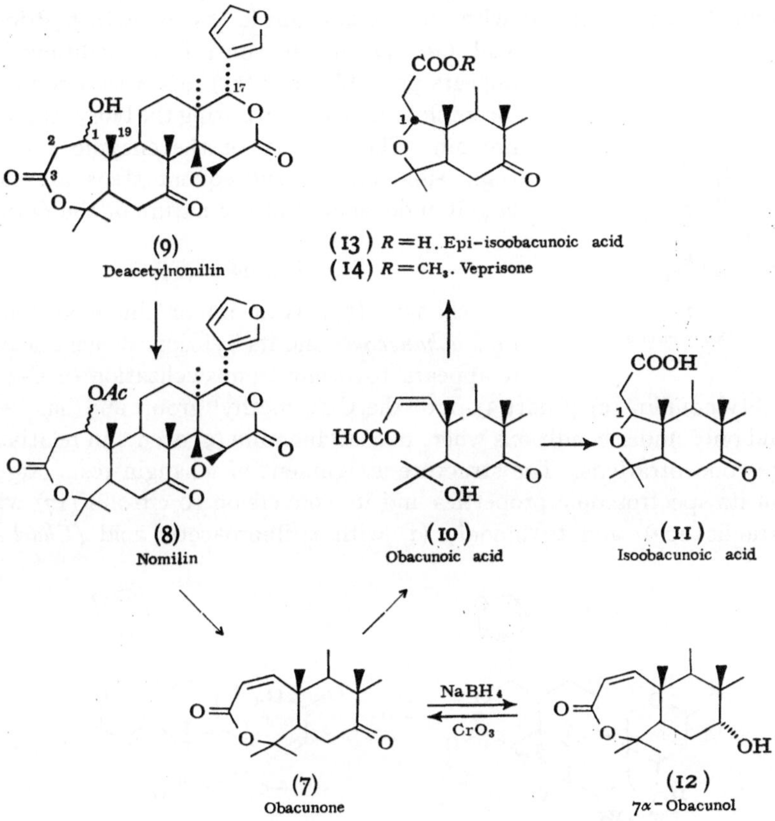

Chart 3. Interconversions of Limonin Congeners

7α-Obacunol (**12**) occurs in *Casimiroa edulis* Llave et Lex (*61a*) and had previously been prepared by DEAN and GEISSMAN (*54*) by reduction of obacunone (**7**).

Veprisone (**14**) (*72*) is the methyl ester of epi-isoobacunoic acid (**13**), which is the ring-closed product of obacunoic acid (**10**) having stereochemistry at C-1 opposite from that of limonin and its derivatives. The isolation from natural sources of the other C-1 epimer, isoobacunoic acid (**11**) appears to be only a matter of time. Although obacunone is easily converted with base to obacunoic acid (**10**), the natural occurrence of the latter has been reported only recently (*108*).

3. Deoxylimonin (15)

Deoxylimonin has been found in very small amounts in grapefruit seeds (57) (*Citrus paradisi* MacF.) but its isolation from this source demands tedious chromatography. It is best prepared directly from limonin (1) by treatment with chromous chloride (*15*) or with hydriodic acid (*20, 54*) under controlled conditions. It appears probable that (15) arises as a result of incomplete epoxidation during the biosynthesis of limonin. The activity of the enzyme systems responsible for the subsequent steps are then largely independent of the nature of the *D*-ring.

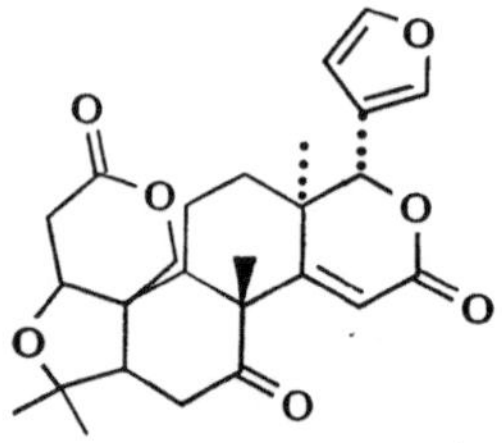

(15)
Deoxylimonin

4. Ichangin (16)

Ichangin (59) is a minor limonoid found in *C. ichangensis* and its hybrids. Biogenetically, it appears to result from cyclization of C-3 in deacetylnomilin (9) (Chart 3) onto the C-19 methyl group and has been found only under conditions where deacetylnomilin (9) occurs in relatively large concentrations. The structure assignment of ichangin rests largely upon its spectroscopic properties and its conversion to citrolin (17) with hydriodic acid, and to limonin (1) with trifluoroacetic acid *(Chart 4)*.

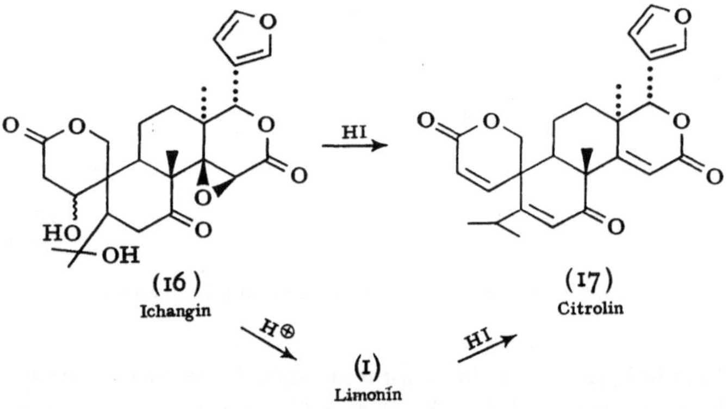

(16)
Ichangin

(17)
Citrolin

(1)
Limonin

Chart 4. Reactions of Ichangin

The occurrence of small amounts of deoxylimonin (15) and ichangin (16) in citrus seeds raises some questions concerning the biogenesis of limonoids. Deoxylimonin (15) appears to have undergone all the enzymatic conversions leading to limonin except the epoxidation step. This suggests that the structural requirements for the substrate are relatively non-specific for many of the later biogenetic steps leading to limonin (*33*).

This is further illustrated by the occurrence of ichangin (**16**), arising presumably from deacetylnomilin (**9**) by oxidation of C-3 onto the C-19 methyl group. Thus, the enzyme system responsible for oxidation of C-19 operates on deacetylnomilin as on obacunone derivatives.

5. *Limonin Diosphenol (Evodol)* (**2**) *and Rutaevin* (**18**)

These two highly oxidized limonoids have been found in only one subfamily of the Rutaceae (*58*). Limonin diosphenol (**2**) is easily prepared directly from limonin (**1**) by a reaction developed by BARTON and co-workers (*18, 20*). It was later recognized by HIROSE (*79*) that limonin diosphenol is also a natural product. Rutaevin (**18**), limonin (**1**) and limonin diosphenol (**2**) have been found co-occurring in three genera of the Rutaceae (*58*) (see *Table 2*, p. 236).

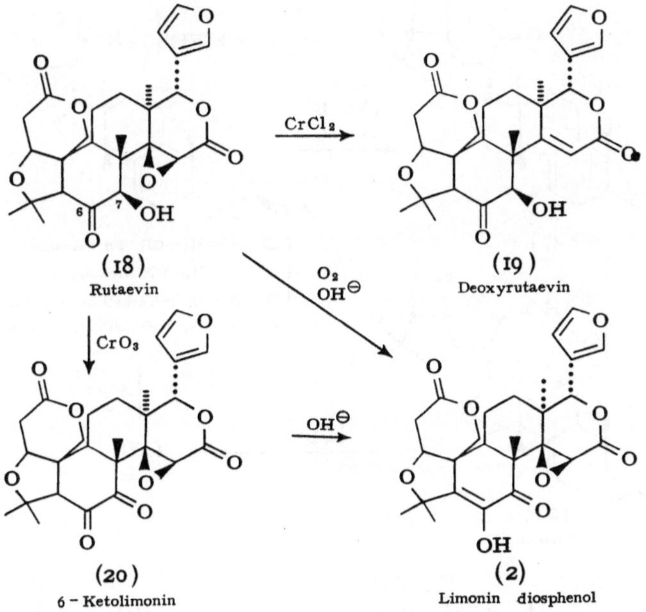

Chart 5. Reactions of Rutaevin

Rutaevin (**18**) has been known for about thirty years (*43, 69*) and proved to be closely related to limonin (**1**). It contains an α-ketol group which easily undergoes base-catalyzed air oxidation to limonin diosphenol (**2**). Reduction of rutaevin with chromous chloride *(Chart 5)* gives the expected deoxy derivative (**19**). Rutaevin is converted to a yellow, enolizable α-diketone (**20**) with chromic acid. Consideration of these conversions with the related spectroscopic properties allows rutaevin to be formulated as 6-keto-epilimonol (**18**) (*60*).

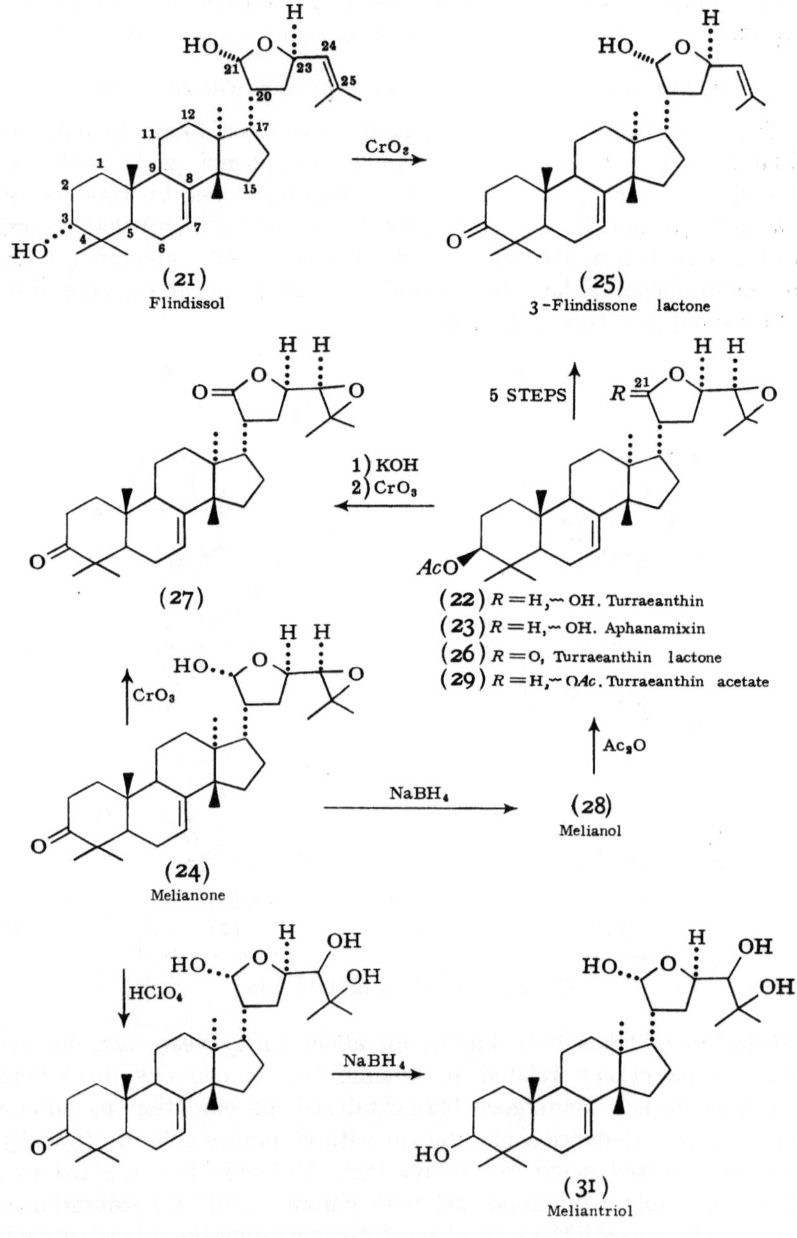

Chart 6. Interconversions of Flindissol Congeners

References, pp. 238—244

6. *Flindissol* (**21**), *Turraeanthin* (**22**), *Aphanamixin* (**23**) and *Melianone* (**24**)

Flindissol (*29, 32*), turraeanthin (*22*), aphanamixin (*42*) and melianone (*82, 87, 89*) are normal C_{30}-triterpenes which are believed to be proto-limonoids by their botanical origin, stereochemistry and tetrahydrofuran system at C-17. Turraeanthin (**22**) is an epoxy flindissol and has been correlated with flindissol by removal of the epoxy group and oxidation of the 3- and 21-hydroxyl groups to give 3-flindissone lactone (**25**) *(Chart 6)*. BIRCH and co-workers (*29*) have in turn related flindissol (**21**) to elema-dienolic acid. The stereochemistry at C-23 in turraeanthin (**22**) was established by application of Klyne's modification of Hudson's lactone rule (*22*). The stereochemistry at C-23 in (**21**) and (**23**) follows from this result. CHATTERJEE and KUNDU (*42*) consider aphanamixin to be a C-21 epimer of turraeathin (**22**), although (**22**) in solution it appears to be a C-21 epimeric mixture (*22*). Chromic acid oxidation of aphanamixin (**23**) gives turraeanthin lactone (**26**) showing the C-21 epimeric relationship between (**22**) and (**23**). Melianone (**24**) (*89*) is a 3-keto triterpene and its ORD curve is superimposable on that of 3-flindissone lactone (**25**), establishing the *A/B*-ring stereochemistry and the position of the double bond in the former. Chromic acid oxidation of melianone (**24**) gives a γ-lactone (**27**) which is presumably identical with 3-turraeanthinone lactone (**27**). Melianone is thus 24,25-epoxyflindissone. Borohydride reduction of melianone yields melianol (**28**). The derived acetate, melianol acetate, is identical with turraeanthin acetate (**29**).

Treatment of melianone (**24**) with hydrobromic acid in acetic acid gave a bromoacetal formulated as (**30**) (*89*). The stereochemistry at C-23 and C-24 follows from the coupling constants of these protons in the nuclear magnetic resonance spectrum and supports previous stereochemical assignments made on turraeanthin (**22**). The hydroxy group at C-21 is α-oriented as was postulated for flindissol (**21**) (*29*). LAVIE and co-workers (*90*) have reported the isolation of a locust antifeeding principle from two

Melia species and identified it as meliantriol (**31**); direct structural proof was obtained by conversion of melianone (**24**) to meliantriol (**31**) in two steps (Chart 6, p. 198).

Turraeanthin (**22**) is found in a genus very closely related to *Aglaia* in which a related tetracyclic triterpene, aglaiol (**32**), occurs (*119*). Ritchie and co-workers (*31*) have reported the structures of three closely related triterpenes, bourjotinolone A (**33**), bourjotinolone B (**34**) and bourjotone (**35**) *(Chart 7)*, as well as of further artifacts

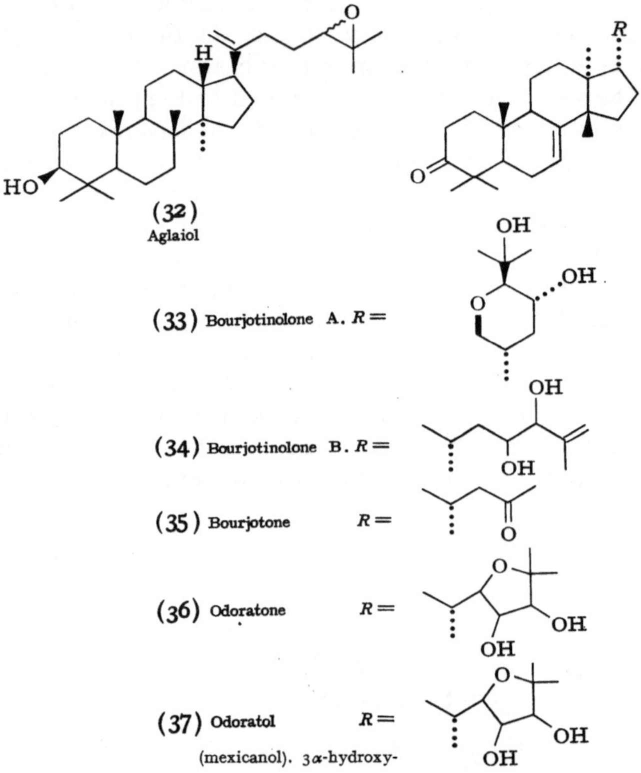

(32)
Aglaiol

(33) Bourjotinolone A. R =

(34) Bourjotinolone B. R =

(35) Bourjotone   R =

(36) Odoratone   R =

(37) Odoratol   R =
(mexicanol). 3α-hydroxy-

Chart 7. Protolimonoids

which occur in the extensively studied genus *Flindersia*. Chan and co-workers (*41*) as well as Overton and his group (*46*) have recently reported the structure determination of two further triterpenes, odoratone (**36**) and odoratol (**37**) which co-occur with limonoids in *Cedrela odorata* L. These triterpenes have the prerequisite stereochemistry of a limonoid precursor; and the stereochemistry at C-23 and C-24 is identical in melianone, melianol, turraeanthin, flindissol, boujotinolone A and B.

References, pp. 238—244

*7. Cedrelone (38), Anthothecol (46), Havanensin-1,7-diacetate (58),
Havanensin-3,7-diacetate (57), Havanensin-1,3,7-triacetate (56),
Heudelottin (66), Hirtin (74), Deacetylhirtin (75), Grandifolione (81),
Grandifolione Acetate (Khayanthone) (84), Azadirone (86),
Azadiradione (87) and Epoxyazadiradione (88)*

The thirteen limonoids of this group have the intact carbocyclic
system of the tetracyclic triterpene precursor. Acid-catalyzed epoxide

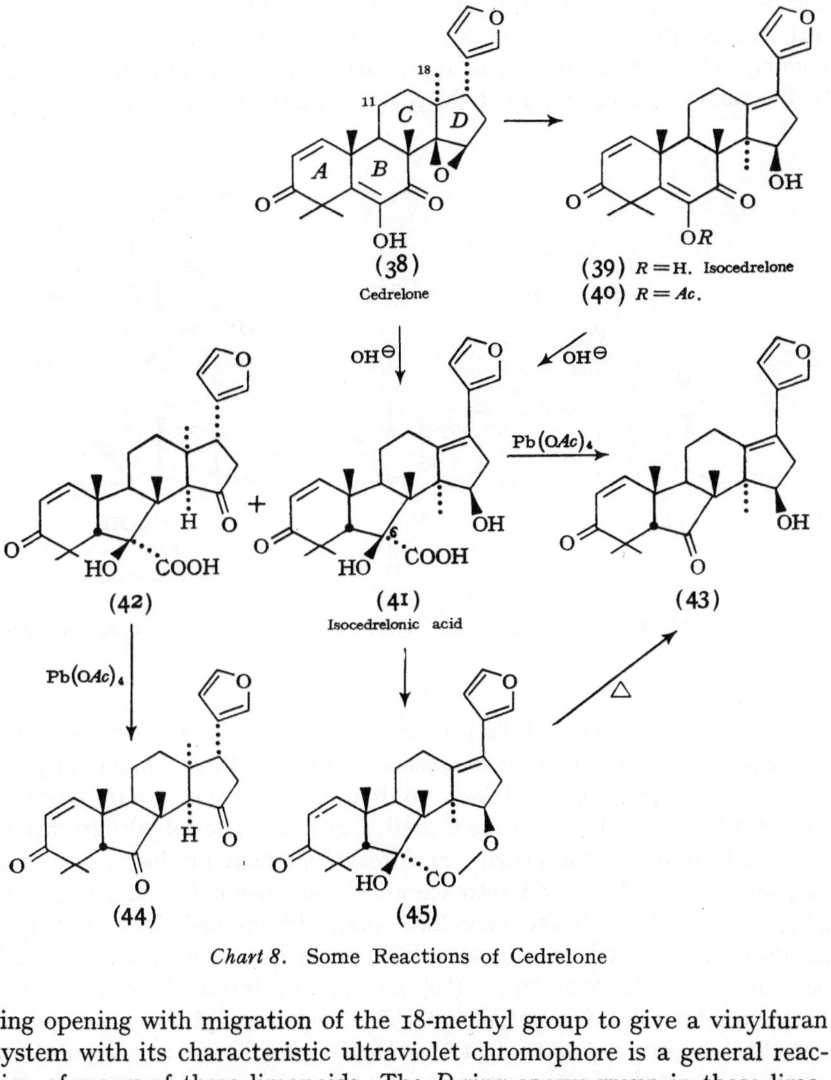

Chart 8. Some Reactions of Cedrelone

ring opening with migration of the 18-methyl group to give a vinylfuran
system with its characteristic ultraviolet chromophore is a general reaction of many of these limonoids. The *D*-ring epoxy group in these limo-

noids appears to undergo acid-catalyzed ring opening more easily than in the α,β-epoxy-δ-lactone system of limonin (1) and its congeners. Such acid-catalyzed reactions of limonin are best performed on its tetrahydro derivative where decomposition of the acid sensitive furan ring is not a problem. In limonin simple epoxide ring opening occurs without migration of the C-18 methyl group (20). Both cedrelone (38) and dihydrocedrelone (A-ring saturated) undergo this conversion to the iso series (39), (41) (71, 73, 80) in which the boat C-ring becomes a chair. Benzilic acid rearrangement of the diosphenol group in cedrelone (38) (*Chart 8*) and anthothecol (46) (Chart 10, p. 203) is also accompanied by epoxide ring opening. Oxide ring opening in these latter cases (46) appears to occur in the course of acidification of the reaction mixture during the work-up.

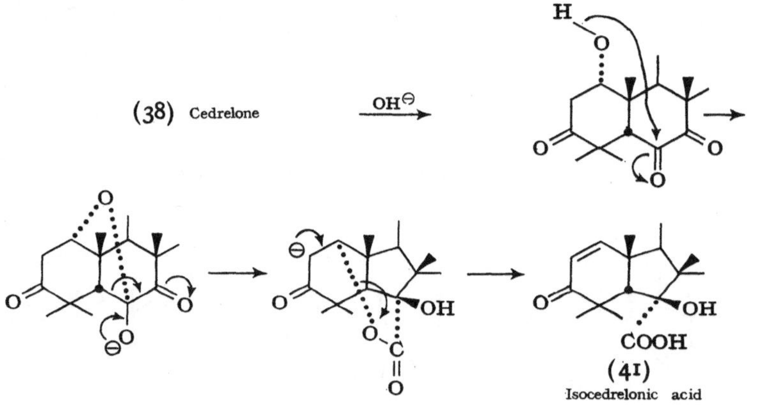

Chart 9. Probable Mechanism of the Conversion of Cedrelone to Isocedrelonic Acid

The products (41) and (42) from the benzilic acid rearrangement of cedrelone (38) appear to have a *cis* A/B ring juncture. Models suggest that the corresponding A/B *trans* system suffers from severe 1,3-non-bonded interactions between the C-methyl groups. The α-hydroxy carboxylic acid groups in the benzilic acid rearrangement products, (41) and (42), are converted by lead tetraacetate to the B-nor ketones, (43) and (44), respectively. Unlike cedrelone (38), dihydrocedrelone and iso-dihydrocedrelone (39) (A-ring saturated) fail to undergo a benzilic acid rearrangement. Hodges et al. (80) suggest an intramolecular process to account for these facts *(Chart 9)*. This sequence (38) → (41) requires that the 6-hydroxyl group of (41) and (42) would have a β-orientation and that the A/B ring juncture is *cis*.

References, pp. 238—244

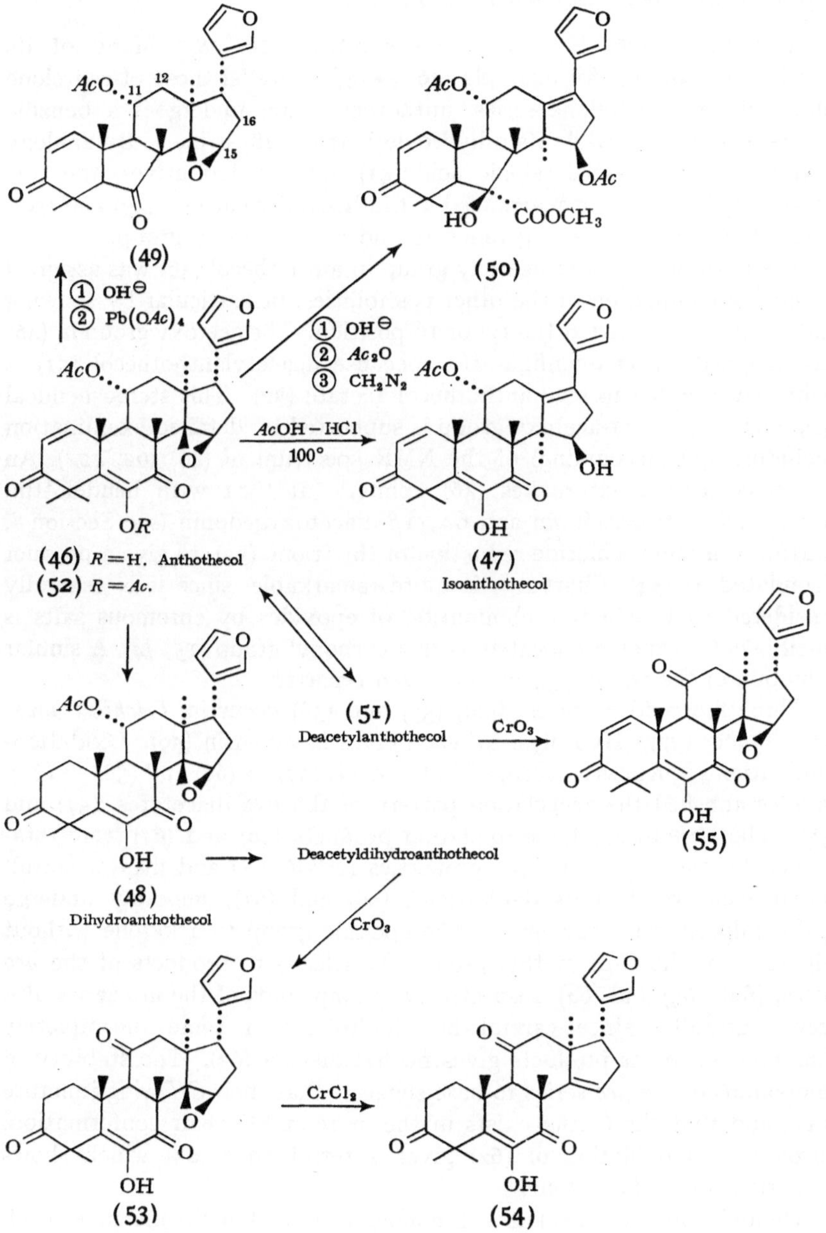

(49)

① OH⊖
② Pb(OAc)₄

(50)

① OH⊖
② Ac₂O
③ CH₂N₂

AcOH – HCl
100°

(46) R = H. Anthothecol
(52) R = Ac.

(47)
Isoanthothecol

(51)
Deacetylanthothecol

CrO₃

(55)

(48)
Dihydroanthothecol

Deacetyldihydroanthothecol

CrO₃

(53)

CrCl₂

(54)

Chart 10. Some Reactions of Anthothecol

The constitution of cedrelone (**38**) is supported by an X-ray crystallographic structure determination of its iodoacetate (*74*). The IR spectra of cedrelone and its derivatives have been discussed in detail (*34*).

Anthothecol (**46**) is 11α-acetoxycedrelone (*25, 28*). Many of its reactions *(Chart 10)*, for example (**46**) → (**47**), parallel those of cedrelone (**38**). Like the cedrelone series, anthothecol (**46**) undergoes a benzilic acid rearrangement while its dihydro derivative (**48**) is inert. By analogy with the data on isocedrelonic acid (**41**) and its derivatives, the corresponding benzilic acid-rearranged anthothecol derivatives, (**49**) and (**50**), should have an *A/B cis* ring juncture and a 6β-hydroxy group.

The position of the 11-acetoxy group in anthothecol (**46**) was assigned largely by elimination of the other possibilities, in particular by showing that it was not located at the 12- or 16-position. The acetoxy group in (**46**) was assigned an 11α-configuration because deacetylanthothecol (**51**) is easily reacetylated to give anthothecol acetate (**52**). This stereochemical assignment of the 11-acetoxy group is supported by detailed examination (including spin decoupling) of the NMR spectrum of (**46**) (*62, 122*). An α-acetoxy assignment makes (**46**) epimeric at C-11 with heudelottin, hirtin, 11β-acetoxygedunin and 6α,11β-diacetoxygedunin (see Section 8, p. 210). Chromous chloride reduction of the trione (**53**), to give a product formulated as (**54**) (Chart 10), is quite remarkable, since it is generally considered that reductive elimination of epoxides by chromous salts is possible only if they are located α- to a carbonyl group (*15, 44*). A similar reduction of the trione (**55**) has not been reported.

Three limonoid acetates, (**56**), (**57**) and (**58**) occur in *Trichilia havanensis* Jacq. (*37*). Hydrolysis of each gives havanensin (**59**). Oxidation-elimination sequences as outlined in *Chart 11*, (**57**) → (**60**) and (**58**) → (**61**), have established the acetylation pattern of the two diacetates, (**57**) and (**58**). The structures of the oxidation products (**60**) and (**61**) were established by their spectroscopic properties (NMR, CD and mass spectra). Havanensin (**59**) and its derivatives, (**56**) and (**61**), smoothly undergo acid-catalyzed rearrangement of the epoxide group to a ketone without migration of the C-18 methyl group. This leads to products of the *neo* series, (**62**), (**63**) and (**64**), respectively. Compounds of the *neo* series also occur naturally, since careful base-hydrolysis of some incompletely characterized plant products gives neohavanensin (**62**). The stability of compounds of the *neo* series to base suggests that their *C/D* ring juncture is *cis* and that the *C*-ring exists in the more stable chair conformation. Chromic acid oxidation of (**62**) gives a tetraketone (**65**) which shows properties of a β-diketone.

Heudelottin (**66**) (*111*) is a limonoid esterified with one mole each of acetic, α-methylbutyric and a structurally unclarified dihydroxylic C_9-carboxylic acid. Base hydrolysis of dihydroheudelottin (**67**) yields a

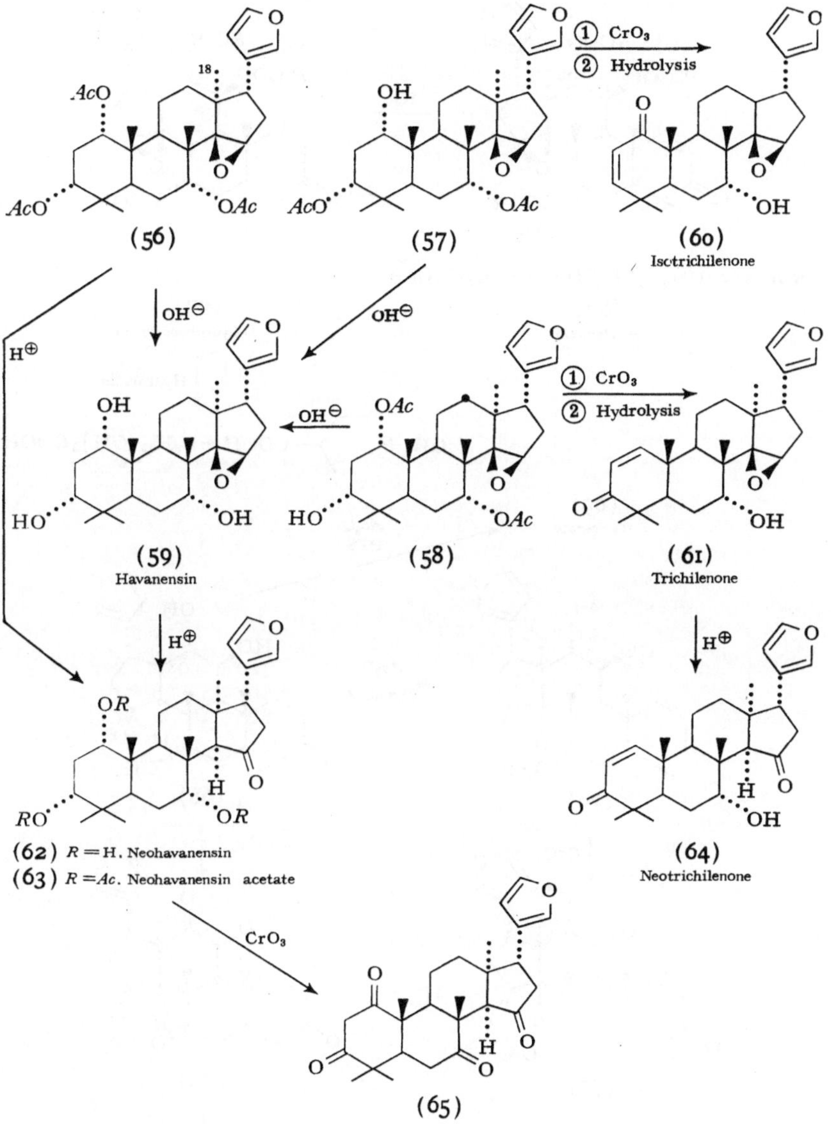

(56)

(57)

① CrO₃
② Hydrolysis

(60)
Isotrichilenone

OH⊖

OH⊖

H⊕

OH⊖

① CrO₃
② Hydrolysis

(59)
Havanensin

(58)

(61)
Trichilenone

H⊕

H⊕

(62) R = H. Neohavanensin
(63) R = Ac. Neohavanensin acetate

(64)
Neotrichilenone

CrO₃

(65)

Chart 11. Interconversions of Havanensin Derivatives

product initially formulated by OKORIE and TAYLOR (*111*) as a 7,14-oxide. Structure (68) has been suggested by CHAN and co-workers (*37*), by analogy with havanensin, as being more consistent with the published data; the conversions of heudelottin outlined in *Chart 12* have been so formu-

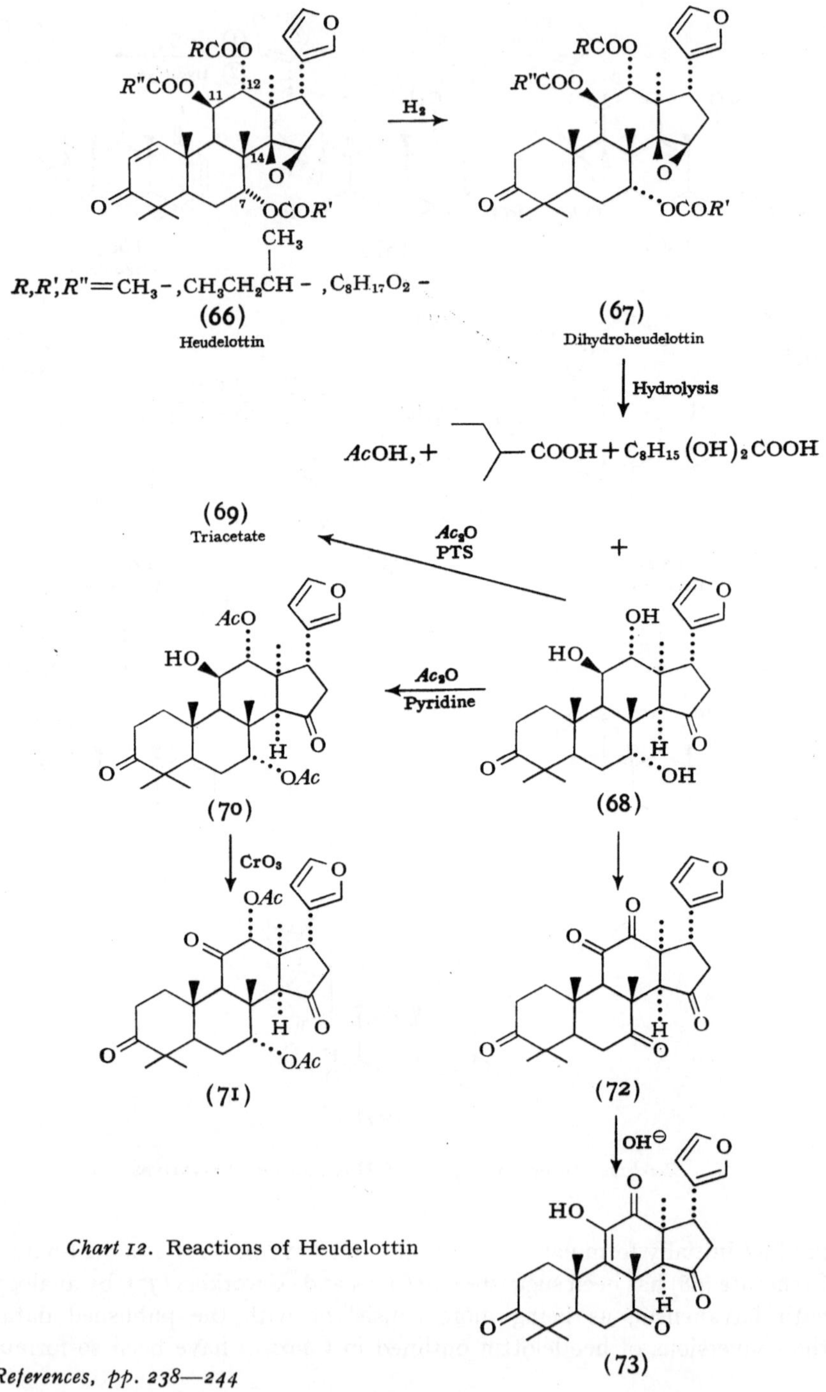
RCOO
R"COO
11
12
14
7
OCOR'
H₂
CH₃
R,R',R" = CH₃-, CH₃CH₂CH-, C₈H₁₇O₂-
(66)
Heudelottin
(67)
Dihydroheudelottin
Hydrolysis
AcOH, + COOH + C₈H₁₅(OH)₂COOH
(69)
Triacetate
Ac₂O
PTS
+
AcO
HO
OH
Ac₂O
Pyridine
H
OAc
(70)
(68)
CrO₃
OAc
(71)
(72)
OH⊖
HO
(73)

Chart 12. Reactions of Heudelottin

lated. The triol (68) gives either a tri- (69) or a diacetate (70). The latter can be oxidized to an α-acetoxyketone formulated as (71). Oxidation

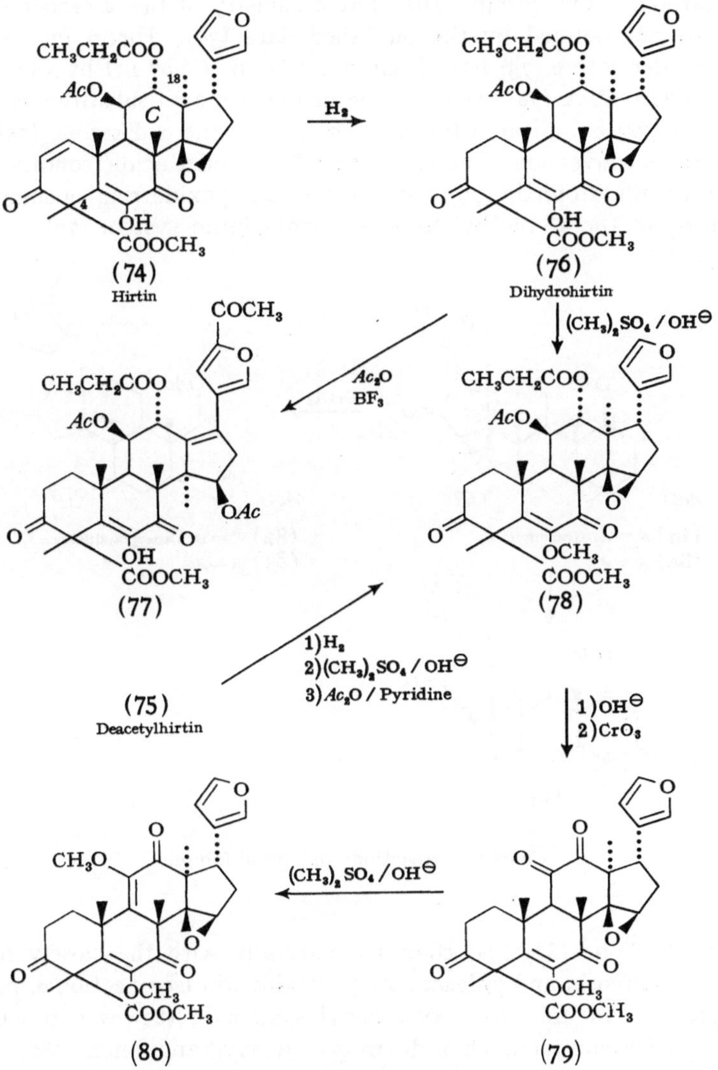

Chart 13. Reactions of Hirtin and Deacetylhirtin

of the triol (68) gives an α-diketone (72) which can be converted to a diosphenol (73) with base. The assigned stereochemistry at C-11 and C-12 in heudelottin (66) is the same as that of hirtin (74).

Hirtin (**74**) and its co-occurring deacetyl derivative (**75**), recently reported by CHAN and TAYLOR (*39*), are further cedrelone derivatives in which one of the C-4 *gem* dimethyl groups has been biogenetically oxidized to a carbomethoxy group. The stereochemistry of this 4-carbomethoxy group is not defined by the published data (*39*). Hirtin (**74**) and its deacetyl derivative (**75**) have been interrelated *(Chart 13)* by way of the dihydro derivative (**78**). The location of the two acyl substituents in the C-ring of (**74**) is indicated by the formation of the α-diketone (**79**) from (**78**) and its conversion to an enol ether (**80**). Under acidic conditions the D-ring of dihydrohirtin (**76**) shows the usual epoxide ring opening with migration of the 18-methyl to give a vinyl furan system (**77**).

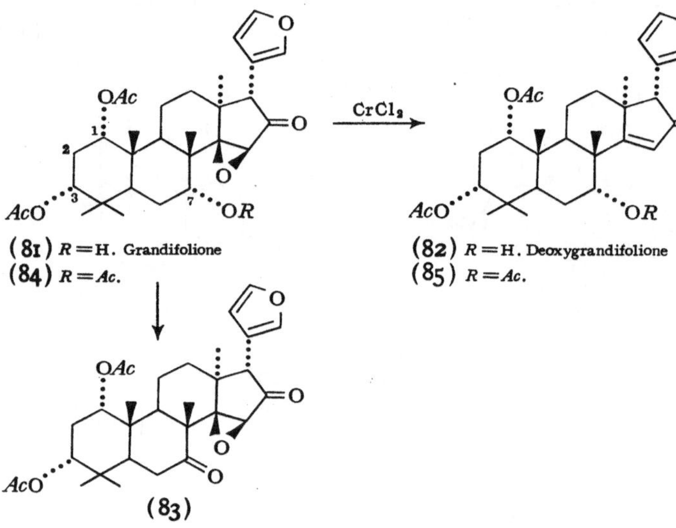

(**81**) R = H. Grandifolione
(**84**) R = Ac.

(**82**) R = H. Deoxygrandifolione
(**85**) R = Ac.

(**83**)

Chart 14. Reactions of Grandifolione

Grandifolione (**81**) (*45*) co-occurs naturally with the closely related 7-deacetylkhivorin and 7-deacetoxy-7-oxokhivorin (see Section 8, p. 210). The presence of an α,β-epoxycarbonyl system in (**81**) was proved by reaction with chromous chloride to give deoxygrandifolione (**82**) which shows the expected spectroscopic properties of an α,β-unsaturated keto group. The 7-hydroxyl in (**81**) was acetylated to give (**84**) and could be oxidized to a 7-keto derivative (**83**). The products, (**82**) and (**83**), showed the expected spectral characteristics. Stereochemistry at the 1-, 3- and 7-positions of (**81**) was assigned from the half-line widths of the methine resonances in the NMR spectrum (*45, 75*).

The 7-acetate of grandifolione (**84**) was found (*8*) co-occurring with khivorin and 3-deacetylkhivorin (Section 8, p. 210) and was named khayanthone *(Chart 14)**. Chromous chloride reduction of (**84**) leads to deoxygrandifolione acetate (**85**) (*8*).

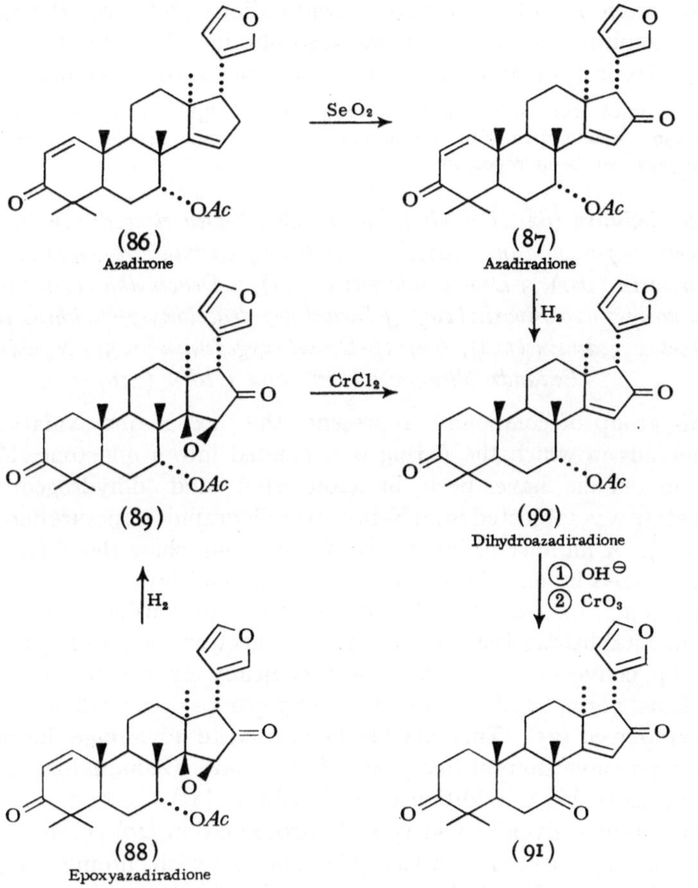

Chart 15. Interconversions of Azadirone and Congeners

Recently, LAVIE and JAIN (*87*) have reported the isolation from *Melia azadirachta* L. and structure determination of an impressive series of limonoids of increasing oxidation level. The co-occurrence of this series of limonoids, azadirone (**86**), azadiradione (**87**) and epoxyazadira-

---

* In view of the simple relationship of khayanthone to grandifolione and in an effort to keep down the proliferation of trivial names, the use of grandifolione acetate instead of khayanthone would be preferable.

dione (**88**), as well as gedunin and 7-deacetylgedunin (Section 8, below) in the same plant provides circumstantial evidence for the original ideas on the biogenesis of limonin (*15*). Azadirone (**86**), azadiradione (**87**) and epoxyazadiradione (**88**) have been interrelated by the conversions outlined in *Chart 15*. Thus, selenium dioxide oxidation of (**86**) gives (**87**). Chromous chloride reduction of dihydroepoxyazadiradone (**89**) gives the expected dihydroazadiradone (**90**) which was also obtained by catalytic reduction of (**87**). Hydrolysis and oxidation of (**90**) yields the triketone (**91**).

No chemical correlation between the cedrelone (**38**), anthothecol (**46**), havanensin (**59**), heudelottin (**66**), hirtin (**74**), grandifolione (**81**) and azadirone (**86**) systems has yet been reported.

*8. Gedunin (**92**), Dihydrogedunin (**98**), 7-Deacetylgedunin (**96**),*
*7-Deacetoxy-7-oxogedunin (**97**), 7-Deacetoxy-7-oxodihydro-α-gedunol (**100**),*
*Khivorin (**101**), 3-Deacetylkhivorin (**111**), 7-Deacetylkhivorin (**104**),*
*7-Deacetoxy-7-oxokhivorin (**105**), 3-Deacetyl-7-deacetoxy-7-oxokhivorin (**113**),*
*11β-Acetoxygedunin (**114**), 6α,11β-Diacetoxygedunin (**115**), Nyasin (**116**),*
*Entandrophragmin (**117**) and Utilin (**118**)*

This group of compounds represents the next major oxidation level of limonoids in which the *D*-ring is expanded into a δ-lactone. Many of these limonoids have been interconverted and dihydrogedun-3β-yl iodoacetate was subjected to an X-ray crystallographic structure determination (*120*). A number of limonoids of this group show the characteristic base-catalyzed limonol (**3**, p. 193) to merolimonol (**4**) conversion (*53, 95*) (see Section 1, p. 192) which has proven to be of considerable diagnostic value in establishing the presence of a 7α-acetoxy or 7α-ol system. The (**3**) → (**4**) conversion appears to be specifically hydroxide ion-catalyzed (*25*). Thus, normal hydrolysis of 7-acetoxy groups occurs when methoxide ion is employed (*25*). This fact has been used to advantage during work on the interconversions of this group of limonoids. Gedunin (**92**) (*10*) gives with hydroxide ion in addition to merogedunin (**93**) a second product (**94**) resulting from a glycidic acid type decarboxylation (*10*) *(Chart 16)*. This hemiacetal (**94**) was oxidized to a γ-lactone (**95**) with chromic acid. Merogedunin (**93**) was dehydrated to its anhydro derivative (**96**) with phosphorous oxychloride.

The initial interrelationships established among the naturally occurring gedunin congeners are illustrated in *Chart 17* (p. 212).

Thus, hydrolysis of gedunin (**92**) gives 7-deacetylgedunin (**96**) (*11*) which upon chromic acid oxidation leads to 7-deacetoxy-7-oxogedunin (**97**) (*25*). Both (**96**) and (**97**) occur naturally. Catalytic reduction of (**92**) gives the naturally occurring dihydrogedunin (**98**) (*25, 81*). Hydrolysis and oxidation of (**98**) yields (**99**) which is also obtained from the naturally occurring 7-deacetoxy-7-oxodihydro-α-gedunol (**100**) (*81*).

Initially, the location of the acetoxy groups in the *A*-ring of khivorin (101) rested on the formation of the triketone (102), which shows spectroscopic properties consistent with those of a β-diketone (*24*). Partial acetylation of the triol (103) gave the naturally occurring 7-deacetylkhivorin (104). Oxidation of (104) gave the 7-keto derivative (105) also found in nature (*25*).

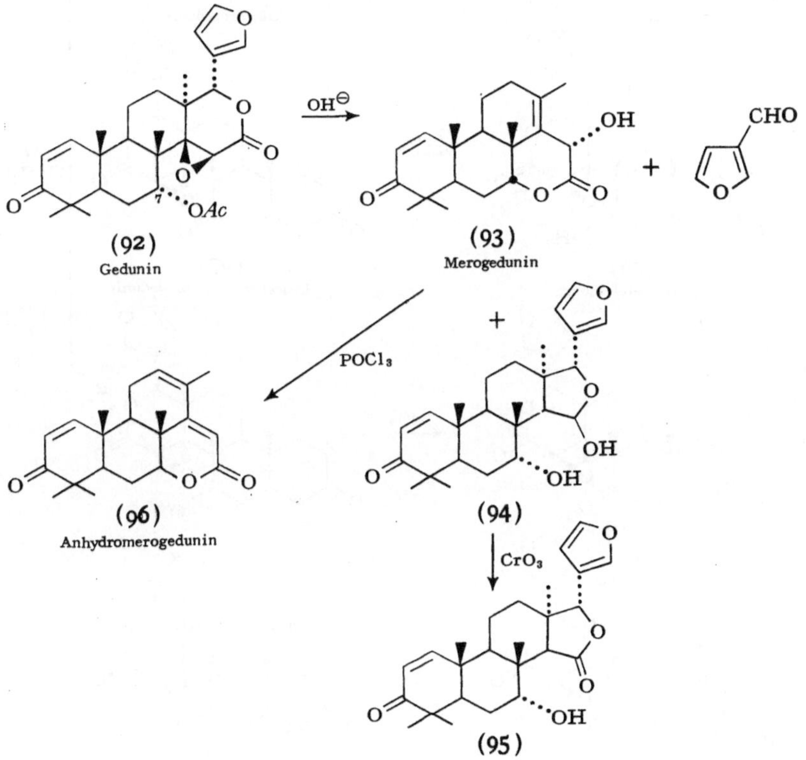

(92) Gedunin

(93) Merogedunin

(96) Anhydromerogedunin

(94)

(95)

Chart 16. Some Reactions of Gedunin

Gedunin (92) and khivorin (101) have been interrelated (*10*) through a isogedunin derivative (106) by the sequence shown in *Chart 18* (p. 213). A key step in this interconversion was the transformation of the α,β-epoxyketone (107) to the allyl alcohol (108) with hydrazine, by the method of WHARTON and BOHLEN (*125*).

The sequence illustrated in Chart 18 does not establish the stereochemistry of the 1- and 3-acetoxy groups in khivorin (101). The half-line widths of the 1- and 3-protons in the NMR spectrum suggests that both acetoxy substituents are axial and in the α-configuration (*6, 10, 45, 75*).

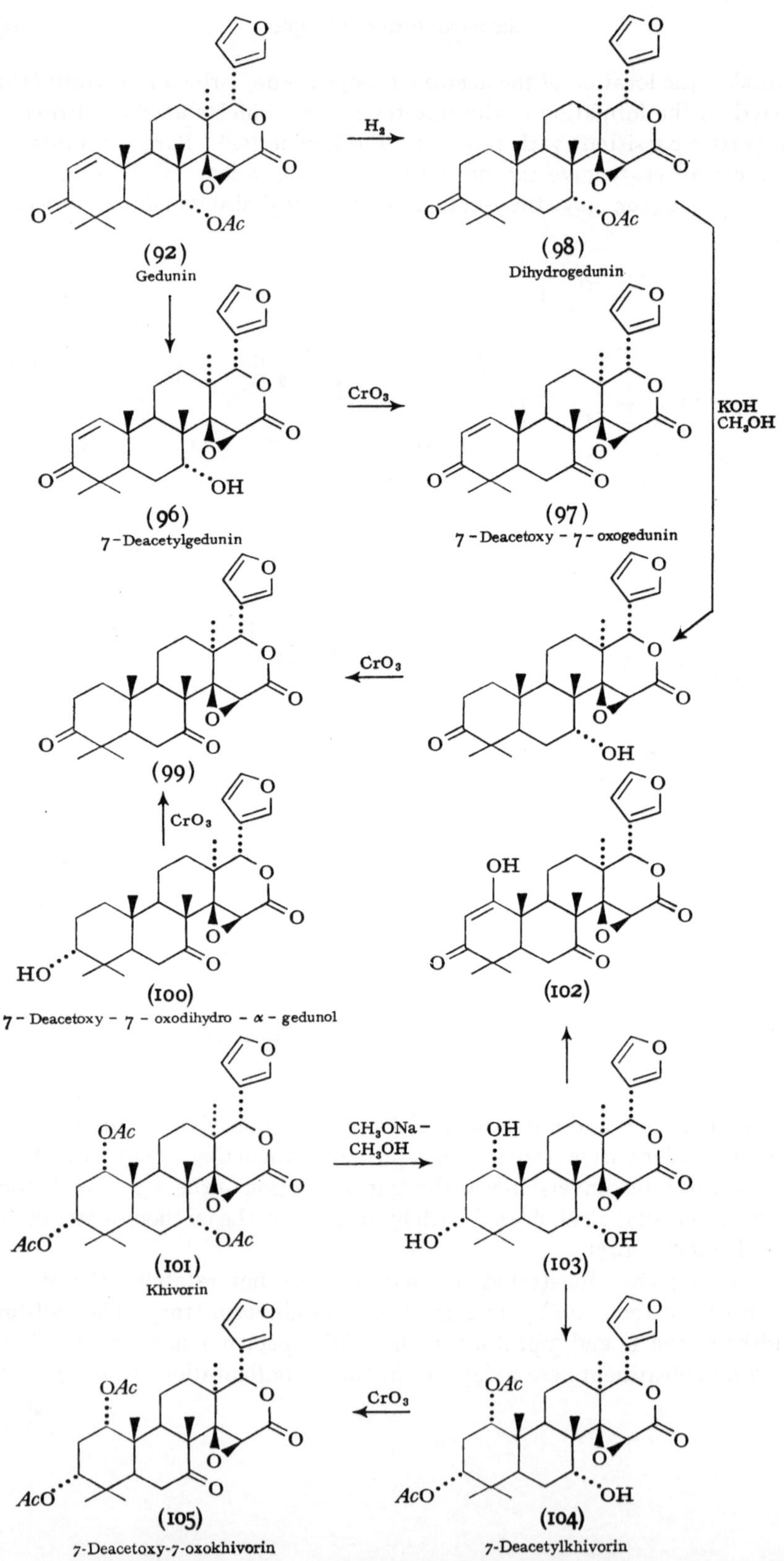

Chart 17. Interconversions of Gedunin Congeners

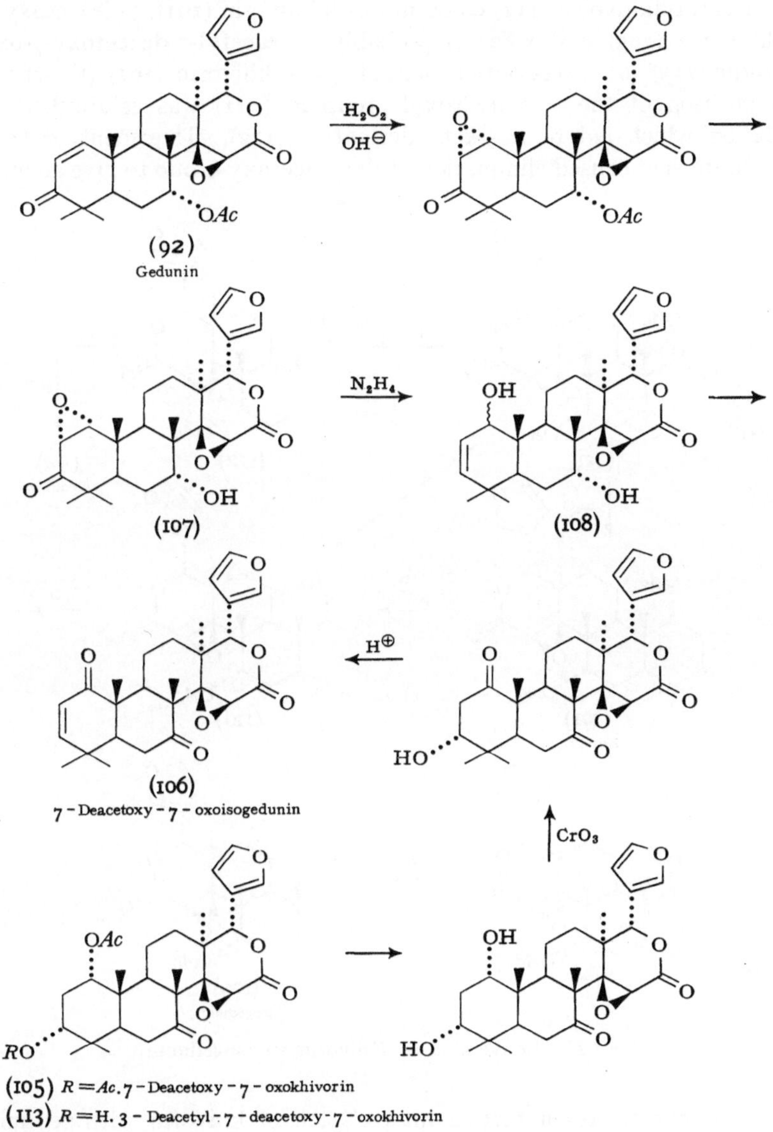

(92)
Gedunin

(107)

(108)

(106)
7 - Deacetoxy - 7 - oxoisogedunin

(105) $R = Ac$. 7 - Deacetoxy - 7 - oxokhivorin
(113) $R = H$. 3 - Deacetyl - 7 - deacetoxy - 7 - oxokhivorin

Chart 18. Interconversions of Gedunin and Khivorin Systems

Khivorin (**101**) itself has also been converted to isogedunin (**109**) *(Chart 19)* by selective hydrolysis of the 1- and 3-acetoxy groups of (**101**) to give (**110**) followed by oxidation *(10)*.

3-Deacetylkhivorin (**111**) co-occurs with khivorin (**101**), 7-deacetoxy-7-oxokhivorin (**105**) and what is probably 3-deacetyl-7-deacetoxy-7-oxokhivorin (**113**) (*6*). Acetylation of (**111**) gave khivorin (**101**) (Chart 19). The position of the free hydroxyl group in (**111**) was established by oxidation which led to a 3-keto derivative (**112**). Treatment of (**112**) with dilute acid caused elimination of the β-acetoxy group to give gedunin

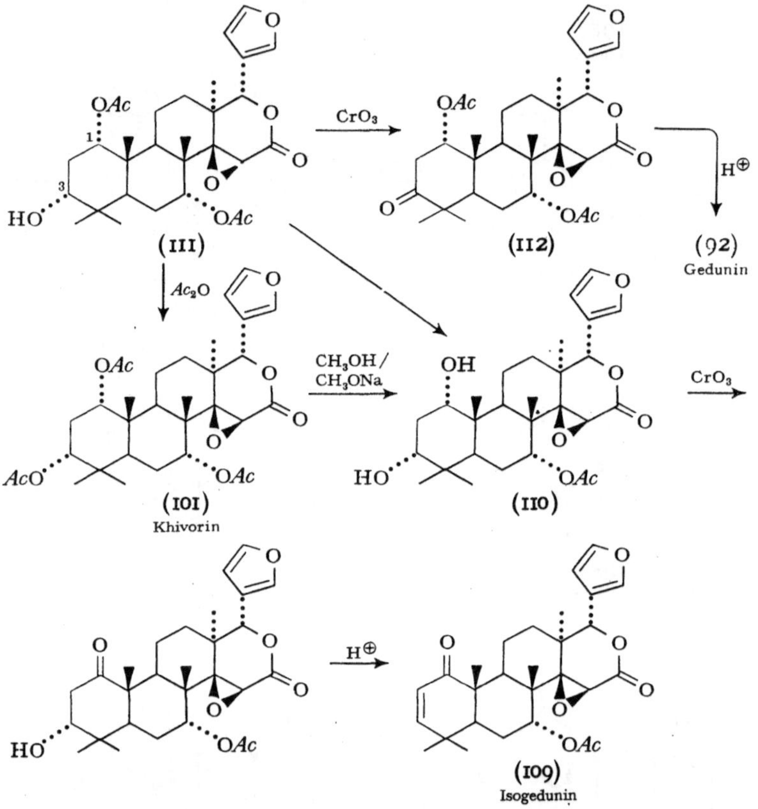

Chart 19. Conversion of Khivorin to Isogedunin

(**92**). This constitutes a further interrelation of khivorin with gedunin. Acetylation of the 3-deacetyl-7-deacetoxy-7-oxokhivorin (**113**) gave a substance which was identical with 7-deacetoxy-7-oxokhivorin (**105**) by thin-layer chromatography (Chart 18) (*6*).

The structural proof of 11β-acetoxygedunin (**114**) and of 6α,11β-diacetoxygedunin (**115**) rests primarily upon their NMR spectra with spin decoupling and their conversion to merogedunin derivatives (*99*). The stereochemical assignment of the 11-acetoxy substituents depends upon the half-line widths of the 11-methine protons. The stereochemical assignment (*49, 75*) of the 11β-acetoxy group in (**114**) and (**115**) is the same as of the 11-acetoxy group in hirtin (**74**) (*39*) and heudelottin (**66**) (*111*), but opposite to that of the 11-acetoxy group in anthothecol (**46**) (*62, 122*). No chemical correlation has been reported of (**114**) and (**115**) with any other gedunin congener.

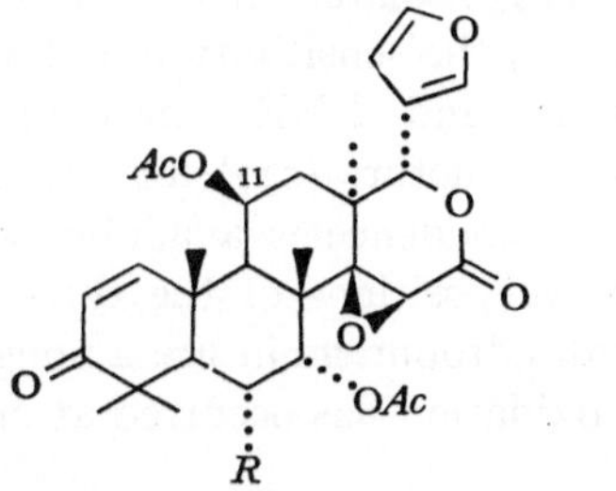

(**114**) R = H. 11β – Acetoxygedunin
(**115**) R = AcO. 6α,11β – Diacetoxygedunin

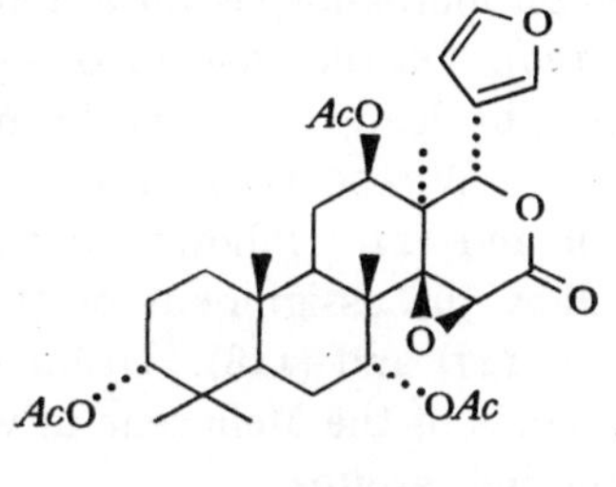

(**116**)
Nyasin

Nyasin is closely related to khivorin (**101**) and has been formulated as (**116**) by TAYLOR (*122*) on spectroscopic grounds.

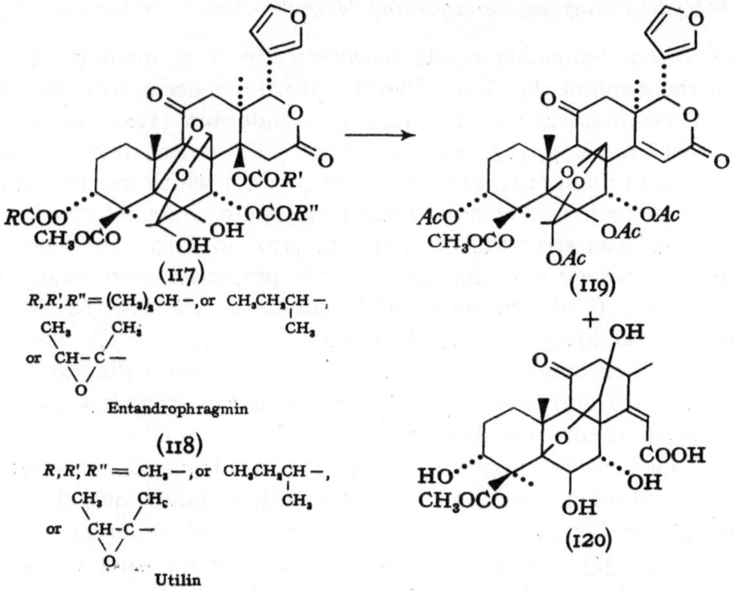

Chart 20. Reactions of Utilin and Entandrophragmin

ARENE et al. (*14*) have reported that the NMR spectra of utilin and entandrophragmin are identical except for the C-methyl region. Both utilin and entandrophragmin *(Chart 20)* which are methyl esters, give furan-3-aldehyde on pyrolysis or base hydrolysis (*12*), suggesting the presence of a 7-α-ol or 7-acyloxy group. Entandrophragmin (**117**) is esterified with one mole each of 2,3-epoxy-2-methylbutyric, 2-methyl-butyric and isobutyric acids while utilin (**118**) is esterified with one mole each of 2,3-epoxy-2-methylbutyric, 2-methylbutyric and acetic acids. Entandrophragmin (**117**) and utilin (**118**) give the same two complex products on methanolysis formulated as (**119**) (isolated as a tetraacetate) and as (**120**). On the basis of this chemistry, the sensitivity of entandro-phragmin to lead tetraacetate oxidation, and NMR considerations, TAYLOR and WRAGG (*124*) have suggested structure (**117**) for entandro-phragmin and (**118**) follows for utilin. No evidence is available which would allow the assignment of the relative positions of the three acyl groups in (**117**) and (**118**). Utilin and entandrophragmin are among the few limonoids of the Meliaceae in which oxidation has occurred at one of the C-methyl groups.

Oxidation at C-19 is common among limonoids of the Rutaceae while oxidation of the methyl at C-8 is common among the simaroubolides (*114a*).

9. *Andirobin* (**121**), *Deoxyandirobin* (**122**), *Methyl Angolensate* (**125**), *Methyl 6-Hydroxyangolensate* (**130**) *and Methyl 6-Acetoxyangolensate* (**131**)

These *B-seco* limonoids result biogenetically from opening of the *B*-ring in the gedunin skeleton. They frequently co-occur with the cor-responding 7-ketolimonoids. For example, andirobin (**121**) and mexi-canolide (Section 11, p. 223) each co-occur with 7-deacetoxy-7-oxoge-dunin (**97**). Andirobin (**121**) (*112*) is a methyl ester which can be hydro-lyzed with base to give an acid from which andirobin can be reformed with diazomethane. The structure of andirobin (**121**) initially rested largely on biogenetic analogy and the spectroscopic properties, particularly on NMR. Treatment of andirobin with chromous chloride *(Chart 21)* afforded deoxyandirobin (**122**) which had the expected UV spectral characteristics. Hydrogenation of andirobin (**121**) gave a dihydro (**123**) and a tetrahydro derivative (**124**). Deoxyandirobin (**122**) has recently been found in nature by ADESOGAN and TAYLOR (*9*).

Methyl angolensate (**125**) is formally a further biogenetic conversion product of andirobin (**121**) and co-occurs with gedunin (**92**), dihydro-gedunin (**98**) and 7-deacetoxy-7-oxogedunin (**97**) in different species (see *Table 1*, p. 234). Much of the work on the structure determination of methyl angolensate (*27, 38*) followed the classical approaches used in the field of the triterpenes. Mild hydrolysis of methyl angolensate (**125**) gave angolensic acid from which (**125**) could be reformed by treatment with diazomethane. Methyl angolensate (**125**) has the same carbon skeleton as andirobin (**121**) with an ether bridge between C-1 and C-14.

This bridge can be opened by strong base to give (**126**) and reversibly closed by acid (*38*) *(Chart 22)*. The stereochemistry at the 1-position in (**125**) is uncertain but a 1β-H is favored on the slight evidence available (*27*). The 1β-H assignment is also supported by NMR arguments (*123a*).

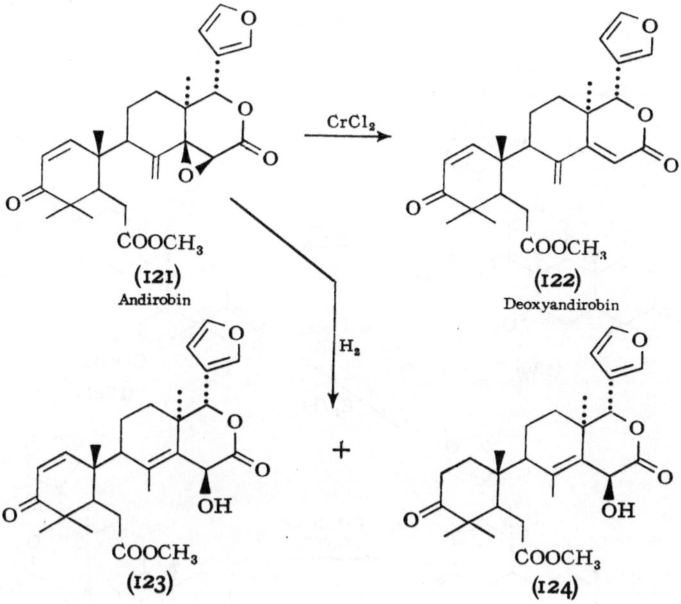

Chart 21.   Some Reactions of Andirobin

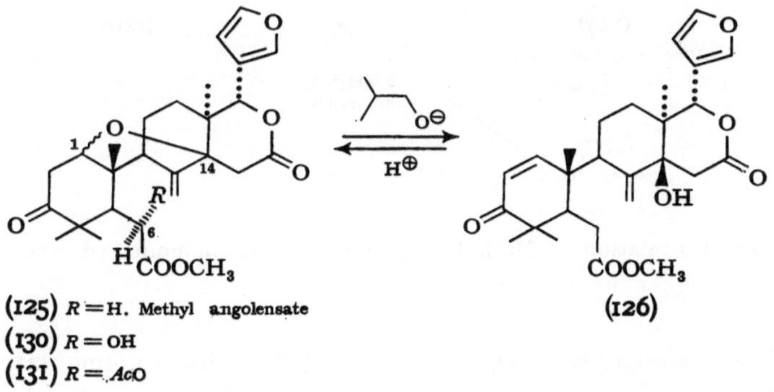

Chart 22.   Ring Opening of Methyl Angolensate

EKONG and OLAGBEMI (*64*) have chemically correlated gedunin (**92**) with methyl angolensate (**125**) and andirobin (**121**) by a route modeled after a sequence of reactions leading from limonin to deoxylimonic acid

(*20*). The ring-opened product (**126**) from methyl angolensate was dehydrated to give deoxyandirobin (**122**) or the corresponding acid (**127**) *(Chart 23)*. Partial hydrogenation of (**127**) yielded a dihydro derivative (**128**) which was also obtained by treatment of deoxy-7-deacetoxy-7-oxogedunin (**129**) with dilute base (*64*).

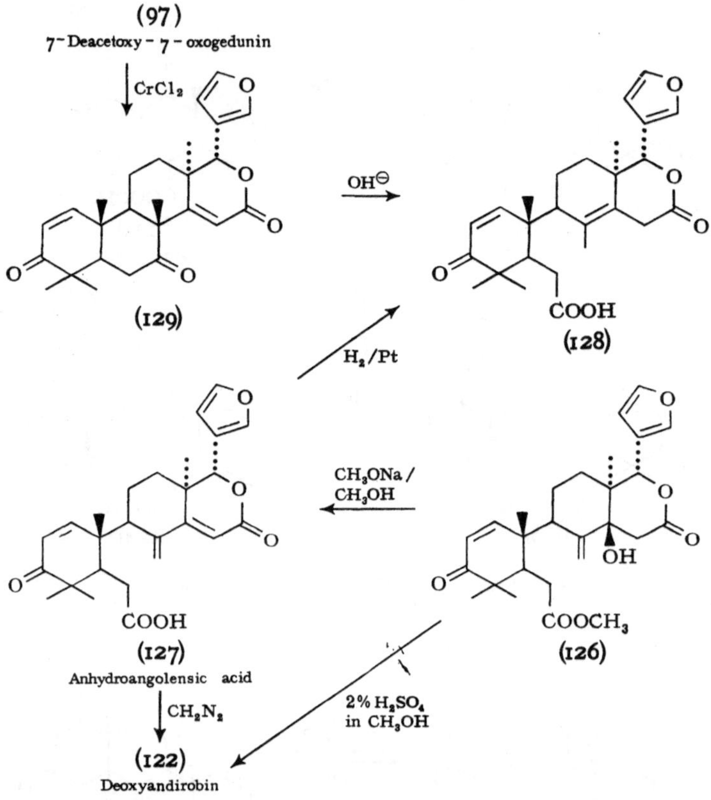

Chart 23. Correlation of Methyl Angolensate with Gedunin and Andirobin

Methyl angolensate (**125**) and methyl 6-hydroxyangolensate (**130**) (Chart 22) co-occur in two *Khaya* species (*2, 45*). Connolly et al. (*51*) have reported the isolation and structure determination of both 6-hydroxy (**130**) and 6-acetoxy methyl angolensate (**131**) *(Chart 22)*. These structures were assigned largely by 100 MHz NMR studies including extensive spin decoupling. Methyl 6-hydroxy-angolensate (**130**) has the same stereochemistry at C-6 as swietenine (*123a*) (see Section 11, p. 223).

10. *Nimbin* (**132**), *Nimbolide* (**150**) *and Salannin* (**152**)

Nimbin (*76, 101, 103*) and salannin (*77*) are two closely related limonoids which co-occur in the seed oil of *Melia azadirachta* L. and possess

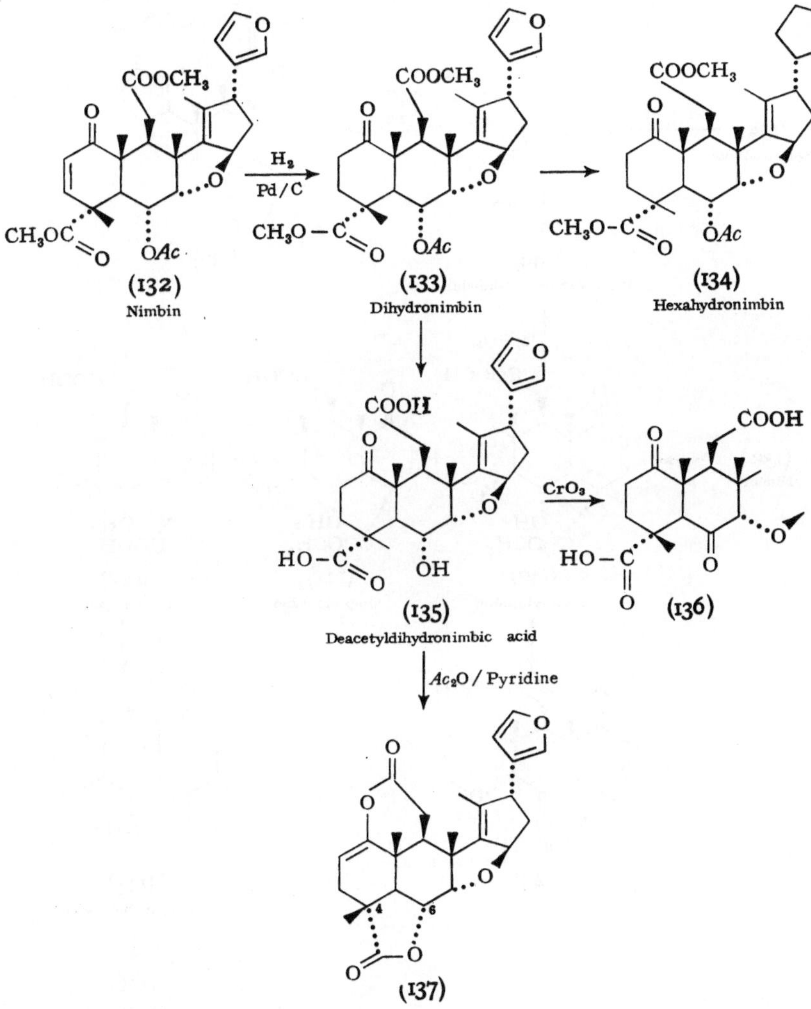

Chart 24. Hydrogenation Products of Nimbin

a *seco* C-ring. The major component, nimbin (**132**), has been the subject of extensive chemical studies (*99, 107, 118*). It can be catalytically reduced to either a dihydro- (**133**) or a hexahydronimbin (**134**) *(Chart 24)*. Since hexahydronimbin (**134**) shows a vinyl methyl resonance in its NMR spectrum, nimbin (**132**) must contain a highly hindered double bond. Hydrolysis of dihydronimbin (**133**) yields the corresponding dihydronimbic acid (**135**) which gives with chromic acid a diketone formulated

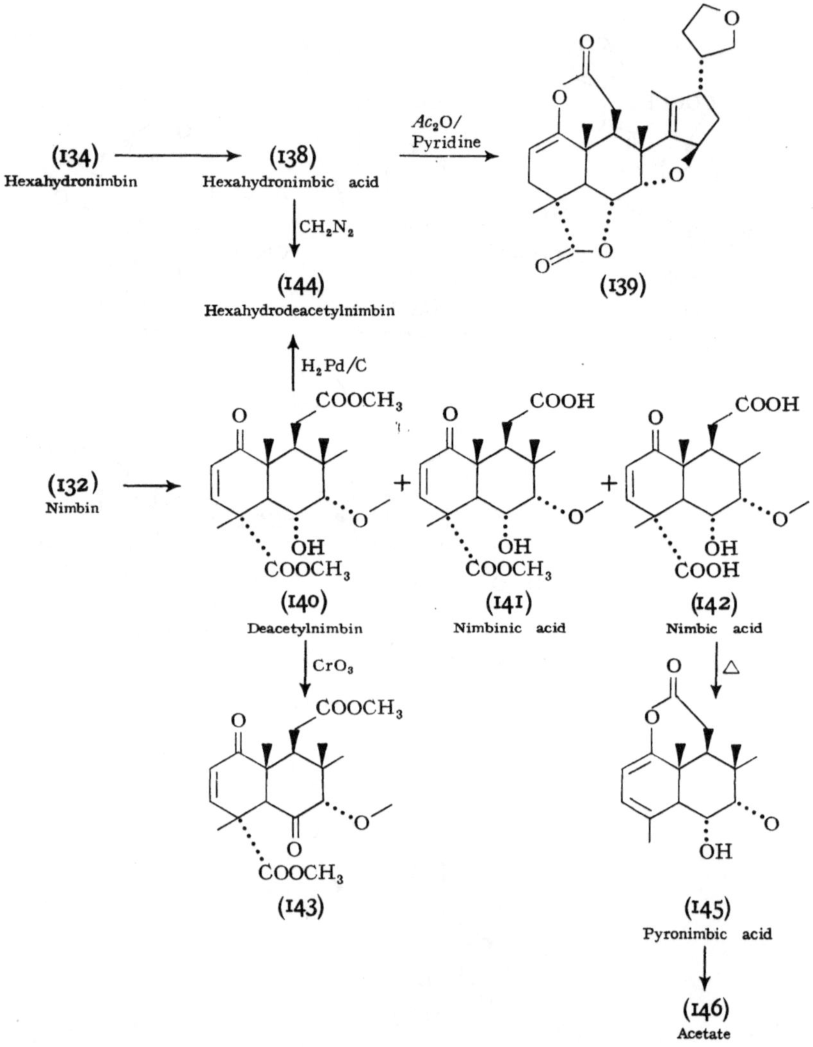

(**134**)   ⟶   (**138**) $\xrightarrow{Ac_2O/Pyridine}$
Hexahydronimbin   Hexahydronimbic acid

$\downarrow CH_2N_2$

(**144**)
Hexahydrodeacetylnimbin

(**139**)

$\uparrow H_2 Pd/C$

(**132**) ⟶ (**140**) + (**141**) + (**142**)
Nimbin   Deacetylnimbin   Nimbinic acid   Nimbic acid

$\downarrow CrO_3$ $\downarrow \triangle$

(**143**)

(**145**)
Pyronimbic acid

$\downarrow$

(**146**)
Acetate

Chart 25. Hydrolysis Products of Nimbin

as (**136**). Dehydration of dihydronimbic acid (**135**) gives (**137**) which shows spectroscopic properties of both a γ-lactone and an enol lactone. A similar product (**139**) was obtained by dehydration of hexahydronimbic acid (**138**) *(Chart 25).*

Mild hydrolysis of nimbin (132) (Chart 25) gives three products, deacetylnimbin (140), nimbinic acid (141) and nimbic acid (142). Deacetylnimbin (140) contains an oxidizable hydroxy group leading to (143) and has been related to hexahydronimbic acid (138) through hexahydrodeacetylnimbin (144). Upon heating, nimbic acid (142) gives carbon dioxide and pyronimbic acid (145) which has characteristic spectroscopic properties and forms the monoacetate (146). NARAYANAN and PACHAPURKAR (*102*) have shown by hydrolysis studies on various derivatives that nimbinic acid (141) is the C-12 carboxylic acid. Thus, hydrolysis of the 4-carbomethoxy group of (132) to give (142) takes place with the assistance of the 6-hydroxyl group. The formation of an enol δ-lactone in (137), (139) and (145) shows the relationship between the keto group and one of the carboxylic acid groups. The formation of the γ-lactone in (137) and (139) shows the relationship of the 6-hydroxyl to the 4-carboxylic acid.

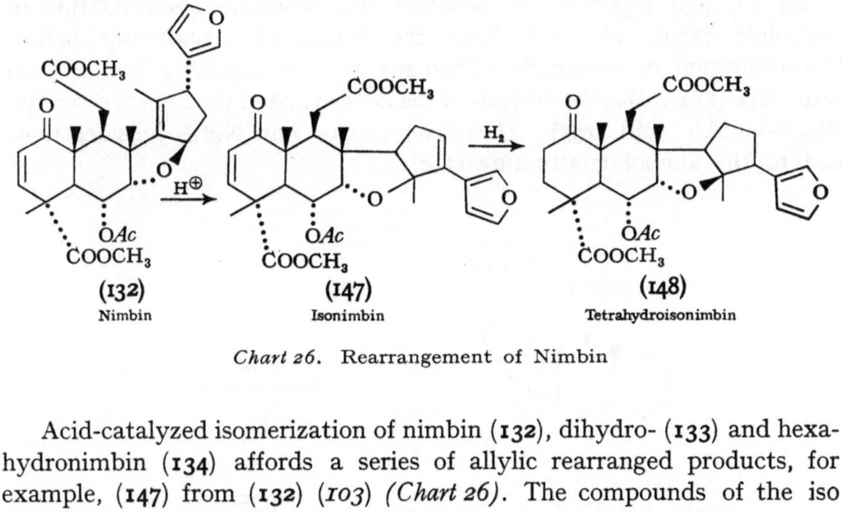

(132) Nimbin

(147) Isonimbin

(148) Tetrahydroisonimbin

Chart 26. Rearrangement of Nimbin

Acid-catalyzed isomerization of nimbin (132), dihydro- (133) and hexahydronimbin (134) affords a series of allylic rearranged products, for example, (147) from (132) (*103*) (*Chart 26*). The compounds of the iso series, e. g., (147) lack a vinyl methyl resonance in their NMR spectra; they show a vinyl proton resonance and four sharp C-methyl resonances. The compounds of the iso series (147) show vinylfuran absorption in the UV which is lacking in the hydrogenation product (148). Many of the stereochemical assignments of nimbin depend on NMR arguments (*77, 100–105*).

The ORD curve of hexahydronimbin (134) is very similar to that of 1-ketocholestane (*103*). It has been suggested (*103*) on this basis that the absolute configuration of nimbin (132) is the same as that of the other limonoids. However, there may be some ambiguity in this approach due to the complex nature of the curve. More recently (*127*), correlation of the ORD curve for the *cis*oid-diene chromophore in pyronimbic acid (**145**), with the chirality of the system, showed that the absolute stereochemistry was the same as that of the other limonoids. Narayanan et al. (*103*) rationalize the biogenetic formation of nimbin by way of (**149**):

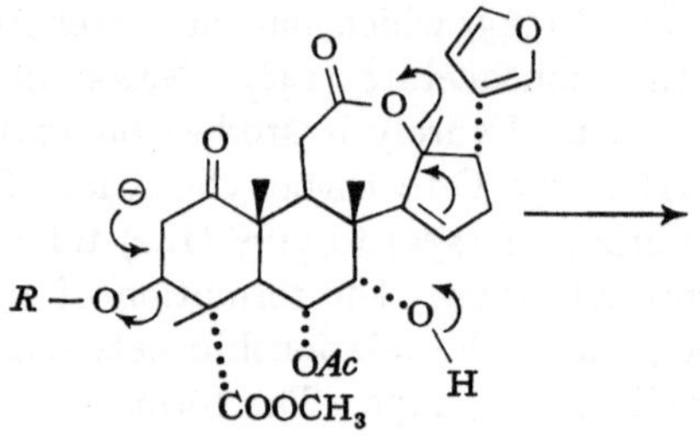

$$\longrightarrow \text{(132)} \quad \text{Nimbin}$$

(**149**)

Ekong (*63*) reported the isolation and structure determination of nimbolide '(**150**), obtained from the leaves of *Azadirachta indica*. Hydrogenation of nimbolide *(Chart 27)* gave a noncrystalline dihydro derivative (**151**). Base hydrolysis of the latter yielded the known deacetyl-dihydronimbic acid (**135**). These conversions and NMR considerations lead to the nimbolide structure (**150**).

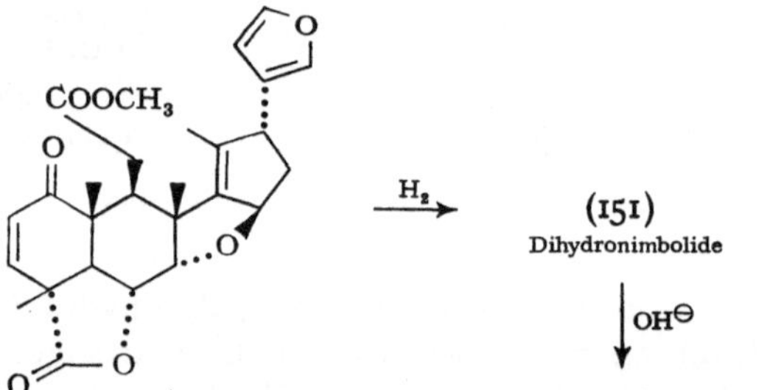

$$\xrightarrow{\text{H}_2} \text{(151)} \quad \text{Dihydronimbolide}$$

$$\downarrow \text{OH}^{\ominus}$$

(**150**) Nimbolide   (**135**) Deacetyldihydronimbic acid

Chart 27. Conversions of Nimbolide

The structure of salannin (**152**) (*77*), a tiglate ester, is based largely on its spectroscopic properties and on biogenetic analogy. NMR played a key role in the structural proposal, particularly the extensive use of spin decoupling. Hydrolysis of salannin afforded two products, (**153**) and (**154**) *(Chart 28)*. Chromic acid oxidation of the 1,3-diol (**153**) gave only a monoketone (**155**) whereas khivorin (**101**) gave a β-diketone (**102**) (Section 8). The assigned stereochemistry to the 1- and 3-substituents of

(**152**) was made on the basis of the half-band widths of the methine protons in the NMR. This stereochemistry is the same as that in grandifolione (**81**), khivorin (**101**) and their congeners (Sections 7 and 8, pp. 201, 210).

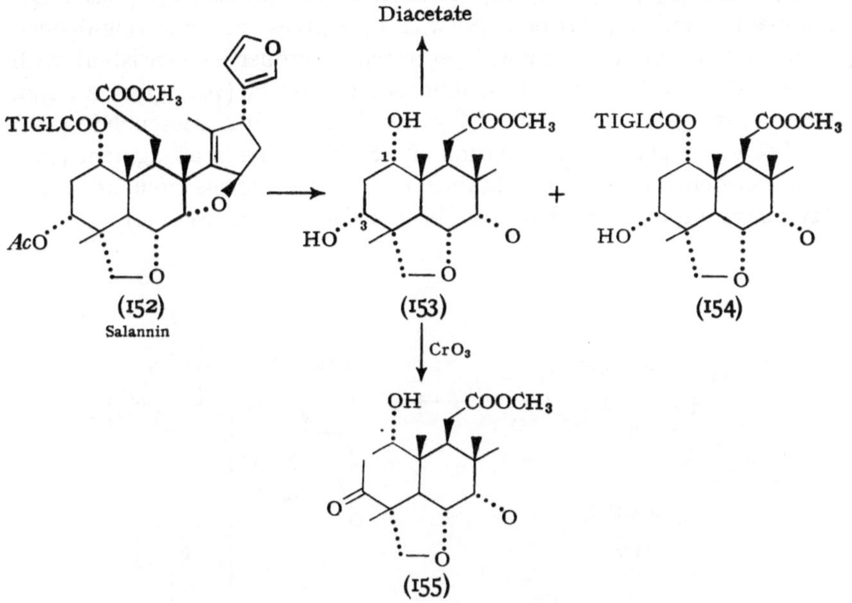

Diacetate

COOCH₃
TIGLCOO

AcO

(**152**)
Salannin

OH   COOCH₃

HO   O

(**153**)

+

TIGLCOO   COOCH₃

HO   O

(**154**)

CrO₃

OH   COOCH₃

O

(**155**)

Chart 28.   Reactions of Salannin

*11. Mexicanolide (Substance B) (**157**), Carapin (**158**),
6-Hydroxycarapin (**159**), Swietenolide (**160**), Fissinolide (**164**),
Khayasin (**165**), 3β-Dihydromexicanolide (**167**), Swietenine (**173**),
6-Deoxydestigloylswietenine Acetate (**185**) and 6-Deoxy-12β-acetoxy-
swietenine Acetate (**186**)*

These compounds are structurally the most complex limonoids presently known. They appear to be formed biogenetically from a 7,8-*seco* system, for example, androbin (**121**) (Section 9, p. 216) by reclosure at C-2 of the *A*-ring of (**156**) onto the exocyclic methylene at C-8, giving a [3,3,1] bicyclic system (**157**) (*47*) *(Chart 29)*. Double-bond isomerization then could lead to the other limonoids of this type for which the general name, bicyclononanolides has been proposed (*47*). Carapin (**158**) and andirobin (**121**) co-occur in the same genus while mexicanolide (**157**) often co-occurs with 7-deacetoxy-7-oxogedunin (**97**), providing some circumstantial evidence for this biogenetic relationship.

Structure determination in this group of limonoids has depended heavily on X-ray crystallographic structure proof; and the structures of both mexicanolide (**157**) (*1*) and swietenolide (**160**) (*94*) have been confirmed in this manner. All of these limonoids contain a highly hindered carbonyl group in the 1-position. They form two isomeric pairs which differ in the position of the double bond. Carapin (**158**) has been isomerized to mexicanolide (**157**) (*25, 26, 48*) under mild conditions (*13*) *(Chart 30)*. Treatment of either (**157**) or (**158**) with base gives the same ring-opened product (**161**); the latter shows spectroscopic properties consistent with an enolizable β-diketone. Thus, mild acetylation of (**161**) gives an enol acetate (**162**) (*5*). Correlation of (**157**) and (**158**) with swietenolide (**160**) was effected by periodiate oxidation of (**161**) resulting in a fragment (**163**) which was identical with that obtained by a similar process from (**160**) (*50*). 6-Hydroxycarapin (**159**) has also been found in nature (*46*).

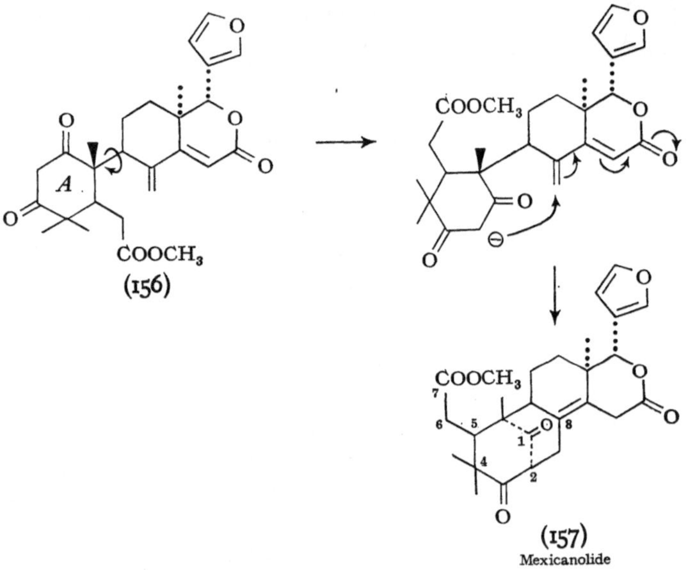

Chart 29. Proposed Biogenetic Formation of Bicyclononanolides

Comparison of the NMR spectra of swietenolide (**160**) and its derivatives with those of mexicanolide (**157**) showed many analogies (*50*). The acetate (fissinolide) (**164**) and isobutyrate (khayasin) (**165**) esters of 3β-dihydromexicanolide were initially found as a difficultly separable mixture (*4*). They were related to mexicanolide (**157**) by way of (**166**) (*4, 5, 48*) *(Chart 31)*. Zelnik and Rosito (*126*) found *Cedrela fissilis* Vell. to be a good source of the pure acetate of 3β-dihydromexicanolide (**164**) which they named fissinolide. They showed that fissinolide was identical with (**164**), previously prepared from mexicanolide (**157**) by borohydride reduction to give (**167**), followed by acetylation (*48*). 3β-Dihydromexicanolide (6-deoxyswietenolide) (**167**) has also been found naturally

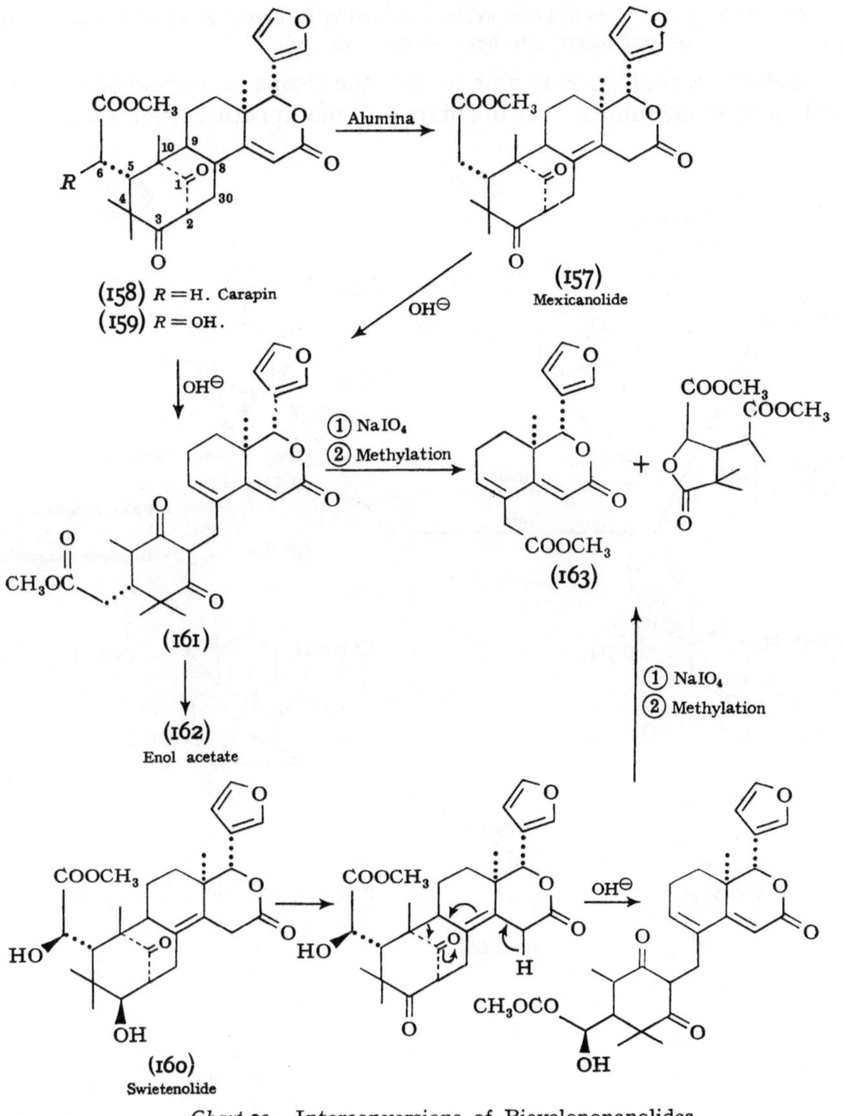
Alumina
OH⊖
① NaIO₄
② Methylation
(158) R=H. Carapin
(159) R=OH.
(157)
Mexicanolide
(161)
(162)
Enol acetate
(163)
(160)
Swietenolide

Chart 30. Interconversions of Bicyclononanolides

occurring (*122*)*. ADESOGAN and TAYLOR (*7*) reported later the isolation of (**164**), which they named grandifoliolin**.

* The tiglate and benzoate esters of 3β-dihydromexicanolide (**167**) have also been found in *Khaya senegalensis* (*123a*).

** In view of the prior work of ZELNIK and ROSITO (*126*) and the similarity of the name to grandifolione (p. 208) the name fissinolide should be retained and grandifoliolin abandoned for (**164**).

The *A/B* [3,3,1] bicyclo ketol system occurring in bicyclononanolides has been the subject of several recent synthetic studies (*78, 93*).

Adesogan et al. (*5*) were able to open the *D*-ring of mexicanolide (**157**) with methanolic sulfuric acid to obtain a 17-methyl ether (**168**) (Chart 31).

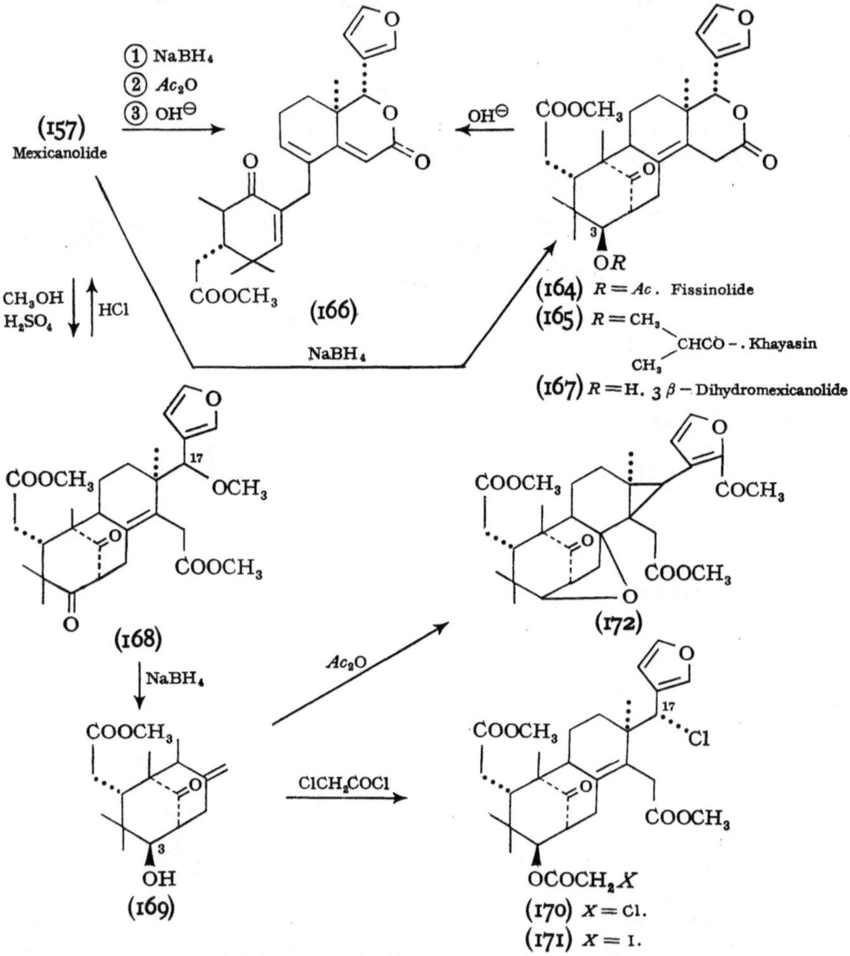

Chart 31. Some Reactions of Mexicanolide

Borohydride reduction of (**168**) gave mainly the 3β-hydroxy derivative (**169**) which was converted to a chloroacetate (**170**), concurrent displacement by chloride ion at C-17 also taking place. The derived iodoacetate (**171**) was used for an X-ray crystallographic structure determination by Adeoye and Bekoe (*1*). Treatment of (**169**) with acetic anhydride and

p-toluenesulfonic acid gave an unusual rearrangement product formulated as (172).

Swietenine was shown by CONNOLLY et al. (47) to have structure (173). Swietenine (173) and its derivatives contain an inert, highly hindered

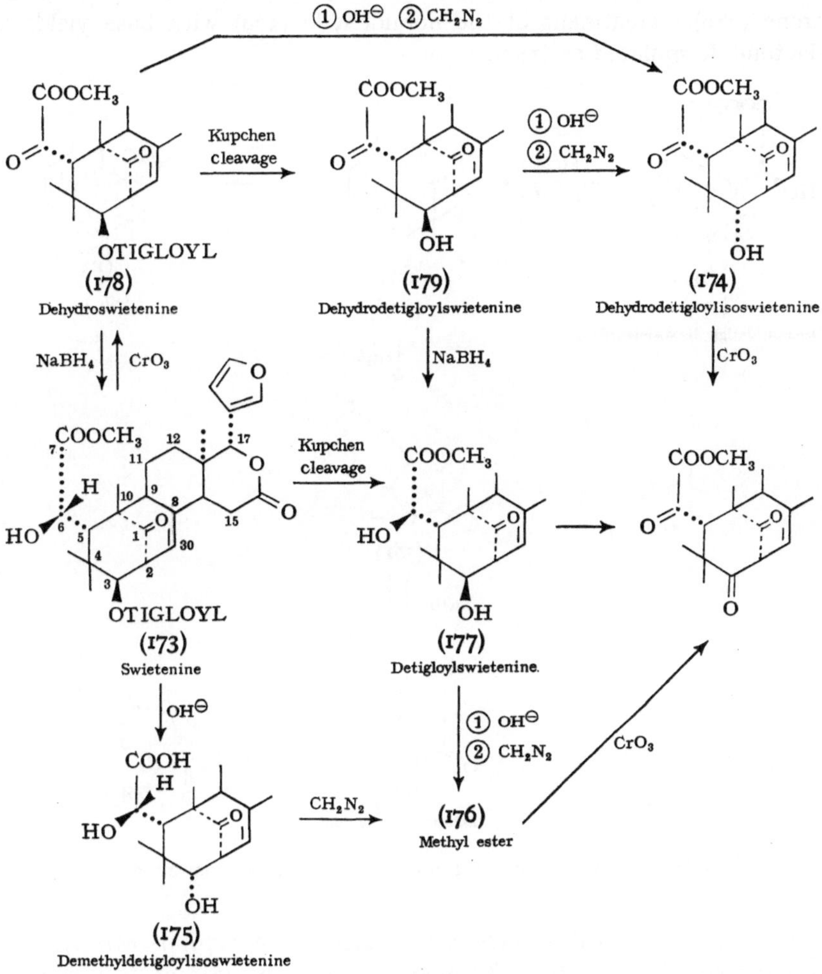

Chart 32. Some Reactions of Swietenine

double bond which resisted all attempts at reduction. Chemical studies on swietenine (47) have given rise to two series of derivatives. The isoswietenine series, (174), (175) and (176), are epimeric at C-3. Conversion of the swietenine derivatives (177), (178) and (179) to the isoswietenine series appears to occur via a dealdolization-realdolization process. The reactions outlined in *Chart 32* illustrate that the swietenine → isoswietenine conversion involves the center at C-3.

Lead tetraacetate oxidation of the α-hydroxy acid (**175**) gives the noraldehyde (**179**) *(Chart 33)*. The aldehyde group in (**179**) is highly hindered and resists oxidation to an acid instead of giving only the β-diketone (**180**). Treatment of the noraldehyde (**179**) with base yields a γ-lactone, formulated as (**181**).

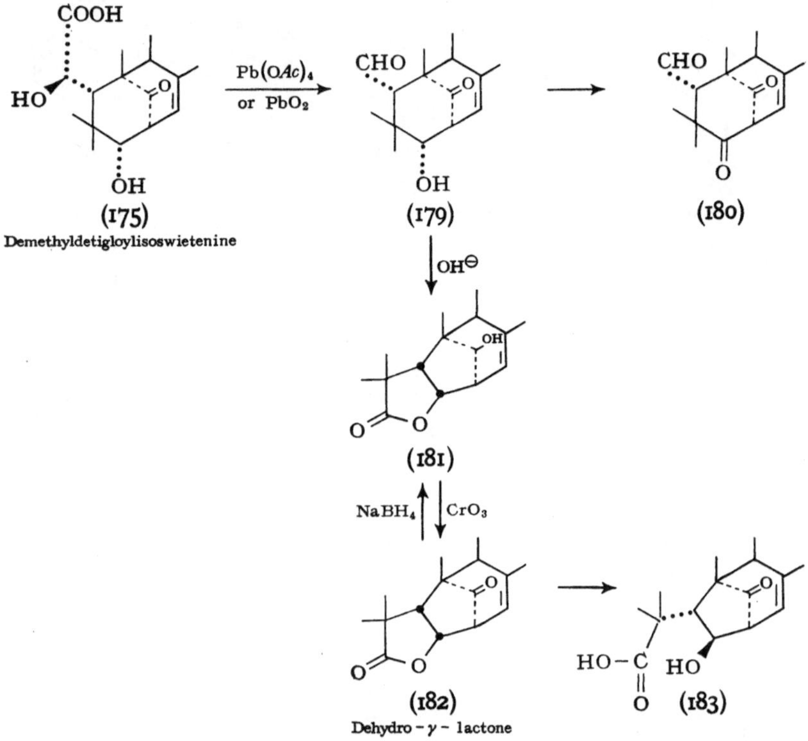

COOH

HO

OH

(**175**)
Demethyldetigloylisoswietenine

$Pb(OAc)_4$
or PbO_2

CHO

OH

(**179**)

CHO

O

(**180**)

$OH^{\ominus}$

OH

O O

(**181**)

$NaBH_4$ CrO_3

O O

(**182**)
Dehydro-γ-lactone

HO-C   HO

O   (**183**)

Chart 33. Reactions of Demethyldetigloyl Isoswietenine

The γ-lactone (**181**) is believed to arise from (**179**) *(Chart 34)* via (a) a reverse aldol reaction, (b) aldol condensation leading to inversion at C-5, (c) an intramolecular Cannizzaro reaction followed by (d) lactonization. The hydride transfer in the intramolecular Cannizzaro reaction of (**179**) (Chart 33) leads to a product (**181**) in which the 1-hydroxy group is *anti* to the γ-lactone ring. Chromic acid oxidation of the γ-lactone (**181**) gives a bicyclic ketone (**182**) which undergoes inversion with base at C-6 to give a hydroxy acid (**183**). This acid (**183**) is also believed to arise from a reverse aldol-realdolization process (*47*).

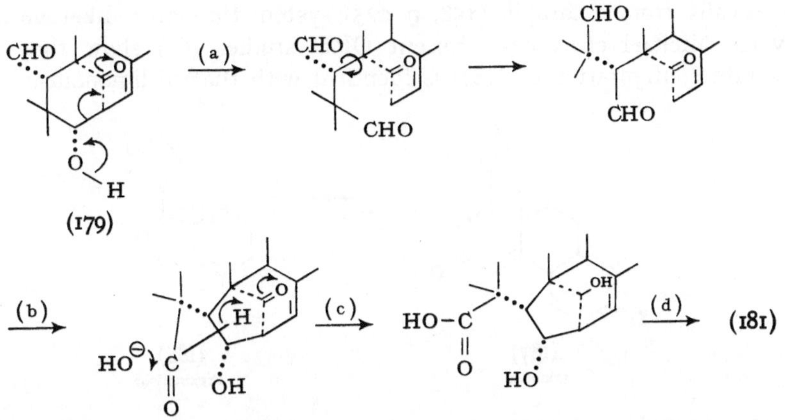

Chart 34. Proposed Mechanism for the Formation of γ-Lactone (175)

An X-ray crystallographic determination (94) of the p-iodobenzoate of detigloylswietenine showed that the A-ring is in a twisted boat conformation and established several of the stereochemical features of swietenine summarized in formula (184).

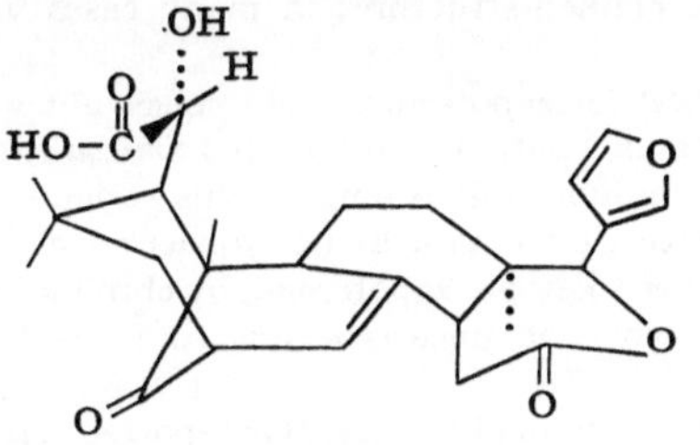

(184)
Swietenine

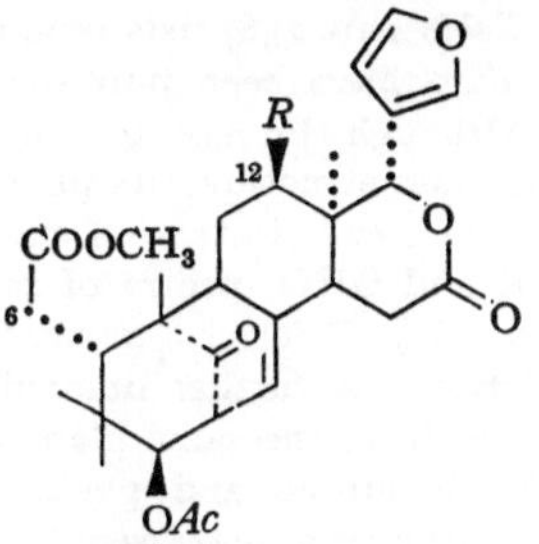

(185) $R = $ H.
(186) $R = $ AcO.

TAYLOR and co-workers (2) have recently reported the isolation of 6-deoxydetigloylswietenine acetate (185) and 6-deoxy-12β-acetoxy-swietenine acetate (186) (originally reported as the 6-epidetigloyl derivative) (123a).

12. Odoratin (187) and Fraxinellone (188)

Studies on the chemical constituents of *Cedrela odorata* L. have given variable results. Some samples have yielded a highly degraded limonoid, odoratin, which, on the basis of spectroscopic studies, has been formulated by CHAN et al. (40) as (187). Odoratin can be imagined as arising biogenetically from a carapin (**158**, p. 225) system through β-diketone and reverse Michael cleavages. Recent ORD studies (*61*) show that the absolute configuration of (**187**) is identical with that of limonoids.

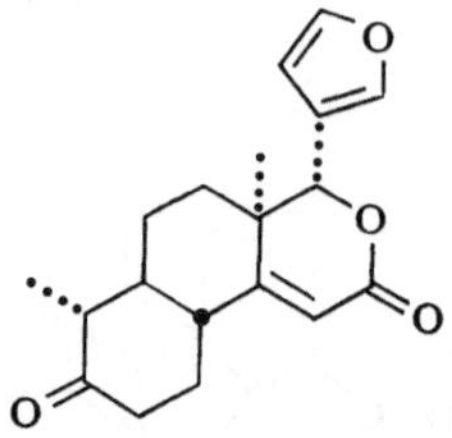

(**187**)
Odoratin

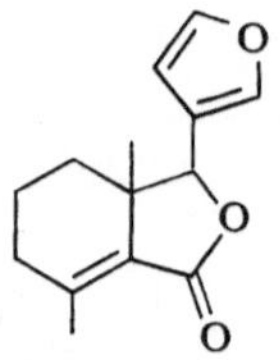

(**188**)
Fraxinellone

Moss (*98*) has raised the interesting possibility that fraxinellone (**188**) (*114*) may be a highly degraded limonoid in which C-16 and the *A*- and *B*-rings have been lost. Fraxinellone (**188**) is found in *Dictamnus albus* L. from which limonoids have previously been isolated. Unfortunately, nothing has been published about the relative or absolute stereochemistry of fraxinellone.

13. Limonoids of Unknown Constitution

Table 4 (p. 238) lists limonoids of undetermined structure; in many cases very few data have been published.

Although the reported analytical data (*84*) for zapoterin are not consistent with a limonoid structure, its physical properties and botanical origin (*Casimiroa edulis* Llave et Lex) (Rutaceae) suggested that it might well belong to this series. The NMR and ORD spectra of zapoterin indicate that it is a hydroxyobacunone derivative (*62*). It forms a monoacetate and yields a ketone, zapoterone, by chromic acid oxidation. A further limonoid (m. p. 249—250°) of uncertain structure has been isolated from the same plant (*58*).

Phellendrone and phellandrine are two limonoids recently reported (*109*), originating from *Phellodendron sacchalinense* (Rutaceae), a genus from which limonoids had been isolated repeatedly (*58*). Although the physical constants of these compounds resemble to some extent those of obacunone (**7**, p. 195) and limonin (**1**, p. 193), respectively, Nikonov and Veremei (*109*) suggest that these are new limonoids.

Mahoganin, isolated from *Swietenia mahogani* appears to be a methyl ester of a C_{28}-limonoid (*36*).

Bussein ($C_{44}H_{58}O_{18}$) (*123*) appears to have many structural features in common with entandrophragmin (**117**) and utilin (**118**) (Section 8, p. 210). Base hydrolysis gives three moles of acetic acid and one mole each of isobutyric and D-(—)-α-methylbutyric acids (*35*). The UV spectrum of bussein indicates that it is an enolizable β-diketone; this is supported by the formation of an enol ether. The NMR spectrum of bussein suggests the presence of an isobutyryl cyclohexanone fragment.

Candollein (*3*) is a further limonoid which appears to be very closely related to entandrophragmin. It is the 3-hydroxy-2-methyl butyrate ester corresponding to entandrophragmin (Section 8, p. 210) and is identical with the chromous chloride reduction product of entandrophragmin (*123a*).

Two limonoids, pseudrelone A_1 and A_2 have been obtained from *Pseudocedrela kotschyii* (*65*). They appear to be closely related to bussein. The UV spectrum of pseudrelone A_1, $C_{38}H_{50}O_{14}$, suggests the presence of an enolizable β-diketone which is supported by the formation of a methyl ether with diazomethane. The NMR spectrum of pseudrelone A_1, which is very similar to that of pseudrelone A_2, indicates the presence of a β-substituted furan ring, seven C-methyls, two acetyl groups, one carbomethoxy group and one tertiary hydroxyl group. Base hydrolysis of pseudrelone A_1 gives two moles of acetic acid and one mole of isobutyric acid.

III. Conformational Problems in Limonoids

There are a number of observations which suggest that limonoids possessing 3-keto groups have A-rings which exist in an "abnormal" conformation relative to normal triterpenes. These conversions include:

1. Borohydride treatment of gedunin (**92**) leads to complete reduction of the 1-en-3-one system to a saturated 3-ol (*11*).

2. The 3-keto group in dihydrocedrelone is resistant to borohydride reduction and fails to form an oxime, while the $\Delta^{1,2}$ double bond of cedrelone (**38**) is reducible with borohydride (*73, 80*).

3. Cedrelone (**38**) gives a Michael addition product with hydroxylamine instead of the expected oxime (*73, 80*).

4. Borohydride reduction of anthothecol methyl ether yields a 1,2-dihydro product (*28*).

The origin of these "abnormal" results can be explained by the 1,3-interactions present in limonoids. The limonoid system is thus warped to relieve these non-bonded interactions (*102*). The presence of a boat C-ring and an axial 8-methyl group add much more strain to limonoids than is normally present in most tetracyclic triterpenes.

The 8-methyl group suffers a severe 1,3-diaxial interaction with the 10-methyl group which is also located 1,3-diaxially to the 4β-methyl when the A-ring is in a boat conformation. The 1,3-diaxial interactions

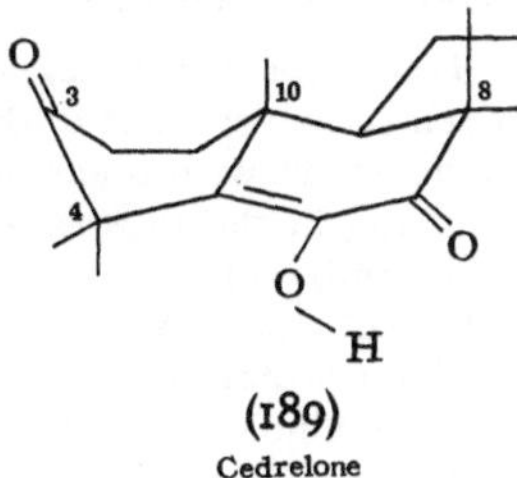

(189)
Cedrelone

can best be minimized by warping the A-ring into a boat conformation. In the case of cedrelone, HODGES et al. (*80*) suggest that in order to relieve these nonbonded interactions, the A-ring exists in a boat conformation. This results in extensive shielding of the 3-keto group by the 4- and 10-methyls and thus leads to low reactivity of the keto group.

X-ray analysis of cedrelone iodocetate (*74*) shows that the A-ring exists in the boat conformation (**189**) that minimizes the nonbonded interactions between the 4,4-*gem* dimethyls and the 6-hydroxyl group (*70*).

There has been some controversy over the preferred conformation of the A-ring in dihydronimbin (**133**) (*100, 102*).

Several publications have dealt exclusively with the IR (*34*), NMR (*56, 110, 115*), ORD, CD (*61*), and mass (*19*) spectra of limonoids.

IV. Some Biological Properties of Limonoids

Many limonoids of the Rutaceae occur in plants which have had use in various systems of folk medicine (*Evodia, Casimiroa, Calodendrum* and *Phellodendron*). In most cases it is unknown whether the biological activity of the plant material is due to limonoids or to other compounds present. For example, alkaloids are widespread in Rutaceae and in many cases alkaloids co-occur with limonoids as in *Phellodendron, Casimiroa* and *Evodia* species. Limonin itself appears to be somewhat toxic to fowl but relatively nontoxic to mammals (*85*). Studies on anthothecol indicate that it may have skin irritant properties in certain cases (*96*). Two species of the genus *Melia* (Meliaceae) are known to contain a locust antifeeding principle which has been identified as meliantriol by LAVIE and co-workers (*90*).

There appears to be no general correlation between taste and structure of the limonoid bitter principles and many compounds of this series are tasteless.

In general, as with many secondary plant metabolites, the role of limonoids in the plant metabolism is not well understood (*33*).

V. Botanical Distribution and Chemotaxonomy of Limonoids

The botanical distribution and chemotaxonomy of limonoids in the Rutaceae and Meliaceae has been discussed in detail (*23, 58, 98, 121*). Limonoids which occur in Rutaceae vary, for the most part, in structure of the *A*-ring and are reasonable intermediates in the biogenesis of limonin itself or are closely related to it biogenetically. All limonoids occurring in the Rutaceae have opened *A*-rings while those in the Meliaceae all have carbocyclic *A*-rings.

Thin-layer chromatography, using Ehrlich's reagent for a detecting spray, has proven to be very useful for routine survey work on limonoids, since it gives characteristic colors with limonoids (*57*). A positive test depends upon the presence of a furan ring in the substrate and the reagent is thus a very selective indicator.

A number of generalizations can be made regarding the distribution of limonoids in the Rutaceae (*58*). These rules, aside from their possible chemotaxonomic value, may be a useful guide in searching for additional limonoids.

1. Limonoids are about equally distributed throughout the three major subfamilies of the Rutaceae. This is equally true of the Meliaceae where limonoids have been reported from each of the three subfamilies.

2. Limonoids show a low frequency of occurrence in the Rutaceae and have been reported in only about 1% of its genera. Rutaceae is a relatively well investigated family and, considering the ease with which most limonoids crystallize, a higher frequency of occurrence would certainly be reflected in more reports of their isolation. This is not true of the less extensively investigated Meliaceae where limonoids have been reported to occur in about 25% of its genera.

3. If limonoids occur in one species of a genus all species of that genus contain them. Thus, limonoids have been found in all species examined of the following genera; *Citrus* (15 examined out of 16 known species), *Phellodendron* (5 of 7 known species), *Casimiroa* (2 of 5 known species) and *Evodia* (4 of 145 known species). This regularity does not apply to Meliaceae where some species of a genus produce limonoids whereas others of the same genus have been reported as devoid of them (*121*).

References, pp. 238—244

4. The limonoids of any given limonoid-producing genus or species are all of about at the same oxidation level. Keeping in mind the proposed biogenetic sequence of steps, *Vespris* and *Casimiroa* show ability to synthesize only limonin precursors. *Phellodendron* species produce mainly obacunone with relatively little limonin. These three genera belong to the subfamily Toddalioideae (24 genera). *Citrus* and its relatives [subfamily Aurantioideae (33 genera)] have the ability of efficiently converting these materials to limonin and they accumulate relatively small amounts of nomilin, deacetylnomilin and obacunone. The last major subfamily, the Rutoideae (86 genera) does not accumulate limonin precursors but produces only limonin and its oxidation products (limonin diosphenol and rutaevin).

5. The subfamilies of the Rutaceae can be ranked by their increasing ability to effect C-19 oxidation of limonoids, e. g. Toddalioideae, Aurantioideae and Rutoideae.

In general, the distribution and structural relationships of limonoids in the Rutaceae support the chemotaxonomic conclusions first drawn by PRICE (*116*), after consideration of the distribution of alkaloids and coumarins in this family.

ADESIDA and TAYLOR (*3*) have discussed in detail the distribution of limonoids in the genus *Entandrophragma*.

VI. Tables

Table 1. Limonoids Occurring in the Meliaceae

	m. p.	$[\alpha]_D$*	Occurrence	References
Cedrelone (38)	209–214°	— 64.5°	*Cedrela toona* Roxb.	(74, 80)
Anthothecol (46)	225°	— 63°	*Khaya anthotheca* (Welw.) C. DC.	(28)
Havanensin 1,7-diacetate (58)	non-crystalline		*Trichilia havanensis* Jacq.	(37)
Havanensin 3,7-diacetate (57)	174–175°		*Trichilia havanensis* Jacq.	(37)
Havanensin 1,3,7-triacetate (56)	188–191°		*Trichilia havanensis* Jacq.	(37)
Heudelottin (66)	178–180°		*Trichilia heudelottii* Planch *ex* Oliv.	(111)
Hirtin (74)	159–161°	+ 26°	*Trichilia hirta* L.	(39)
Deacetylhirtin (75)	non-crystalline		*Trichilia hirta* L.	(39)
Grandifolione (81)	233–235°	— 35°	*Khaya grandifoliola* C. DC.	(45)
Grandifolione acetate (84)	215–217°	— 57°	*Khaya anthotheca* (Welw.) C. DC.	(8)
Azadirone (86)	non-crystalline	+ 26°	*Melia azadirachta* L.	(87)
Azadiradione (87)	non-crystalline	— 24°	*Melia azadirachta* L.	(87)
Epoxyazadiradione (88)	199–200°	— 75° (+ 45°)	*Melia azadirachta* L.	(87, 106)
Gedunin (92)	157°/218°	+ 44°	*Entandrophragma angolense, E. utile, E. delevoyi* De Wild, *Xylocarpus granatum* Koen., *Trichillia trifolia, Melia azadirachta* L., *Cedrela glaziovii*	(11)
7-Deacetylgedunin (96)	250–272°	+ 75°	*Melia azadirachta* L., *Pseudocedrala kotschyii*	(11, 65)
Dihydrogedunin (98)	236°	+ 8°	*Guarea thompsonii*	(11, 12, 81)
7-Deacetoxy-7-oxogedunin (97)	262–265°	— 50°	*Cedrela odorata* L., *C. glaziovii, Carapa guayanensis* Aubl., *Pseudocedrela kotschyii*	(25)
7-Deacetoxy-7-oxo-α-dihydro gedunol (100)	300–301°	— 126°	*Guarea thompsonii*	(81)
Khivorin (101)	256–263°	— 42°	*Khaya ivorensis, K. nyasica* Stapf., *K. anthotheca* (Welw.) C. DC., *K. senegalensis* A. Juss.	(7, 24)
7-Deacetylkhivorin (104)	non-crystalline	— 32°	*Khaya ivorensis, K. grandifoliola* C. DC.	(45)
7-Deacetoxy-7-oxokhivorin (105)	228–230°	— 106°	*Khaya senegalensis* (Desr.) A. Juss., *K. grandifoliola* C. DC.	(45)
3-Deacetylkhivorin (111)	180°/245°	— 38°	*Khaya senegalensis* A. Juss., *K. anthotheca* (Welw.) C. DC.	(6, 7, 122)

Compound	m.p.	[α]*	Source	Ref.
3-Deacetyl-7-deacetoxy-7-oxokhivorin (**113**)	210–214°/240°		*Khaya senegalensis* A. Juss.	(*6*)
11β-Acetoxygedunin (**114**)	176–178°	+ 33°	*Carapa guianensis* Aubl.	(*49*)
6α,11β Diacetoxygedunin (**115**)	184–188°/248–251°	+ 120°	*Carapa guianensis* Aubl.	(*49*)
Nyasin (**116**)	302°	— 42°	*Khaya nyasica* Stapf.	(*122*)
Entandrophragmin (**117**)	256°	— 4°	*Entandrophragma cylindricum, E. bussei* Harms, *E. utile, E. caudatum, E. spicatum*	(*14*)
Utilin (**118**)	278°	— 355°	*Entandrophragma utile*	(*14*)
Andirobin (**121**)	195–197°		*Carapa guayanensis* Aubl.	(*112*)
Deoxyandirobin (**122**)	171–174°		*Khaya grandifoliola* C. DC.	(*9, 64, 112*)
Methyl angolensate (**125**)	197°	— 43°	*Entandrophragma angolense, F. utile, Cedrela odorata* L., *C. glaziovii, Guarea thompsonii, Khaya ivorensis, K. grandifoliola* C. DC., *K. senegalensis* A. Juss.	(*27, 38*)
Methyl 6-hydroxyangolensate (**130**)	252° (237–239°)	— 85°	*Khaya grandifoliola* C. DC., *K. senegalensis* A. Juss.	(*2, 45*)
Methyl 6-Acetoxyangolensate (**131**)	172–174°	— 82°	*Khaya grandifoliola* C. DC.	(*51*)
Nimbin (**132**)	204–205°	+ 168°	*Melia azadirachta* L.	(*103*)
Nimbolide (**150**)	245–247°	+ 206°	*Azadirachta indica*	(*63*)
Salannin (**152**)	167–170°	+ 167°	*Melia azadirachta* L.	(*77*)
Carapin (**158**)	175–178°	+ 64°	*Carapa procera, Cedrela glaziovii*	(*13*)
6-Hydroxycarapin (**159**)			*Cedrela glaziovii*	(*46*)
Mexicanolide (**157**)	226–230°	— 100°	*Cedrela mexicana, C. odorata* L., *C. glaziovii, Carapa procera, Khaya ivorensis*	(*48*)
Swietenolide (**160**)	221–225°	— 136°	*Swietenia macrophylla*	(*50*)
Fissinolide (**164**)	168–171°	— 157°	*Cedrela fissilis* Vell., *Khaya senegalensis* (Desr.) A. Juss, *K. grandifoliola* C. DC., *Guarea trichilioides* L.	(*7, 126*)
Khayasin (**165**)	114–116°	— 165°	*Khaya senegalensis* (Desr.) A. Juss.	(*5*)
3β-Dihydromexicanolide (**167**)	194–196°	— 141°	*Khaya ivorensis*	(*48, 122*)
Swietenine (**173**)	272–276°	— 182°	*Swietenia macrophylla*	(*47*)
6-Deoxydetigloylswietenine acetate (**185**)	223–225°	— 150°	*Khaya senegalensis* A. Juss.	(*2*)
6-Deoxy-12β-acetoxyswietenine acetate (**186**)	250–252°	— 131°	*Khaya senegalensis* A. Juss.	(*2*)
Odoratin (**187**)	216–223°	+ 155°	*Cedrela odorata* L.	(*40*)

* Rotations have been determined in chloroform when specified in the original work.

Table 2. Limonoids Occurring in the Rutaceae

	m. p.	$[\alpha]_D$	Occurence	References
Deacetylnomilin (9)	263–265°	— 112° (dioxane)	*Citrus, Casimiroa*	(*57, 62*)
Nomilin (8)	278–279°	— 95.7° (acetone)	*Citrus, Casimiroa*	(*58, 66*)
Obacunone (7)...........	229–230°	— 50.2° (CHCl$_3$)	*Citrus, Casimiroa, Phellodendron*	(*58*)
7α-Obacunol (12)	224–245°	+ 91° (dioxane)	*Casimiroa*	(*62*)
Obacunoic Acid (10)	205–208°	— 99° (acetone)	*Dictamnus*	(*108*)
Veprisone (14)	180–181°	— 18° (CHCl$_3$)	*Vepris*	(*72*)
Ichangin (16)............	209–212°	— 78° (dioxane)	*Citrus*	(*59*)
Limonin (1)	298°	— 125° (acetone)	*Citrus, Evodia, Phellodendron, Luvunga, Calodendrum, Dictamnus, Ponicrus*	(*20, 58*) (*70*)
Deoxylimonin (15)	323–325°	— 39° (CHCl$_3$)	*Citrus*	(*57*)
Limonin diosphenol (2) ..	273–288°	— 200° (acetone)	*Evodia, Calodendrum*	(*60, 79*)
Rutaevin (18)	301–304°	— 130° (CH$_3$CN)	*Evodia, Dictamnus, Calodendrum*	(*43, 58, 60, 69*)
Fraxinellone (188)	116°	— 44° (ethanol)	*Dictamnus*	(*98, 114*)

Table 3. Protolimonoids Occurring in Rutaceae and Meliaceae

	m. p.	$[\alpha]_D$*	Occurrence	References
In Rutaceae				
Flindissol (**21**)	198°	— 46°	*Flindersia dissosperma* Domin., *F. maculosa* Lindl	(*29*)
Bourjotinolone A (**33**)	176°	— 34°	*Flindersia bourjotiana* F. Muell.	(*31*)
Bourjotinolone B (**34**)	202°	— 89°	*Flindersia bourjotiana* F. Muell.	(*31*)
Bourjotone (**35**)	151°	— 37°	*Flindersia bourjotiana* F. Muell.	(*31*)
In Meliaceae				
Turraeanthin (**22**)	218–220°	+ 3°	*Turraeanthus africanus*	(*22*)
Melianone (**24**)	223–224°/232–233°	— 62°	*Melia azedarach* L.	(*89*)
Aphanamixin (**23**)	232–234°	— 45°	*Aphanamixis polystachya* Wall and Parker	(*42*)
Aglaiol (**32**)	218–220°	+ 3°	*Aglaia odorata* Lour.	(*119*)
Odoratone (**36**)	230–231°	— 90°	*Cedrela odorata* L., *C. glaziovii*	(*41*)
Odoratol (mexicanol) (**37**).....	236–238°	— 45°	*Cedrela odorata* L., *C. glaziovii, C. mexicana*	(*41, 46*)
Meliantriol (**31**)...............	176–178°	— 23°	*Melia azadirachta* L., *M. azedarach* L.	(*90*)

* Rotations have been determined in chloroform when specified in the original work.

Table 4. Limonoids of Unknown Structure

	m. p.	$[\alpha]_D$*	Occurrence	References
In Rutaceae				
Zapoterin	257–259°	— 51°	*Casimiroa edulis* Llave et Lex	*(84)*
Unnamed	249–250°		*Casimirosa edulis* Llave et Lex	*(58)*
Phellandrone	224°	— 80°	*Phellodendron sacchalinense*	*(108)*
Phellandrine	280–281°	— 34°	*Phellodendron sacchalinense*	*(108)*
In Meliaceae				
Bussein	300–304°		*Entandrophragma bussei* Harms, *E. caudatum* Sprague	*(123)*
Prieurianin			*Trichilia prieuriana*	*(23)*
Ekebergolactone A .			*Ekebergia senegalensis* (Desr.) A. Juss.	*(23)*
Ekebergolactone B .			*Ekebergia senegalensis* (Desr.) A. Juss.	*(23)*
Ekebergolactone C .			*Ekebergia senegalensis* (Desr.) A. Juss.	*(23)*
Mahoganin.........	225–228°	— 72°	*Swietenia mohogani*	*(36)*
Dregeanin	278°		*Trichilia dregeana, T. heudelottii* Planch *ex* Oliv.	*(111, 121)*
Pseudrelone A₁.....	265–268°		*Pseudocedrela kotschyii*	*(65)*
Pseudrelone A₂.....			*Pseudocedrela kotschyii*	*(65)*
Candollein	254–256°		*Entandrophragma candollei* Harms	*(3)*

* Rotations have been determined in chloroform when specified in the original work.

References

1. Adeoye, S. A. and D. A. Bekoe: The Molecular Structure of *Cedrela odorata* Substance B. Chem. Commun. **1965**, 301.

2. Adesida, G. A., E. K. Adesogan and D. A. H. Taylor: Extractives from *Khaya senegalensis* A. Juss. Chem. Commun. **1967**, 790.

3. Adesida, G. A. and D. A. H. Taylor: The Chemistry of the Genus *Entandrophragma*. Phytochem. **6**, 1429 (1967).

4. Adesogan, E. K., C. W. L. Bevan, J. W. Powell and D. A. H. Taylor: Extractives from West African Timbers. XV. The Structure of the Low-melting Compound from *Khaya senegalensis*. Chem. Commun. **1966**, 27.

5. — — — — West African Timbers. XVIII. Some Reactions of *Cedrela odorata* Substance "B" and Khayasin. J. Chem. Soc. (London) C **1966**, 2127.

6. Adesogan, E. K., J. W. Powell and D. A. H. Taylor: Extractives from the Seed of *Khaya senegalensis*. J. Chem. Soc. (London) C **1967**, 554.

7. Adesogan, E. K. and D. A. H. Taylor: Grandifoliolin, A New Limonoid from *Khaya grandifoliola* C. DC. Chem. Commun. **1967**, 225.

8. — — Extractives from the Seed of *Khaya anthotheca* (Welw.) C. DC. Chem. Commun. **1967**, 379.

9. — — Isolation of a Steroid Hormone from *Khaya grandifoliola*. Chem. and Ind. **1967**, 1365.

10. AKISANYA, A., E. O. ARENE, C. W. L. BEVAN, D. E. U. EKONG, M. N. NWAJI, J. I. OKOGUN, J. W. POWELL and D. A. H. TAYLOR: West African Timbers. XII. The Inter-relation of Gedunin and Khivorin. J. Chem. Soc. (London) C 1966, 506.

11. AKISANYA, A., C. W. L. BEVAN, T. G. HALSALL, J. W. POWELL and D. A. H. TAYLOR: West African Timbers. IV. Some Reactions of Gedunin. J. Chem. Soc. (London) 1961, 3705.

12. AKISANYA, A., C. W. L. BEVAN, J. HIRST, T. G. HALSALL and D. A. H. TAYLOR: West African Timbers. III. Petroleum Extracts from the Genus *Entandrophragma.* J. Chem. Soc. (London) 1960, 3827.

13. ARENE, E. O., C. W. L. BEVAN, J. W. POWELL and D. A. H. TAYLOR: West African Timbers. XI. The Structure of Carapin, an Extractive from *Carapa procera.* Chem. Commun. 1965, 302.

14. ARENE, E. O., C. W. L. BEVAN, J. W. POWELL, D. A. H. TAYLOR and K. WRAGG: The Inter-relation of Utilin and Entandrophragmin. Chem. Commun. 1966, 627.

15. ARIGONI, D., D. H. R. BARTON, E. J. COREY, O. JEGER, L. CAGLIOTI, S. DEV, P. G. FERRINI, E. R. GLAZIER, A. MELERA, S. K. PRADHAN, K. SCHAFFNER, S. STERNHELL, J. F. TEMPLETON and S. TOBINAGA: The Constitution of Limonin. Experientia 16, 41 (1960).

16. ARNOTT, S., A. W. DAVIE, J. M. ROBERTSON, G. A. SIM and D. G. WATSON: The Structure of Limonin. Experientia 16, 49 (1960).

17. — — — — — The Structure of Limonin: X-Ray Analysis of Epilimonol Iodoacetate. J. Chem. Soc. (London) 1961, 4183.

18. BAILEY, E. J., D. H. R. BARTON, J. ELKS and J. F. TEMPLETON: Compounds Related to Steroid Hormones. IX. Oxygenation of Steroid Ketones in Strongly Basic Medium: A New Method of Preparation of 17α-Hydroxypregnan-20-ones. J. Chem. Soc. (London) 1962, 1578.

19. BALDWIN, M. A., A. G. LOUDON, A. MACCOLL and C. W. L. BEVAN: The Mass Spectra of Meliacins and Related Compounds. I. Gedunin and Related Compounds. J. Chem. Soc. (London) C 1967, 1026.

20. BARTON, D. H. R., S. K. PRADHAN, S. STERNHELL and J. F. TEMPLETON: Triterpenoids. XXV. The Constitution of Limonin and Related Bitter Principles. J. Chem. Soc. (London) 1961, 255.

21. BERNAYS: Limonin. Liebigs Ann. Chem. 40, 317 (1841).

22. BEVAN, C. W. L., D. E. U. EKONG, T. G. HALSALL and P. TOFT: West African Timbers. XX. The Structure of Turraeanthin, an Oxygenated Tetracyclic Triterpene Monoacetate. J. Chem. Soc. (London) C 1967, 820.

23. BEVAN, C. W. L., D. E. U. EKONG and D. A. H. TAYLOR: Extractives from West African Members of the Family Meliaceae. Nature 206, 1323 (1965).

24. BEVAN, C. W. L., T. G. HALSALL, M. N. NWAJI and D. A. H. TAYLOR: West African Timbers. V. The Structure of Khivorin, A Consitutent of *Khaya ivorensis.* J. Chem. Soc. (London) 1962, 768.

25. BEVAN, C. W. L., J. W. POWELL and D. A. H. TAYLOR: West African Timbers. VI. Petroleum Extracts from Species of the Genera *Khaya, Carapa* and *Cedrela.* J. Chem. Soc. (London) 1963, 980.

26. — — — West African Timbers. X. The Structure of *Cedrela odorata* Substance B. Chem. Commun. 1965, 281.

27. BEVAN, C. W. L., J. W. POWELL, D. A. H. TAYLOR, T. G. HALSALL, P. TOFT and M. WELFORD: West African Timbers. XIX. The Structure of Methyl Angolensate, a Ring-B-seco Tetranor-tetracyclic Triterpene of the Meliacin Family. J. Chem. Soc. (London) C 1967, 163.

28. Bevan, C. W. L., A. H. Rees and D. A. H. Taylor: West African Timbers. VII. Anthothecol, an Extractive from *Khaya anthotheca.* J. Chem. Soc. (London) **1963,** 983.

29. Birch, A. J., D. J. Collins, S. Muhammad and J. P. Turnbull: The Structure of Flindissol. Some Remarks on the Elemi Acids. J. Chem. Soc. (London) **1963,** 2762.

30. Bredenberg, J. B.: Biogenesis of the Simarubaceae Bitter Compounds. Chem. and Ind. **1964,** 73.

31. Breen, G. J. W., E. Ritchie, W. T. L. Sidwell and W. C. Taylor: The Chemical Constituents of Australian *Flindersia* Species. XIX. Triterpenoids from the Leaves of *F. bourjotiana* F. Muell. Austral. J. Chem. **19,** 455 (1966).

32. Brown, R. F. C., P. T. Gilham, G. K. Hughes and E. Ritchie: The Chemical Constituents of Australian *Flindersia* Species. V. The Constituents of *Flindersia maculosa* Lindl. Austral. J. Chem. **7,** 181 (1954).

33. Bu'Lock, J. D.: The Biosynthesis of Natural Products, p. 82. London: McGraw-Hill. 1965.

34. Cairns, T., G. Eglinton and S. G. McGeachin: Infrared Studies of Terpenoid Compounds. I. Hydrogen Bonding in Cedrelone and Related Compounds. J. Chem Soc. (London) **1965,** 1235.

35. Calam, D. H. and D. A. H. Taylor: Extractives of East African Timbers. III. The Identification of Volatile Acids from Complex Natural Esters. J. Chem. Soc. (London) **C 1966,** 949.

36. Chakraborty, D. P. and B. K. Barman: Chemical Taxonomy. III. Isolation of Mahoganin, a Non-bitter Constituent of *Swietenia mahogani.* Science and Cult. (India) **31,** 241 (1965) [Chem. Abstr. **64,** 537 (1966)].

37. Chan, W. R., J. A. Gibbs and D. R. Taylor: The Limonoids of *Trichilia havanensis* Jacq.: an Epoxide Rearrangement. Chem. Commun. **1967** 720.

38. Chan, W. R., K. E. Magnus and B. S. Mootoo: Extractives from *Cedrela odorata* L. The Structure of Methyl Angolensate. J. Chem. Soc. (London) **C 1967,** 171.

39. Chan, W. R. and D. R. Taylor: Hirtin and Deacetylhirtin: New "Limonoids" from *Trichilia hirta.* Chem. Commun. **1966,** 206.

40. Chan, W. R., D. R. Taylor and R. T. Aplin: Odoratin, an Undecanortriterpenoid from *Cedrela odorata* L. Chem. Commun. **1966,** 576.

41. Chan, W. R., D. R. Taylor, G. Snatzke and H.-W. Fehlhaber: The Tetracyclic Triterpenes from *Cedrela odorata* L. Chem. Commun. **1967,** 548.

42. Chatterjee, A. and A. B. Kundu: Isolation, Structure and Stereochemistry of Aphanamixin — a New Triterpene from *Aphanamixis polystachya* Wall and Parker. Tetrahedron Letters **1967,** 1471.

43. Chu, J. H.: Constituents of the Chinese Drug Wu-Chu-Yu, *Evodia rutaecarpa.* Science Record (China) **4,** 279 (1951) [Chem. Abstr. **46,** 11589 (1952)].

44. Cole, W. and P. L. Julian: Sterols. XIV. Reduction of Epoxy Ketones by Chromous Salts. J. Organ. Chem. (USA) **19,** 131 (1954).

45. Connolly, J. D., K. L. Handa, R. McCrindle and K. H. Overton: Grandifolione: a Novel Tetranortriterpenoid. Chem. Commun. **1966,** 867.

46. — — — — Mexicanol. Tetrahedron Letters **1967,** 3449.

47. Connolly, J. D., R. Henderson, R. McCrindle, K. H. Overton and N. S. Bhacca: Tetranortriterpenoids. I. [Bicyclononanolides. I.] The Constitution of Swietenine. J. Chem. Soc. (London) **1965,** 6935.

48. Connolly, J. D., R. McCrindle and K. H. Overton: The Constitution of Mexicanolide. A Novel Cleavage Reaction in a Naturally Occurring Bicyclo-[3,3,1]nonane Derivative. Chem. Commun. **1965,** 162.

49. CONNOLLY, J. D., R. McCRINDLE, K. H. OVERTON and J. FEENEY: Tetranortriterpenoids. II. Hartwood Constituents of *Carapa guianensis* Aubl. Tetrahedron **22**, 891 (1966).

50. CONNOLLY, J. D., R. McCRINDLE, K. H. OVERTON and W. D. C. WARNOCK: Swietenolide. Tetrahedron Letters **1965**, 2937.

51. — — — — Tetranortriterpenoids. III. 6-Hydroxy- and 6-Acetoxy-methyl Angolensate from the Heartwood of *Khaya grandifolia*. Tetrahedron **23**, 4035 (1967).

52. COURTNEY, J. L.: The Terpenoid Bitter Principles. Rev. Pure Appl. Chem. **11**, 118 (1961).

53. CROSS, A. D.: The Chemistry of Naturally Occurring 1,2-Epoxides. Quart. Rev. (Chem. Soc. London) **14**, 317 (1960).

54. DEAN, F. M. and T. A. GEISSMAN: The Functional Groups of Nomilin and Obacunone. J. Organ. Chem. (USA) **23**, 596 (1958).

55. DREYER, D. L.: Citrus Bitter Principles. I. A Biogenetic Proposal for the Simaroubaceous Bitter Principles. Experientia **20**, 297 (1964).

56. — Citrus Bitter Principles. II. Application of NMR to Structural and Stereochemical Problems. Tetrahedron **21**, 75 (1965).

57. — Citrus Bitter Principles. III. Isolation of Deacetylnomilin and Deoxylimonin. J. Organ. Chem. (USA) **30**, 749 (1965).

58. — Citrus Bitter Principles. V. Botanical Distribution and Chemotaxonomy in the Rutaceae. Phytochem. **5**, 367 (1966).

59. — Citrus Bitter Principles. VI. Ichangin. J. Organ. Chem. (USA) **31**, 2279 (1966).

60. — Citrus Bitter Principles. VII. Rutaevin. J. Organ. Chem. (USA) **32**, 3442 (1967).

61. — Citrus Bitter Principles. VIII. Application of ORD and CD to Conformational and Stereochemical Problems. Tetrahedron (in press).

61 a. — Citrus Bitter Principles. IX. Extractives of *Casimiroa edulis* Llave et Lex: The Structure of Zapoterin. J. Organ. Chem. (USA) (in press).

62. — unpublished results.

63. EKONG, D. E. U.: Chemistry of the Meliacins (Limonoids). The Structure of Nimbolide, a New Meliacin from *Azadirachta indica*. Chem. Commun. **1967**, 808.

64. EKONG, D. E. U. and E. O. OLAGBEMI: West African Timbers. XVII. Correlation of Gedunin, Methyl Angolensate and Andirobin. J. Chem. Soc. (London) C **1966**, 944.

65. — — Novel Meliacins (Limonoids) from the Wood of *Pseudocedrela kotschyii*. Tetrahedron Letters **1967**, 3525.

66. EMERSON, O. H.: The Bitter Principles of Citrus Fruit. I. Isolation of Nomilin, a New Bitter Principle from the Seeds of Oranges and Lemons. J. Amer. Chem. Soc. **70**, 545 (1948).

67. — The Bitter Principle in Navel Oranges. Food Technol. **3**, 248 (1949).

68. — Bitter Principles of Citrus. II. Relation of Nomilin and Obacunone. J. Amer. Chem. Soc. **73**, 2621 (1951).

69. FUJITA, A. and M. AKATSUKA: Obakulactone. V. Evodin. J. Pharmac. Soc. Japan **69**, 322 (1949) [Chem. Abstr. **44**, 1954 (1950)].

70. GANGULY, A. K., T. R. GOVINDACHARI, A. MANMADE and P. A. MOHAMED: Chemical Investigation of *Luvunga eleutherandra*. Indian J. Chem. **4**, 292 (1966).

71. GOPINATH, K. W., T. R. GOVINDACHARI, P. C. PARTHASARATHY, N. VISWANATHAN, D. ARIGONI and W. C. WILDMAN: The Structure of Cedrelone. Proc. Chem. Soc. (London) **1961**, 446.

72. GOVINDACHARI, T. R., B. S. JOSHI and V. N. SUNDARARAJAN: Structure of Veprisone, a Constituent of *Vepris bilocularis*. Tetrahedron **20**, 2985 (1964).

73. GRANT, I. G., J. A. HAMILTON, T. A. HAMOR, R. HODGES, S. G. MCGEACHIN, R. A. RAPHAEL, J. M. ROBERTSON and G. A. SIM: The Structure of Cedrelone. Proc. Chem. Soc. (London) **1961**, 444.

74. GRANT, I. G., J. A. HAMILTON, T. A. HAMOR, J. M. ROBERTSON and G. A. SIM: The Structure of Cedrelone: X-Ray Analysis of Cedrelone Iodoacetate. J. Chem. Soc. (London) **1963**, 2506.

75. HASSNER, A. and C. HEATHCOCK: Reaction of Nitrosyl Chloride with Steroid 5-Enes. Nuclear Magnetic Resonance as a Stereochemical Tool in Steroids. J. Organ. Chem. (USA) **29**, 1350 (1964).

76. HENDERSON, R., R. MCCRINDLE, K. H. OVERTON, M. HARRIS and D. W. TURNER: The Constitution of Nimbin. Proc. Chem. Soc. (London) **1963**, 269.

77. HENDERSON, R., R. MCCRINDLE, K. H. OVERTON and A. MELERA: Salannin. Tetrahedron Letters **1964**, 3969.

78. HICKMOTT, P. W. and J. R. HARGREAVES: Reaction of α,β-Unsaturated Acid Chlorides with Enamines. I. Preparation of Bicyclo(3,3,1)nonane-2,9-diones. Tetrahedron **23**, 3151 (1967).

79. HIROSE, Y.: The Structure of Evodol, a Principle of *Evodia rutaecarpa*. Chem. and Pharm. Bull. Japan **11**, 535 (1963).

80. HODGES, R., S. G. MCGEACHIN and R. A. RAPHAEL: The Chemistry of Cedrelone. J. Chem. Soc. (London) **1963**, 2515.

81. HOUSLEY, J. R., F. E. KING, T. J. KING and P. R. TAYLOR: The Chemistry of Hardwood Extractives. XXXIV. Constituents of *Guarea* Species. J. Chem. Soc. (London) **1962**, 5095.

82. JAIN, M. K., I. KIRSON and D. LAVIE: The Complete Structure of Melianone. Israel J. Chem. **4**, 32p (1966).

83. KEFFORD, J. F.: The Chemical Constituents of Citrus Fruits. Adv. Food Res. **9**, 285 (1959).

84. KINCL, F. A., J. ROMO, G. ROSENKRANZ and F. SONDHEIMER: The Constituents of *Casimiroa edulis* Llave et Lex. I. The Seed. J. Chem. Soc. (London) **1956**, 4163.

85. KINGSBURY, J. M.: Poisonous Plants of the United States and Canada, p. 205. Englewood Cliffs: Prentice-Hall. 1964.

86. KUBOTA, T., T. MATSUURA, T. TOKOROYAMA, T. KAMIKAWA and T. MATSUMOTO: Establishment of the Correlation of Obacunone and Limonin. Tetrahedron Letters **1961**, 325.

87. LAVIE, D. and M. K. JAIN: Tetranortriterpenoids from *Melia azadirachta* L. Chem. Commun. **1967**, 278.

88. LAVIE, D., M. K. JAIN and I. KIRSON: Terpenoids. V. Melianone from *Melia azedarach* L. Tetrahedron Letters **1966**, 2049.

89. — — — Terpenoids. VI. The Complete Structure of Melianone. J. Chem. Soc. (London) C **1967**, 1347.

90. LAVIE, D., M. K. JAIN and S. R. SHPAN-GABRIELITH: A Locust Phagorepellent from two *Melia* Species. Chem. Commun. **1967**, 910.

91. MAIER, V. P. and G. D. BEVERLY: Limonin Monolactone, the Non-bitter Precursor Responsible for Delayed Bitterness of Certain Citrus Juices. J. Food Sci. (in press).

92. MAIER, V. P. and D. L. DREYER: Citrus Bitter Principles. IV. Occurrence of Limonin in Grapefruit Juice. J. Food Sci. **30**, 874 (1965).

93. Martin, J., W. Parker, B. Shroot and T. Stewart: Bridged Ring Systems. X. The Reductive Rearrangement of δ-Enol-lactones. J. Chem. Soc. (London) C 1967, 101.

94. McPhail, A. T. and G. A. Sim: The Structure of Swietenine: X-Ray Analysis of Detigloylswietenine *p*-Iodobenzoate. J. Chem. Soc. (London) B 1966, 318.

95. Melera, A., K. Schaffner, D. Arigoni und O. Jeger: Zur Konstitution des Limonins. I. Über den Verlauf der alkalischen Hydrolyse von Limonin und Limonol. Helv. Chim. Acta 40, 1420 (1957).

96. Morgan, J. W. W. and D. S. Wilkinson: Sensitization to *Khaya anthotheca*. Nature 207, 1101 (1965).

97. Moron, J., J. Rondest et J. Polonsky: Sur la biosynthèse des constituents amers des Simarubacées. Experientia 22, 511 (1966).

98. Moss, G. P.: Some Aspects of Triterpene Bitter Principle Biosynthesis. Planta Med., Suppl. 1966, 86.

99. Narasimhan, N. S.: Konstitution des Nimbins. I. Natur der funktionellen Gruppen. Chem. Ber. 92, 769 (1959).

100. — Stereochemistry of Nimbin. Chem. and Ind. 1966, 944.

101. Narayanan, C. R. and R. V. Pachapurkar: Ring D in Nimbin. Tetrahedron Letters 1963, 4333.

102. — — The Structure of Nimbinic Acid. Tetrahedron Letters 1966, 553.

103. Narayanan, C. R., R. V. Pachapurkar, S. K. Pradhan, V. R. Shah and N. S. Narasimhan: Structure of Nimbin. Indian J. Chem. 2, 108 (1964).

104. — — — — — — Structure of Nimbin. Chem. and Ind. 1964, 322.

105. — — — — — Stereochemistry of Nimbin. Chem. and Ind. 1964, 324.

106. Narayanan, C. R., R. V. Pachapurkar and B. M. Sawant: Nimbinin: a New Tetranortriterpenoid. Tetrahedron Letters 1967, 3563.

107. Narayanan, C. R., S. K. Pradhan, R. V. Pachapurkar and N. S. Narasimhan: The Molecular Formula of Nimbin. Chem. and Ind. 1962, 1283.

108. Nikonov, G. K.: Chemical Study of *Dictamnus dasycarpus*. Med. Prom. SSSR 18, 15 (1964) [Chem. Abstr. 62, 12157 (1965)].

109. Nikonov, G. K. and R. K. Veremei: Lactones of *Phellodendron sacchalinense*. Doklady Akad. Nauk. SSSR 148, 850 (1963) [Chem. Abstr. 59, 3785 (1963)].

110. Ohochuku, N. S. and J. W. Powell: Nuclear Magnetic Resonance Spectroscopy, the Bandwidths of the Singlets in Gedunin and its Simple Derivatives. Chem. Commun. 1966, 422.

111. Okorie, D. A. and D. A. H. Taylor: The Structure of Heudelottin, an Extractive from *Trichilia heudelottii*. Chem. Commun. 1967, 83.

112. Ollis, W. D., A. D. Ward and R. Zelnik: Andirobin. Tetrahedron Letters 1964, 2607.

113. Ourisson, G., P. Crabbe and O. Rodig: Tetracyclic Triterpenes. San Francisco: Holden-Day. 1964.

114. Pailer, M., G. Schaden, G. Spiteller und W. Fenzl: Die Konstitution des Fraxinellons. Monatsh. Chem. 96, 1324 (1965).

114a. Polonsky, J.: Les principes amers des Simarubacées. Planta Med., Suppl. 1966, 107.

115. Powell, J. W.: Nuclear Magnetic Resonance Spectroscopy. A Correlation of the Spectra of some Meliacins and their Derivatives. J. Chem. Soc. (London) C 1966, 1794.

116. Price, J. R.: In: T. Swain (ed.), Chemical Plant Taxonomy 429. London: Academic Press. 1963.

117. Ritchie, E.: Chemistry of *Flindersia* Species. Rev. Pure Appl. Chem. 14, 47 (1964).

118. SENGUPTA, P., S. K. SENGUPTA and H. N. KHASTGIR: Terpenoids and Related Compounds. II. Investigations on the Structure of Nimbin. Tetrahedron **11**, 67 (1960).

119. SHIENGTHONG, D., A. VERASARN, P. NANONGGAI-SUWANRATH and E. W. WARNHOFF: Constituents of Thai Medicinal Plants. I. Aglaiol. Tetrahedron **21**, 917 (1965).

120. SUTHERLAND, S. A., G. A. SIM and J. M. ROBERTSON: The Structure of Gedunin. Proc. Chem. Soc. (London) **1962**, 222.

121. TAYLOR, D. A. H.: Extractives from East African Timbers. I. J. Chem. Soc. (London) **1965**, 3495.

122. — Nyasin, an Extractive from *Khaya nyasica* Stapf. Chem. Commun. **1967**, 500.

123. — Functional Groups of Bussein. Chem. and Ind. **1967**, 582.

123 a. — unpublished.

124. TAYLOR, D. A. H. and K. WRAGG: The Structure of Entandrophragmin. Chem. Commun. **1967**, 81.

125. WHARTON, P. S. and D. H. BOHLEN: Hydrazine Reduction of α,β-Epoxyketones to Allylic Alcohols. J. Organ. Chem. (USA) **26**, 3615 (1961).

126. ZELNIK, R. et C. M. ROSITO: Le fissinolide. Tetrahedron Letters **1966**, 6441.

127. ZIFFER, H., U. WEISS, C. R. NARAYANAN and R. V. PACHAPURKAR: Absolute Stereochemistry of Nimbin. "Complex" Optical Rotatory Dispersion of Pyronimbic Acid. J. Organ. Chem. (USA) **31**, 2691 (1966).

(Received, November 28, 1967)

Proaporphin-Alkaloide

Von **K. Bernauer** und **W. Hofheinz**, Basel

Inhaltsübersicht

I. Einleitung

Barton und Cohen haben 1957 in einer Arbeit „Some Biogenetic Aspects of Phenol Oxidation" (3) Dienone der allgemeinen Formel (1) als biogenetische Vorstufen bestimmter Aporphin-Alkaloide postuliert (S. 260). Strukturen dieses Typs wurden erstmals 1963 für die Alkaloide D-(+)-Pronuciferin und D-(+)-Crotonosin bewiesen [Bernauer (11); Haynes, Stuart, Barton und Kirby (35)]. Der für solche Verbindungen vorgeschlagene Sammelname „Proaporphine" (23, 65) bringt den inzwischen experimentell bewiesenen biogenetischen Zusammenhang mit der Gruppe der Aporphin-Alkaloide zum Ausdruck, ist aber auch vom präparativ-chemischen Standpunkt sinnvoll, da sich Proaporphine (1) leicht in Aporphine (2) umwandeln lassen (S. 250).

Außer den Alkaloiden mit Dienongruppierung sind auch solche bekannt geworden, in welchen eine oder mehrere Doppelbindungen des Dienonsystems aushydriert sind; sie werden sinnvollerweise den Proaporphinen zugerechnet. Dieser Aufsatz berücksichtigt alle natürlichen Alkaloide und ihre wichtigsten Derivate sowie alle synthetischen Verbindungen, die das Skelett (3) besitzen (*Tabellen 1—3*, SS. 270—279). Er macht ausschließlich von der in (3) angegebenen Bezifferung Gebrauch, welche der Nomenklatur der IUPAC und der Chemical Abstracts entspricht.

Verbindung (1, $R = H$) ist wie folgt zu bezeichnen: 2',3',8',8'a-Tetrahydro-5',6'-dihydroxy-spiro[2,5-cyclohexadien-1,7'(1'H)-cyclopent[ij]-isochinolin]-4-on.

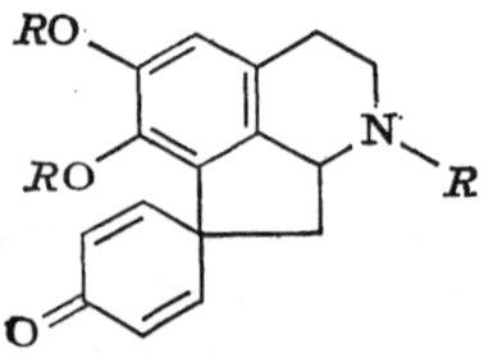

(1) $R = H$ oder Alkyl
Proaporphine

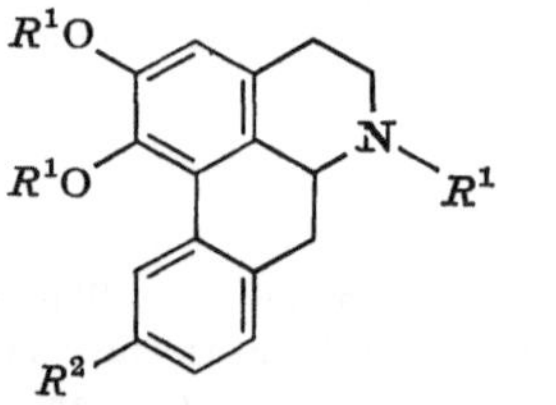

(2) $R^1 = H$ oder Alkyl
$R^2 = H$, OH oder O–Alkyl
Aporphine

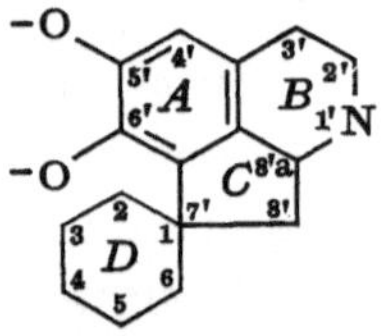

(3)

In den Originalarbeiten sind verschiedene, von der in (3) angegebenen abweichende Bezifferungsarten verwendet worden. Keine hat sich allgemein durchgesetzt. Auch die von Slavík (69) für das sauerstofffreie Grundgerüst der Proaporphine vorgeschlagene Bezeichnung Mecambran hat keinen Eingang in die Literatur gefunden.

Alle bis jetzt bekannt gewordenen Proaporphin-Alkaloide besitzen wie (1) ein Asymmetriezentrum in C-8'a. Sie sind über die 1-Benzyl-tetrahydro-isochinolin-Alkaloide mit den natürlichen Aminosäuren verknüpft (vgl. S. 251). Es ist daher sinnvoll, die Enantiomeren als D- bzw. L-Proaporphine zu bezeichnen. D-Konfiguration ist bei allen Proaporphinen gleichbedeutend mit (8'aR)-Konfiguration, L-Konfiguration mit (8'aS)-Konfiguration. Von der R,S-Nomenklatur wird im folgenden nur bei Verbindungen Gebrauch gemacht, die mehr als 1 Asymmetriezentrum besitzen.

Eine erste Übersicht über die Proaporphine findet sich in einer Arbeit von KÜHN und PFEIFER (46) aus dem Jahre 1965. Die Biosynthese und Synthese von Proaporphinen und Aporphinen durch Phenoloxidation hat BATTERSBY (4a) zusammenfassend dargestellt.

II. Vorkommen der Proaporphin-Alkaloide

Bis jetzt sind 16 verschiedene Proaporphin-Alkaloide isoliert worden, und zwar aus Pflanzen folgender Familien und Gattungen:

Euphorbiaceae, Croton (1 Art).
Lauraceae, Ocotea (1 Art), Neolitsea (1 Art).
Menispermaceae, Stephania (2 Arten), Pericampylus (1 Art).
Monimiaceae, Laurelia (1 Art).
Nymphaeaceae, Nelumbo (1 Art).
Papaveraceae, Papaver (11 Arten), Meconopsis (1 Art).

Tabelle 1, S. 270, führt diese Verbindungen auf.

Von einer systematischen Suche nach Proaporphin-Alkaloiden im Pflanzenreich kann bis jetzt nicht die Rede sein. Lediglich der Tribus Papavereae der Familie Papaveraceae ist recht gut durchforscht. Berücksichtigt man die biosynthetische Verwandtschaft der Proaporphine mit den Aporphinen, so kann man annehmen, daß Proaporphine in allen Familien zu erwarten sind, die Aporphine führen (*19, 65*), d. h. auch in den Familien Anonaceae, Berberidaceae, Ranunculaceae und Rutaceae. Ein Beispiel für die erfolgreiche gezielte Suche nach einem Proaporphin als biosynthetischer Vorstufe eines Aporphins ist die Isolierung des L-(—)-Orientalinons aus *Papaver orientale* L., in welchem das Aporphin L-(+)-Isothebain vorkommt (*8*).

Zieht man als Substituenten an C-5' und C-6' OH- und OCH$_3$-Gruppen, resp. eine Methylendioxygruppe, als Substituenten an N-1' H oder CH$_3$ in Betracht, und läßt man eine mögliche Substitution am Ring *D* außer acht, so errechnet sich die Zahl der denkbaren *p*-Cyclohexadienon-Proaporphine (*1a*) zu 20. Gefunden wurden bisher erst acht, nämlich 4 Alkaloide der D-Reihe und 4 Alkaloide der L-Reihe. Von 220 denkbaren hydrierten Proaporphinen (vgl. S. 248) der gleichen Substitutionstypen sind bisher ebenfalls nur 7 entdeckt worden.

Cyclohexadienon-Proaporphine und hydrierte Proaporphine können nebeneinander in der gleichen Pflanze vorkommen (z. B. in *Croton linearis* Jacq. und *Papaver orientale* L.), ebenso Proaporphine der D- und L-Reihe (z. B. in *C. linearis* Jacq.).

III. Konstitution und Eigenschaften
der Proaporphin-Alkaloide

1. Allgemeines

Neben den eigentlichen Proaporphinen mit p-Cyclohexadienon-Gruppierung (4), sind Cyclohexenon- (6), Cyclohexenol- (7) und Cyclohexanol-Proaporphine (9) als natürliche Alkaloide aufgefunden worden. Dienole vom Typus (5), obgleich wichtige Zwischenprodukte der biosynthetischen Umwandlung von Proaporphinen in Aporphine, konnten bisher nur bei der Reduktion von Dienonen (4) mit komplexen Hydriden gefaßt werden. Cyclohexanon-Proaporphine (8) haben, da sie aus allen natürlichen Proaporphinen erhalten werden können, erhebliche Bedeutung für Korrelation und Konstitutionsaufklärung erlangt.

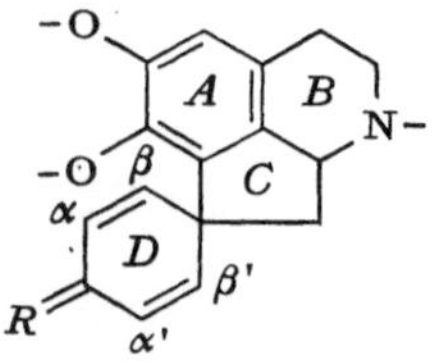

(4) $R = O$
(5) $R = H, OH$

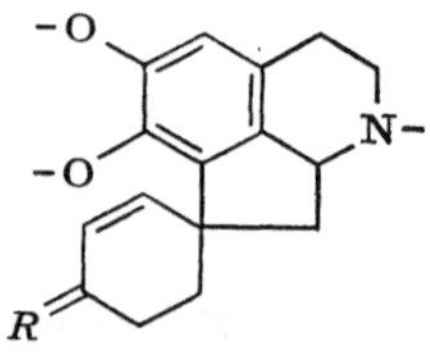

(6) $R = O$
(7) $R = H, OH$

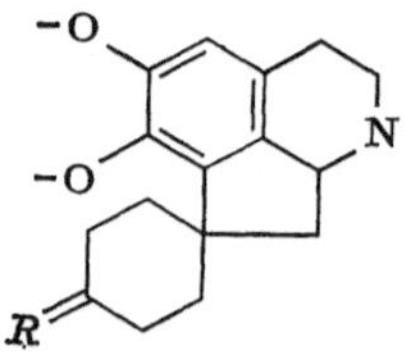

(8) $R = O$
(9) $R = H, OH$

Bemerkenswerte Zusammenhänge bestehen zwischen dem Hydrierungsgrad und dem Circulardichroismus der Proaporphine (*72*). Sie zeigen wie alle Tetrahydroisochinolin-Alkaloide zwischen 225 und 300 mμ zwei Cottoneffekte. Deren Vorzeichen sind bei (8'aS)- oder (7'R, 8'aS)-Konfiguration negativ bei den Typen (4), (7), (8) und (9); positiv dagegen bei dem Typ (6) (wie auch bei den 1-Benzyltetrahydroisochinolinen mit L-Konfiguration).

Der Hydrierungsgrad beeinflußt auch deutlich die Basizität der Proaporphine: Mit zunehmender Sättigung des Ringes D nimmt die Basizität zu.

Die pK-Werte der N-Methylverbindungen betragen bei (4) 6,1—6,2, bei (6) und (7) 6,5—6,6, bei (8) zirka 6,8 und bei (9) zirka 7,2. Die pK-Werte der sekundären Basen liegen um etwa 1 bis 1,5 Einheiten höher.

2. Cyclohexadienon-Proaporphine

a. Allgemeine Eigenschaften

Als Substitutionsvarianten der gleichen Grundverbindung zeigen die Dienon-Proaporphine (4) weitgehende Übereinstimmung in ihren spektroskopischen* und chemischen Eigenschaften. Die Maxima der UV-Absorption liegen bei 226—235 mμ (log ε 4,3—4,5) und bei 282—294 mμ (log ε 3,2—3,7). Die Art der Sauerstoff-Substituenten übt keinen großen Einfluß auf die Lage der Maxima aus. Überlagert man die Spektren von p-Cyclohexadienonen und von dem Ring A entsprechenden, alkylierten Brenzkatechinäthern, so erhält man ganz ähnliche Spektralkurven (*15, 30*). Dies deutet darauf hin, daß sich Cyclohexadienon- und Arylchromophor nur wenig beeinflussen.

In den IR-Spektren sind die intensiven Valenzschwingungsbanden des Cyclohexadienon-Systems bemerkenswert: ν (C=O)-Bande zwischen 1656 und 1673 cm^{-1}, ν (C=C)-Bande(n) zwischen 1605 und 1650 cm^{-1}.

Sehr charakteristisch sind die NMR-Spektren. Wegen des Asymmetriezentrums in C-8'a sind die vier Dienonprotonen nicht paarweise äquivalent. Ihre Signale erscheinen in zwei Gruppen**: bei 2,7—3,3 die der β,β'-Protonen an C-2 und C-6, bei 3,4—3,9 die der α,α'-Protonen an C-3 und C-5.

Die Spektren zweier Alkaloide wurden vollständig analysiert (*37*). Es ergaben sich folgende Kopplungskonstanten: $J_{\alpha\beta} = J_{\alpha'\beta'} = 10$ Hz; $J_{\alpha\alpha'} = 1,5$ bzw. 1,6 Hz; $J_{\beta\beta'} = 2,5$ Hz. Das Signal des isolierten aromatischen Protons an C-4' ist stets ein scharfes Singlett bei 3,35—3,46. Methoxylgruppen an C-5' sind als Singlette bei 6,15—6,23 zu erkennen, während die Signale von Methoxylgruppen an C-6' infolge der diamagnetischen Abschirmung durch den Cyclohexadienonring um zirka 0,25 ppm nach höherem Feld verschoben sind. Die Signale von N-Methylgruppen liegen stets bei etwa 7,6.

Im Massenspektrometer zerfallen alle Dienon-Proaporphine im wesentlichen nach *Schema 1* (*1, 8, 30, 66, 73*), das am Beispiel des Pronuciferins erläutert sei: Der Basispeak entspricht dem Molekülion (**10**). Stark ausgeprägt ist der M-1-Peak [Ion (**11**)]. Die Hauptfragmentierung erfolgt über einen Dien-Zerfall zu dem Ion (**13**), das sich durch Eliminierung eines Methyl-(→ **14**) oder eines Methoxylradikals (→ **16**) stabilisieren kann. Alle Fragmente, welche die Carbonylgruppe des Dienonsystems enthalten, können unter CO-Abspaltung weiter zerfallen (Ionen **12**, **15** und **17**). Die hydrierten Proaporphine zerfallen in grundsätzlich gleicher Weise (*1, 8, 64, 66*). Das Fragmentierungsverhalten der Proaporphine gleicht weitgehend dem der Aporphine (*21*).

Bei der klassischen Polarographie scheinen sich die Cyclohexadienon-Proaporphine einheitlich zu verhalten, ohne sich jedoch von anderen Cyclohexadienon-Alkaloiden zu unterscheiden (*50*).

---

* Eine Ausnahme macht das Orientalinon, dessen Cyclohexadienonring substituiert ist (vgl. S. 254).

** Alle chemischen Verschiebungen werden im folgenden in τ-Werten angegeben.

Die in der zitierten Arbeit (*50*) für das Alkaloid Amurin vorgeschlagene Proaporphin-Struktur ist inzwischen zugunsten einer Morphinan-Struktur revidiert worden (*25*).

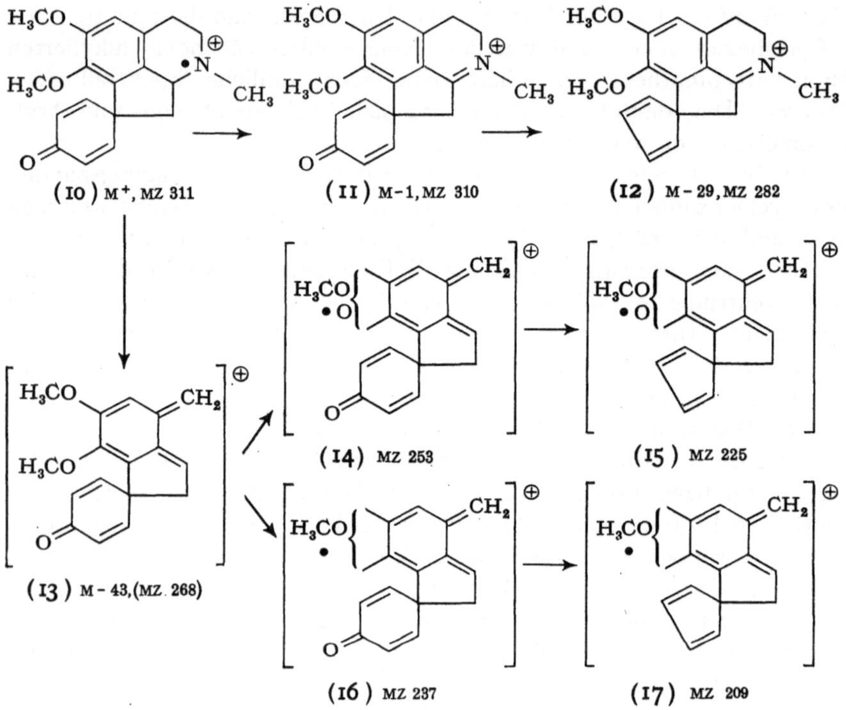

(10) M+, MZ 311   (11) M-1, MZ 310   (12) M-29, MZ 282

(13) M-43, (MZ 268)

(14) MZ 253   (15) MZ 225

(16) MZ 237   (17) MZ 209

Schema 1. Verhalten des Pronuciferins im Massenspektrometer (*73*)

Alle Cyclohexadienon-Proaporphine erleiden bei der Einwirkung von Säuren Dienon-Phenol-Umlagerung (**18 → 20**). Mit komplexen Hydriden

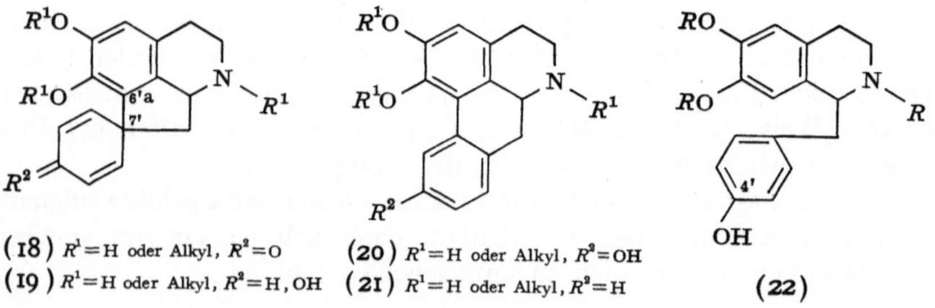

(18) R^1=H oder Alkyl, R^2=O
(19) R^1=H oder Alkyl, R^2=H,OH

(20) R^1=H oder Alkyl, R^2=OH
(21) R^1=H oder Alkyl, R^2=H

(22)

werden sie zu Dienolen (**19**) reduziert, die ihrerseits mit Säure sehr rasch und glatt Dienol-Benzol-Umlagerung (*29, 63*) eingehen (**19 → 21**). Beide Reaktionsfolgen führen zu Aporphinen und sind im Zusammenhang mit

Literaturverzeichnis: SS. 279—283

deren Biosynthese von Interesse (vgl. S. 260 ff.). Alle bekannt gewordenen Umlagerungen können mit einer Phenyl-Wanderung erklärt werden.

Wichtig für die Festlegung der absoluten Konfiguration an C-8'a ist die von Cava et al. (23) entdeckte reduktive Spaltung der 6'a,7'-Bindung durch Natrium in flüssigem Ammoniak, wobei 4'-Hydroxy-benzyl-tetrahydro-isochinolin-Alkaloide (22) entstehen.

b. D-(+)-Pronuciferin und L-(—)-Pronuciferin (23)

Von Pronuciferin kommen beide Enantiomere natürlich vor (11, 61). Die (+)-Verbindung (23) ist das erste Alkaloid, das als Proaporphin erkannt worden ist (11, 15), und zwar auf Grund der folgenden Befunde: Pronuciferin enthält eine N-Methyl- und zwei O-Methyl-Gruppen. Sein IR-Spektrum zeigt keine NH- oder OH-Banden, dagegen die starken Valenzschwingungsbanden einer Dienongruppierung. Das UV-Spektrum besitzt große Ähnlichkeit mit der Spektralkurve, die durch Addition der Spektren von Homoveratrylamin und 4-Allyl-4-methyl-cyclohexa-2,5-dien-1-on erhalten wird. Bei der katalytischen Hydrierung mit Platin in Eisessig entsteht aus (23) das Cyclohexanol (31) mit axialer Hydroxyl-gruppe, dessen Konfiguration an C-4, zunächst an Hand von Modell-betrachtungen abgeleitet, später durch synthetische Arbeiten bestätigt worden ist (vgl. S. 268 und 258). Durch Natriumborhydrid wird (+)-Pro-nuciferin zu einem Gemisch der epimeren Dienole (33) und (34) reduziert. Die Konfiguration dieser Dienole folgt aus ihrer Hydrierung zu den epimeren Cyclohexanolen (31) bzw. (32). Beide Dienole liefern bei Säure-einwirkung durch Dienol-Benzol-Umlagerung das Aporphin (—)-Nuci-ferin (26). Durch Dienon-Phenol-Umlagerung entsteht aus (+)-Pro-nuciferin das (—)-10-Hydroxy-1,2-dimethoxy-aporphin [(—)-Nuciferolin (60)] (28).

Mit Natrium in flüssigem Ammoniak geht (+)-Pronuciferin in D-(—)-Armepavin (30) über, woraus seine Zugehörigkeit zur D-Reihe [(8'a R)-Konfiguration] folgt (23).

D-(+)-Pronuciferin ist identisch mit „Base A" (35) und Miltanthin (60).

c. D-(+)-Stepharin (25)

(+)-Stepharin wird durch N-Methylierung in D-(+)-Pronuciferin (23) übergeführt und ist demnach D-(+)-Norpronuciferin. Sein N-Acetyl-derivat kann in gleicher Weise wie Pronuciferin reduktiv zu einem Benzylisochinolin gespalten werden, dessen Struktur durch Vergleich mit einem synthetischen (racemischen) Produkt gesichert worden ist (23).

d. D-(+)-Glaziovin (24)

Die Cyclohexadienon-Struktur dieses Alkaloids ist auf Grund von IR-, UV- und NMR-Daten und der Hydrierung zu dem Cyclohexanon (35) abgeleitet worden. Dienon-Phenol-Umlagerung führt zu D-(—)-

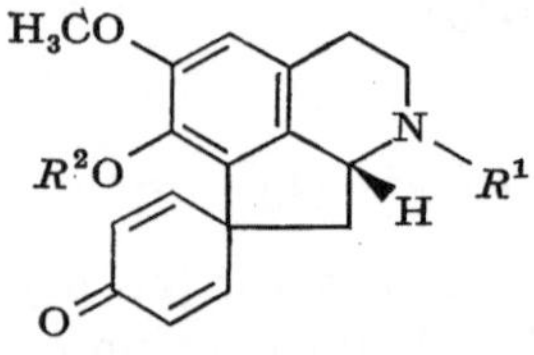

(23) $R^1=R^2=CH_3$. D-(+)-Pronuciferin
(24) $R^1=CH_3$, $R^2=H$. D-(+)-Glaziovin
(25) $R^1=H$, $R^2=CH_3$. D-(+)-Stepharin

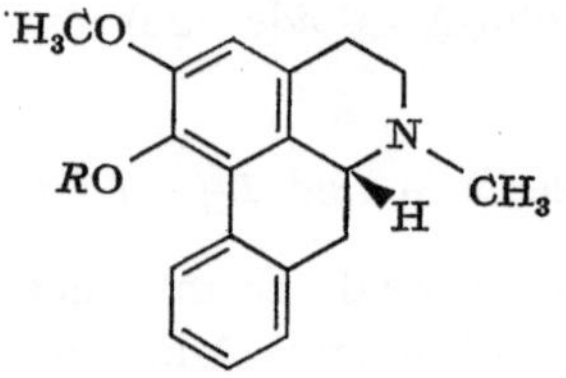

(26) $R=CH_3$. (-)-Nuciferin
(27) $R=H$

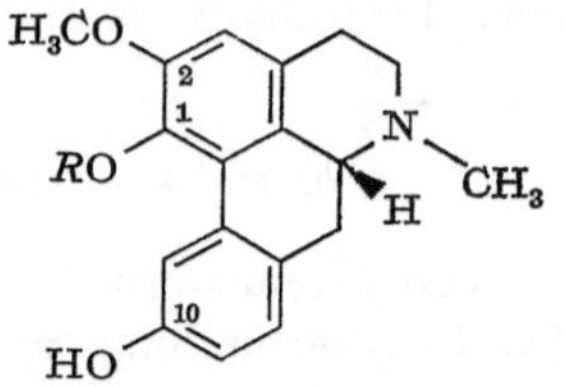

(28) $R=CH_3$. Nuciferolin
(29) $R=H$. Apoglaziovin

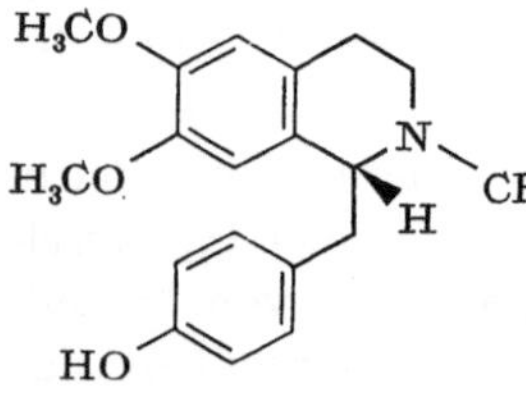

(30) D-(-)Armepavin

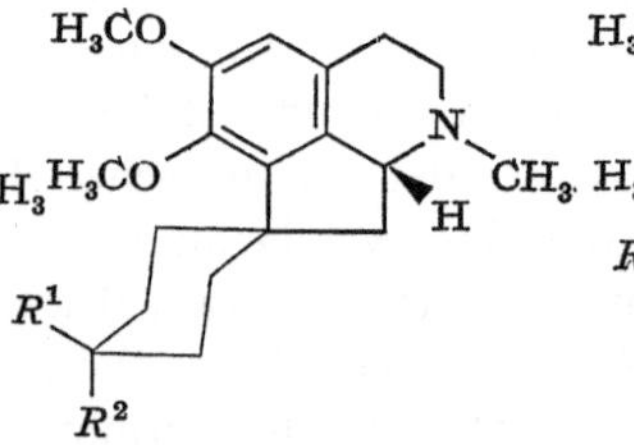

(31) $R^1=OH$, $R^2=H$
(32) $R^1=H$, $R^2=OH$

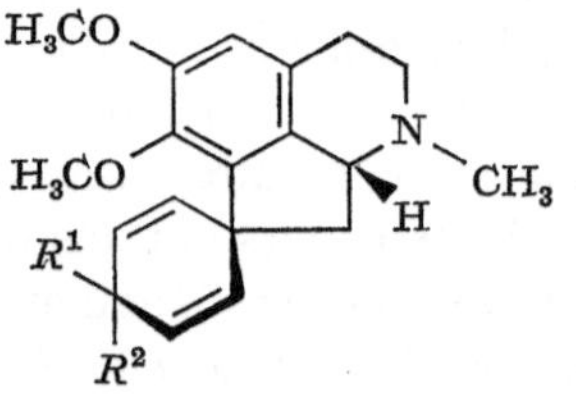

(33) $R^1=H$, $R^2=OH$
(34) $R^1=OH$, $R^2=H$

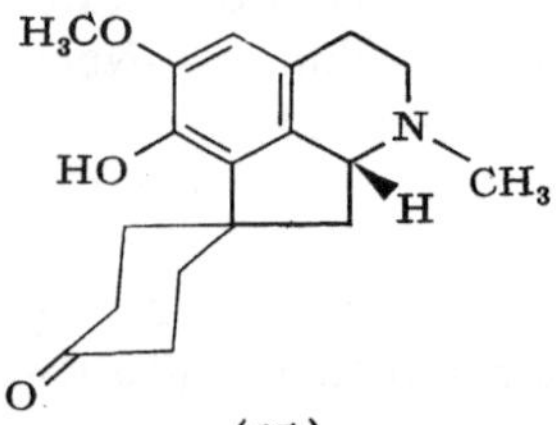

(35)

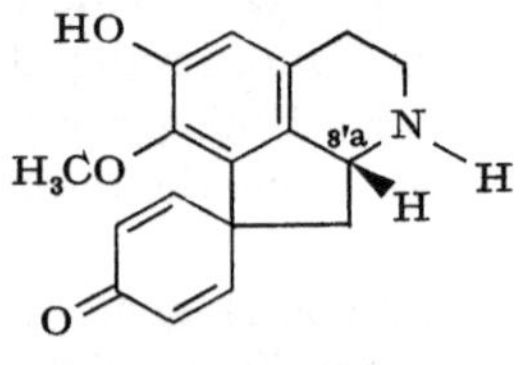

(36) D-(+)-Crotonosin

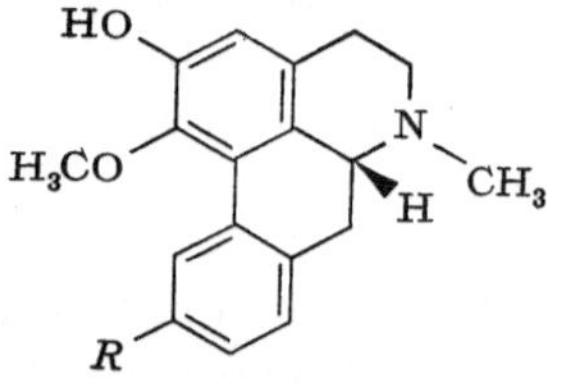

(37) $R=OH$
(38) $R=H$

1,10-Dihydroxy-2-methoxy-aporphin (29), das neben Glaziovin in der gleichen Pflanze vorkommt. Die Struktur dieser Verbindung folgt aus UV- und NMR-Spektren sowie der Methylierung zu Aporphinen bekannter Konstitution.

Borhydrid-Reduktion von Glaziovin gefolgt von Dienol-Benzol-Umlagerung lieferte ein Aporphin (27), das durch Diazomethan in D-(—)-Nuciferin (26) umgewandelt wird. Damit ist (+)-Glaziovin mit D-(+)-Pronuciferin verknüpft und die Struktur (24) gesichert. [Gilbert et al. (30)].

Literaturverzeichnis: SS. 279—283

e. D-(+)-Crotonosin (36)

Für die sekundäre Base (+)-Crotonosin war ursprünglich eine Morphinandienon-Struktur angenommen worden (*33*). Eingehende NMR-Studien an Crotonosin, seinem Tetrahydroderivat und einem durch Dienon-Phenol-Umlagerung von N,O-Diacetylcrotonosin erhaltenen Aporphin haben später zu einer Korrektur zugunsten einer Proaporphin-Struktur geführt (*35*). Die absolute Konfiguration geht aus der Identität von N,O-Dimethyl-crotonosin und D-(+)-Pronuciferin (**23**) hervor (*35*).

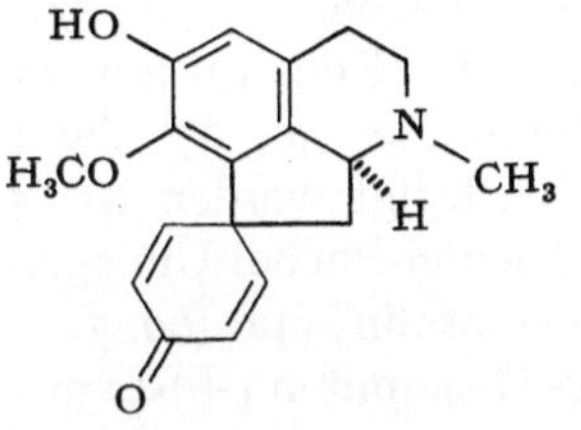

(39) L-(−)-N-Methylcrotonosin

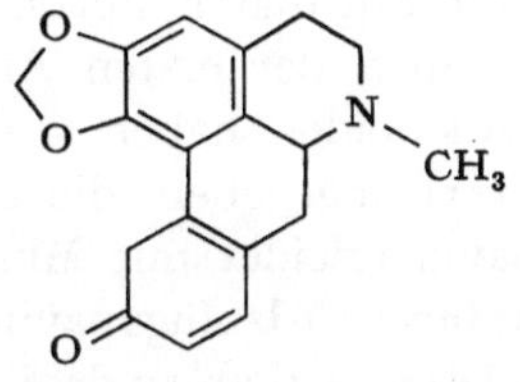

(40)

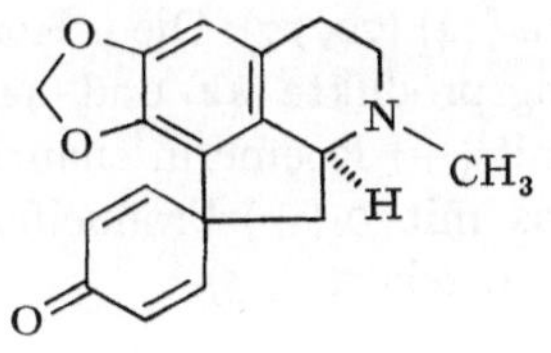

(41) L-(−)-Fugapavin
(Mecambrin)

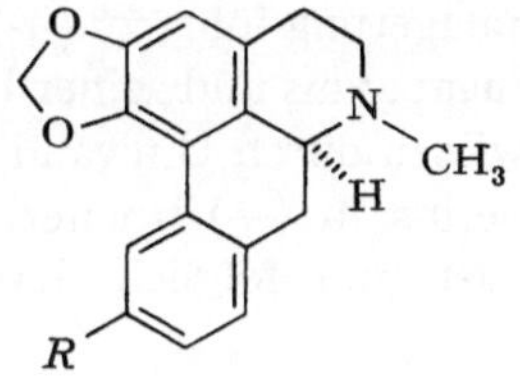

(42) $R = $OH. L-(+)-Isofugapavin
L-(+)-Mecambrolin

(43) $R = OCH_3$. L-(+)-Laurelin

(44) $R = $H. L-(+)-Roemerin

Die Stellung der Hydroxy- und Methoxygruppe ergab sich aus einem Vergleich des aus N-Methyl-crotonosin durch Dienon-Phenol-Umlagerung dargestellten sogenannten Apo-N-methyl-crotonosins (**37**) mit dem analogen Apoglaziovin (**29**) (*36, 37*). Das erstere (**37**) tauscht mit alkalischem D$_2$O drei Protonen orthoständig zu den phenolischen Hydroxylgruppen aus, Apoglaziovin (**29**) hingegen nur zwei. Dienol-Benzol-Umlagerung des Borhydrid-Reduktionsproduktes des N-Methylcrotonosins ergibt das natürliche Alkaloid (—)-2-Hydroxy-1-methoxyaporphin (**38**) (*37*).

Einen unabhängigen Beweis für die D-Konfiguration des Crotonosins kann man aus den Versuchen über die Biosynthese des Alkaloids ableiten (*2, 34*; vgl. S. 260).

f. L-(—)-N-Methylcrotonosin (39)

L-(—)-N-Methylcrotonosin wurde als Begleiter des D-(+)-Crotonosins (**36**) isoliert und anfänglich „Homolinearisin" benannt (*32*). Seine Struktur folgt aus dem direkten Vergleich mit D-(+)-N-Methyl-crotonosin (*36, 37*).

g. L-(—)-Fugapavin (Mecambrin) (41)

Die Geschichte dieses Alkaloids ist recht verworren. Erwähnt sei, daß es anscheinend schon 1907 aus *Papaver dubium* L. isoliert worden ist (*58, 59*; vgl. *68*). 1961 wurde im Anschluß an seine Gewinnung aus *P. fugax* für Fugapavin Formel (**40**) vorgeschlagen (*53, 76, 77*). Nach dem Erscheinen der ersten Arbeiten über die Konstitution von Proaporphin-Alkaloiden haben mehrere Autoren (*18, 30, 65*) die richtige Struktur (**41**) angegeben, die inzwischen bestätigt worden ist (*52, 69*).

Fugapavin erleidet mit Mineralsäure Dienon-Phenol-Umlagerung zu dem Aporphin (+)-Isofugapavin [(+)-Mecambrolin] (**42**) (*69, 71, 53, 77*), das durch Diazomethan zu dem schon länger bekannten (+)-Laurelin (**43**) methyliert wird (*53, 69*).

Reduktion mit Lithiumaluminiumhydrid und anschließende Dienol-Benzol-Umlagerung führt zu (+)-Roemerin (**44**) (*70, 77*). Die L-Konfiguration des Fugapavins und seiner Umlagerungsprodukte (**42**) und (**44**) wurde streng bewiesen durch Umwandlung des mit (+)-Roemerin enantiomeren (—)-Roemerins in (—)-Nuciferin (**26**), das mit D-(+)-Pronuciferin (**23**) korreliert ist und folglich der D-Reihe angehört (*55*).

h. L-(—)-Orientalinon (45)

Orientalinon ist zuerst synthetisiert und dann als natürliches Alkaloid aufgefunden worden (*8, 9, 38*). Wegen der Methoxylgruppe an C-3, die das Spirozentrum zu einem Asymmetriezentrum macht, nimmt es eine Sonderstellung unter den Cyclohexadienon-Proaporphinen ein. Seine Konstitution und die S-Konfiguration an C-8′a ergaben sich aus der ersten Synthese (*9*; vgl. S. 264). Eine zweite Synthese (*41*; vgl. SS. 265, 266) macht (7′S)-Konfiguration wahrscheinlich.

Die Substitution am Dienon-chromophor äußert sich im UV-Spektrum durch das Auftreten eines zusätzlichen Maximums bei 242 mμ (neben Maxima bei 231 und 284 mμ). Das NMR-Spektrum ist einfacher als bei den übrigen Cyclohexadienon-Proaporphinen, da infolge des Methoxylsubstituenten an C-3 nur noch drei olefinische Protonen vorhanden sind und das Proton an C-2 eine andere chemische Verschiebung zeigt als das Proton an C-6. Die Signale der drei Protonen liegen bei 4,08 (Dublett, H an C-2), 3,67 (Dublett, H an C-5) und 3,21 (Quartett, H an C-6; $J_{2.6} =$ = 2,5 Hz und $J_{5.6} =$ 10 Hz) (*9, 38*).

Literaturverzeichnis: SS. 279—283

Bei der Dienol-Benzol-Umlagerung wandert die Phenylgruppe in die
o-Stellung zur Methoxylgruppe und es entsteht L-(+)-Isothebain (**46**)
(*9, 38*).

Das Papaver-Alkaloid Bractein (*44*) ist möglicherweise D-(+)-Orientalinon (*38*).

i. L-(—)-Crotsparin (46a)

Aus *Croton sparsiflorus* Morong ist neben D-(+)-Pronuciferin (**23**) ein
neues Cyclohexadienon-Alkaloid, Crotsparin (**46a**), isoliert worden.
Die Struktur wurde anhand von IR-, NMR- und Massenspektren und
Vergleich des N-Methylderivates mit Glaziovin abgeleitet (*17a*).

(45) L-(-)-Orientalinon   **(46)** L-(+)-Isothebain   **(46a)** L-(-)-Crotsparin

3. Cyclohexenon-Proaporphine

a. L-(+)-Linearisin (47)

Linearisin kommt zusammen mit D-(+)-Crotonosin, D-(+)-Pronuci-
ferin und L-(—)-N-Methylcrotonosin in der gleichen Pflanze vor. Sein
IR-Spektrum, das große Ähnlichkeit mit dem des L-(—)-N-Methyl-
crotonosins („Homolinearisin") besitzt, wies es als α,β-ungesättigtes
Keton aus (*32*). Hydrierung führte zu einem gesättigten Keton, das mit
dem aus D-(+)-Crotonosin erhaltenen Tetrahydro-N-methyl-crotonosin
enantiomer ist. Daraus folgte die Konstitution (**47**) und die L-Konfiguration des Asymmetriezentrums 8'a (*36, 37*).

Die absolute Konfiguration des zweiten Asymmetriezentrums, des
Spirozentrums, wurde aus dem Circulardichroismus abgeleitet (*72*):
Modellbetrachtungen zeigen, daß jede der beiden diastereomeren Formen
des Linearisins in nur einer Konformation vorliegen kann. Aus dem
positiven Cottoneffekt bei 230 mμ ergibt sich dann an Hand der Regeln
für die Korrelation des Circulardichroismus und der Stereochemie von
Enon-Chromophoren eine R-Konfiguration des Spirozentrums. (+)-Line-
arisin besitzt somit (7'R, 8'aS)-Konfiguration.

b. L-(+)-Amuronin (48)

Die Auswertung der NMR-Spektren führte in Verbindung mit den anderen spektroskopischen und analytischen Daten zu der Annahme, daß (+)-Amuronin ein partiell hydriertes Pronuciferin sei. Der direkte Vergleich mit synthetischem (±)-Dihydropronuciferin (vgl. S. 268) bestätigte diese Vermutung (26). Die relative und die absolute Konfiguration wurde in analoger Weise wie bei Linearisin bestimmt. Sie ist wie bei diesem (7'R, 8'aS) (26, 72).

c. (+)-Dihydroorientalinon (49)

Das Massenspektrum dieses Alkaloids zeigt große Ähnlichkeit mit dem des Orientalinons, alle Schlüsselbruchstücke sind jedoch um zwei Einheiten nach höheren Massenzahlen verschoben. Die IR- und NMR-Spektren weisen es als α,β-ungesättigtes Keton mit einer Methoxylgruppe in α-Stellung aus. Die Bestätigung der Struktur (49) (mit der relativen Konfiguration des Orientalinons) ergab sich aus der Reduktion der Ketogruppe mit Borhydrid. Es entstand dabei ein Paar epimerer Cyclohexanole, das auch bei der Borhydrid-Reduktion des Orientalinons hatte gefaßt werden können. Die absolute Konfiguration ist noch unbestimmt (8).

4. Cyclohexenol-Proaporphine

a. L-(+)-Amurolin (50)

Das als Nebenalkaloid des L-(+)-Amuronins (48) isolierte (+)-Amurolin wird durch Mangandioxid zu L-(+)-Amuronin oxydiert (20), aus dem es umgekehrt als Hauptprodukt der Borhydrid-Reduktion entsteht. Es besitzt somit die gleiche relative und absolute Konfiguration wie L-(+)-Amuronin (7'R, 8'aS) (26).

Dihydroamurolin stimmt in allen vergleichbaren Daten mit demjenigen der beiden racemischen Hexahydropronuciferine überein, dessen 4-Hydroxylgruppe äquatorial ist (17; vgl. S. 269). Daraus folgt die angegebene Konfiguration des L-(+)-Amurolins.

b. Alkaloid $C_{18}H_{23}NO_3$ (51)

Dieses bisher nicht benannte Alkaloid wurde als O,O-Diacetat (52) isoliert (31). Katalytische Hydrierung führte unter Hydrogenolyse der allylständigen 4-Acetatgruppe zu (53). Das Reduktionsprodukt (53) wurde nach Verseifung der phenolischen Acetylgruppe zu einem Methojodid (55) methyliert, das auch aus D-(+)-Tetrahydropronuciferin (56) durch Wolff-Kishner-Reduktion und anschließende Methylierung erhalten werden konnte. Verbindung (53) ist nicht identisch mit dem Acetat (54) des Wolff-Kishner-Produktes des N-Methyltetrahydrocrotonosins,

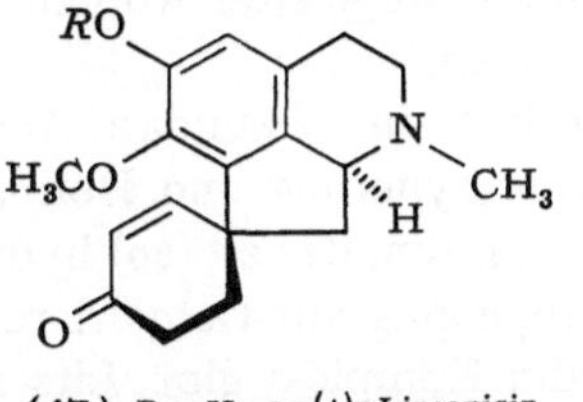

(47) $R = H$. L-(+)-Linearisin
(48) $R = CH_3$. L-(+)-Amuronin

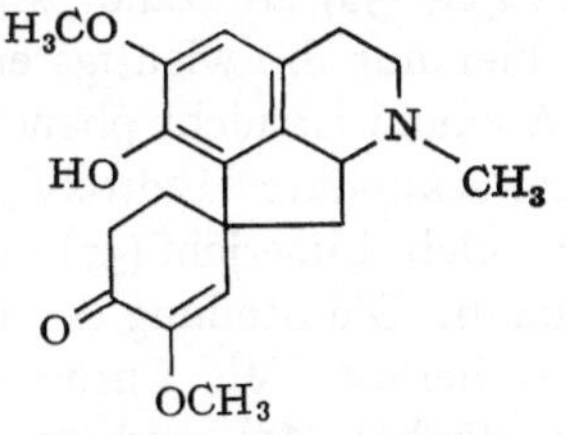

(49) (+)-Dihydroorientalinon

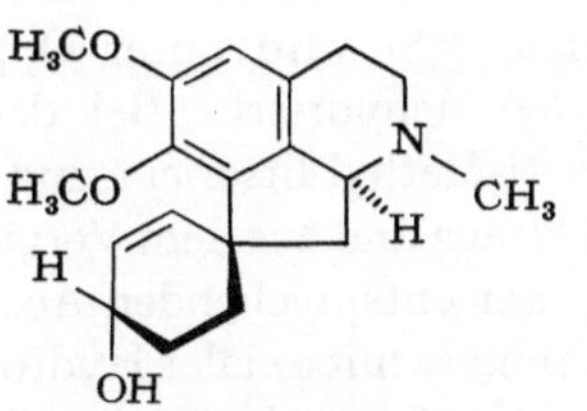

(50) L-(+)-Amurolin

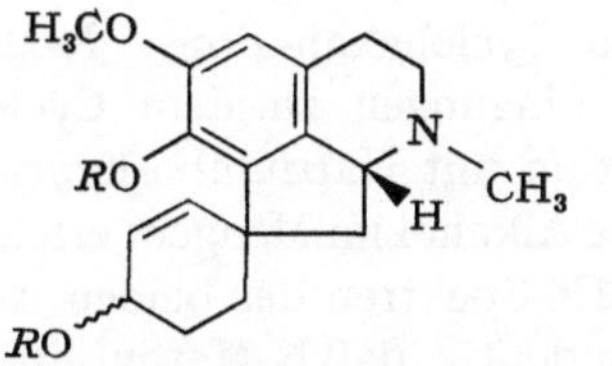

(51) $R = H$. Alkaloid $C_{18}H_{23}NO_3$
(52) $R = COCH_3$

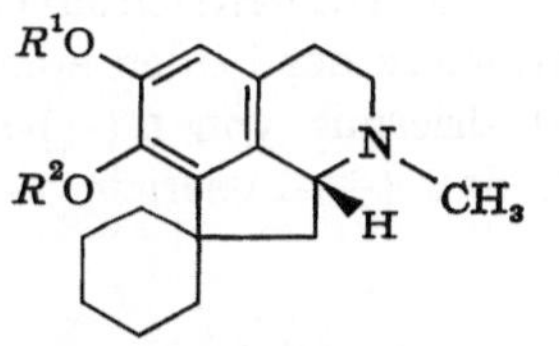

(53) $R^1 = CH_3 . R^2 = COCH_3$
(54) $R^1 = COCH_3 . R^2 = CH_3$

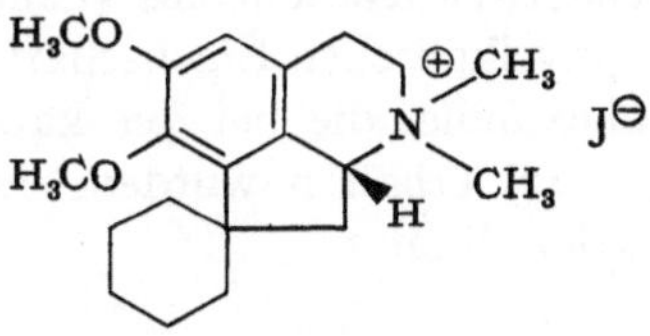

(55)

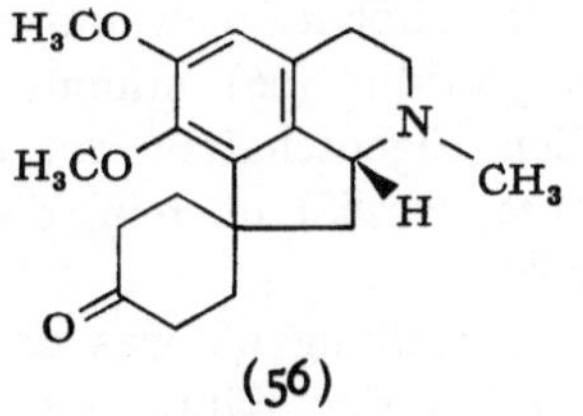

(56)

von dem es sich nur durch die Stellung der Methoxyl- bzw. der Acetoxygruppe unterscheiden kann. Aus diesen Vergleichen folgt für das Alkaloid die Struktur (51), die durch das NMR-Spektrum bestätigt wird (31).

5. Cyclohexanol-Proaporphine

a. D-(+)-Litsericin (57)

Die Konstitution dieses Alkaloids ist von Nakasato, Asada und Koezuka (*54—56*) an Hand zahlreicher Befunde abgeleitet worden, von welchen hier nur die wichtigsten angeführt seien.

Das Alkaloid ist nicht-phenolisch und besitzt eine sekundäre Amino- sowie eine sekundäre Hydroxylgruppe. Permethylierung und Hofmann-Abbau wandeln Litsericin (57) in das Methin (59) um, das zu (60) hydriert werden kann. Die Stellung der Hydroxylgruppe ging aus Dehydrierungsversuchen hervor: Mit Chromtrioxyd-Pyridin-Komplex gibt Litsericin den Ketoalkohol (61) und das Diketon (62). Für die Oxidation der C-8′-Methylengruppe gibt es Analogien in der Aporphin-Reihe (*75a*). Verbindung (62) zeigt sehr klar das A_2B_2-Spektrum eines 1,1-disubstituierten Cyclohexan-4-ons. N-Methyl-litsericin (58) wird unter Oppenauer-Bedingungen zu dem Cyclohexanon (63) dehydriert. Bei dessen Reduktion mit Natriumborhydrid entstehen N-Methyl-litsericin und der epimere Alkohol im Mengenverhältnis 1 : 3. Daraus und aus dem Vergleich der NMR-Spektren der beiden Alkohole und der entsprechenden Acetate folgt eindeutig, daß N-Methyl-litsericin das Epimere mit axialer Hydroxylgruppe ist; und da die energetisch ungünstige Konformation (64) nicht in Betracht gezogen werden muß, ergibt sich auch die angegebene Konfiguration der Hydroxylgruppe in (57) und (58).

N-Methyl-litsericin und N-Methyl-4-epilitsericin erwiesen sich auf Grund ihrer physikalischen Eigenschaften als Enantiomere der beiden epimeren Cyclohexanole, die bei der katalytischen Hydrierung von L-(—)-Fugapavin (41) erhalten wurden. Daraus folgt, daß (+)-Litsericin D-Konfiguration besitzt.

b. L-(—)-Oridin (Oreolin) (65)

L-(—)-Oridin (*51*), das mit dem kurze Zeit später isolierten Oreolin (*62*) identisch ist, ist eine sekundäre Base mit einer aliphatischen und einer phenolischen Hydroxylgruppe. N,O-Dimethyloridin (66) stimmt abgesehen von der optischen Drehung in allen Eigenschaften mit dem synthetischen (±)-Hexahydropronuciferin (**120**, S. 268) überein, dessen Hydroxylgruppe axial ist (*49, 64*; vgl. S. 268).

Die phenolische Hydroxylgruppe steht in Stellung 6′, was daraus hervorgeht, daß aus Oridin unter dem Einfluß starker Säuren ein cyclischer Äther (68) entsteht (*49*). Die L-Konfiguration wurde mit Hilfe der optischen Drehung und der Rotationsdispersion in Korrelation zu D-(+)-Pronuciferin (**23**, S. 252), L-(—)-Fugapavin (41) und L-(—)-Hexahydrofugapavin ermittelt (*64*).

Literaturverzeichnis: SS. 279—283

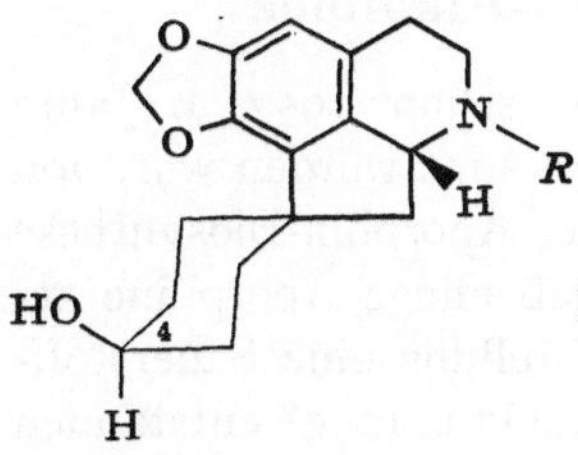

(57) $R=$ H. D-(+)-Litsericin
(58) $R=$ CH$_3$

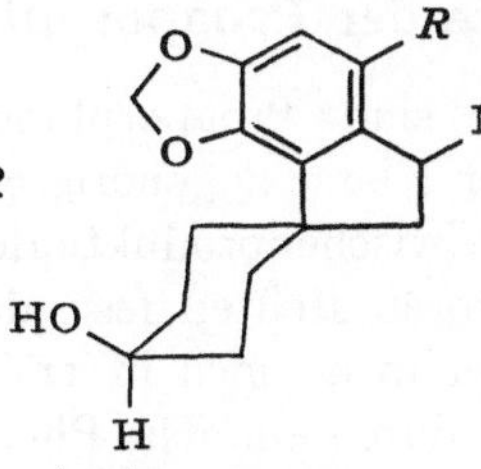

(59) $R=$ CH$=$CH$_2$
(60) $R=$ C$_2$H$_5$

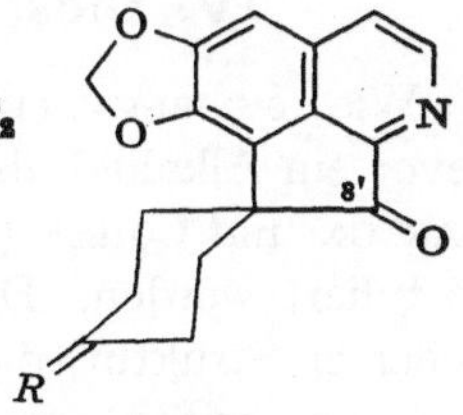

(61) $R=$ H,OH
(62) $R=$ O

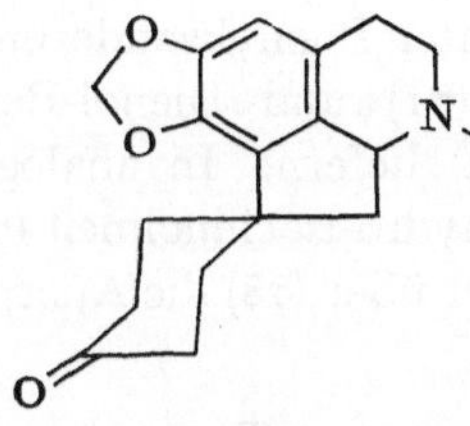

(63)

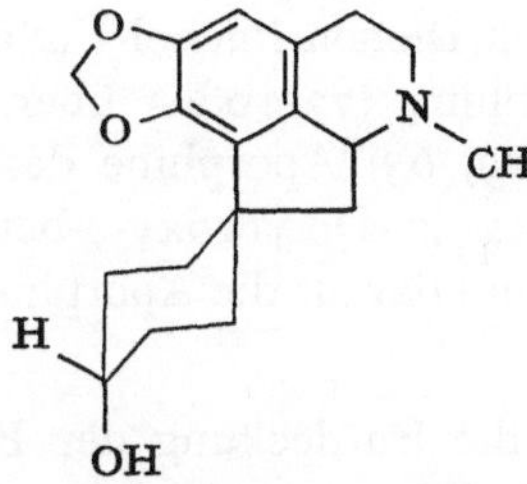

(64)

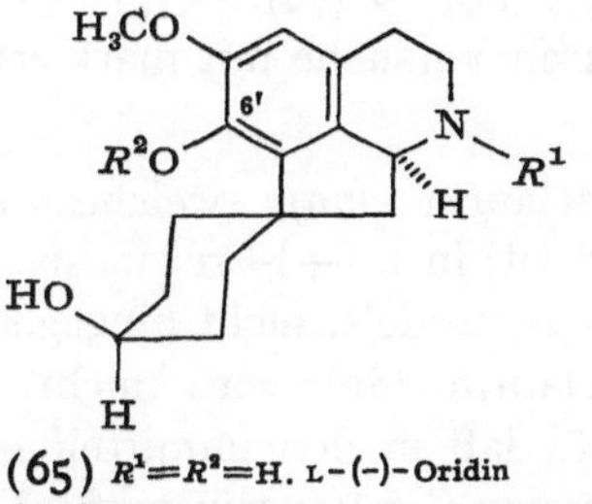

(65) $R^1=R^2=$ H. L-(-)-Oridin
(66) $R^1=R^2=$ CH$_3$. N,O-Dimethyloridin
(67) $R^1=$ CH$_3$, $R^2=$ H. N-Methyloridin

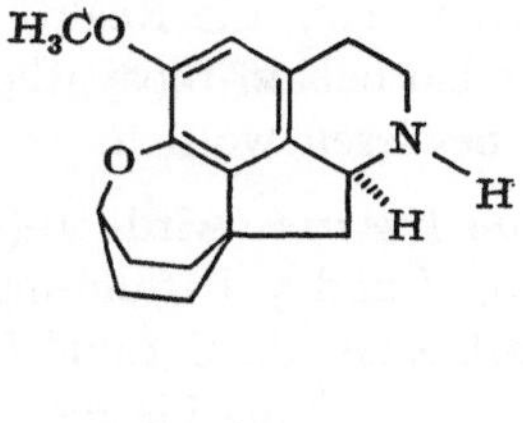

(68)

c. N-Methyloridin

N-Methyloridin kommt als Nebenalkaloid des L-(—)-Oridins (65) vor (62) und „...kann aus Oreolin durch N-Methylierung erhalten werden..." (49). Nach der zitierten Angabe dürfte ihm Struktur (67) und L-Konfiguration zukommen.

17*

IV. Biosynthese der Proaporphin-Alkaloide

Wie eingangs erwähnt, sind Proaporphine schon 1957, 6 Jahre bevor ein Alkaloid als dieser Klasse zugehörig erkannt worden war, von Barton und Cohen (*3*) als Zwischenprodukte der Aporphin-Biosynthese postuliert worden. Die Autoren stellten fest, daß einige Aporphine gesicherter Struktur, die weder in 9- noch in 11-Stellung eine Sauerstofffunktion besitzen (**75**), nicht durch einfache Phenolkupplung* entstanden sein können.

Die Biosynthese dieser Alkaloide läßt sich nach Barton und Cohen dennoch zwanglos mit dem Konzept der Phenolkupplung vereinbaren, wenn man annimmt, daß 7,4'-Dihydroxy-1-benzyl-tetrahydro-isochinoline (**69**) zunächst zu p-Dienonen (**70**) dehydriert werden, welche dann entweder durch Dienon-Phenol-Umlagerung unter Phenylwanderung 10-Hydroxyaporphine (**73**) oder über Dienole (**71**) und Dienol-Benzol-Umlagerung (*29, 63*) Aporphine des Typs (**74**) liefern. In analoger Weise könnten aus 7,2'-Dihydroxy-1-benzyl-tetrahydro-isochinolinen (**76**) o-Dienone (**77**) und daraus die Aporphine (**72**) oder über (**78**) die Aporphine (**74**) entstehen.

Durch die Entdeckung der Proaporphine vom Typus (**70**), überdies teilweise in Pflanzen, welche auch Aporphine der Typen (**73**) und (**74**) führen (vgl. *8, 11, 30, 71*), hat die geschilderte Theorie, was die Schritte (**69**) → (**70**) → (**73**) und (**69**) → (**70**) → (**71**) → (**74**) anbelangt, eine gute Stütze erhalten. Die Reaktionsfolge (**69**) → (**70**) ist am Beispiel des Crotonosins (**36**, S. 252), die Reaktionsfolge (**69**) → (**70**) → (**71**) → (**74**) am Beispiel der Isothebain-Biosynthese durch Versuche mit markierten Verbindungen bewiesen worden.

In *Croton linearis* wird D-(—)-Coclaurin (**79**), welches an den C-Atomen 8, 3' und 5' tritium-markiert ist, in D-(+)-Crotonosin (**36**) mit Tritium-Markierung an C-3 und C-5 umgewandelt, nicht hingegen L-(+)-Coclaurin. Tritium-markiertes Isococlaurin (**80**) wird nicht in Crotonosin übergeführt, was dafür spricht, daß in den unmittelbaren Proaporphin-Vorläufern die Hydroxylgruppe an C-7 frei sein muß. Versuche, mit [Methyl-^{14}C; 8,3',5'-^{3}H$_3$]-coclaurin die Herkunft der O-Methylgruppe des Crotonosins (**36**) zu klären, brachten keine eindeutigen Resultate. Die Methylgruppe des Coclaurins (**79**) wird nur teilweise (und in von Versuch zu Versuch wechselndem Ausmaß) in Crotonosin eingebaut. Dies kann mit einem Demethylierungs-Remethylierungs-Mechanismus erklärt werden. Man kann jedoch nicht ausschließen, daß daneben eine intramolekulare Methylwanderung stattfindet (*2, 34*).

---

* Eine Diskussion dieses Reaktionstypus findet man in (*1a, 4a*).

Literaturverzeichnis: SS. 279—283

Das Aporphin L-(+)-Isothebain (46) könnte, da es eine Sauerstofffunktion an C-11 besitzt, durch direkte Phenolkupplung entstanden sein.
BATTERSBY (4) hat für dieses Alkaloid einen alternativen Biosyntheseweg

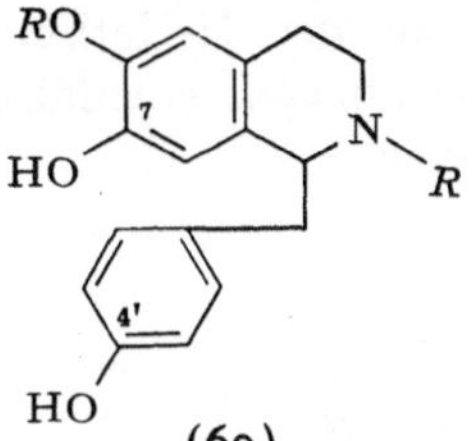

(69) 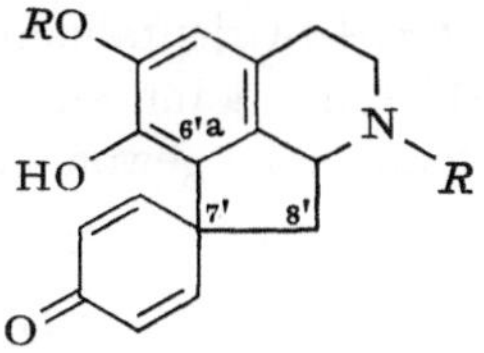   (70) 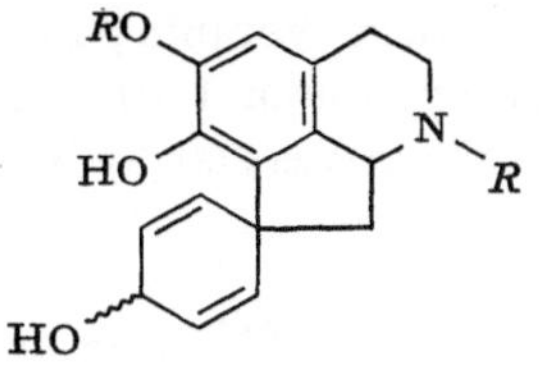   (71)

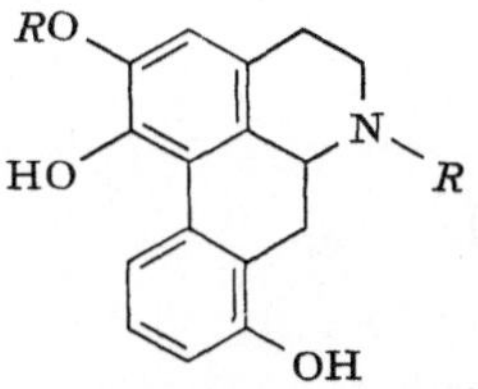

(72) 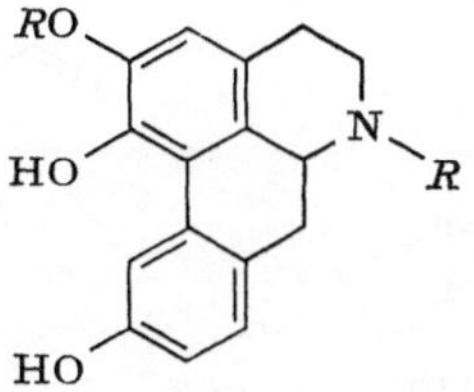   (73) 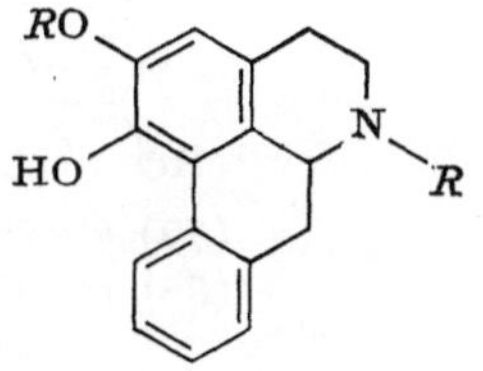   (74)

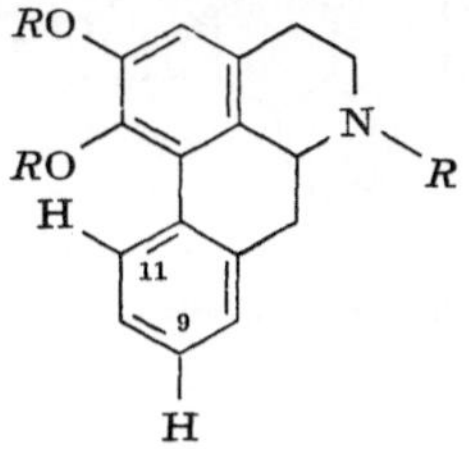

(75)

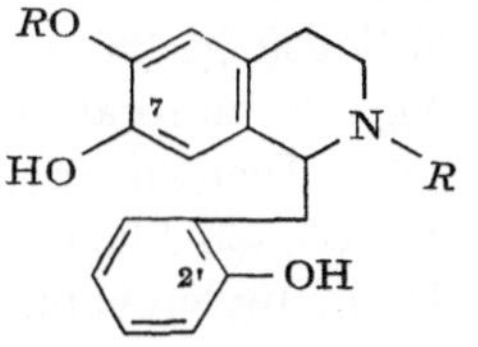

(76) 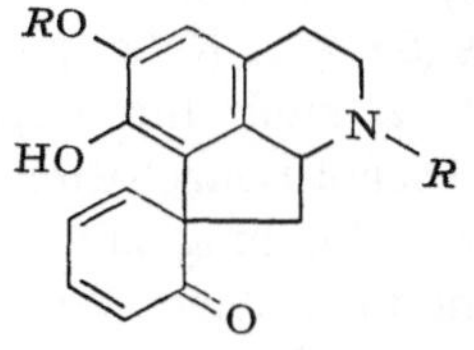   (77) 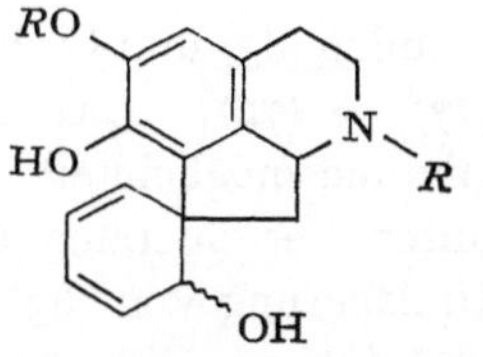   (78)

[(81) → (45) → (82) → (46)] in Erwägung gezogen und später bewiesen
(6, 8): Versuche mit *Papaver orientale* zeigten, daß markiertes Orientalin
(81) spezifisch in L-(−)-Orientalinon (45) und L-(+)-Isothebain (46) übergeführt wird. Dabei wurde erstmals das Vorkommen des Orientalinons (45)
nachgewiesen. Bei der Umwandlung von (81) in (46) bleibt die Methoxyl-

gruppe an C-3' erhalten, wie aus Versuchen mit doppelmarkiertem (±)-Orientalin hervorging. Von den beiden enantiomeren Orientalinen wird der L-Antipode sehr viel besser als der D-Antipode in L-Isothebain umgewandelt.

Nach Verabfolgung von (±)-Orientalinon, das an der N-Methylgruppe durch Tritium markiert ist, kann aus *Papaver orientale* ebenfalls radioaktives L-(+)-Isothebain (46) gewonnen werden.

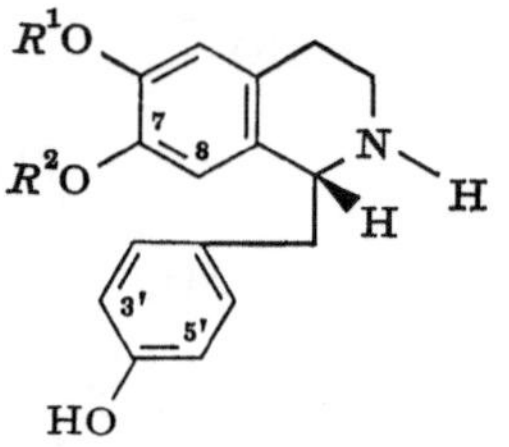

(79) R^1=CH$_3$, R^2=H
(80) R^1=H, R^2=CH$_3$

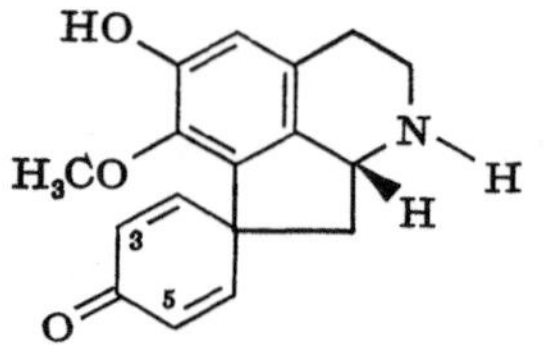

(36)

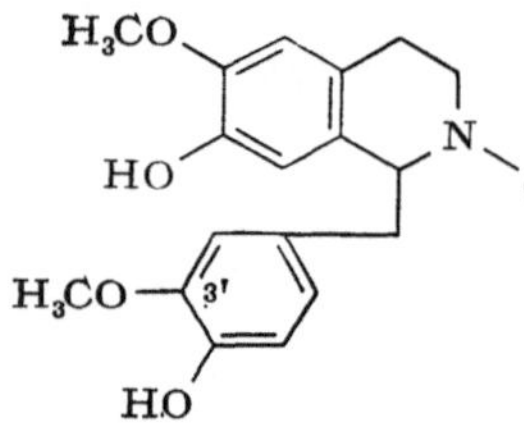

(81)

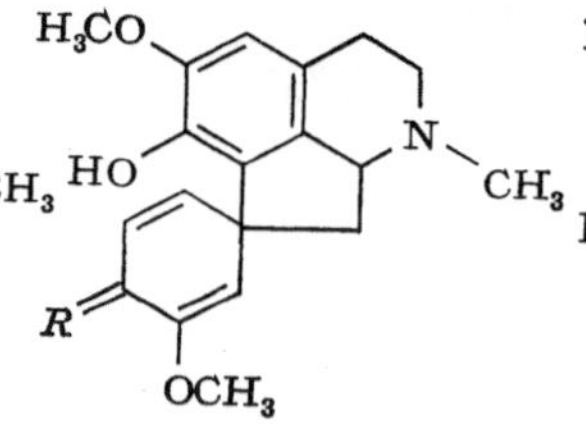

(45) R=O
(82) R=H,OH

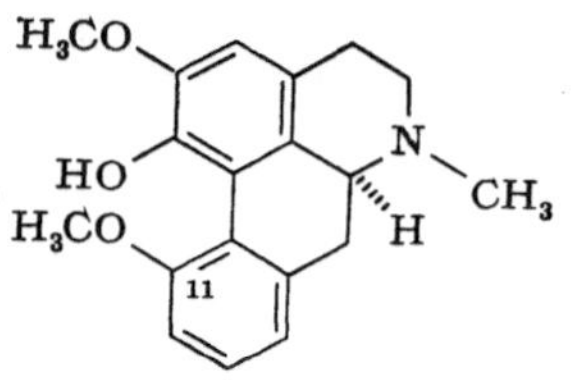

(46) L-(+)-Isothebain

o-Dienone des Typs (77) könnten bei der Biosynthese des Stephanins (85) und Crebanins (87) eine Rolle spielen [Reaktionsfolge (76) → (77) → (72)]. Battersby (4) stellt für Stephanin (85) als alternative Bildungsmöglichkeit die Dienol-Benzol-Umlagerung des Dienols (84) unter Verschiebung der 7',8'-Bindung zur Diskussion. Dienon-Phenol-Umlagerung von (83) würde in analoger Weise zu O-Desmethylcrebanin (86) führen. Für keine der beiden Alternativen liegen experimentelle Argumente vor.

Versuche an *Meconopsis cambrica*- und *Papaver dubium*-Pflanzen haben bewiesen, daß L-(−)-Mecambrin (41) aus (±)-Coclaurin (79) und (±)-N-Methylcoclaurin entsteht und in L-(+)-Römerin (44) bzw. L-(+)-Mecambrolin (42) übergeht. Mit der Umwandlung (41) → (42) ist erstmals die Dienon-Phenol-Umlagerung eines Proaporphins in vivo nachgewiesen (*1 b*).

Literaturverzeichnis: SS. 279—283

Es sei hier noch darauf hingewiesen, daß aus *Kreysigia multiflora* Reichb. kürzlich Vertreter einer neuen Alkaloidgruppe, der Homoaporphine, isoliert worden sind (5). Diese Alkaloide sind über Homo-

(83) $R = O$.
(84) $R = H, OH$

(85) $R = H$. Stephanin
(86) $R = OH$. O-Desmethylcrebanin
(87) $R = OCH_3$. Crebanin

(88) $R = OCH_3$
(89) $R = H$. Kreysiginon

(90) (±)-Multifloramin

(91)

(92)

(93)

proaporphine als Zwischenstufen synthetisiert worden (5, 42), so z. B. (±)-Multifloramin (90) über das Dienon (88). Kurz darauf sind aus der gleichen Pflanze die beiden ersten natürlichen Homoproaporphine, das Kreysiginon (89) und ein Dihydrokreysiginon isoliert worden (10).

V. Synthese der Proaporphin-Alkaloide

Bei racemischen Verbindungen wird die Struktur nur eines Antipoden angegeben.

Bisher sind drei verschiedene Synthesewege bekannt geworden:

1. Die intramolekulare Phenolkupplung von 7,4'-Dihydroxy-1-benzyl-1,2,3,4-tetrahydro-isochinolinen (91), formal eine Nachahmung der Biosynthese *(Weg A)*.

2. Die intramolekulare Phenolkupplung von 7,2'-Dihydroxy-4'-alkoxy-1-benzyl-1,2,3,4-tetrahydro-isochinolinen (92) zu Alkoxy-ortho-Dienonen und deren weitere Umwandlung *(Weg B)*.

3. Die Totalsynthese über 2,3,8,8a-Tetrahydrocyclopent[ij]-isochinolin-7(1 H)-one (93) als typische Zwischenprodukte *(Weg C)*.

Nach Weg C sind auch hydrierte Proaporphine dargestellt worden.

Alle synthetisch erhaltenen Proaporphine sind in *Tabelle 3* (S. 278) zusammengestellt.

Die Synthese der Proaporphin-Alkaloide ist nicht nur an und für sich von Interesse, sondern auch deshalb, weil diese Verbindungen in guten bis sehr guten Ausbeuten durch Dienon-Phenol-Umlagerung oder — nach Reduktion mit komplexen Hydriden — durch Dienol-Benzol-Umlagerung in Aporphine umgewandelt werden können.

1. Proaporphine durch intramolekulare Phenolkupplung
von 7,4'-Dihydroxy-1-benzyl-1,2,3,4-tetrahydroisochinolinen *(Weg A)*

Die nächstliegende Synthese der Proaporphine besteht in der intramolekularen Phenolkupplung von Verbindungen des Typs (91). Auf diesem Weg wurde das erste synthetische Proaporphin, (±)-Orientalinon (96), von Battersby und Brown (7) erhalten. Diese Autoren dehydrierten racemisches Orientalin (94) in einem zweiphasigen, aus Chloroform und wäßriger Ammoniumacetat-Lösung bestehenden System mit Kaliumferricyanid und faßten (96) in etwa 2% Ausbeute. Natriumborhydrid-Reduktion und anschließende säurekatalysierte Dienol- Benzol-Umlagerung führte (96) in das Aporphin (±)-Isothebain (98) über. Diese Reaktionsfolge ist später auch mit den beiden Antipoden von (94) durchgeführt worden (9).

Ebenfalls durch intramolekulare Phenolkupplung wurde (±)-Glaziovin (97) aus (±)-N-Methylcoclaurin (95) erhalten (24, 43). (97) wurde mit Diazomethan zu (±)-Pronuciferin (117, S. 268) methyliert (43).

2. Proaporphine durch intramolekulare Phenolkupplung
von 7,2'-Dihydroxy-1-benzyl-1,2,3,4-tetrahydroisochinolinen *(Weg B)*

Jackson und Martin *(39—41)* haben gefunden, daß die Verbindung (99) bei der Kaliumferricyanid-Dehydrierung im System Chloroform-wäßrige Ammoniumacetat-Lösung die zwei diastereomeren 2,4-Dienone (**100**) (6,9% als Picronolat) und (**101**) (10,7% als Picronolat) liefert. Die Zuteilung der Strukturen beruht auf der Annahme, daß die weniger polare —

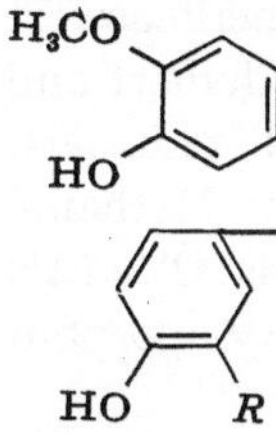

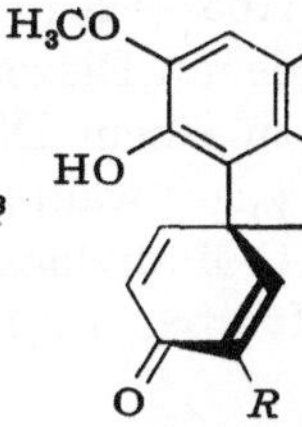

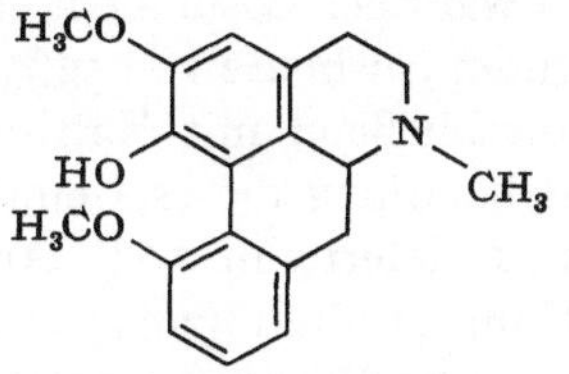

(**94**) R = OCH$_3$. (±)-Orientalin (**96**) R = OCH$_3$. (±)-Orientalinon (**98**) (±)-Isothebain
(**95**) R = H. (±)-N-Methylcoclaurin (**97**) R = H. (±)-Glaziovin

im Dünnschichtchromatogramm schneller laufende — der beiden Verbindungen die Struktur (**100**) haben muß, weil, wie Modellbetrachtungen zeigen, nur bei dieser eine Wasserstoffbrücke zwischen phenolischer

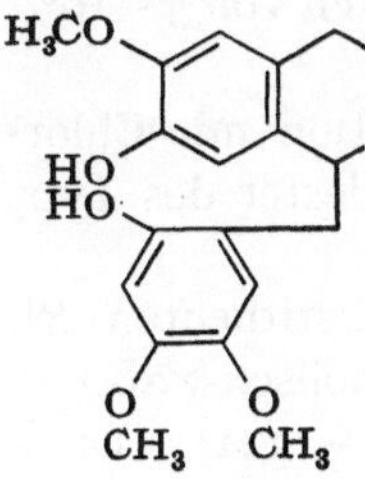

(**99**) 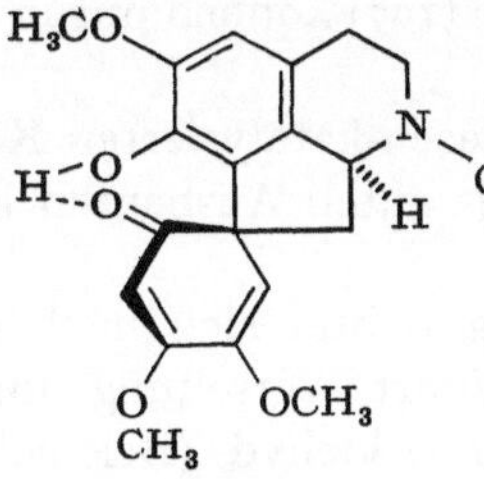 (**100**) 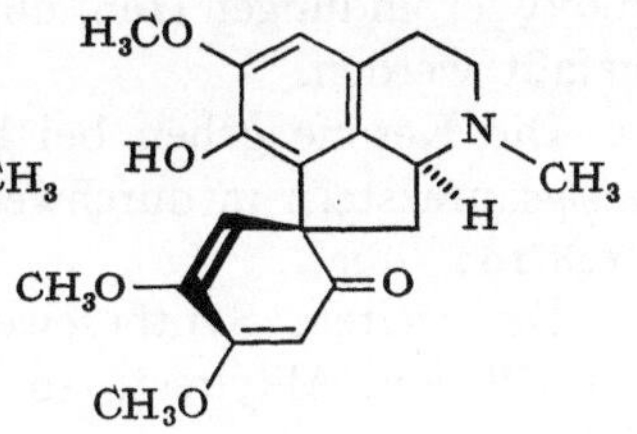 (**101**)

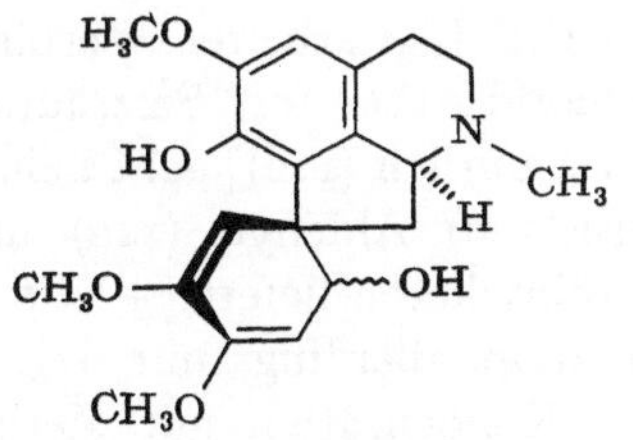

(**102**)

Hydroxylgruppe und Carbonylgruppe möglich ist. (**101**) läßt sich mit Natriumborhydrid zu (**102**) reduzieren. (**102**) ist wahrscheinlich ein Gemisch der beiden möglichen diastereomeren Dienole. Beim Behandeln mit verdünnter methanolisch-wäßriger Salzsäure gibt (**102**) in 50% Ausbeute (±)-Orientalinon (**96**). Unter der Voraussetzung, daß die

Konfigurationszuteilung für (100) und (101) richtig ist, ist die Frage der Konfiguration des Orientalinons im Sinne der Formel (96) gelöst.

Shamma und Slusarchyk (*66*) haben ebenfalls die Dehydrierung von (99) studiert. Sie erhielten mit Kaliumferricyanid-Ammoniumacetat in wäßriger Lösung ein öliges Dienon (100 + 101 ?) in der bemerkenswert guten Ausbeute von 52%. Dieses wurde mit Diazomethan methyliert und anschließend mit Natriumborhydrid zu einem Dienolgemisch reduziert. Behandlung dieses Gemisches mit 10%iger wäßriger Salzsäure-Methanol 1:1 lieferte in 10% Ausbeute (±)-O-Methylorientalinon als Öl. (Als Hauptprodukt (29%) wurde das 10-Hydroxy-1,2,11-trimethoxy-aporphin (±)-Pseudocorydin gefaßt.)

3. Proaporphine über 2,3,8,8a-Tetrahydrocyclopent[ij]-isochinolin-7(1H)-one *(Weg C)*

Eine Totalsynthese von Proaporphinen über Ketone des Typs (93) als Zwischenprodukte ist von Bernauer (*12, 13, 17*) ausgearbeitet worden. Diese Ketone erhält man durch Cyclisierung von 1,2,3,4-Tetrahydro-isochinolin-1-essigsäuren (103) mit Polyphosphorsäure; so z. B. (104) in 44% und (105) in 60% Ausbeute (vgl. auch *22, 22a*). Die Methylen-dioxyverbindungen (106) und (107) können nur in Ausbeuten von 5—10% gefaßt werden.

Die Ketone gehen bei basen-katalysierter Kondensation mit Chlor-essigsäureestern in durchwegs guten Ausbeuten in Glycidester des Typs (108/109) über.

Der weitere Syntheseweg richtet sich nach dem Substituenten R^1. Ist R^1 ein Alkylrest, so führt Verseifung mit äthanolisch-wäßriger Lauge zu dem gewünschten Aldehyd (Beispiel: 108 → 111). Diese Methode versagt, wenn R^1 ein Acylrest ist. In diesem Fall kommt man aber zum Ziel, indem man mit einem t-Butyl-ester ($R^4 =$ $= C(CH_3)_3$) arbeitet: Verbindung (109), zum Beispiel, wird in Methylen-chlorid-Äther mit Salzsäuregas behandelt. Das rohe Reaktionsprodukt [vermutlich (110)] geht beim Erwärmen mit Pyridin in den (nicht kristal-lisierten) Aldehyd (112) über. Kondensation von (111) mit Methyl-äthinylketon liefert (±)-Pronuciferin (117). Die Ausbeute in dieser Stufe beträgt allerdings nur 6%.

Kondensation mit Methylvinylketon führt von den Aldehyden des Typs (111/112) zu Dihydroproaporphinen des Typs (113) (Ausbeute an 113 zirka 40%). Diese können mit 2,3-Dichlor-5,6-dicyanbenzochinon zu den entsprechenden Dienonen dehydriert werden. Diese Umwandlung ist auch durch Bromierung-Dehydrobromierung möglich. So läßt sich (113) in Eisessiglösung in Gegenwart von Bromwasserstoffsäure als Katalysator in die Bromverbindung (114) überführen (Ausbeute zirka 30%). Als Nebenprodukte entstehen je nach der verwendeten Brom-

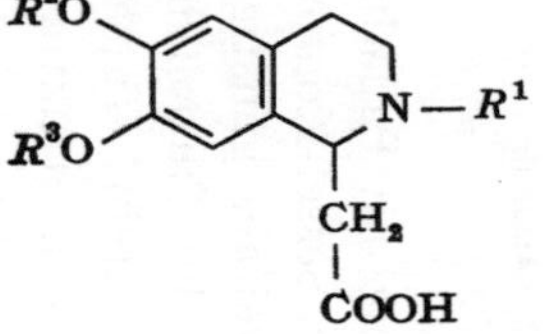

(110)

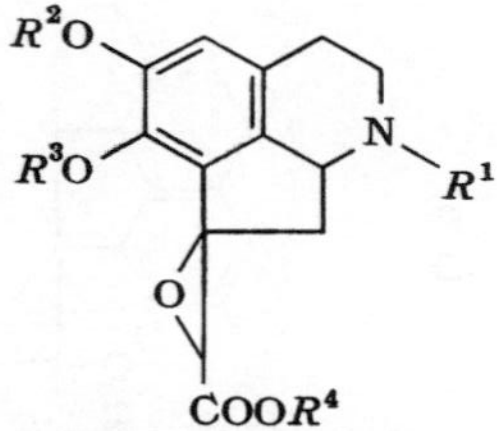

(108) $R^1=R^2=R^3=R^4=CH_3$
(109) $R^1=CHO, R^2=R^3=CH_3, R^4=C(CH_3)_3$

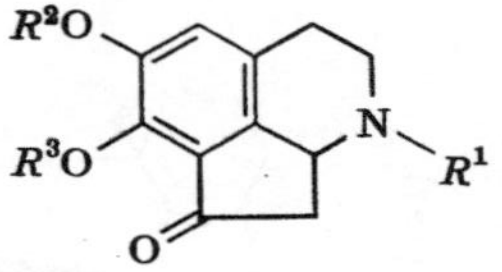

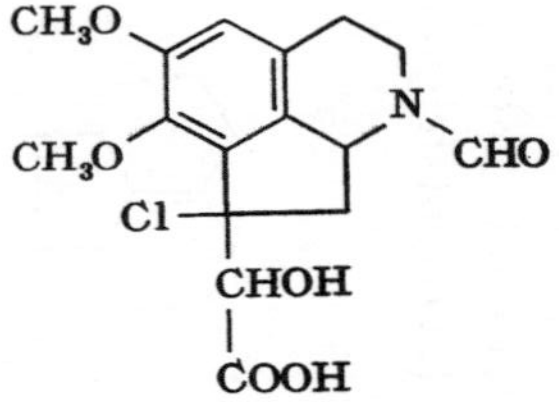

(104) $R^1=R^2=R^3=CH_3$
(105) $R^1=H, R^2=R^3=CH_3$
(106) $R^1=H, R^2+R^3=CH_2$
(107) $R^1=CH_3, R^2+R^3=CH_2$

(103)

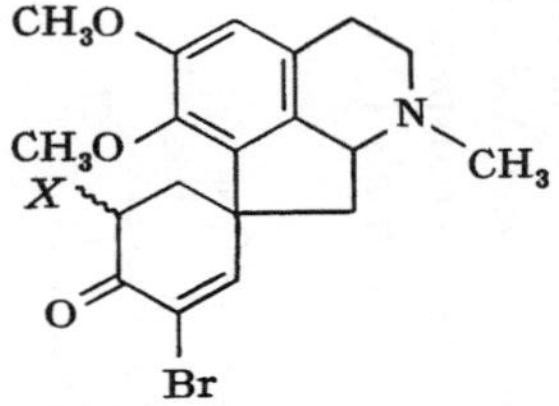

(115) $X=H$
(116) $X=Br$

(114)

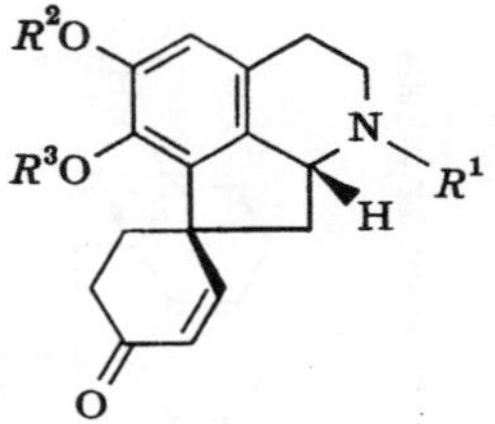

(113) $R^1=R^2=R^3=CH_3$
(±)-Amuronin

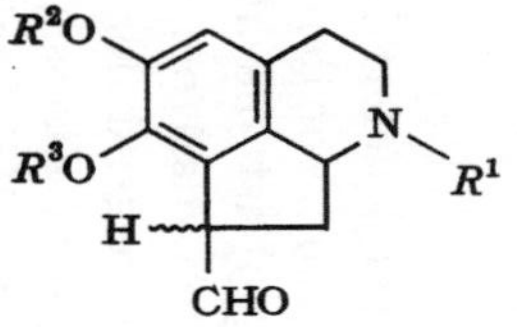

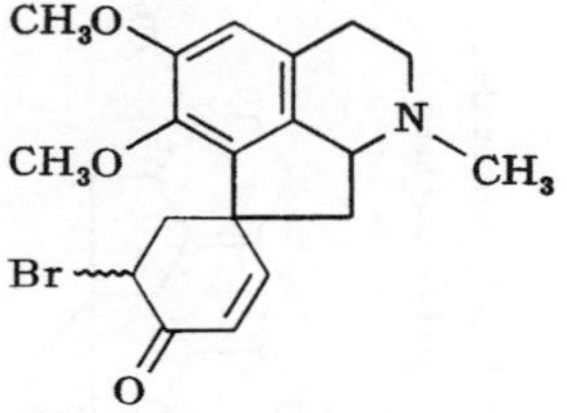

(111) $R^1=R^2=R^3=CH_3$
(112) $R^1=CHO, R^2=R^3=CH_3$

menge entweder die Monobromverbindung (**115**) oder die Dibromverbindung (**116**). (**114**) liefert bei kurzem Erwärmen mit 1,5-Diazabicyclo-[4.3.0]-5-nonen (57) in 60% Ausbeute (±)-Pronuciferin (**117**).

Die Kondensation der Aldehyde (**111**) und (**112**) mit Methylvinylketon verläuft fast völlig stereospezifisch. Die aus (**111**) entstehende Verbindung (**113**) erwies sich als (±)-Amuronin.

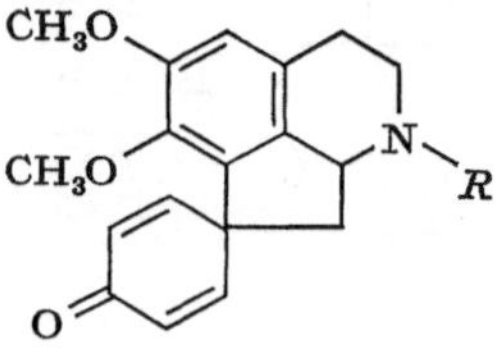

(**117**) $R = CH_3$. (±)-Pronuciferin
(**118**) $R = H$. (±)-Stepharin

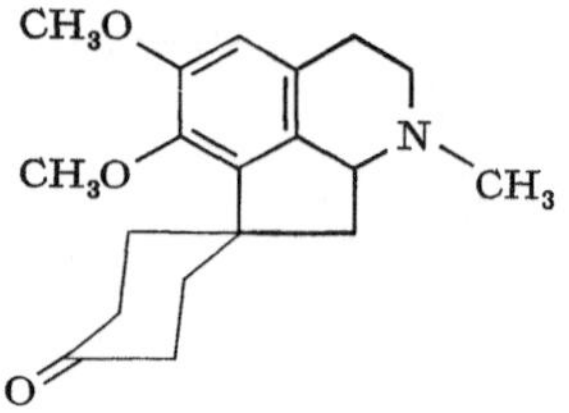

(**119**)

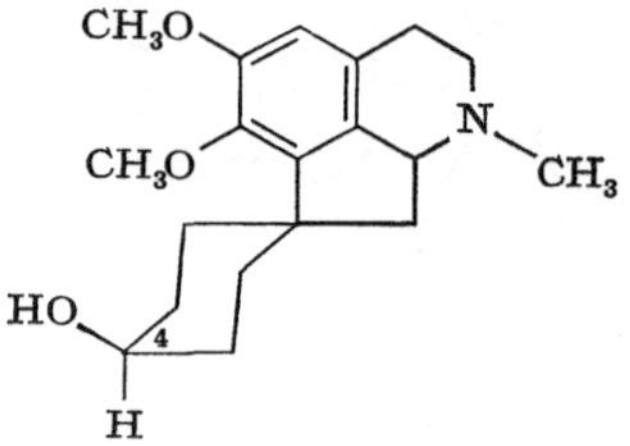

(**120**) (±)-N, O-Dimethyloridin

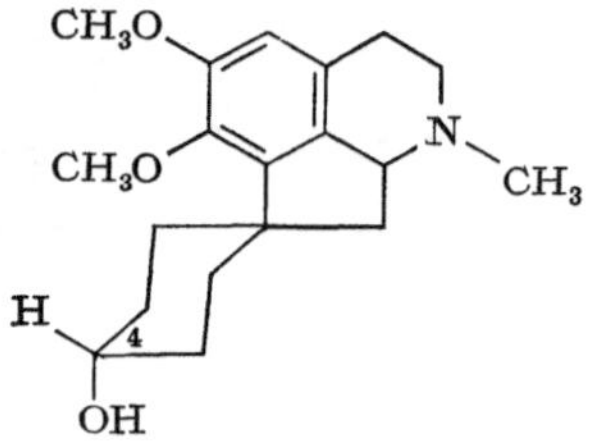

(**121**)

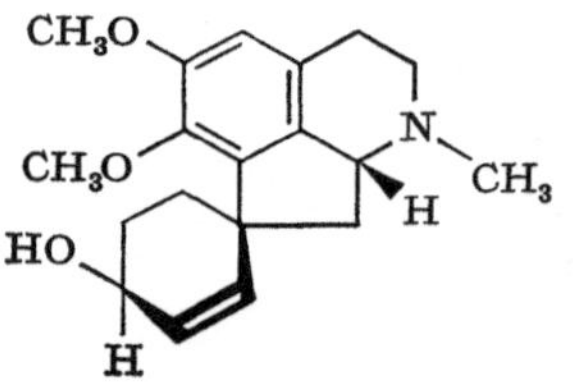

(**122**)

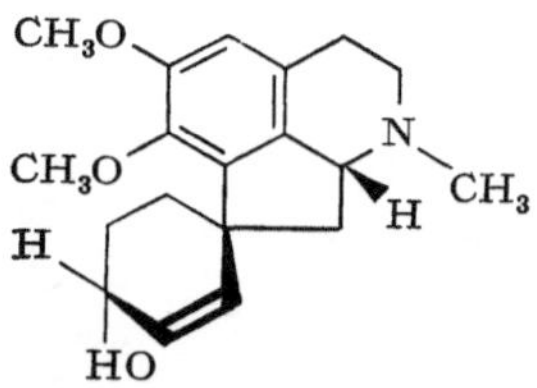

(**123**) (±)-Amurolin

(±)-Amuronin (**113**) liefert bei der Hydrierung mit Palladium-Kohle das Cyclohexanon (**119**). Von den beiden möglichen Sessel-Konformationen der Verbindung (**119**) ist die abgebildete die weitaus günstigere. Man kann daher erwarten, daß bei der Hydrierung der Carbonylgruppe in saurer Lösung überwiegend der Alkohol (**120**) entsteht. Tatsächlich erhält man mit Platin in Eisessig einen Alkohol, dem Struktur (**120**) zuzuordnen ist, da sein NMR-Spektrum bei 5,9 ein relativ scharfes Signal für ein äquatoriales Proton an C-4 zeigt. Die epimere Verbindung (**121**)

Literaturverzeichnis: SS. 279—283

entsteht neben wenig (**120**) bei der Natriumborhydrid-Reduktion von
(**119**). Im NMR-Spektrum von (**121**) findet man eine breite Bande für
das axiale Proton an C-4 bei zirka 6,4 (man vergleiche die Konstitutionsaufklärung des Litsericins, S. 258). Verbindung (**120**) stimmt hinsichtlich
des IR-Spektrums in Lösung, des Massenspektrums und des chromatographischen Verhaltens mit N,O-Dimethyloridin überein (vgl. S. 258).

($\pm$)-Amuronin (**113**) liefert bei der Reduktion mit Natriumborhydrid
die zwei epimeren Alkohole (**122**) und (**123**), deren relative Konfiguration
aus der Hydrierung zu (**120**) resp. (**121**) folgt. (**123**) hat sich als ($\pm$)-Amu-
rolin erwiesen.

VI. Pharmakologische Eigenschaften

Es liegen nur spärliche Angaben über pharmakologische Eigenschaften
von Proaporphin-Alkaloiden vor, und zwar nur für D-($+$)-Pronuciferin (**23**),
D-($+$)-Crotonosin (**36**) und D-($+$)-Glaziovin (**24**).

Die LD 50 an der Maus beträgt bei intraperitonealer Verabfolgung für Pronuciferin 120 mg/kg, für Crotonosin 1,2 g/kg. Beide Alkaloide wirken im Meer-
schweinchen-Test nach Bülbring und Wajda stärker lokalanästhetisch als Lig-
nocain und Procain. Am Meerschweinchen-Ileum verstärken 0,001—1,0 mcg/ml
Pronuciferin und 0,1—10,0 mcg/ml Crotonosin die durch Acetylcholin induzierten
Kontraktionen; höhere Dosen (5—200 mcg/ml bzw. 100—200 mcg/ml) inhibieren
die Wirkung von Acetylcholin und Nicotin. Skelettmuskelpräparate werden durch
beide Verbindungen inhibiert. Die Wirkung von D-($+$)-Crotonosin (**36**) ähnelt
derjenigen eines depolarisierenden neuromuskulären Blockers, die von D-($+$)-
Pronuciferin (**23**) der eines kompetitiven Blockers (*28*). Glaziovin (**24**) wirkt nach
Angaben in einem Patent (*67*) antidepressiv und anxiolytisch.

VII. Tabellen

Tabelle 1. Natürliche Proaporphin-Alkaloide

No.	Name (Synonyma) Bruttoformel	Struktur	Schmelzpunkt	$[\alpha]_D$ (Lösungsmittel)	Isoliert aus	Konstitution	Spektren
1	D-(+)-Pro-nuciferin (Base A, Miltan-thin) $C_{19}H_{21}NO_3$	CH_3O, CH_3O, N—CH_3, H, O	127—129°	+ 99° $(CHCl_3)$ *(11)* + 86° $(CHCl_3)$ *(23)* + 106° (C_2H_5OH) *(11)* + 111° (C_2H_5OH) *(32)*	*Nelumbo nucifera* Gaertn. *(11, 14, 27)* *Croton linearis* Jacq. *(32)* *Stephania glabra* Miers *(23)*	*(11, 15, 23)*	IR, UV, NMR: *(15)* MS: *(73)*
2	L-(—)-Pro-nuciferin $C_{19}H_{21}NO_3$	CH_3O, CH_3O, N—CH_3, H, O	126—128°	— 109° (C_2H_5OH) *(61)*	*Papaver persicum* Lindl. *(61, 45)* *P. caucasicum* Marsch.-Bieb. *(48, 60,* vgl. aber *61)* *P. fugax* Poir. *(45)** *P. triniaefolium* Briss. *(45)** *P. polychaetum* Schott et Kotschy *(45)** *P. armeniacum* L. *(45)*	*(61)*	
3	D-(+)-Stepharin $C_{18}H_{19}NO_3$	CH_3O, CH_3O, N—H, H, O	179—181°	+ 143° $(CHCl_3)$ *(23)*	*Stephania glabra* Miers *(23)* *St. rotunda* Loureiro *(75)* *Pericampylus formo-sanus* Diels *(74)* *Laurelia novaezelan-*	*(23)*	MS: *(73)*

4	D-(+)-Glaziovin $C_{18}H_{19}NO_3$		235—237° Zers.	+ 7° ($CHCl_3$) (30)	*Ocotea glaziovii* Mez. (30) *Papaver caucasicum* Marsch.-Bieb. (47)*	IR, UV, NMR, MS: (30)
5	D-(+)-Crotonosin $C_{17}H_{17}NO_3$		Zers. ab 197°	+ 180° (CH_3OH) (32)	*Croton linearis* Jacq. (32) *C. discolor* Willd. (37) (35, 37)	IR, UV: (32) NMR: (37)
6	L-(—)-N-Methylcrotonosin (Homolinearisin) $C_{18}H_{19}NO_3$		218—220° Zers.	— 116,5° (CH_3OH) (32)	*Croton linearis* Jacq. (32) *Papaver caucasicum* Marsch.-Bieb. (47) (36, 37)	IR, UV, NMR: (37)

* Absolute Konfiguration unsicher.

(Fortsetzung der Tabelle 1)

No.	Name (Synonyma) Bruttoformel	Struktur	Schmelzpunkt	$[\alpha]_D$ (Lösungsmittel)	Isoliert aus	Konstitution	Spektren
7	L-(—)-Fuga-pavin (Mecam-brin) $C_{18}H_{17}NO_3$	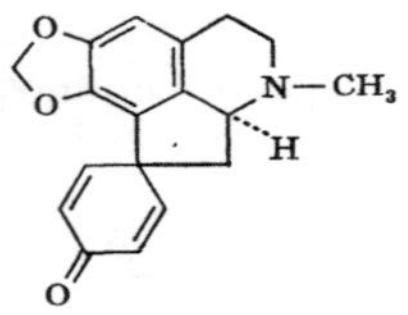	179—180°	— 116° (CHCl₃) (76) — 94° (CHCl₃) (71)	Meconopsis cambrica (L.) Vig. (71) Papaver dubium L. (68) P. caucasicum Marsch.-Bieb. (48, 63a) P. triniaefolium Boiss. (69, 48) P. armeniacum L. DC. (69, 48) P. persicum Lindl. (69, 48, 63a) P. fugax Poir. (76, 77, 48) P. polychaetum Schott et Kotschy (48)	(18, 30, 52, 69, 55)	IR, UV: (71)
8	L-(—)-Orien-talinon $C_{19}H_{21}NO_4$	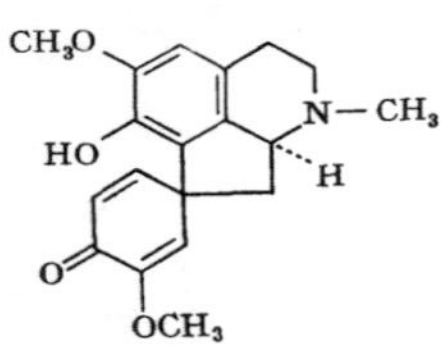	230—232° Zers. (38)	— 76° (CHCl₃) (38)	Papaver orientale (8) P. bracteatum Lindl. (38)	(9, 38)	IR, NMR: (9) MS: (8) UV: (38)
9	L-(+)-Amuronin $C_{19}H_{23}NO_3$	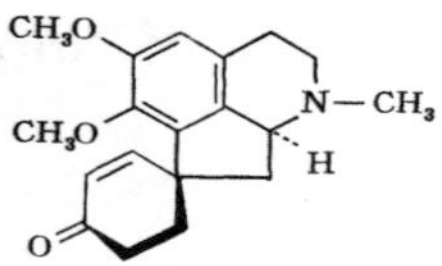	119—120° (20) 120° und 131—132° (26)	+ 140° (CHCl₃) (20) + 124° (CHCl₃) (26)	Papaver nudicaule, var. amurense (20)	(26, 72)	IR, UV, NMR: (26)

			Smp.	$[\alpha]_D$	Vorkommen		
10	L-(+)-Li-nearisin $C_{18}H_{21}NO_3$	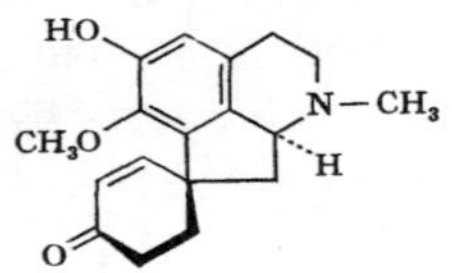 [Struktur]	219—222° Zers.	+ 116° (CH_3OH) (*32*)	*Croton linearis* Jacq. (*32*)	(*36, 37, 72*)	IR, UV: (*32*) NMR: (*37*)
11	(+)-Di-hydro-orienta-linon $C_{19}H_{23}NO_4$	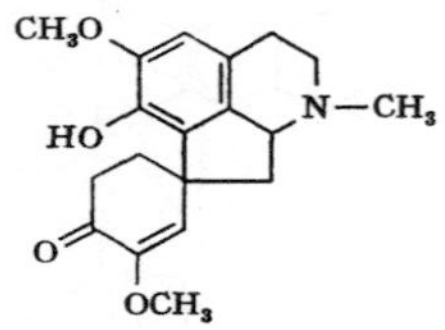 [Struktur]		+ 50° ($CHCl_3$)	*Papaver orientale* (*8*)	(*8*)	IR, NMR, MS: (*8*)
12	L-(+)-Amurolin $C_{19}H_{25}NO_3$	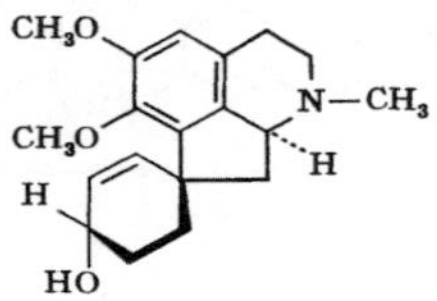 [Struktur]	169—170°	+ 106° ($CHCl_3$) (*26*)	*Papaver nudicaule,* var. *amurense* (*20*)	(*26, 17*)	IR, UV: (*26*)
13	$C_{18}H_{23}NO_3$	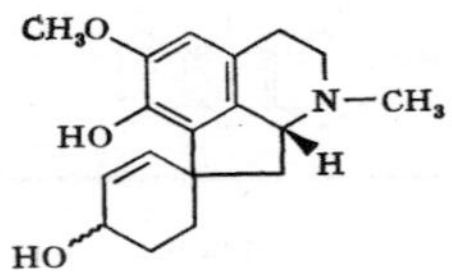 [Struktur]	isoliert als Diacetat Smp. 190—193°	Diacetat — 18° (CH_3OH)	*Croton linearis* Jacq. (*31*)	(*31*)	IR, UV, NMR: (*31*)

(Fortsetzung der Tabelle 1)

No.	Name (Synonyma) Bruttoformel	Struktur	Schmelzpunkt	$[\alpha]_D$ (Lösungsmittel)	Isoliert aus	Konstitution	Spektren
14	D-(+)-Lit-sericin $C_{17}H_{21}NO_3$	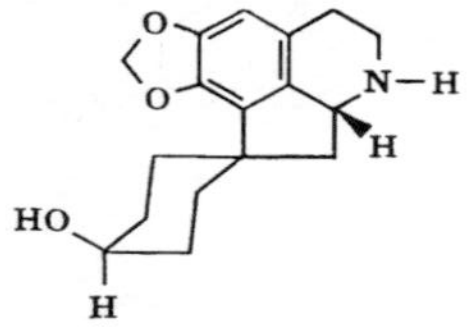	156°	+67° (C_2H_5OH) (56)	*Neolitsea sericea* (Blume) Koidz. (56)	(54, 55)	IR, UV, NMR: (54)
15	L-(−)-Oridin (Oreolin) $C_{17}H_{23}NO_3$	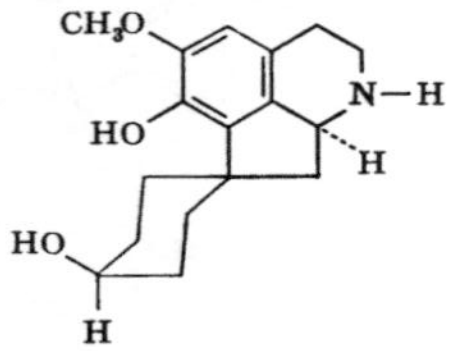	234—236°	−87° (CH_3OH) (49)	*Papaver oreophilum* F. J. Rupr. (51, 62)	(49, 64)	IR: (51), UV: (49), NMR, MS: (64)
16	N-Methyl-oreolin $C_{18}H_{25}NO_3$	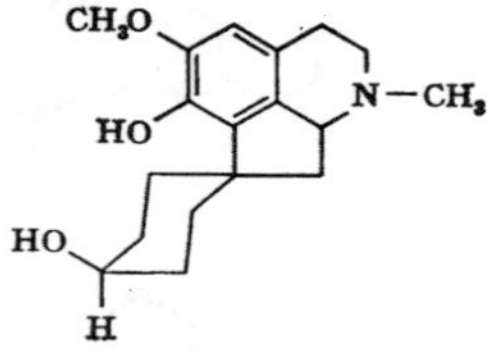	192—193°		*Papaver oreophilum* F. J. Rupr. (62)	(49)	UV: (49)
8a	L-(−)-Crot-sparin	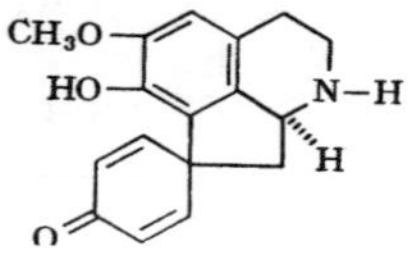	193—195°	−30° ($CHCl_3$) (17a)	*Crofon sparsiflorus* Morong (17a)	(17a)	IR, UV, NMR, MS: (17a)

Tabelle 2. Derivate natürlicher Proaporphin-Alkaloide

No.	Name/Bruttoformel	Struktur	Schmelzpunkt	$[\alpha]_D$ (Lösungsmittel)	Gewonnen aus
1	D-(+)-Methyl-crotonosin $C_{18}H_{19}NO_3$		216—218° (Zers.)	+ 122° (CH_3OH)	Crotonosin (*37*)
2	L-(+)-Epiamurolin $C_{19}H_{25}NO_3$		144—145,5°	+ 37° ($CHCl_3$)	Amuronin (*26*)
3	L-(—)-Dihydro-amuronin $C_{19}H_{25}NO_3$		125—126°	— 40° ($CHCl_3$)	Amuronin (*26*)
4	L-(—)-Dihydro-amuronin-metho-jodid $C_{20}H_{28}NO_3J$		242—244°	— 23° (CH_3OH)	Amuronin (*26*)

18*

No.	Name/Bruttoformel	Struktur	Schmelzpunkt	$[\alpha]_D$ (Lösungsmittel)	Gewonnen aus
5	D-Tetrahydro-glaziovin $C_{18}H_{23}NO_3$	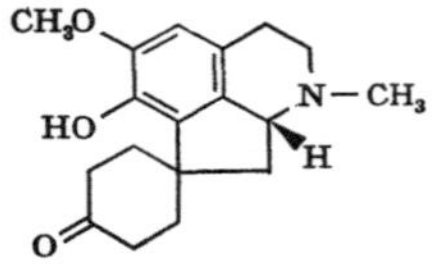	112—116°		Glaziovin (30)
6	D-(+)-Tetrahydro-N-methyl-crotonosin $C_{18}H_{23}NO_3$	8′aR	Zers. ab 210°, Smp. 225—228°	+ 59° (CH_3OH)	D-(+)-N-Methyl-crotonosin (37)
7	L-(—)-Dihydrolinearisin	8′aS 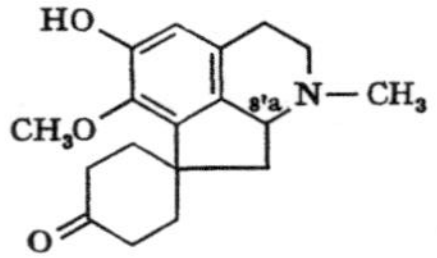	Zers. ab 210°, Smp. 225—227°	— 60,6° (CH_3OH)	Linearisin (37, 36)
8	D-(+)-N-Methyl-litsericinon $C_{18}H_{21}NO_3$	8′aR	148°	+ 91° (C_2H_5OH)	N-Methyl-litsericin (55)
9	L-Tetrahydro-fugapavin	8′aS 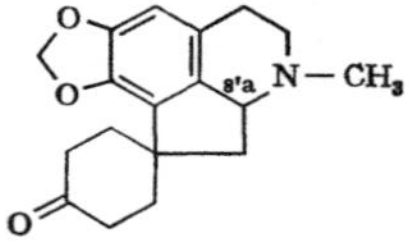	144—156°		Fugapavin (55)
10	D-(+)-N-Methyl-litsericin $C_{18}H_{23}NO_3$	8′aR	185°	+ 53° (CH_3OH)	Litsericin (54, 55)
11	L-(—)-„cis“-Hexahydro-fugapavin	8′aS 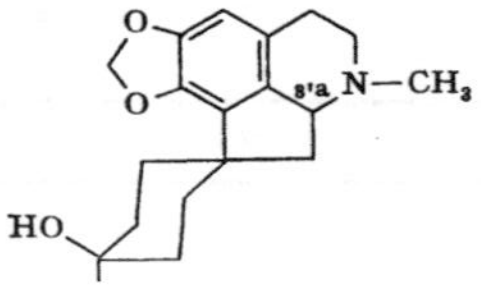	267—269° (53) 186° (55)	— 38,2° (CH_3OH) (53) — 47° (CH_3OH) (55)	Fugapavin (53), Tetrahydro-fugapavin (55)

Literaturverzeichnis: SS. 279—283

12	D-(+)-N-Methyl-epilitsericin $C_{18}H_{23}NO_3$	8'aR	258°	+ 59° (CH_3OH)	Litsericin (55)
13	L-(—)-Hexahydro-fugapavin	8'aS 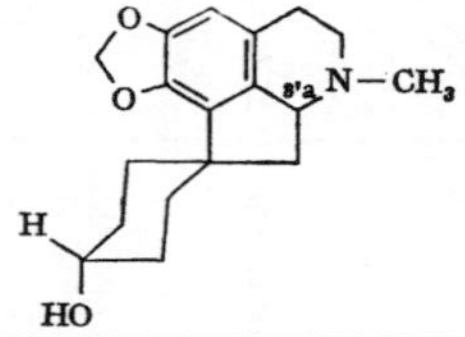	256°	— 44° (CH_3OH)	Fugapavin (55)
14	D-,,cis''-Hexahydro-pronuciferin $C_{19}H_{27}NO_3$	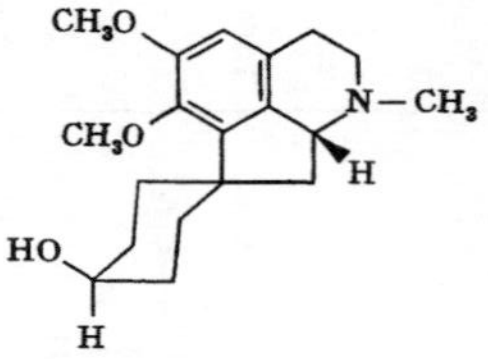	165—168° Hydrochlorid 253—254°		Pronuciferin (15)
15	D-,,trans''-Hexahydro-pronuci-ferin $C_{19}H_{27}NO_3$	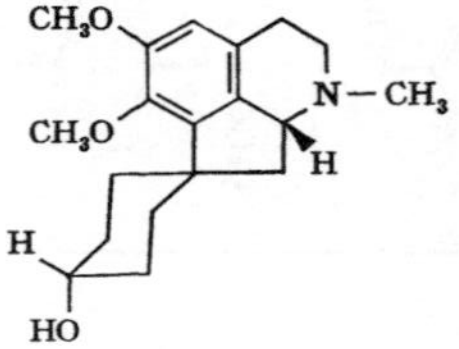	Hydrochlorid 205—208°		Pronuciferin (15)
16	D-Hexahydro-desoxy-crotonosin-methojodid $C_{20}H_{30}NO_2J$	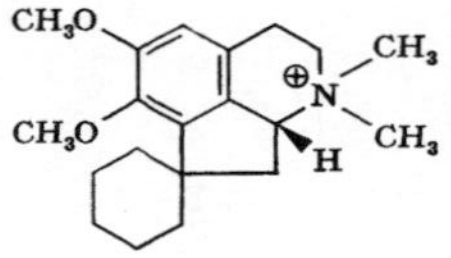	231—235°	+ 2,3° (CH_3OH)	Crotonosin (31)

Tabelle 3. Synthetische Proaporphine*

Name/Bruttoformel	Struktur	Schmelzpunkt	Synthese-weg**
(±)-Pronuciferin $C_{19}H_{21}NO_3$	No. 1, Tab. 1	148—150°	C (*12, 17, 43*)
(±)-Stepharin $C_{18}H_{19}NO_3$	No. 3, Tab. 1		C (*17*)
(±)-Glaziovin $C_{18}H_{19}NO_3$	No. 4, Tab. 1	227—228°	A (*24, 43*)
D-(+)-Orientalinon	8'aR	184—186° (Zers.)	A (*9*)
L-(—)-Orientalinon	8' aS	183—184° (Zers.)	A (*9*)
(±)-Orientalinon $C_{19}H_{21}NO_4$	No. 8, Tab. 1	203° (Zers.)	A (*7, 9*)
(±)-O-Methyl-orientalinon $C_{20}H_{23}NO_4$	(Struktur) CH₃O, CH₃O, N—CH₃, O, OCH₃	Oel	B (*66*)
(±) $C_{20}H_{23}NO_5$	(Struktur) CH₃O, HO, N—CH₃, H, O, CH₃O, OCH₃	Picronolat 230°	B (*41*)
(±) $C_{20}H_{23}NO_5$	(Struktur) CH₃O, HO, N—CH₃, H, CH₃O, O, CH₃O	Picronolat 193°	B (*41*)
(±)-Amuronin $C_{19}H_{23}NO_3$	No. 9, Tab. 1	121—123°	C (*17*)

* Es werden nur einheitliche, gut charakterisierte Verbindungen aufgeführt.
** Vgl. S. 264 ff.

(Fortsetzung der Tabelle 3)

Name/Bruttoformel	Struktur	Schmelzpunkt	Synthese-weg*
$(\pm)$-2,3-Dihydro-stepharin $C_{18}H_{21}NO_3$		134—$136°$ N-Formyl-derivat 178—$179°$	C (17)
$(\pm)$-Amurolin $C_{19}H_{25}NO_3$	No. 12, Tab. 1	Hydrochlorid 235—$237°$ Zers.	C (17)
$(\pm)$-Epiamurolin $C_{19}H_{25}NO_3$	No. 2, Tab. 2	133—$134°$	C (17)
$(\pm)$-2,3,5,6-Tetrahydro-pronuciferin $C_{19}H_{25}NO_3$	No. 3, Tab. 2	124—$126°$	C (17)
$(\pm)$-,,cis''-2,3,4,5,6,o-Hexahydro-pronuciferin $C_{19}H_{27}NO_3$	No. 14, Tab. 2	165—$168°$	C (17)
$(\pm)$-,,trans''-2,3,4,5,6,o-Hexahydro-pronuciferin $C_{19}H_{27}NO_3$	No. 15, Tab. 2	84—$86°$ oder 146—$149°$	C (17)
$(\pm)$-3,5-Dibrom-2,3-di-hydro-pronuciferin $C_{19}H_{21}NO_3Br_2$		Zers. ab $180°$	C (17)

* Vgl. S. 264 ff.

Literaturverzeichnis

1. BALDWIN, M., A. G. LOUDON, A. MACCOLL, L. J. HAYNES and K. L. STUART: Alkaloids from Croton Species. VI. Mass Spectrometric Studies of the Croto-nosine Alkaloids. J. Chem. Soc. (London) **C 1967**, 154.

1a. BARTON, D. H. R.: A Region of Biosynthesis. Chem. in Britain **1967**, 330.

1b. BARTON, D. H. R., D. S. BHAKUNI, G. M. CHAPMAN and G. W. KIRBY: Phenol Oxidation and Biosynthesis. XV. The Biosynthesis of Roemerine, Anonaine, and Mecambrine. J. Chem. Soc. (London) **C 1967**, 2134.

2. Barton, D. H. R., D. S. Bhakuni, G. M. Chapman, G. W. Kirby, L. J. Haynes and K. L. Stuart: Phenol Oxidation and Biosynthesis. XIV. (Alkaloids from Croton Species. VII.) The Biosynthesis of Crotonosine. J. Chem. Soc. (London) C 1967, 1295.

3. Barton, D. H. R. and T. Cohen: Some Biogenetic Aspects of Phenol Oxidation. Festschrift Arthur Stoll, p. 117. Basel: Birkhäuser. 1957.

4. Battersby, A. R.: The Biosynthesis of Alkaloids. Proc. Chem. Soc. (London) 1963, 189.

4a. — Phenol Oxidations in the Alkaloid Field. In: W. I. Taylor and A. R. Battersby (eds.), Oxidative Coupling of Phenols. New York: Dekker. 1967.

5. Battersby, A. R., R. B. Bradbury, R. B. Herbert, M. H. G. Munro and R. Ramage: Structure and Synthesis of Homoproaporphines: A New Group of 1-Phenethylisoquinoline Alkaloids. Chem. Commun. 1967, 450.

6. Battersby, A. R., R. T. Brown, J. H. Clements and G. G. Iverach: On the Biosynthesis of Isothebaine. Chem. Commun. 1965, 230.

7. Battersby, A. R. and T. H. Brown: Synthesis of (±)-Isothebaine. Proc. Chem. Soc. (London) 1964, 85.

8. — — Orientalinone, Dihydro-orientalinone, and Salutaridine from *Papaver orientale*: Related Tracer Experiments. Chem. Commun. 1966, 170.

9. Battersby, A. R., T. H. Brown and J. H. Clements: Synthesis along Biosynthetic Pathways. I. Synthesis of (+)-Isothebaine. J. Chem. Soc. (London) 1965, 4550.

10. Battersby, A. R., E. McDonald, M. H. G. Munro and R. Ramage: Homoaporphine Systems and Related Dienones: Isolation, Structure, and Synthesis. Chem. Commun. 1967, 934.

11. Bernauer, K.: Pronuciferin, ein Benzylisochinolin-Alkaloid mit *para*-Cyclohexadienon-Gruppierung. Helv. Chim. Acta 46, 1783 (1963).

12. — Eine Totalsynthese des D,L-Pronuciferins. Experientia 20, 380 (1964).

13. — Synthesen in der Proaporphin- und Aporphin-Reihe. Chimia 18, 407 (1964).

14. — Über die Isolierung von (+)-Pronuciferin und (—)-Anonain aus den Keimlingen von *Nelumbo nucifera* Gaertn. 2. Mitt. über natürliche und synthetische Isochinolinderivate. Helv. Chim. Acta 47, 2119 (1964).

15. — Konstitution und Reaktionen des (+)-Pronuciferins. 3. Mitt. über natürliche und synthetische Isochinolinderivate. Helv. Chim. Acta 47, 2122 (1964).

16. — Über Alkaloide aus *Laurelia novae-zelandiae* A. Cunn. 5. Mitt. über natürliche und synthetische Isochinolinderivate. Helv. Chim. Acta 50, 1583 (1967).

17. — Über die Synthese des Pronuciferins und einiger weiterer Proaporphin-Alkaloide. 6. Mitt. über natürliche und synthetische Isochinolinderivate. Helv. Chim. Acta 51, 1119 (1968).

17a. Bhakuni, D. S. and M. M. Dhar: Crotsparine, a New Proaporphine Alkaloid from *Croton sparsiflorus* Morong. Experientia 24, 10 (1968).

18. Bick, I. R. C.: The Structure of Fugapavine. Experientia 20, 362 (1964).

19. Boit, H.-G.: Ergebnisse der Alkaloid-Chemie bis 1960. Berlin: Akademie-Verl. 1961.

20. Boit, H.-G. und H. Flentje: Neue Alkaloide aus *Papaver amurense*. Naturwiss. 46, 514 (1959).

21. Budzikiewicz, H., C. Djerassi and D. H. Williams: Structure Elucidation of Natural Products by Mass Spectrometry. Vol. 1, Alkaloids. San Francisco: Holden-Day. 1964.

22. Carlson, R. M. and R. K. Hill: Intramolecular Cyclization of N-Formyl-1-carboxymethyl-6,7-dimethoxy-1,2,3,4-tetrahydroisoquinoline. J. Organ. Chem. (USA) 31, 2385 (1966).

22 a. CARSON, J. R.: U. S. Patent 3247210 (1966).

23. CAVA, M. P., K. NOMURA, R. H. SCHLESSINGER, K. T. BUCK, B. DOUGLAS, R. F. RAFFAUF and J. A. WEISSBACH: A Chemical Determination of the Absolute Configuration of Aporphines; the Reductive Cleavage of Proaporphines. Chem. and Ind. **1964**, 282.

24. CHAPMAN, G. M.: Ph. D. Thesis, London 1966. [Zitiert nach (*1*)].

25. FLENTJE, H., W. DÖPKE und D. W. JEFFS: Über die Konstitution des Amurins und Nudaurins. Naturwiss. **52**, 259 (1965).

26. — — — Amuronin und Amurolin, zwei neue Dihydroproaporphin-Alkaloide. Pharmazie **21**, 379 (1966).

27. FURUKAWA, H.: On the Alkaloids of *Nelumbo nucifera* Gaertn. XII. Alkaloids of Loti Embryo. J. Pharmac. Soc. Japan **86**, 75 (1966).

28. GASKIN, R. ST. C. and P. C. FENG: Some Pharmacological Activities of Crotonosine and Pronuciferine. J. Pharm. Pharmacol. **19**, 195 (1967).

29. GENTLES, M. J., J. B. MOSS, H. L. HERZOG and E. B. HERSHBERG: The Dienol-Benzene Rearrangement. Some Chemistry of 1,4-Androstadiene-3,17-dione. J. Amer. Chem. Soc. **80**, 3702 (1958).

30. GILBERT, B., M. E. A. GILBERT, M. M. DE OLIVEIRA, O. RIBEIRO, E. WENKERT, B. WICKBERG, U. HOLLSTEIN and H. RAPOPORT: The Aporphine and Isoquinolinedienone Alkaloids of *Ocotea glaziovii*. J. Amer. Chem. Soc. **86**, 694 (1964).

31. HAYNES, L. J., G. E. M. HUSBANDS and K. L. STUART: Alkaloids from Croton Species. IV. The Isolation and Structure of a New Proaporphine Derivative from *Croton linearis* Jacq. J. Chem. Soc. (London) C **1966**, 1680.

32. HAYNES, L. J. and K. L. STUART: Alkaloids from Croton Species. I. The Isolation of Alkaloids from *C. linearis* Jacq., and the Detection of Alkaloids in *C. glabellus* L., *C. humilis* L., and *C. flavens* L. J. Chem. Soc. (London) **1963**, 1784.

33. — — Alkaloids from Croton Species. II. Structural Determination of Base A, Linearisine and Crotonosine from *C. linearis* Jacq. J. Chem. Soc. (London) **1963**, 1789.

34. HAYNES, L. J., K. L. STUART, D. H. R. BARTON, D. S. BHAKUNI and G. W. KIRBY: On the Biosynthesis of Crotonosine. Chem. Commun. **1965**, 141.

35. HAYNES, L. J., K. L. STUART, D. H. R. BARTON and G. W. KIRBY: The Constitution of Crotonosine. Proc. Chem. Soc. (London) **1963**, 280.

36. — — — — The Constitutions of Crotonosine, Linearisine, and Homolinearisine. Proc. Chem. Soc. (London) **1964**, 261.

37. — — — — Alkaloids from Croton Species. III. The Constitution of the Proaporphines Crotonosine, Homolinearisine, Base A, and the Dihydroproaporphine Linearisine. J. Chem. Soc. (London) C **1966**, 1676.

38. HEYDENREICH, K. und S. PFEIFER: Über Alkaloide der Gattung Papaver. 11. Mitt. Isolierung von (—)-Orientalinon, Salutaridin und Oreophilin aus *Papaver bracteatum* Lindl. Pharmazie **21**, 121 (1966).

39. JACKSON, A. H. and J. A. MARTIN: Phenol-coupling Reactions: The Synthesis of Corydine and Isocorytuberine. Chem. Commun. **1965**, 142.

40. — — Biogenetic Type Syntheses of Aporphine Alkaloids: Isoboldine and Corydine. Chem. Commun. **1965**, 420.

41. — — Phenol Oxidation. II. Synthesis of Orientalinone, Corydine and Isocorytuberine. J. Chem. Soc. (London) C **1966**, 2222.

42. KAMETANI, T., K. FUKUMOTO, H. YAGI and F. SATCH: Syntheses of Homoproaporphine-type Compounds by Phenolic Oxidative Coupling. Chem. Commun. **1967**, 878.

43. Kametani, T. and H. Yagi: Total Syntheses of ($\pm$)-Glaziovine and ($\pm$)-Pronuciferine by Phenolic Oxidative Coupling. J. Chem. Soc. (London) C **1967,** 2182.

44. Kiselev, V. V. and R. A. Konovalova: Alkaloids of Wild Species of the Poppy. VIII. Alkaloids of *Papaver bracteatum.* J. Gen. Chem. (USSR) **18,** 142 (1948) [Chem. Abstr. **42,** 5073 (1948)].

45. Kühn, L. und S. Pfeifer: Über Alkaloide der Gattung Papaver. 8. Mitt. *Papaver fugax* Poir., *P. triniaefolium* Boiss., *P. persicum* Lindl., *P. polychaetum* Schott et Kotschy. Pharmazie **20,** 520 (1965).

46. — — Theorien und Befunde zur Biosynthese von Aporphin- und Morphinan-Alkaloiden unter besonderer Berücksichtigung von Dienonstrukturen. Pharmazie **20,** 659 (1965).

47. — — Über Alkaloide der Gattung Papaver. 19. Mitt. Isolierung von Porphyroxin, Salutaridin, (—)-N-Methylcrotonosin und Glaziovin aus *Papaver caucasicum* Marsch.-Bieb. Pharmazie **22,** 58 (1967).

48. Kühn, L., S. Pfeifer, J. Slavík und J. Appelt: Über Alkaloide von *Papaver caucasicum* Marsch.-Bieb. Naturwiss. **51,** 556 (1964).

49. Mann, I. und S. Pfeifer: Zur Struktur von Oreolin und N-Methyl-oreolin. 20. Mitt. über die Gattung Papaver. Pharmazie **22,** 124 (1967).

50. Maturová, M., L. Hruban, F. Šantavý und W. Wiegrebe: Beiträge zur Konstitution des Alkaloids Amurin und zur Polarographie der Cyclohexadienon-Verbindungen. Arch. Pharmaz. **298,** 209 (1965).

51. Maturová, M., D. Pavlásková und F. Šantavý: Isolierung der Alkaloide aus einigen Arten der Gattung *Papaver.* Isolierung und Chemie der Alkaloide einiger Papaverarten. XXXIV. Planta med. **14,** 22 (1966).

52. Mnatsakanyan, V. A. and A. R. Mkrtchyan: Structure of the Alkaloid Fugapavine. Armyansk. Khim. Zh. **19,** 466 (1966) [Chem. Abstr. **66,** 38091 f (1967)].

53. Mnatsakanyan, V. A. and S. Yu. Yunusov: The Constitution of Fugapavine. Dokl. Akad. Nauk. Uz.SSR **1961,** 36 [Chem. Abstr. **58,** 1503 (1963)].

54. Nakasato, T. and S. Asada: Alkaloids of Lauraceae Plants. VI. Structure of Litsericine. J. Pharmac. Soc. Japan **89,** 134 (1966) [Chem. Abstr. **64,** 19696 (1966)].

55. — — Structure of Litsericine. II. J. Pharmac. Soc. Japan **86,** 1205 (1966) [Chem. Abstr. **66,** 65671 b (1967)].

56. Nakasato, T., S. Asada and Y. Koezuka: Alkaloids of Lauraceae Plants. V. Alkaloids Isolated from the Trunk Bark of *Neolitsea sericea* (Blume) Koidz. J. Pharmac. Soc. Japan **86,** 129 (1966) [Chem. Abstr. **64,** 19695 (1966)].

57. Oediger, H., H.-J. Kabbe, F. Möller und K. Eiter: 1,5-Diazabicyclo-[4.3.0]nonen-(5). Ein neues Reagenz zur Einführung von Doppelbindungen. Chem. Ber. **99,** 2012 (1966).

58. Pavesi, V.: Nochmals über das Aporhein und die anderen Alkaloide von *Papaver dubium.* Gazz. chim. ital. **37** I, 629 (1907) [Chem. Zbl. **1907,** II 820].

59. — Neue Bemerkungen über das Aporhein und seine Salze. Gazz. chim. ital. **44** I, 398 (1914) [Chem. Zbl. **1914,** II 837].

60. Pfeifer, S. und L. Kühn: Zur Kenntnis der Alkaloide von *Papaver caucasicum* Marsch.-Bieb. Pharmazie **20,** 394 (1965).

61. — — Über Alkaloide der Gattung Papaver, 22. Mitt. Isolierung von L-(—)-Pronuciferin und L-(+)-Nuciferin aus *Papaver persicum* Lindl. Pharmazie **22,** 221 (1967).

62. Pfeifer, S. und I. Mann: Oreolin, ein neues Papaveralkaloid. Pharmazie **21,** 251 (1966).

63. Plieninger, H. und G. Keilich: Die Dienol-Benzol-Umlagerung. Chem. Ber. **91,** 1891 (1958).

63a. PREININGER, V., J. APPELT, L. SLAVÍKOVÁ und J. SLAVÍK: Alkaloide der Mohngewächse (Papaveraceae), XXXVII. Alkaloide aus *Papaver persicum* und *P. caucasicum*. Collect. Czech. Chem. Comm. **32**, 2682 (1967).

64. ŠANTAVÝ, F. und M. MATUROVÁ: Isolierung und Chemie der Gattung Papaver. XLI. Beitrag zur Konstitution der Oridin-Alkaloide. Planta med. **15**, 311 (1967).

65. SHAMMA, M. and W. A. SLUSARCHYK: The Aporphine Alkaloids. Chem. Rev. **64**, 59 (1964).

66. — — The Synthesis of Glaucine, O-Methylcorydine and Pseudocorydine via Phenolic Oxidative Coupling. Chem. Commun. **1965**, 528.

67. SIPHAR, S. A.: Belg. Patent 688814 (1966).

68. SLAVÍK, J.: Alkaloide der Mohngewächse (Papaveraceae). XXIII. Über die Alkaloide aus *Papaver dubium* L. und über die Konstitution des Aporheins. Collect. Czech. Chem. Comm. **28**, 1738 (1963).

69. — Alkaloide der Mohngewächse (Papaveraceae), XXVII. Über die Konstitution und Konfiguration des Mecambrins. Collect. Czech. Chem. Comm. **30**, 914 (1965).

70. — Alkaloide der Mohngewächse (Papaveraceae), XXXV. Überführung des (—)-Mecambrins in (+)-Roemerin (Aporhein) und Zusammensetzung des Mecambridins. Collect. Czech. Comm. **31**, 4184 (1966).

71. SLAVÍK, J. und L. SLAVÍKOVÁ: Alkaloide der Mohngewächse (Papaveraceae), XXI. Über die Alkaloide aus *Meconopsis cambrica* (L.) Vig. Collect. Czech. Chem. Comm. **28**, 1720 (1963).

72. SNATZKE, G. and G. WOLLENBERG: Alkaloids from Croton Species. V. Configurations of Linearisine and Circular Dichroism of Proaporphine-type Alkaloids. J. Chem. Soc. (London) **C 1966**, 1681.

73. TOMITA, M., A. KATO, T. IBUKA, H. FURUKAWA and M. KOZUKA: Mass Spectra of Pronuciferine and Stepharine. Tetrahedron Letters **1965**, 2825.

74. TOMITA, M., M. KOZUKA and S.-T. LU: Studies on the Alkaloids of Menispermaceous Plants. CCXXVIII. Alkaloids of *Pericampylus formosanus* Diels. J. Pharmac. Soc. Japan **87**, 315 (1967) [Chem. Abstr. **67**, 32854f (1967)].

75. TOMITA, M., M. KOZUKA and S. UYEO: Studies on the Alkaloids of Menispermaceous Plants. CCXXIII. Alkaloids of *Stephania rotunda* Loureiro. J. Pharmac. Soc. Japan **86**, 460 (1966) [Chem. Abstr. **65**, 10633 (1966)].

75a. TOMITA, M., T.-H. YANG, H. FURUKAWA and H.-M. YANG: Oxidation of Aporphine Alkaloids with CrO_3-Pyridine Complex. J. Pharmac. Soc. Japan **82**, 1574 (1962) [Chem. Abstr. **58**, 14012 (1963)].

76. YUNUSOV, S. YU., V. A. MNATSAKANYAN and S. T. AKRAMOV: Alkaloids from *Papaver fugax*. Isoremerins. Dokl. Akad. Nauk. Uz.SSR **1961**, 43 [Chem. Abstr. **57**, 9900 (1962)].

77. — — — Alkaloids of Some Species of Papaver and Roemeria and the Structure of Fugapavine. Izv. Akad. Nauk. SSSR, Ser. Khim. **1965**, 502 [Chem. Abstr. **63**, 642 (1965)].

(Eingelaufen am 31. Oktober 1967)

Chemie der Chlorine und Porphyrine

Von **H. H. Inhoffen, J. W. Buchler** und **P. Jäger,** Braunschweig

Mit 1 Abbildung

Inhaltsübersicht

Einleitung

Der vorliegende Bericht über die Chemie der Chlorophylle und Hämine resp. der Chlorine und Porphyrine soll einen Überblick über die wichtigen und interessanten Fortschritte etwa während der letzten 10 Jahre geben. Das unsubstituierte Porphin (52) bildet den Grundkörper der Porphyrine, deren Hauptvertreter der rote Blutfarbstoff Hämin ist. Die in (52) nach H. Fischer mit 1 bis 8 numerierten C-Atome, S. 286, heißen „peripher" (kurz: „peri"), die mit α bis δ gekennzeichneten C-Atome „meso"-ständig bzw. „Methinbrücken". [Beim Übergang auf andere Tetrapyrrolpigmente ist eine rationelle Nomenklatur zweckmäßiger (108).] Hydrierung einer peripheren Doppelbindung in (52), z. B. zwischen C-7 und C-8, ergibt das Chlorin (60), den Grundkörper der grünen Blattpigmente.

Die Totalsynthese des Chlorophylls a von Woodward und Mitarbeitern aus dem Jahre 1960, mit der Fülle ihrer neuartigen, grundsätzlich wichtigen Ergebnisse, wird als so impulsgebend betrachtet, daß sie an den Anfang gestellt wird (*212—215*).

(52)   (60)

Die seitdem erarbeiteten Ergebnisse sind so zahlreich — zum Teil auch noch nicht hinreichend erschlossen — daß nur eine Auswahl referiert werden kann, um den gegebenen Umfang nicht allzusehr zu überschreiten.

I. Abschluß der Strukturermittlung der Chlorophylle a und b

1. Totalsynthese des Chlorophylls a

a) Der Grundgedanke

Die Bewältigung einer derartig schwierigen Aufgabe, wie sie mit dem Chlorophyll-Molekül gestellt wurde, erfordert verständlicherweise die Lösung einer ganzen Reihe von Einzelproblemen, die nacheinander behandelt werden sollen. Unter der Fülle der gestellten Einzelaufgaben ragt jedoch eine einzige derartig heraus, daß sie zuerst und gesondert behandelt wird: Die Einführung der beiden Wasserstoffe an den C-Atomen 7 und 8 des Ringes IV, und zwar in *trans*-Stellung. Diese ehemals „überzähligen", auch heute „zusätzlichen" Wasserstoffe des Chlorin-Moleküls bleiben noch immer eines der schwierigsten und zugleich interessantesten Probleme der Chlorophyll-Chemie.

Wäre das Chlorophyll ein Protochlorophyll, d. h. ein voll aromatisches Porphyrin, also praktisch ohne Asymmetrie-Zentren (das asymmetrische C-10 wird hier nicht als Aufgabe angesehen), so wäre eine Synthese des 2-Vinyl-phäoporphyrins a$_5$ eine relativ leichte Aufgabe.

Die Erkenntnis des Vorhandenseins sowie zugleich der Konsequenzen der sterischen Verdichtung in der unteren Hälfte des Chlorophyll-Moleküls, d. h. an den C-Atomen 5, 6, γ, 7 und 8, hat Woodward zu seinem Grundgedanken geführt: das diffizile, differenzierte Wechselspiel zwischen sterischer Verdichtung und Entspannung nicht nur zu überwinden, sondern geradezu als synthetische Möglichkeit zu nutzen, und zwar mehrfach.

Literaturverzeichnis: SS. 345—355

Daß der γ-Substituent im meso-γ-Phylloporphyrin die Hydrierung der C-Atome 7 und 8 begünstigt, hatten schon STOLL und WIEDEMANN (*197*) erkannt und kausal gedeutet, aber es wurden keine Konsequenzen daraus gezogen.

Bereits hier gestalteten sich WOODWARDS Gedankengänge grundlegend anders als die seiner Vorgänger: Der an den C-Atomen 7 und 8 benötigte Wasserstoff sollte nicht unmittelbar von außen an die Molekel herangebracht werden — auf dem Wege einer unübersichtlichen und schwierig steuerbaren Zweiphasen-Reaktion — sondern von anderen, leichter bestimmbaren Zentren an den gewünschten Ort wandern. Das heißt, eine empfindlichere Vorstufe sollte sich letztlich selbst, quasi aus sich heraus, in einer thermodynamisch kontrollierten Reaktion zum stabileren Endprodukt orientieren. Die folgenden Bilder veranschaulichen die Situation.

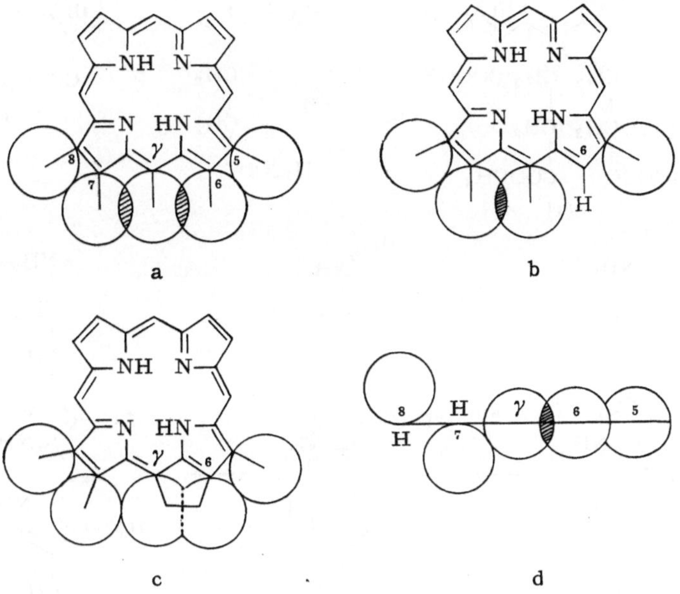

Bei voller Substitution (a) kommt es zu starker sterischer Verdichtung zwischen 6, γ und 7, die bei Fehlen eines Substituenten, z. B. in 6-Stellung (b), verringert wird. Gleichfalls wird Entspannung durch Ringbildung zwischen C-6 und C-γ bewirkt (c). Die wichtige Entspannung zwischen den C-Atomen γ, 7 und 8 (a) durch Bildung der beiden tetraedrischen Kohlenstoffatome in 7- und 8-Stellung wird durch d veranschaulicht.

b) Die neue Porphyrin-Synthese

Ein neues Porphyrin war das erste Ziel: (**1**). In der 2-Stellung trägt es einen β-Aminoäthyl-Rest, der später in die Vinyl-Gruppe umzuwandeln war. Insbesondere enthält es in der γ-Position einen Propionsäure-Rest.

Nicht nur würde eine dort kohlenstoffzahlmäßig stimmende Essigsäure unter mancherlei Reaktionsbedingungen aus dem Molekül abgestoßen werden, der Wasserstoff der Propionyl-Gruppe könnte sich im Verlauf

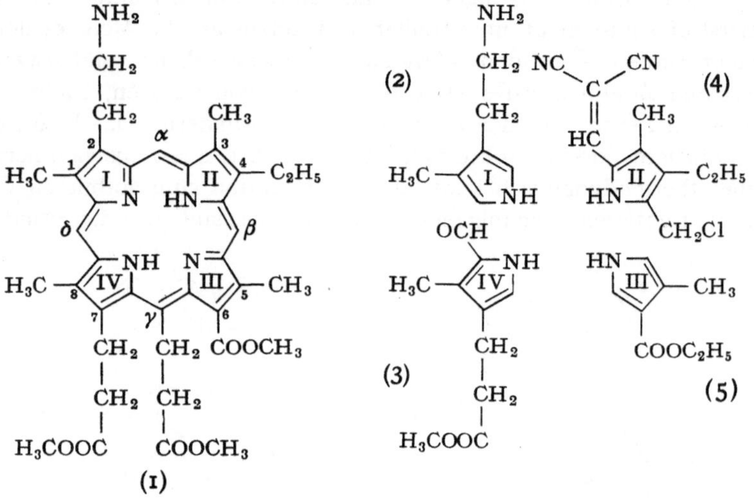

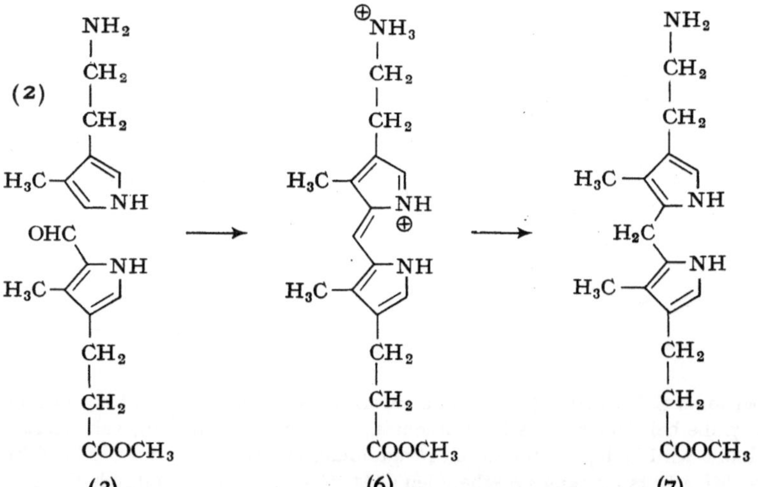

einer intramolekularen Verschiebung nützlich verwerten lassen. Die vier benötigten Pyrrol-Derivate für die Ringe I, II, III und IV waren praktisch bis auf wenige Modifikationen bekannt: (2)—(5). Die Komponenten (2) und (3) ließen sich zum Dipyrromethen (6) kondensieren und dieses mittels $NaBH_4$ zum Dipyrromethan (7) reduzieren (Ausbeute 71%).

Literaturverzeichnis: SS. 345—355

Für den Aufbau der zweiten Hälfte wurden (4) und (5) zu (8) kondensiert (Ausbeute insgesamt 44%). Die noch freie 5'-Stellung in (8) konnte glatt mit β-Carbomethoxy-propionylchlorid unter der Einwirkung von wasserfreiem Zinkchlorid zu (9) zur Reaktion gebracht werden.

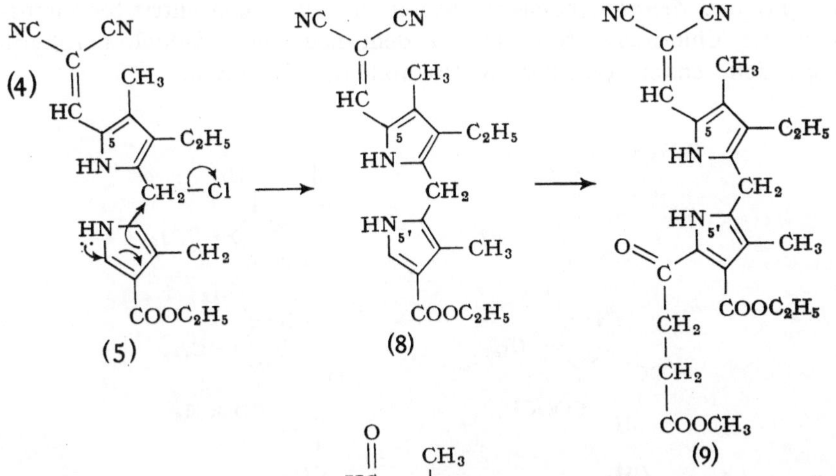

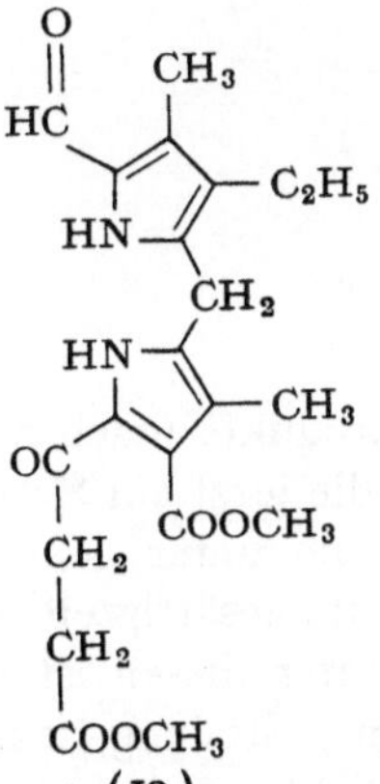

(4)

(5)

(8)

(9)

(10)

Nun konnte die Dicyanovinyl-Schutzgruppe hydrolysiert und damit die benötigte Aldehyd-Funktion freigelegt werden (10).

Eine direkte Kondensation der beiden unsymmetrischen Dipyrromethan-Hälften mit zwei Carbonyl-Funktionen einerseits und zwei freien α-Positionen anderseits würde zu einem Gemisch zweier strukturisomerer Porphyrine führen. Außer dem Materialverlust könnte die chromatographische Trennung des wahrscheinlich gleichartig polaren Gemisches ein Problem darstellen.

Um diesen Mißstand zu beseitigen, wurde die nicht steuerbare intermolekulare Kondensation zu einer eindeutig, d. h. einseitig verlaufenden intramolekularen Ringschlußreaktion umgestaltet, und zwar über die

Schiff-Base (13). Hierzu erwies es sich als notwendig, die nur schwach elektrophile Aldehyd-Gruppe in den reaktionsbereiten Thioaldehyd zu überführen, was in einfachster Weise durch Bildung des Azomethins (11) und dessen Umsetzung mit H_2S zum Thioaldehyd (12) gelang. Amin (7) und Thioaldehyd (12) vereinigen sich zunächst rasch und ohne Katalysator zu (13) und dieses kondensiert weiter in kalter, konzentrierter methanolischer Chlorwasserstoffsäure zu dem neuartigen Dihydroporphyrin (14), dem ersten isolierbaren Produkt der Reaktion.

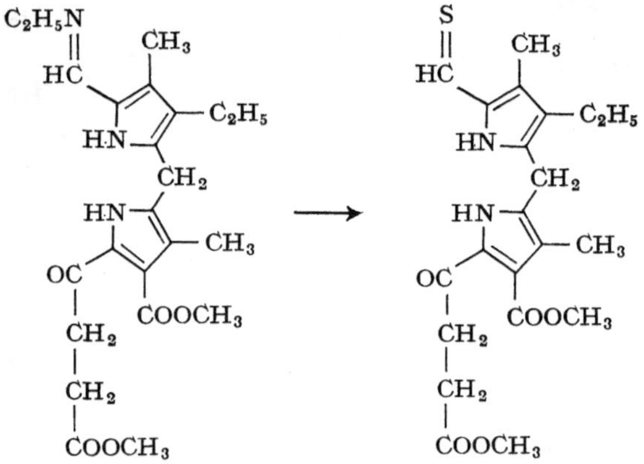

$$\text{(11)} \longrightarrow \text{(12)}$$

Hierbei sind Zwischenprodukte (nach der Wasser- und „Ammoniak"-Abspaltung) anzunehmen, die jetzt ein System von konjugierten Doppelbindungen enthalten und die unter dem Einfluß der Säure zu einem Isomeren-Gleichgewicht mit gesättigten meso-Positionen in α, β, γ oder δ tautomerisieren. Unter diesen ist (14) das stabilste, denn seine Propionsäure-Seitenkette in γ-Stellung ist aus der Ebene herausgetreten, womit die sterische Verdichtung zwischen C-6 und C-7 aufgehoben wird.

Das hiermit gewonnene Dihydro-porphyrin (14) ist der erste Vertreter einer neuen Verbindungsklasse, nämlich der *Phlorine*; ihre Salze geben grüne Lösungen, während die freien Phlorine sich in organischen Lösungsmitteln mit leuchtend tiefblauer Farbe lösen. Phlorine sind recht beständige Verbindungen und können aus konzentrierter Schwefelsäure unzerstört zurückgewonnen werden. Jedoch lassen sie sich sehr leicht zu Porphyrinen oxidieren (s. auch S. 315).

Dies geschieht zum Abschluß der eingangs gekennzeichneten Porphyrin-Synthese mittels Jods. Nach weiterer Acetylierung der Amin-Gruppe wurde schließlich das acetylierte Porphyrin (15), frei von anderen Porphyrinen, in einer Gesamtausbeute von 50% isoliert.

Literaturverzeichnis: SS. 345—355

c) Der Weg vom Porphyrin (**15**) *zum Chlorin* e_6-*trimethylester*

Im Porphyrin (**15**), mit seinem resonanz-stabilisierten System kon-
jugierter Doppelbindungen, wurde zwischen den Seitenketten in 5, 6,

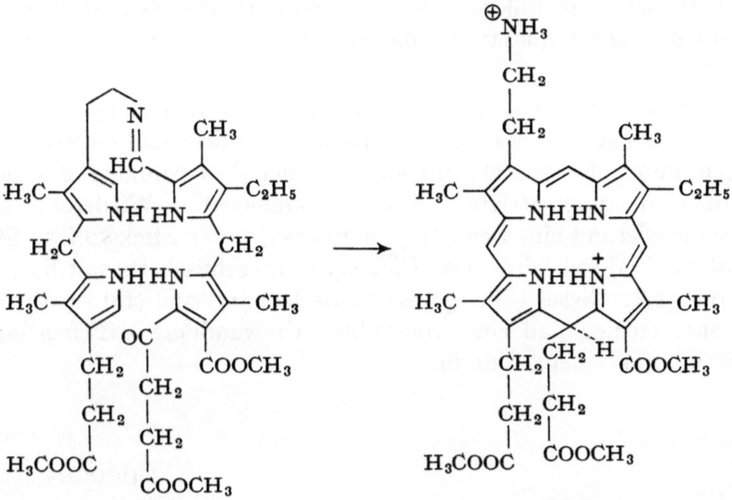

γ, 7 und 8 eine sehr starke sterische Verdichtung erzwungen. Diese sterische Situation ermöglicht, resp. erleichtert eine äußerst bemerkenswerte
Isomerisierung. Wird (**15**) in Eisessig unter Ausschluß von Sauerstoff
erhitzt, tautomerisiert die Verbindung zu dem neuen Phlorin (**16**), indem

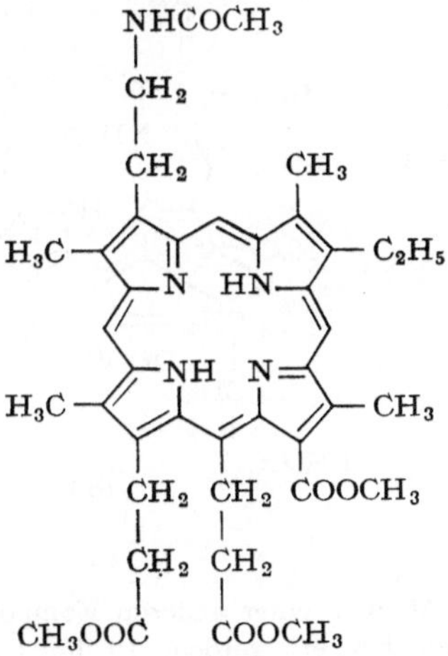

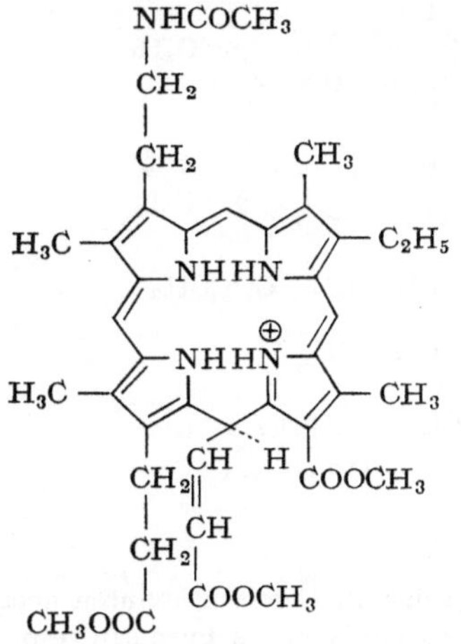

unter Wanderung von Wasserstoff aus der Propionsäurekette das γ-C-Atom wiederum tetraedrisch wird. Trotzdem läßt sich auch hier wieder — unter Überwindung der sterischen Verdichtung — die Ausbildung eines resonanz-stabilisierten konjugierten Doppelbindungs-Systems erzwingen, wenn man das neue Phlorin (16) in Gegenwart von Sauerstoff erhitzt. Hierbei wird nahezu quantitativ das neue Porphyrin (17) mit Acrylsäurekette in γ erhalten.

Jetzt realisiert sich — zum letzten, entscheidenden Mal — das faszinierende Wechselspiel zwischen sterischer Verdichtung einerseits, erzwungen durch Resonanzbestreben, und der Entspannung des hochsubstituierten, ungesättigten Systems anderseits: Wiederum setzt eine Isomerisierung ein, wenn (17) in Eisessig unter Stickstoff 30 Stunden auf 110° erhitzt wird. Das Gleichgewichtsgemisch enthält 63% des Purpurins (18). Dieses Gleichgewicht zwischen (17) und (18) ist der erste und bisher einzige Fall einer reversiblen Umwandlung zwischen einem Porphyrin und einem Chlorin.

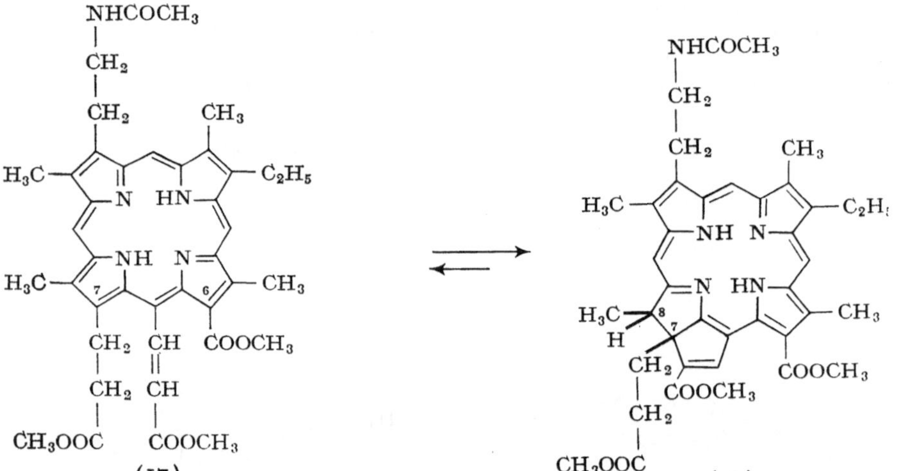

(17) $\rightleftarrows$ (18)

In diesem Gemisch konnten noch geringe Mengen einer anderen Komponente, nämlich ein isomeres Purpurin resp. Chlorin identifiziert werden. In ihm hat sich der neue Ring zur Position 6 gebildet. Hierbei muß die 6-Carbomethoxy-Gruppe aus ihrer Konjugation zum Aromatensystem herausgedrückt werden, wohl der Hauptgrund dafür, daß sich mengenmäßig nur wenig dieser Komponente bilden kann.

Da es sich um eine thermodynamisch kontrollierte Gleichgewichts-Reaktion handelt, müssen die Seitenketten an den C-Atomen 7 und 8 die stabilere *trans*-Konfiguration einnehmen.

Literaturverzeichnis: SS. 345—355

An dieser Stelle wird die β-Acetylamino-Seitenkette in bekannter einfacher Weise (Hofmann-Abbau) in die Vinyl-Gruppe zurückverwandelt (**19**).

Bezüglich der weiteren Reaktionen ist zu bemerken, daß glückliche Umstände eine bedeutende, nicht vorhersehbare Schwierigkeit überwinden halfen.

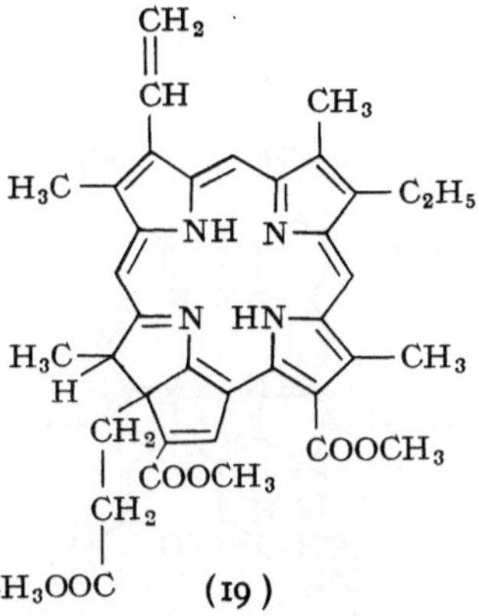

(**19**) und (**20**): Chlorin-Strukturformeln mit den Seitenketten $CH_2{=}CH$, CH_3, H_3C, C_2H_5, NH, N, HN, $COOCH_3$, CH_2, CH_3OOC; bei (**20**) zusätzlich CO, CHO und die Positionsbezeichnungen 7 und γ.

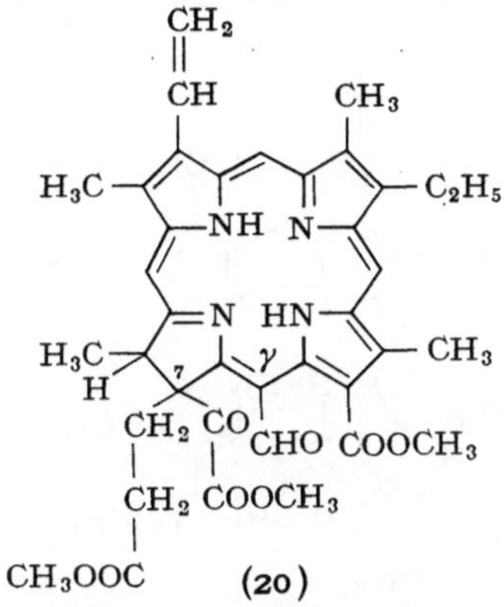

Gemäß dem ursprünglichen Konzept sollten zwei Wasserstoff-Atome ihre Wanderung aus anderen Positionen an die C-Atome 7 und 8 antreten. Aber nur eines war diesem Befehl gefolgt. Am C-8 hatte sich jedoch eine unerwünschte (zunächst) und stabil aussehende Kohlenstoff-Kohlenstoff-Bindung gebildet. In einer sehr bemerkenswerten Photo-Oxidation, die weder die räumlich exponierte Vinyl-Gruppe noch das recht empfindliche π-Elektronen-System angriff, konnte die im neuen Fünf-ring von (**19**) vorhandene Doppelbindung gesprengt werden, und zwar in 74%iger Ausbeute. Die primär entstandene Dicarbonyl-Verbindung (**20**) besitzt eine Aldehyd-Gruppe in γ-Stellung.

Die basenkatalysierte Abspaltung des Oxalyl-Restes zu (**21**) als Racemat bot keine Schwierigkeiten, so daß hiermit endlich auch das zweite „zusätzliche" H-Atom in seine Endposition gebracht war. Die Stereochemie dieser Reaktion wird wieder kontrolliert durch Abstoßungs-kräfte zwischen den Seitenketten, so daß die *trans*-7"-Carbomethoxy-äthyl-8-methyl-Verbindung vorliegt. (**21**) reagiert sogleich weiter zum Aldehyd-lacton-methyläther (**22**), dem Racemat des seit H. FISCHER bekannten Isopurpurin 5-methylesters, nunmehr richtig formuliert. Vor-sichtige (partielle) alkalische Hydrolyse des synthetischen, racemischen Isopurpurin 5-methylesters (**22**) ergab das entsprechende freie Aldehyd-lacton, dessen Chininsalze sich durch fraktionierte Kristallisation trennen ließen. Das aus dem rechtsdrehenden Chininsalz zurückgewonnene freie

Chlorin 5 besaß die Drehung $[\alpha]^{23}_{546} = + 1810°$, während das natürliche Vergleichspräparat den Drehwert $[\alpha]^{23}_{546} = + 1823°$ aufwies. Die Messung der optischen Aktivität der tiefgefärbten Verbindung wurde durch ihr Absorptionsminimum bei 546 mμ sehr erleichtert.

Behandlung mit Diazomethan lieferte totalsynthetischen, optisch aktiven Purpurin 5-dimethylester (21), der in jeder Hinsicht mit dem natürlichen Präparat identisch war.

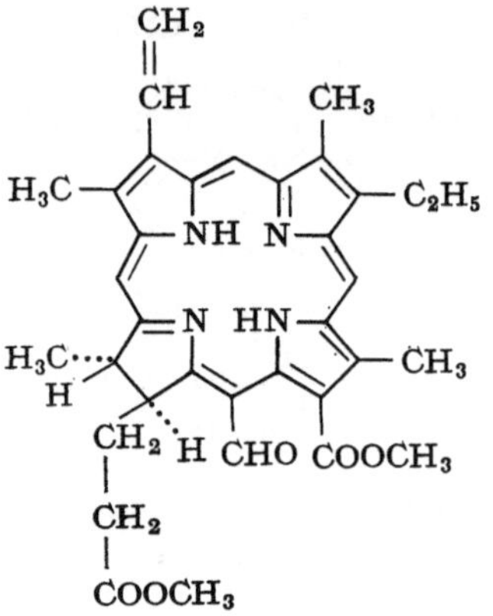

(21) 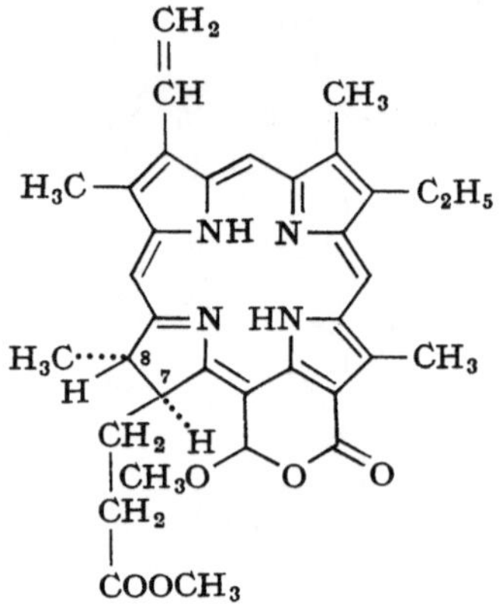   (22)

Die weiteren und letzten Schritte weisen keine Besonderheiten mehr auf. Behandlung von (21) mit Blausäure und Reduktion des Cyanolactons mit Zink und Eisessig führten (nach Wiederveresterung) zur γ-Cyanomethyl-Verbindung, die sich schließlich glatt in den Chlorin e_6-trimethylester (47, S. 304), gleichfalls identisch mit natürlichem Material, als Endprodukt der Totalsynthese überführen ließ.

Die letzten drei Stufen zum totalsynthetischen Chlorophyll a sind historisch bekannt: durch Willstätter und Fischer die Bildung des Fünfringes sowie die Wiedereinführung von Phytol und Magnesium, und schließlich die Totalsynthese optisch aktiven Phytols durch Weedon.

Als letztes Problem blieb noch die absolute Konfiguration an den asymmetrischen Kohlenstoffatomen 7 und 8.

2. Die absolute Konfiguration der Chlorophylle a und b

Die von H. Fischer postulierte, von Linstead wahrscheinlich gemachte und von Woodward bewiesene *trans*-Stellung der H-Atome an C-7 und C-8 ließ noch eine Frage offen, nämlich die nach der absoluten Konfiguration. Woodward hatte ohne nähere Begründung einem der beiden Antipoden den Vorzug gegeben. Fleming (98) ist es kürzlich gelungen, auch diese Frage einwandfrei zu klären.

(+)-threo-Dihydrohämatinsäure-tri-p-bromphenacylester (**23**) konnte aus (—)-α-Santonin, dessen Struktur und absolute Konfiguration mit großer Genauigkeit festgelegt ist, durch Abbau erhalten werden. (**23**) kommt also die absolute Konfiguration 2 S, 3 S zu. (—)-threo-Dihydrohämatinsäure-tri-p-bromphenacylester ließ sich erstmals durch Hydrolyse des rechtsdrehenden, synthetischen threo-Dihydrohämatinsäureimids ($[\alpha]_D^{20} = + 67°$) und Veresterung der erhaltenen (+)-threo-Dihydrohämatinsäure gewinnen. Der aus (—)-α-Santonin gewonnene rechtsdrehende threo-Dihydrohämatinsäure-tri-p-bromphenacylester (**23**) ($[\alpha]_D^{20} =$

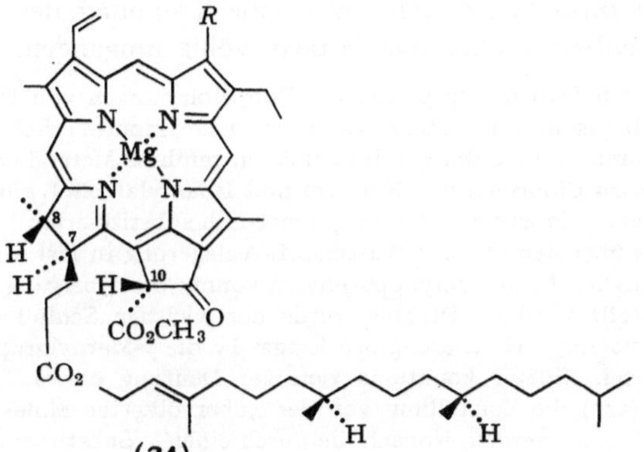

$R = CH_2COC_6H_4 - p - Br$

(23)

(24)

Chlorophyll b : R = CH₃

Chlorophyll a : R = CHO

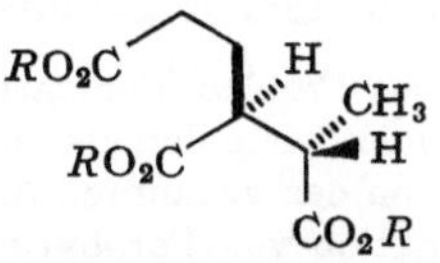

= + 3°) wurde durch eine Wiederholung des oxidativen Abbaus von Methylphäophorbid a dargestellt (*25*). Beide Präparate erwiesen sich als identisch. Unter weiterer Berücksichtigung der Arbeiten über die relative Konfiguration an C-10 und C-7 (*39, 211*) (siehe S. 342) kommt somit den Chlorophyllen a und b die Konstitution (**24**) zu. (Chlorophyll a und Chlorophyll b lassen sich wechselseitig ineinander überführen; siehe auch S. 305.)

3. Strells Publikation

In einer Kurzmitteilung aus dem Jahre 1960 haben STRELL und Mitarb. (*198*), aufbauend auf den Grundlagen von H. FISCHER, eine Reaktionsfolge angegeben, die im Prinzip zu einem zweiten künstlichen Aufbau von Chlorophyll a führen könnte. Ausgangspunkt der STRELL-schen Versuche ist das 2-Desäthyl-γ-phylloporphyrin von TREIBS und SCHMIDT (*201*), das aus zwei entsprechend substituierten Dipyrromethenen durch Kondensation gebildet wird. Unter den energischen Reaktionsbedingungen der Brenzweinsäure-Schmelze bei 140—160° entstanden mindestens acht verschiedene Porphyrine mit einer Gesamtausbeute von 2%. Offensichtlich müssen auch rückläufige Spaltungsprozesse an Zwischenprodukten und damit nicht vorhergesehene, unerwünschte Kondensationen stattfinden, so daß insgesamt der Porphyrin-Bildungsprozeß als unübersichtlich zu bezeichnen ist. Das in 0,6% entstandene 2-Desäthyl-γ-phylloporphyrin wurde durch Vergleich mit natürlich-analytischem Abbaumaterial identifiziert.

Ob man eine auf diesem Ergebnis aufbauende Reaktionsfolge eine Synthese nennen soll, ist eine Ermessensfrage.

Den wichtigsten Teil einer jeden Chlorophyll-Synthese, nämlich die Einführung der beiden Wasserstoff-Atome in 7- und 8-Stellung und zwar in der geforderten *trans*-Anordnung sowie eine Trennung des zu erwartenden Racemats, haben STRELL und Mitarb. völlig umgangen.

Das Problem der partiellen Hydrierung peripherer Doppelbindungen von Porphyrinen hatte damals durch FISCHER und HELBERGER (*70*) neue Impulse erhalten. Die von den genannten Autoren in die Chlorophyll-Chemie eingeführte Methode der Reduktion von Porphyrinen zu Chlorinen mit Natrium und Isoamylalkohol wurde auf breiter Basis in die Versuche einbezogen. Bei unsymmetrisch substituierten Porphyrinen ließ sich allerdings über den Ort der Wasserstoff-Anlagerung in fast allen Fällen nichts aussagen. Lediglich beim γ-Phylloporphyrin konnte der Übergang zu einem 7,8-Chlorin sichergestellt werden. Hieraus wurde der wichtige Schluß gezogen (*197*), daß für eine bevorzugte Hydrierung des Ringes IV die γ-Methylgruppe verantwortlich zu machen sei. Später konnte in richtiger Deutung dieses Tatbestandes von WOODWARD (*212*) die Vorstellung von der „übervölkerten Molekülebene" in die Diskussion gebracht werden, wonach die durch einen γ-Substituenten erzeugte Raumspannung einen bevorzugten Übergang benachbarter trigonaler C-Atome in tetraedrische zur Folge hat, so daß hiermit zugleich die Propionsäure-Seitenkette an C-7 aus der übervölkerten Molekülebene unter Entspannung des gesamten Systems heraustreten kann.

Wie weit waren nun die Arbeiten bei H. FISCHER bezüglich der sterischen Probleme der Hydrierung einer peripheren Doppelbindung eines unsymmetrisch substituierten Porphyrins mit Natrium und Isoamylalkohol vorgedrungen? Da die seinerzeitigen Arbeiten zu dieser Frage keine gültige Aussage machen, erscheint es zweckmäßig, an dieser Stelle die inhaltsreiche Arbeit von FISCHER und GIBIAN über die „Racemisierung von Chlorophyllderivaten" (*67*) zu diskutieren, die wegen ihres gleichfalls stark basischen Reaktionsmediums als analog aufgefaßt wird:

Bedingung der Racemisierung von Chlorinen: Erhitzen in Pyridin mit überschüssiger 10%iger Natriummethylatlösung und Hydrazinhydrat 3 bis 7 Stunden im Bombenrohr auf $115 \pm 2°$ (praktisch unter Stickstoff).

Bedingung der Reduktion mit Natrium und Isoamylalkohol: Lösen in viel Isoamylalkohol, Kp. 130°, und einstündiges Kochen mit viel Natrium auf dem Sandbad unter Durchleiten von Wasserstoff.

Der wesentliche Inhalt der Arbeit besagt, daß es möglich ist, Chlorine unter den oben gekennzeichneten, sogenannten Wolff-Kishner-Bedingungen zu racemisieren. Im Verlauf dieser noch nicht voll durchschaubaren Reaktion (wahrscheinlich werden intermediär Protonen von den asymmetrischen C-Atomen abgelöst) finden an den Asymmetriezentren sterische Umlagerungen statt, denn es werden stereoisomere, optisch inaktive Chlorine, respektive Phäophorbide erhalten. Von den von FISCHER und GIBIAN (*67*) diskutierten Möglichkeiten für die sterische Struktur der neugewonnenen Reaktionsprodukte seien die beiden wichtigsten herausgehoben:

1. Es findet 50%ige Racemisierung an beiden Asymmetriezentren C-7 und C-8 statt; dann entstünde ein *trans*-Racemat.

2. Nur jeweils eines der beiden Asymmetriezentren wird betroffen; damit würde ein *cis*-Racemat gebildet.

Es bleibt die Frage zu prüfen, ob unter diesen Bedingungen nur *trans*-Racemate oder daneben teilweise auch *cis*-Racemate entstehen können. Das gleiche sollte auch für die Reduktion mit Natrium und Isoamylalkohol gelten (s. auch SS. 309, 310).

Es wäre wünschenswert, wenn ein Fall eines γ,7,8-substituierten 7,8-Chlorins dahingehend geklärt würde, daß alle vier Stereoisomeren dargestellt und auf ihre chemischen und physikalischen Eigenschaften untersucht werden.

Was die weiteren Reaktionen von STRELL anbetrifft, so ist schließlich noch zu bemerken, daß die anschließende Umwandlung des γ-Methyls in einen Essigsäure-Rest lediglich mit natürlichem Material durchgeführt wurde.

Insgesamt ist (von weiteren noch offenen Fragen abgesehen) fest-zustellen, daß das FISCHERsche Konzept einer Chlorophyll-Synthese im Prinzip als durchführbar erscheint, daß jedoch die Behauptung von STRELL, die FISCHERschen Haupt-Vorarbeiten zum erfolgreichen Ab-schluß gebracht zu haben, nicht den Tatsachen entspricht.

II. Bacteriochlorophyll

Die Struktur des Bacteriochlorophylls ist seit langem von H. FISCHER (*77*) durch Überführung in ein Derivat des Chlorophylls a geklärt, wo-nach Bacteriochlorophyll (**25**) ein 3,4-Dihydro-chlorophyll a Derivat dar-stellt. Die letzte, offene Frage, nämlich die nach der sterischen Anordnung an den C-Atomen 3 und 4 konnte von BROCKMANN jr. geklärt werden (*26*). (+)-threo-2-Methyl-3-äthyl-bernsteinsäure besitzt die absolute Kon-figuration 2 R, 3 R, unter Beziehung auf die absolute Konfiguration des (+)-2,3-Dimethyl-bernsteinsäureanhydrids, die unter Anwendung der Kirkwood-Regel aus der bekannten, durch Röntgenbeugungsmessung bewiesenen absoluten Konfiguration des (2 R)-(+)-Methyl-bernstein-säureanhydrids abgeleitet werden konnte. Aus der rechtsdrehenden 2-Methyl-3-äthyl-bernsteinsäure wurde der linksdrehende Di-*p*-brom-phenacylester vom Schmelzp. 115° und $[\alpha]_D^{20} = -36,5° \pm 3°$ hergestellt. LINSTEAD und Mitarb. (*101*) erhielten den gleichfalls linksdrehenden

Ester vom Schmelzp. 112—113° und $[\alpha]_D^{24} = -37°$ durch oxidativen Abbau von Bacteriophäophorbid, dessen Konfiguration an Ring II damit 3 R, 4 R sein muß.

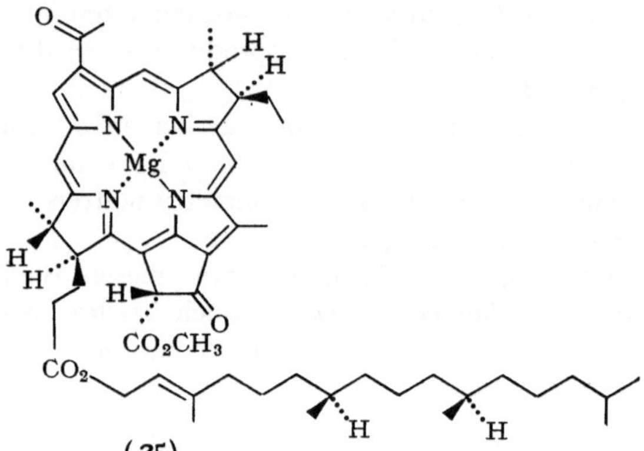

(25)

Bacteriochlorophyll

Weitere Chlorophylle. Neben den bekannten, ,,klassischen" Chlorophyllen a und b sowie dem Bacteriochlorophyll sind in jüngerer Zeit einige weitere verwandte Farbstoffe aufgefunden worden. Erwähnt seien hier die Chlorobium-Chlorophylle, in denen die 7-Propionsäurekette an Stelle von Phytol mit Farnesol verestert ist, die Chlorophylle c und d sowie das Bacteriochlorophyll b. Bezüglich Einzelheiten vgl. (*113*).

III. Porphyrinsynthesen

Der Stand unserer Kenntnis über die Synthesen von Porphyrinen und verwandten makrocyclischen Systemen bis Ende Januar 1965 ist kürzlich von Johnson und Mitarb. (*108*) dargelegt worden. Wir können uns daher auf die eindeutige Synthese unsymmetrischer Porphyrine und deren gegebenenfalls zur Darstellung natürlicher Porphyrine notwendigen Umbau beschränken.

1. Synthese unsymmetrischer Porphyrine

Die von Fischer erfundene Dipyrromethen-Kondensation (*60, 74, 80*) und ihre Variante, die Dipyrromethan-Kondensation nach MacDonald (*7, 59*) sind mit dem Nachteil behaftet, daß die zwei zum Porphyrin-makrocyclus führenden Verknüpfungen gleichberechtigt sind. Daher muß nach (**26**) + (**27**) → (**28**) eines von zwei verschiedenen Dipyrromethanen für sich symmetrisch sein (*200*) oder man ist nach (**29**) + + (**29**) → (**30**) auf die Selbstkondensation zweier identischer Dipyrromethane beschränkt (*140*), wenn die Synthesen eindeutig verlaufen sollen. (Im letzten Fall entstehen drehsymmetrische Porphyrine.) Wie Wood-

WARD die Eindeutigkeit im Zuge der Chlorophyllsynthese erzielte, wurde schon geschildert (S. 289). In den Arbeitsgruppen von JOHNSON (*150*), KENNER (*142*) und ESCHENMOSER (*14*) wurden andere Methoden entwickelt, die die schrittweise Vereinigung zweier verschiedener Dipyrromethane oder -methene über lineare Tetrapyrrolpigmente zu Porphyrinen beinhalten. Dabei wurden die Verknüpfungsstellen so gestaltet, daß die Kondensation nacheinander unter verschiedenen Bedingungen gelang.

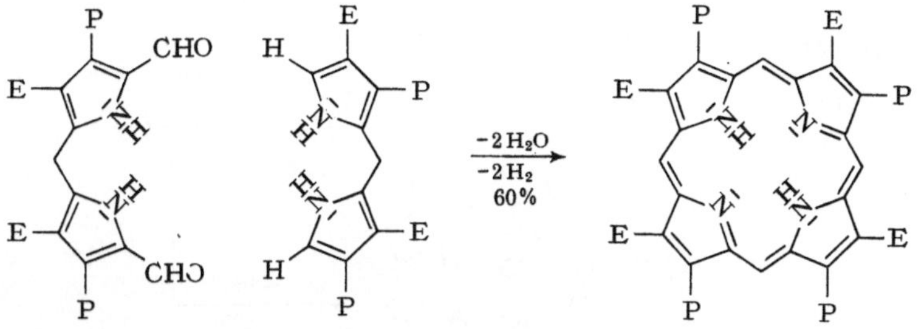

$$E = CH_2COOR$$
$$P = CH_2CH_2COOR$$

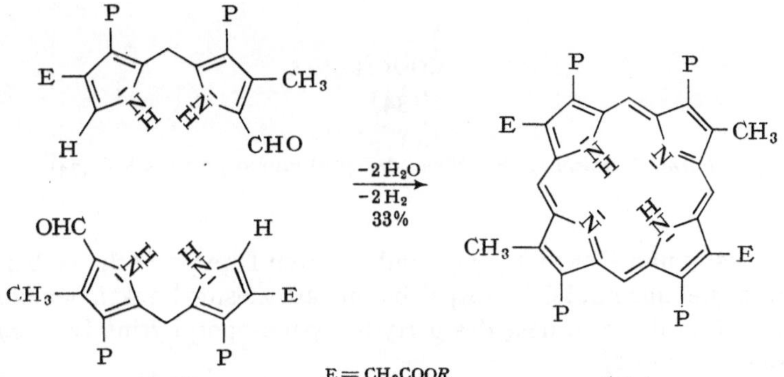

Die Synthese des γ-Phylloporphyrins XV (34) (*109*) möge JOHNSONS Weg schematisch verdeutlichen (*Formelübersicht 1*, S. 300).

Das Dipyrromethen (31) reagiert ohne Selbstkondensation mit der leichter angreifbaren freien α-Stellung des Dipyrromethens (32) zum Biladien-a,c-Salz (33), dessen terminale Äthylgruppe erst bei 180° in den Ringschluß eintritt. Biladien-a,c-Salze konnten auch mit Dimethylsulfoxid in Pyridin bei Raumtemperatur cyclisiert werden (*12*). In Gegenwart von Kupfer(II)-acetat in siedendem Methanol erhält man die Kupferchelate der Porphine, während in Anwesenheit von Nickel(II)- oder Kobalt(II)-salzen Chelate der Tetradehydrocorrine entstehen (*150, 108*).

Die hier skizzierte Synthese ist wegen ihrer Eignung zur Darstellung meso-alkylierter Porphine auch im Hinblick auf Chlorobiumchlorophylle nteressant (S. 298).

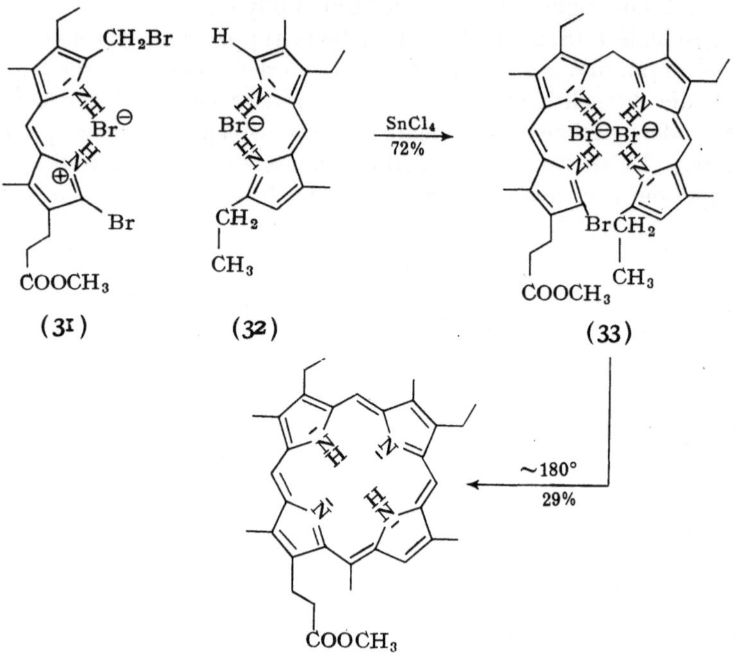

(31)   (32)   (33)

(34)

Formelübersicht 1. Synthese des γ-Phylloporphyrins XV (34)

Jackson und Kenner (*142*) kondensierten Dipyrromethane, bei denen
je eine 5-Stellung mit leicht abspaltbaren Carbonester-Funktionen blockiert
sind, z. B. in der Synthese des β-Hydroxy-mesoporphyrins IX (*144, 145*)
(*Formelübersicht 2*).

Das noch fehlende C-δ wird in Form von Orthoameisensäuremethylester eingeführt. Die Oxo-Gruppe an C-β des hier in der Oxophlorinform (siehe S. 336) ge-
zeichneten (36) kann durch Acetylierung und katalytische Hydrierung unter Ausbildung eines Porphinogens entfernt werden. Dieses läßt sich dann zum Mesoporphyrin IX (95b, S. 326) dehydrieren.

Zwar ist diese über b-Oxo-bilane (35a, b) führende Synthese (*31, 149*)
weniger ergiebig als die von Johnson (*109*) angegebene, aber sie stellt
Oxophlorine mit genau bekannter Position des O-Atoms bereit, die zum
Studium des oxidativen Abbaus nativer Porphyrine zu Gallenfarbstoffen
benötigt werden (S. 336).

Auch a-Oxobilane (*144, 31, 146*) und neuerdings b-Bilene (*42*) wurden
von Kenner auf analoge Weise zu ausbaufähigen Porphyrin-Synthesen
herangezogen.

Literaturverzeichnis: SS. 345—355

$\cdot POCl_3$

$C_6H_5CH_2OOC$ $CO\cdot N(CH_3)_2$

$C_6H_5CH_2OOC$ H

COOCH$_3$ COOCH$_3$
(35)

$C_6H_5CH_2OOC$
$C_6H_5CH_2OOC$ $=O$

COOCH$_3$ COOCH$_3$
(35^a)

① H_2 ($-2\,C_6H_5CH_3$)
② $185°$ ($-2\,CO_2$)

① $HC(OCH_3)_3$
② O_2

δ $=O$

COOCH$_3$ COOCH$_3$
(36)

H
H $=O$

COOCH$_3$ COOCH$_3$
(35^b)

Formelübersicht 2. Synthese des β-Hydroxy-mesoporphyrins IX (36)

Das von ESCHENMOSER hauptsächlich zur Synthese von Corrinen angewendete Prinzip der Iminoester-Kondensation (*14*) kann auch zum Aufbau des Porphyrinmakrocyclus dienen (*52*) (*Formelübersicht 3*, S. 302).

Die Gefahr der Selbstkondensation von (37) und (38) wird durch den Umstand unterdrückt, daß die reaktivere, konjugierte Iminoestergruppierung in der Komponente (38) mit weniger reaktivem Enaminzentrum steht. Das Enaminsystem in (37) läßt sich unter viel milderen Bedingungen aktivieren, so daß die Cyclisierung in die zwei gezeigten Schritte (37) + (38) → (39) → (40) zerlegt werden kann.

(40) ist als Iso-octahydroporphyrin-Derivat mit gesättigter Peripherie sowie endo-ständiger Methylgruppe bemerkenswert.

2. Umbau von Porphyrinen
als Abschluß der Synthese natürlicher Porphyrine

Native Porphyrine tragen die verschiedensten Seitenketten, die die Kondensationsreaktionen des Gerüstaufbaus stören oder dabei irre-

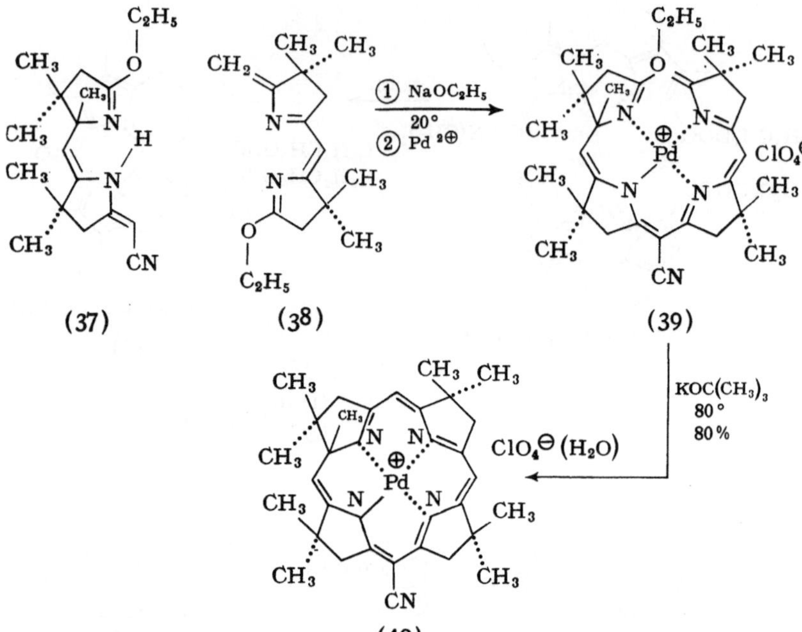

(37)   (38)   (39)

(40)

Formelübersicht 3. Aufbau des Porphyrin-Makrocyclus durch Iminoester-Kondensation

versibel verändert werden. Hierzu gehören Formyl-, Acetyl- und Vinyl-reste. Diese wurden daher schon von FISCHER (*92, 91*) nach dem Ringschluß durch Acylierung (siehe S. 330) und weitere Umwandlung erzeugt

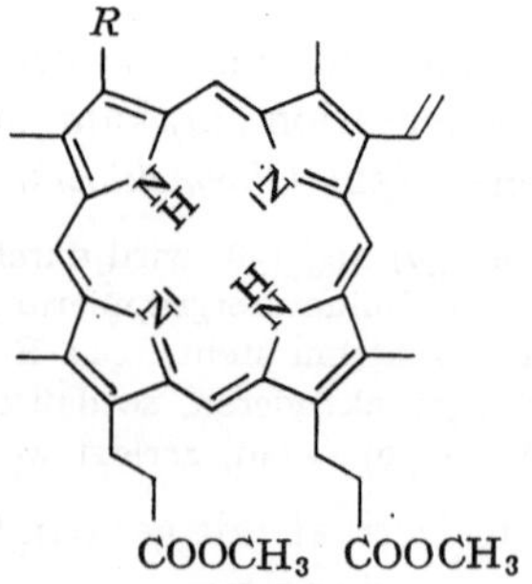

(41a) $R = H$

(41b) $R = CHO$

(41c) $R = CH \dot= CH_2$

(41)

(siehe auch S. 293, 338). Heute pflegt man die Vinylgruppe als 2″-Amino-äthylgruppe (*11, 30, 53, 212*) oder 2″-Acetoxyäthylgruppe (*31, 149*) in die Synthese mit einzuplanen und zum Schluß durch Hofmann-Abbau oder Überführung des Acetats in Chlorid und Chlorwasserstoffabspaltung die Vinylgruppe freizulegen. So wurde das kürzlich aus menschlichen Faeces (*180*) isolierte Pemptoporphyrin (**41a**) synthetisiert und damit in seiner Struktur sichergestellt (*149, 11*).

Spirographis-Porphyrin (**41b**), das in Form seines Eisen(III)-chelats die prosthetische Gruppe des Blutfarbstoffes einiger Polychätenwürmer bildet, konnte durch Formylierung einer Vorstufe von (**41a**) und Freilegung der Vinylgruppe unter vorübergehender Maskierung der Formylgruppe gewonnen werden (*149*). Es ist auch durch Photooxidation des Protoporphyrins IX (**41c**) zum sogenannten „Photoprotoporphyrin IX" (**42**) (*54*) und dessen anschließenden Umbau gemäß den Teilformeln (**43**) und (**44**) zugänglich (*118*) *(Formelübersicht 4).*

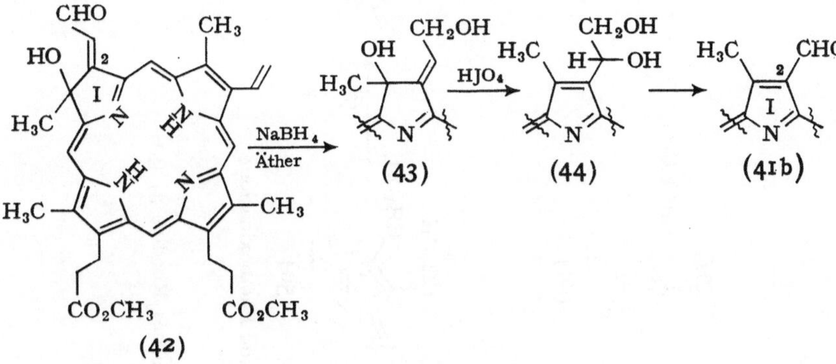

Formelübersicht 4. Umbau des Photoprotoporphyrins (**42**) in *Spirographis*-Porphyrin (**41b**)

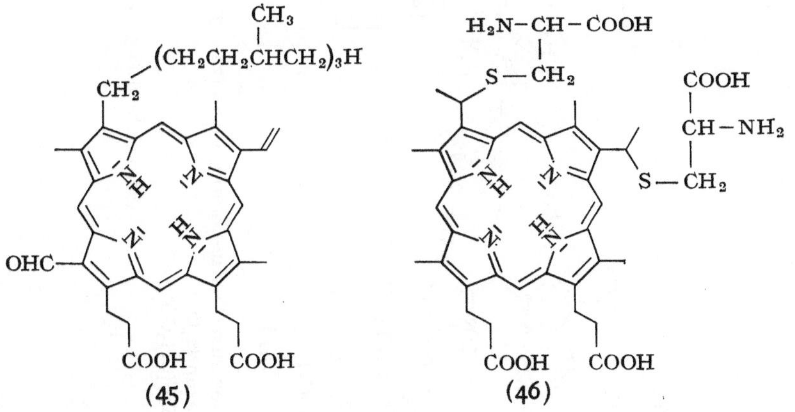

Porphyrin a (**45**) und Porphyrin c (**46**) bilden als Eisen(III)-chelate die prosthetischen Gruppen der Cytochrome a, a$_3$ und c. (**45**) wurde von LYNEN und Mitarb. in seiner Struktur aufgeklärt (*104, 105*); die Synthese ist in Arbeit (*8*, *42*). (**46**) konnte aus 2',4'-Dibrom-mesoporphyrin IX und L-Cystein in kristalliner Form erhalten werden (*172*). Auch einige Abbauporphyrine der *Chlorobium*-Chlorophylle (S. 298) sind kürzlich synthetisiert worden (*5*, *6*, *42*, *113*).

Formelübersicht 5. Teilsynthese des Rhodins g_7 (**51**)

1 Reduktion zum Chlorin-phlorin — vgl. (*76*), (*77*), (*83*) — und Photooxidation;
2 HCl/Dioxan/H_2O;
3 $(CH_3)_2SO/(CH_3CO)_2O$.

Die auf S. 323 beschriebene Funktionalisierung der Alkylgruppen an C-3 und C-4 von Chlorinen eröffnet einen Übergang von der Chlorophyll a-Reihe in die Derivate der b-Reihe. Wie noch näher ausgeführt werden wird, sind durch Photooxydation von Chlorin-phlorinen Derivate des 3,4-Dihydroxybacteriochlorins zugänglich. Die Anwendung dieser Reaktion auf Chlorin e_6-TME (47) in Dioxan-Wasser führt zum Isomerengemisch des *trans*-,,Bacteriodiols" (48), das durch säurekatalysierte Umlagerung in Dioxan-Wasser neben dem 4'-Alkohol (49) des Chlorin e_6-TME auch die 3-Hydroxymethyl-Verbindung des Chlorin e_6-TME (Trimethylester) (50), also den Alkohol des Rhodin g_7-TME, ergibt. Diese Verbindung kann, besonders durch Oxidation mit Dimethylsulfoxid/Acetanhydrid leicht in den Rhodin g_7-TME (51) übergeführt werden *(Formelübersicht 5)*. Die Identität der natürlichen und synthetischen Verbindungen (51) ist durch Dünnschichtchromatographie, NMR- und Massenspektroskopie gesichert. Damit sind die Verbindungen der b-Reihe an die Totalsynthese des Chlorophylls a angeschlossen (*129*).

3. Biosynthese

Vergleicht man die Formeln von Chlorophyll und Hämin, so drängt sich die Vermutung auf, daß ihre Biosynthese weitgehend gleich verlaufen und eine Trennung der Synthesewege erst in einem späteren Stadium erfolgen sollte. Diese Vorstellungen sind experimentell verifiziert worden. Methoden, die zur Aufklärung der Biosynthese herangezogen wurden, waren hauptsächlich: die Verwendung von isotopenmarkierten Verbindungen beim Wachstum lebender Zellen, die Untersuchung von Porphyrinen, die von lebenden Zellen unter speziellen experimentellen Bedingungen erzeugt werden, und das Studium enzymatischer Reaktionen außerhalb der lebenden Zelle.

Für nähere Einzelheiten siehe (*15*, *161*).

IV. Die Bedeutung physikalischer und quantenchemischer Methoden für die präparative Porphyrinchemie

Physikalische Meßtechnik und Quantenchemie haben sich wegen der Schlüsselstellung des Chlorophylls und der Häme in der biochemischen Energetik schon eingehend mit den Porphyrinen befaßt. Wir wollen hier nur die für den präparativ arbeitenden organischen Chemiker besonders interessierenden Ergebnisse hervorheben.

1. Physikalische Methoden

Röntgenstrukturanalysen haben die durch Abbau und Synthese chemisch erschlossenen Strukturformeln voll bestätigt (*94—97*, *99*,

110—112, 160, 194, 206, 207). Die Deutung chemischer Befunde muß in Zukunft auf die hier festgestellten sterischen und elektronischen Besonderheiten des Porphyrinsystems Rücksicht nehmen.

Ein Porphinmolekül ist durch die beiden mesomeren und tautomeren Grenzstrukturen (**52a, 52b**) als Überlagerung sich gegenseitig durchdringender aromatischer Teilsysteme dargestellt:

p
p
e
e
(52a)
(52b)
(e = endo-C-Atome)
(p = peri-C-Atome)

Chlorophyll, Hämin und ihre Abkömmlinge sind an Hand vieler Veröffentlichungen und Übersichtsreferate, z. B. über Elektronenspektren (*33, 34, 54, 100, 102, 183*), Infrarotspektren (*20, 22, 32, 155, 183*), Kernresonanzspektren (*1—4, 24, 30, 39, 155—158*), optische Rotationsdispersion und Circulardichroismus (*23, 46, 159, 204, 205, 210, 211*) sowie Massenspektren (*9, 29, 143, 148*) spektroskopisch gründlich charakterisiert.

Der Nutzen dieser Methoden zeigt sich beispielsweise, wenn man zwei isomere Dihydroporphine — etwa ein Phlorin (**58**, S. 309) von einem Chlorin (**60**, S. 309) — unterscheiden will. Im Phlorin ist das makrocyclische π-Elektronensystem unterbrochen, was sich in einer Abnahme der Extinktion der Soret-Bande im Elektronenspektrum (*103*) und einem Verschwinden des „Ringstromeffektes" in der Protonenresonanz (*13*) bemerkbar macht.

Dagegen ist das makrocyclische π-Elektronensystem im Chlorin erhalten geblieben und damit auch eine normal ausgeprägte Soret-Bande sowie ein „Ringstromeffekt". Die Aufhebung eines pyrrolischen, aromatischen Teilsystems äußert sich durch das Auftreten des typischen Elektronenspektrums (*54*), der „Chlorinbande" im Infrarot bei etwa 1600 cm⁻¹ (*165*) — die übrigens auch bei Pyrromethenen (*51*), nicht aber bei Porphinen zu finden ist — und durch eine Änderung des massenspektrometrischen Fragmentierungsmusters (*148*). Bei allen Messungen in Lösung ist die Assoziationsneigung der Porphyrine (*155, 21*) zu beachten.

Für analytische Zwecke ist in Anbetracht der über 500 liegenden Molmassen die hochauflösende Massenspektrometrie zur exakten Bestimmung der Summenformel von unschätzbarem Wert, während die Dünnschichtchromatographie davor bewahrt, ein Gemisch von Stoffen gleichen Elektronenspektrums als einheitliche Substanz anzusprechen.

Literaturverzeichnis: SS. 345—355

2. Theoretische Methoden

Bei dem ausgedehnten aromatischen π-Elektronensystem interessiert die Abgrenzung der Reaktionsbereitschaft der einzelnen Positionen gegeneinander. Im Einklang mit der Erfahrung an kondensierten Aromaten ist ein Angriff auf eines der in (52a) mit e bezeichneten „endo"-C-Atome recht selten (weswegen diese C-Atome im Fischer-Numerierungsschema unbezeichnet sind), er wird aber um so wahrscheinlicher, je mehr die Pyrrolringe — etwa durch Hydrierung — ihre Eigenreaktivität entfalten. Die größere Bereitschaft der peri-C-Atome („p" in (52b)) gegenüber elektrophilem Angriff im Vergleich zu den meso-C-Atomen beschrieb WOODWARD (214) durch eine Konzentration der π-Elektronen in den pyrrolischen Teilsystemen nach (53), welche die peri-Stellungen negativiert, die meso-Stellungen positiviert.

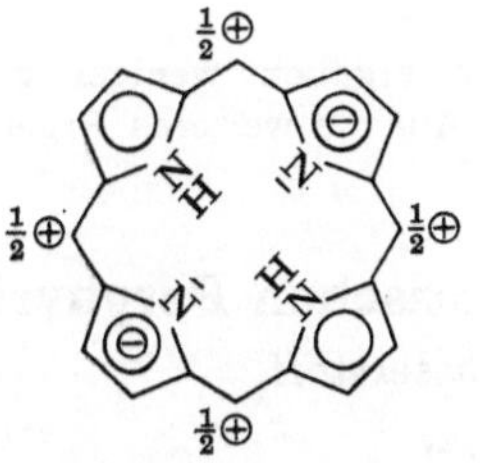

$$q_m = 0,91 \qquad q_p = 0,93$$

(53) 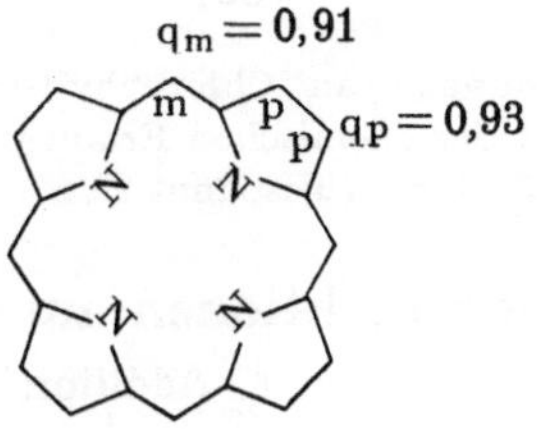   (54)
(q = Elektonendichte)

Zum gleichen Schluß kommt man durch die Auswertung der zahlreich vorliegenden Molekül-Orbital-(MO-)-Berechnungen [Übersicht in (102)] unter dem Gesichtspunkt chemischer Reaktivität. Die Arbeiten von GOUTERMAN (33, 102, 103, 208, 217—221) sind hier hervorzuheben. Sie befassen sich hauptsächlich mit der Deutung der Elektronenspektren. Da im Porphin ein nichtalternierendes heterocyclisches π-Elektronensystem vorliegt, sollten Elektronendichten q und Lokalisierungsenergien L bei der Auslegung von MO-Rechnungen bevorzugt werden. Die in (54) skizzierten, aufgerundeten q-Werte (103) beschreiben den erleichterten elektrophilen Angriff an C-p relativ zu C-m richtig. (Die H-Atome an den N-Atomen bleiben unberücksichtigt.)

Die im Vergleich zum Porphin leichter und selektiv an C-γ und C-δ angreifende elektrophile Substitution am Chlorin folgt nicht nur aus den in (55) angegebenen q-Werten (103), sondern auch aus Lokalisierungsenergien (178). Ferner geht die in der Reihe C-m (Porphin) < C-α,β (Chlorin) < C-γ-δ (Chlorin) < C-β-δ (a-Tetrahydroporphin) < C-γ (a-Tetrahydroporphin) zunehmende Leichtigkeit des Deuteriumaustausches (18) einher mit zunehmenden q-Werten in (54)—(56) (103). Für (55) ist weiter $q_2 = 0,94$ und $q_6 = 0,98$. Falls diese Werte auf das Kupferchelat des Isochlorin e_4-dimethylesters (99a, S. 334) übertragbar sind, ließe sich damit die relativ zu C-2 erleichterte Acetylierung an C-6 deuten.

Die Betrachtung der ψ-Amplitudenwerte der obersten besetzten MOs im Porphin-dianion (*102*) und der q-Werte im obersten besetzten MO des Phlorinanions (*115*) macht schließlich verständlich, daß eine stufenweise Entladung des Octaäthyl-porphinatozink-dianions mit Methyljodid zur α,γ-Addition zweier Methylgruppen führt (*121*).

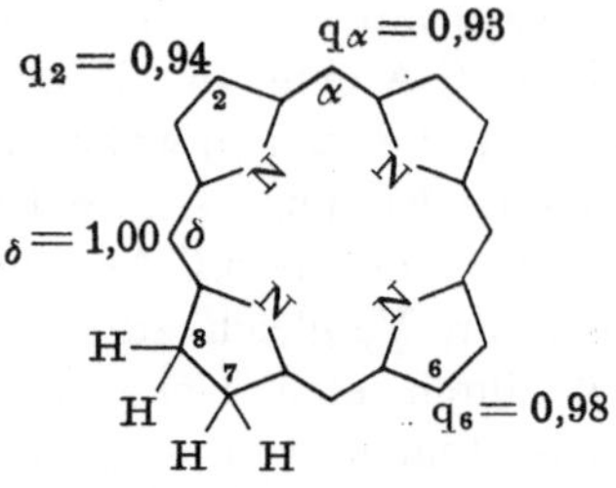

$q_2 = 0,94$ $q_\alpha = 0,93$ $q_\alpha = 0,94$

$q_\delta = 1,00$ $q_\delta = 1,05$

$q_6 = 0,98$

$q_\gamma = 1,08$

(55) 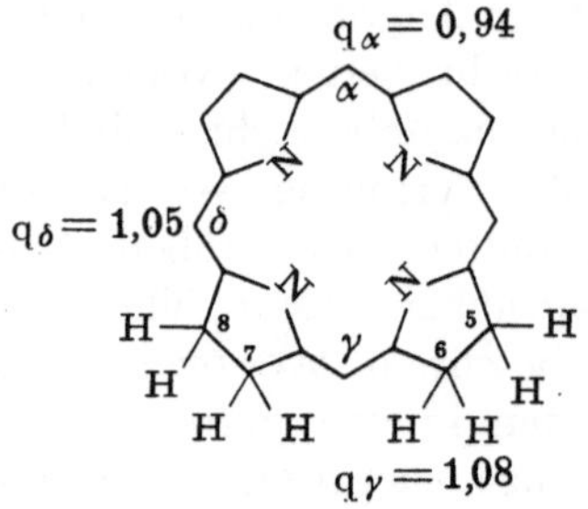   (56)

MO-Rechnungen an Chlorophyll-Derivaten ergaben weniger gute Überein-stimmung mit der chemischen Erfahrung (*45*). Auf die weiteren Ergebnisse solcher Rechnungen darf man gespannt sein.

V. Additionsreaktionen am aromatischen Porphyrinsystem

1. Addition von Wasserstoff

a) Allgemeines

Die am häufigsten durchgeführte Additionsreaktion am Porphyrin-system ist die teilweise Absättigung des π-Elektronensystems mit Wasser-stoff zu Dihydro-, Tetrahydro- und Hexahydroporphinen, die in der Regel die Endstufe der Hydrierung darstellen. Nach Mauzerall (*167*) sind — ausgehend von einem totalsymmetrischen Porphin — allein 30 Dihydroporphine denkbar. Nach der Regel von der Erhaltung einer maximalen Anzahl aromatischer Teilsysteme reduziert sich die Menge der Dihydroporphine auf 7 (*27*), von denen außer Woodwards hypothetischem Isophlorin (57) für die weitere Betrachtung nur das Phlorin (58), das α,γ-Dihydroporphin (59) und das Chlorin (60) in Frage kommen.

Grundsätzlich ist also eine Absättigung peripherer Doppelbindungen (60) und der Methinbrücken (58) möglich. Die nacheinander erfolgende Hydrierung peripherer Doppelbindungen führt über (60) zu den a-Tetra-hydroporphinen (61), den b-Tetrahydroporphinen (62) oder zu Hexa-hydroporphinen (*47, 44*). Fortlaufende Wasserstoffanlagerung an die meso-C-Atome liefert über (58) die α,β,γ,N-Tetrahydroporphine (74) und die farblosen Porphinogene (73, S. 315). Gemischte meso- und peri-Hydrierung liegt bei den Chlorin-phlorinen (76, S. 317) β,δ-Dihydro-chlorinen (78, S. 318) und „Chlorinogenen" (75) vor. Die Stabilität der einzelnen Hydroporphyrine ist unterschiedlich (siehe S. 319); allgemein werden die Brückenstellungen leichter dehydriert als die peri-Stellungen.

Literaturverzeichnis: SS. 345—355

Grad und Richtung der Hydrierung sind nur schwer zu steuern. Da noch keine systematischen Untersuchungen vorliegen, kann man über die Beeinflussung des Hydrierungsgangs durch Reaktionsbedingungen und Zentralmetall nur Vermutungen äußern (S. 343).

In diesem Zusammenhang ist es bemerkenswert, daß WOODWARD zur Absättigung der 7,8-Doppelbindung in (15, S. 291) keine Hydrierung durchgeführt hat.

Die Hydroporphyrine haben charakteristische Elektronenspektren, die zur Identifikation dienen können. Für Einzelheiten sei auf die Originalliteratur verwiesen.

(57)   (58)

(59)   (60)

b) Reduktion peripherer Doppelbindungen

Einwirkung von Metallen in protonenhaltigem Milieu. Die Hydrierung eines Porphyrins läßt sich nach MAUZERALL (*167*) als Ergebnis einer Übertragung zweier Elektronen und nachfolgender Addition zweier Protonen auffassen. Diese Art der Reduktion ist in der von FISCHER (*70, 89*) entwickelten Hydrierung mit Natrium in siedendem Isoamylalkohol verwirklicht, wobei das Metall die Elektronen, der Alkohol die Protonen liefert. Gewöhnlich geht man von einem Porphinato-eisen(III)-chlorid aus. Man erhält als Hauptprodukte Eisen(II)-chelate der Chlorine sowie der a- und b-Tetrahydroporphine [a: aneinanderliegende hydrierte Pyrrolringe wie in (61) und b: Anordnung der hydrierten Pyrrolringe wie im Bacteriochlorophyll, etwa (62)]. Bei der Aufarbeitung mit Säure erhält man unter Abspaltung des Eisens die Hydroporphine selbst. Am übersichtlichsten verläuft dieser Prozeß bei totalsymmetrischen Porphinen, z. B. Octaäthylporphin (*47, 122*). Hier konnte nicht nur das von FISCHER schon bei der Reduktion seiner Porphyrinmonocarbonsäure-VII (*70*) beobachtete a-Tetrahydroporphin (61) gefaßt, sondern auch das b-Tetrahydroporphin (62) isoliert (*47*) und schließlich in kristalliner Form (*122*) erhalten werden.

Ergab sich die Struktur des b-Tetrahydroprodukts (62) auf physikalischem Wege bisher nur durch das Elektronenspektrum (47) und indirekt aus dem ^{1}H-Resonanzspektrum (NMR) von (61) (18), das drei verschiedene Methinprotonen anzeigt, so kann die Struktur (62) nach dem ^{1}H-Resonanzspektrum mit nur einem einzigen Satz von vier Methinprotonen (δ = 8,82 ppm) unabhängig vom Elektronenspektrum als gesichert gelten (122).

(61)   (62)

Mit Natrium in Isoamylalkohol erfolgt die Hydrierung des Octaäthylhämins (OÄP)FeCl überwiegend in dem Sinne, daß die „überzähligen" Wasserstoffe im Octaäthylchlorin anticlinal — einfacher, aber weniger präzis: *trans*-zueinander — stehen (122). Es entsteht also bevorzugt die *trans*-Racemform des Octaäthylchlorins. Die *cis*-Mesoform findet sich nur in Spuren; sie konnte besser durch katalytische Hydrierung (siehe S. 313) und Hydroborierung (S. 314) bereitet werden.

Die Oxidation der a- und b-Tetrahydro-octaäthylporphine (61) und (62) mit Sauerstoff führte ausschließlich zum *trans*-Chlorin zurück, woraus geschlossen wurde, daß die Tetrahydroporphine ihrerseits durch *trans*-Addition des Wasserstoffs entstanden waren (122).

Unübersichtlich ist die Reduktion unsymmetrischer Porphyrine zu Chlorinen, weil hier — abgesehen von Stereoisomeren — vier Strukturisomere möglich sind.

Den Ort der Hydrierung zu bestimmen, ist nur in seltenen Fällen gelungen. So behaupten Corwin und Collins (41), das 2-Vinylchloroporphyrin e$_6$-zinn-dichlorid in ein 7,8-Dihydroprodukt (mit 2-Äthyl) überführt zu haben, während Inhoffen und Samblebe (138) bei einer Nacharbeitung diesen Körper nicht zu isolieren vermochten. Ein klassisches Beispiel für eine strukturspezifische Hydrierung ist die des Meso-phylloporphyrins (34) durch Fischer und Laubereau (76, 197), die nur das 7,8-Dihydroprodukt (63) isolierten.

Wenn man die überwiegende *trans*-Hydrierung bei der Natrium-Isoamylalkohol-Reduktion am Octaäthylporphinato-eisen(III)chlorid verallgemeinert, so soll man vermuten, daß (63) mit der *trans*-Anordnung der überzähligen Wasserstoffatome entstanden war. Eine spezifische Hydrierung ist dagegen beim Pyrrohämin (67), Protohämin und Deuterohämin (91) nicht beobachtet worden. Hier fehlt die sterische Verdichtung durch meso-Substituenten.

Zum Einfluß des Zentralmetalls liegen Untersuchungen an Chelaten des Ätioporphyrins II vor (41, 182).

Die Chlorin-Ausbeuten liegen selten über 20%.

Literaturverzeichnis: SS. 345—355

Der Mechanismus dieser Reduktion ist unbekannt. Die Bildung von Chlorinen aus Metalloporphyrinen bei der Einwirkung von Natrium in höheren Alkoholen scheint hier nur ein Sonderfall zu sein, denn eine Reihe von Metalloporphyrinen geben bei der Behandlung mit Lithium und Äthylendiamin zum Teil Leuko-Verbindungen (*48*), desgleichen Octaäthylporphinato-zink mit Natrium in flüssigem Ammoniak (*121*) und 2-Vinyl-chloroporphyrin e$_6$-trimethylester-zinn-dichlorid mit Eisenpulver in Chlorwasserstoff-Eisessig (*41*).

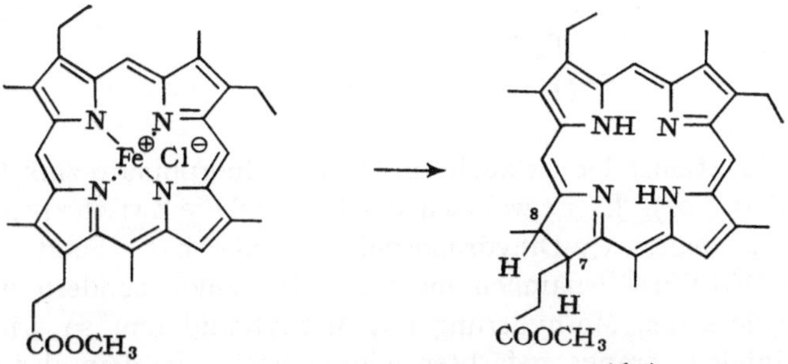

$$(34) \longrightarrow (63)$$

Erwähnenswert ist auch die Bildung von Chlorinen aus Porphyrinen durch Einwirkung von Natriumäthylat bei 170—200° (*203, 202*), wobei in günstigen Fällen 20 bis 25% Ausbeute erreicht werden.

Die zwei auf S. 309 postulierten Teilschritte der Hydrierung eines Metalloporphyrins lassen sich getrennt realisieren, wenn man wasserfrei arbeitet — etwa in Tetrahydrofuran. So haben HUSH und DODD (*43*) aus Natrium und Metalloporphyrinen Metalloporphyrin-anion-Radikale und Metalloporphyrin-dianionen hergestellt.

Präparativ einfacher ist es, Anthracen als Elektronenüberträger zwischenzuschalten. Man bereitet in Tetrahydrofuran das Anionradikal des Anthracens $C_{14}H_{10}$:

$$2\,Na + 2\,C_{14}H_{10} \rightarrow 2\,Na^{\oplus}C_{14}H_{10}{}^{\ominus}$$

und läßt dieses nach CLOSS und CLOSS (*38*) in geringem Überschuß auf Metalloporphyrine einwirken. Nach Zugabe eines Äquivalents $NaC_{14}H_{10}$ schlägt z. B. die rote Farbe des Zinkkomplexes nach grün, nach 2 Äquivalenten wieder nach rot um. Die bei der Reduktion des meso-Tetraphenylporphinato-zinks (TPP)Zn erhaltenen Salze $Na^{\oplus}[(TPP)Zn]^{\ominus}$ und $Na_2{}^{\oplus}[(TPP)Zn]^{\ominus\ominus}$ wurden in Substanz isoliert. Ersteres ist paramagnetisch, letzteres diamagnetisch. Beide sind luftempfindlich, die Reoxidation mit Jod oder Sauerstoff liefert quantitativ das Metalloporphyrin zurück. Bemerkenswert verläuft der zweite Schritt, die Protolyse mit Methanol, die aus dem Dianion $[(TPP)Zn]^{\ominus\ominus}$ zunächst ein Phlorinanion (**64**) erzeugt, das sich über ein nicht näher bezeichnetes „Phlorin" (**65**) in Tetraphenylchlorinato-zink (**66**) umlagert (*38*).

$$[(64)] \quad Na^{\oplus} \longrightarrow (65) \longrightarrow (66)$$

Diese Reaktionsfolge ist auch am Octaäthylporphinato-zink (OÄP)Zn durchführbar (*121*). Es ist wahrscheinlicher, daß die Zwischenverbindung die Struktur eines α,γ-Dihydroporphinato-zinks hat. Setzt man das Dianion [(OÄP)Zn]$^{\ominus\ominus}$ nämlich nicht mit Methanol, sondern nach Art einer reduzierenden Methylierung mit Methyljodid um, so erhält man einen instabilen, orange gefärbten Zinkkomplex, in dem der Ort der Reduktion durch Methylgruppen festgelegt sein sollte. Dieser geht durch Oxidation mit Sauerstoff in Eisessig in α,γ-Dimethyl-octaäthylporphyrin (67), durch Einwirkung von Nickelacetat unter Luftausschluß in α,γ-Dihydro-α,γ-dimethyl-octaäthylporphinato-nickel (68) über, eine im Gegensatz zu den bisher bekannten meso-Hydroporphinen luftbeständige Substanz.

Das Eintreten der Methylgruppen in die meso-Position legt nahe, daß [(TPP)Zn]$^{\ominus\ominus}$, [(OÄP)Zn]$^{\ominus\ominus}$ und (64) die maximale Elektronendichte an den meso-C-Atomen aufweisen. Dies steht im Einklang mit MO-Berechnungen (S. 308).

Die Formel (68) spricht dafür, daß auch (65) eine analoge Struktur besitzt. Die prototrope Verwandlung von (65) in (66) ist daher eher als α,γ-Dihydroporphin-Chlorin-Umlagerung aufzufassen. Auch Pyridin wurde kürzlich als Elektronenüberträger verwendet (*192*, *193*).

Literaturverzeichnis: SS. 345—355

Katalytische Hydrierung. Im Gegensatz zur Hydrierung freier Porphyrine führt die katalytische Hydrierung mit wenigen Ausnahmen (*10, 44*) nur dann zu peripher hydrierten Produkten, wenn die Bildung von Porphinogenen durch Einführung eines Metallions unterbunden wird (*72, 209*). Die Hydrierung wird gleichzeitig erschwert. So ging Octaäthylporphinato-zink (*122*) mit Pd/C und 180 atm H_2 in 90 Stunden bei 90° zu etwa 8% in *cis*-Octaäthylchlorinato-zink über (vgl. S. 310).

Die *cis*-Addition des Wasserstoffs war nach dem Verlauf der Hydrierung des Octamethyl-tetraaza-porphinato-magnesiums, der durch oxidativen Abbau des erhaltenen b-Tetrahydro-Produkts geklärt wurde (*58*), zu erwarten und ergab sich weiter daraus, daß *cis*-Octaäthylchlorinato-zink zum *cis*-Octaäthylchlorin entmetalliert werden konnte. Dieses war vom isomeren, konfigurativ gesicherten „*trans*"-Octaäthylchlorin (S. 310) im NMR-Spektrum, Elektronenspektrum und chromatographischen Verhalten deutlich verschieden.

Mit gutem Erfolg wurde die katalytische Hydrierung von Nickel-Komplexen zur Darstellung peripher hydrierter Verbindungen angewandt (*134*). Hydrierte man den Nickel-Komplex von Pyromethylphäophorbid a (**69a**) mit Raney-Nickel in Äther, so entstanden neben Mesopyromethylphäophorbid a-Ni (**69b**), etwa 20% 1,2-Dihydro-mesopyromethylphäophorbid a-Ni (**70**), sowie 2,5% 1,2,3,4,-Tetrahydro-mesopyromethylphäophorbid a-Ni (**71**), 9-Desoxo-mesopyromethylphäophorbid a-Ni (**69c**) (20%) und 9-Desoxo-9-hydroxy-mesopyromethylphäophorbid a-Ni (**69d**) (zirka 1,5%).

$$(69)\qquad\qquad\qquad (70)$$

$$(69\,a)\ R^1 = C_2H_5;\ \left.\begin{matrix}R^2\\R^3\end{matrix}\right\} = O$$

$$(69\,b)\ R^1 = C_2H_5;\ \left.\begin{matrix}R^2\\R^3\end{matrix}\right\} = O$$

$$(69\,c)\ R^1 = C_2H_5;\ R^2,\ R^3 = H$$

$$(69\,d)\ R^1 = C_2H_5;\ R^2 = OH;\ R^3 = H$$

Die Struktur von (**70**) wurde durch Elektronen-, IR-, NMR- und Massenspektroskopie bewiesen, die Struktur von (**71**) durch Elektronenspektroskopie und Abbau zu (**70**) gesichert. Die Substanzen (**69c**) und (**69d**) ließen sich durch Vergleich mit

Verbindungen identifizieren, die durch gezielten Abbau der Carbonylfunktion am
C-9 dargestellt werden konnten. Zur Aufklärung der Stereochemie der Substanzen
(*70*) und (*71*) kann auf die Arbeiten von Linstead et al. (*58*) verwiesen werden, die für
ähnliche Verbindungen eine *cis*-Hydrierung nachgewiesen haben. Die weitergehen-
den Überlegungen bezüglich der Stellung der eingeführten Wasserstoffatome jeweils
zur Propionsäurekette am C-7 (syn/anti) können hier keine Anwendung finden.

Reduktion mit Metallhydriden. Lewis-acide Metallhydride, etwa
Diboran oder Di-isobutyl-aluminiumhydrid reduzieren das π-Elektronen-
system des Octaäthylporphins unter geeigneten Bedingungen. Man darf
annehmen, daß Bor- oder Aluminium-Komplexe in die Reaktion eintreten (*122, 120*). Die Einwirkung von Diboran in Tetrahydrofuran
liefert nach der Hydrolyse etwa 10% eines trennbaren Gemisches von
cis- und *trans*-Octaäthylchlorin (*122*). Daher dürfte die Reaktion nicht
allein in der Hydroborierung einer peripheren Doppelbindung bestehen.

Ein zu großer Überschuß an Diboran führt zu farblosen bis gelben Verbindungen, die nicht reoxidiert werden können (*122, 145*). Die direkte Anlagerung von
Di-isobutyl-aluminiumhydrid an eine periphere Doppelbindung gelingt nicht. Jedoch
bilden sich in Tetrahydrofuran mit diesem Reagens brückenhydrierte Porphin-
aluminium-chelate, die sich in saurem Milieu zu Chlorin-Komplexen umlagern lassen.
Auf diesem Wege erhielt man Octaäthylchlorinato-aluminiumhydroxid in 19%iger
Ausbeute (*120*).

Reduktion mit Hydrazin und Derivaten. Schon Fischer beobachtete,
daß Hydrazin in Methanol ein Porphin zum Chlorin reduzieren kann (*80*).
Neuerdings wurde Hydrazin in Pyridin von Sidorov zur photochemischen
Reduktion wie zur Dunkelhydrierung eingesetzt (*181, 190, 191*). Die
Dunkelhydrierung verläuft nur in Gegenwart von Sauerstoff. So konnte
Chlorophyll zu einem a-Tetrahydroporphin reduziert werden, jedoch ließ
sich dieses nur spektral charakterisieren. Da Hydrazin auch mit den
Seitenketten der nativen Chlorine reagiert (*66, 68*), ist hier eine unüber-
sichtliche Reaktion zu erwarten.

Octaäthylchlorin wurde mit Hydrazinhydrat, Natriummethylat und
Pyridin unter Luftausschluß im Rohr bei 185° in 8 bis 10% Ausbeute
zum b-Tetrahydro-octaäthylporphin hydriert, das mit dem nach S. 309
erhaltenen identisch war. a-Tetrahydro-octaäthylporphin wurde nicht
gefunden; es ist unter diesen Bedingungen nicht beständig, wie unabhängig geprüft wurde (*122*).

Auch die hydrierende Wirkung des Diimids wurde untersucht (*137*),
und Mesoporphyrin-IX-DME-zinndichlorid und Octaäthylporphinato-
zinndiacetat mit *p*-Toluolsulfhydrazid und etwas KOH in siedendem
Diäthylenglycol-dimethyläther umgesetzt. Im ersten Fall erhielt man ein
Gemisch verschiedener Chlorinkomplexe, im zweiten Fall in 33% der
Theorie Octaäthylchlorinato-zinndihydroxid, das als Diacetat charakterisiert wurde, sowie Tetrahydro-Verbindungen. Octaäthylchlorinato-
zinndiacetat enthält die „überzähligen" Wasserstoffatome in „*trans*"-

Konfiguration, was durch Einführung des Zinns in unabhängig gewonnenes *trans*-Octaäthylchlorin sichergestellt wurde (*122*). Hier wird also nicht die normale *cis*-Hydrierung mit Diimid beobachtet.

Photochemische Reaktionen. Wegen des engen Zusammenhanges mit der Photosynthese sind photochemische Reaktionen der Porphyrine und Chlorine, insbesondere des Chlorophylls selbst, häufig untersucht worden. Bezüglich der Photochemie des Chlorophylls sei hier auf Übersichtsartikel von SEELY (*186*) und SIDOROV (*191*) verwiesen.

In jüngerer Zeit konnten SEELY und TALMADGE (*189*) zeigen, daß bereits der Zn-Komplex des einfachen Porphins mit Ascorbinsäure in Gegenwart von Aminen eine Photoreaktion eingeht, die zu Chlorinen und einem Tetrahydro-Produkt führt. Die Untersuchungen ergaben, daß die Reaktion über ein Zwischenprodukt (72) ($\lambda_{max} = 435$ mμ) verläuft, das nach seinem Hydrierungsgrad isomer mit dem Chlorin ist und sich photochemisch zum Chlorin umlagert. Dieses Zwischenprodukt, das sich im Dunkeln in Gegenwart der oxidierten Ascorbinsäure wieder zum Porphin umwandelt, hat so im Reaktionsschema eine zentrale Stellung. Die Autoren glauben, daß es einem Phlorin sehr ähnlich, aber selbst kein Phlorin sein müsse und schlagen die Struktur (72) dafür vor.

Dieser Vorschlag wurde von SIDOROV (*191*) in Frage gestellt. Nach den Ergebnissen der reduzierenden Methylierung (S. 312) könnte es sich hier um α,γ-Dihydroporphine handeln, deren Struktur (59, S. 309) sich besser mit einer Absorptionsbande bei 435 mμ vereinen ließe.

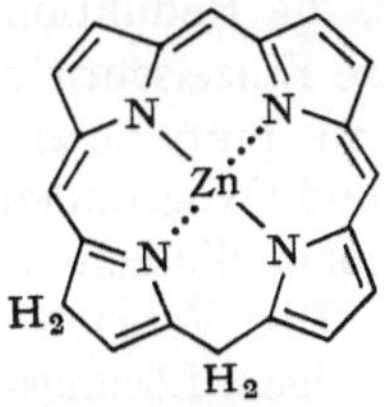

(72) 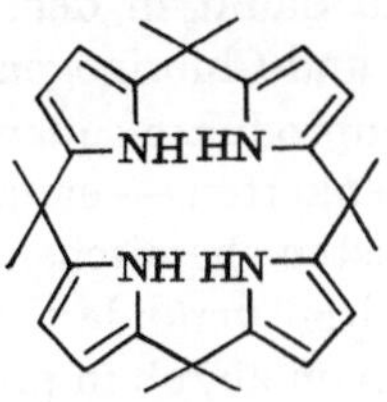   (73) 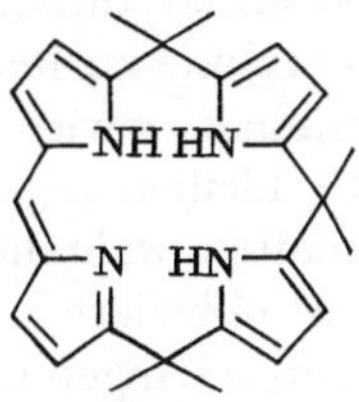   (74)

c) *Reduktion an den Methinbrücken*

Chemische Reduktion. Einwirkung von Reduktionsmitteln wie Natriumamalgam, Eisen in Ameisensäure (*80*), Jodwasserstoff in Eisessig (*83*), Zinkstaub in Eisessig (*85*) oder Alkali überführt Porphyrine in Leuko-Verbindungen, die sogenannten Porphyrinogene (73). Diese Hexahydroporphine dürften sich wohl deshalb bevorzugt ausbilden, da trotz der Anlagerung von sechs H-Atomen nur ein aromatisches Teilsystem aufgehoben wird, im Gegensatz zur Weiterhydrierung der Peripherie des Porphyrinsystems. Dabei treten nach MAUZERALL (*167*) die von WOODWARD (siehe S. 290) entdeckten Phlorine (58, S. 309) und die α,β,γ-N-Tetrahydroporphyrine (74) als Zwischenstufen auf. Letztere kann man spezifisch mit Titan(III)- oder Chrom(II)-Salzen in saurem

Milieu gewinnen. Sie entstehen auch durch Disproportionierung von Phlorinen. Ihre Struktur ergibt sich zwanglos daraus, daß sie Zwischenprodukte der Reduktion von Phlorinen zu Porphinogenen sind.

Die Phlorine bereitet man spezifisch auf elektrochemischem Wege (S. 317). Zur Darstellung der Porphinogene ist Arbeiten in neutralem oder basischem Milieu zu empfehlen, da die Porphinogene in Säuren zur Spaltung und ungeordneten Resynthesen neigen (*85*).

Katalytische Hydrierung. Die katalytische Hydrierung der Porphyrine (*93*) und Chlorine mit Metallkatalysatoren wie Pt, Pd oder Ni führt zu Hexahydro-porphyrinen, nämlich Porphinogenen (*73*) und „Chlorinogenen" (*75*), deren Existenz von Fischer (*62*) nachgewiesen wurde. Chlorinogene gehen an der Luft in neutralem Milieu wieder in Chlorine über. (Zur Hydrierung in saurem Milieu vgl. S. 319.) Die sterische Verdichtung in natürlichen Chlorinen ermöglicht bei vorsichtiger Reaktionsführung die Isolierung von Chlorin-phlorinen (**76**), die zwangsläufig Zwischenprodukte der Hydrierung zu (**75**) sind. Chlorin-phlorine sind 7,8-meso-N-Tetrahydroporphine, z. B. das β-Chlorin-phlorin (**76**).

Polarographie und elektrochemische Reduktion. Die elektrochemische Reduktion der Porphyrine ist im Vergleich zu anderen Methoden deswegen reizvoll, weil man eine genau definierte Anzahl von Reduktionsäquivalenten zuführen kann und damit in der Lage ist, die Reduktion der Porphine zu Porphinogenen und Chlorinogenen in ihre Einzelschritte aufzuteilen sowie die Zwischenprodukte quantitativ zu präparieren. Weiter bleiben ungesättigte Seitenketten — etwa Vinyl- und Oxogruppen — erhalten, während bei den meisten chemischen Reduktionen die Seitenketten reduziert oder anderweitig verändert werden. Die Versuchsergebnisse hängen weitgehend davon ab, ob in protonenhaltigem Lösungsmittel oder protonenfrei gearbeitet wird.

Reduktion in protonenfreien Lösungsmitteln. Bei der Verwendung aprotischer Lösungsmittel, z. B. Dimethylformamid oder Dimethylsulfoxid mit Tetraalkylammoniumsalzen als Elektrolyt, unterbleibt naturgemäß eine Anlagerung von Protonen im potentialbestimmenden Schritt. Unter diesen Bedingungen ergeben Porphyrine, wie Tetraphenyl- und Ätioporphyrine und deren Metallkomplexe (*35, 55*) sowie aus Chlorophyll dargestellte Chlorine und Chlorin-Komplexe im Bereich bis — 2 V zwei polarographische Stufen; eine oder zwei weitere Stufen werden gegebenenfalls bei höheren Potentialen gefunden (*127, 128*). Jedoch resultieren hier weniger gut ausgebildete Grenzströme. Es kann auf verschiedenen Wegen bewiesen werden, daß die Stufen dem Übergang von jeweils einem Elektron entsprechen, so daß die zugehörigen Reduktionsprodukte Mono-, Di- und mehrfach geladene Anionen des Porphyrin- bzw. Chlorin-Systems sind.

Literaturverzeichnis: SS. 345—355

Vergleicht man die Verschiebung der Halbstufenpotentiale, die als Folge der Einführung von Komplexmetallen auftritt, so lassen sich aus den Gesetzmäßigkeiten Rückschlüsse auf die Lokalisierung der an der Kathode aufgenommenen Elektronen ziehen. Danach kann man annehmen, daß in der Regel die Aufnahme der Elektronen in das niedrigste unbesetzte Orbital des Porphyrinsystems erfolgt, nicht in ein dem Komplexmetall zugehöriges Orbital (55).

Weiterhin wurden spezielle elektrochemische Untersuchungen in aprotischem Lösungsmittel über Chlorophyll a und b durchgeführt (56).

Reduktion in protonenhaltigen Lösungsmitteln. In protonenhaltigem Lösungsmittel zeigen Porphyrine zwei bis drei, Chlorine ein bis zwei polarographische Stufen (127, 179). Durch präparative Elektrolyse dieser Substanzen beim Potential der ersten Stufe werden Dihydro-Produkte erzeugt, wie sich aus coulombmetrischen Messungen ergibt. Die aus Porphyrinen erzeugten Dihydro-Produkte sind Phlorine (58, S. 309), die aus Chlorinen erzeugten Chlorin-phlorine (76) (128), z. B. (77) (133). Geht man von einem unsymmetrischen Porphyrin aus, so entsteht bei der Elektrolyse ein Gemisch der vier isomeren Phlorine.

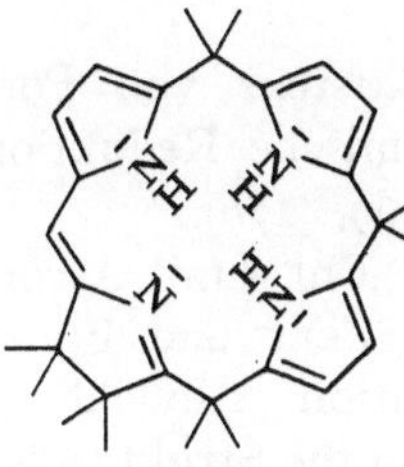

(75) 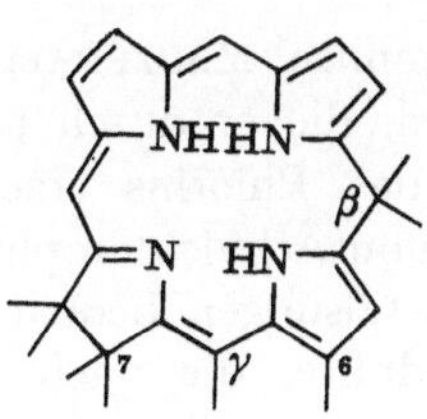   (76) 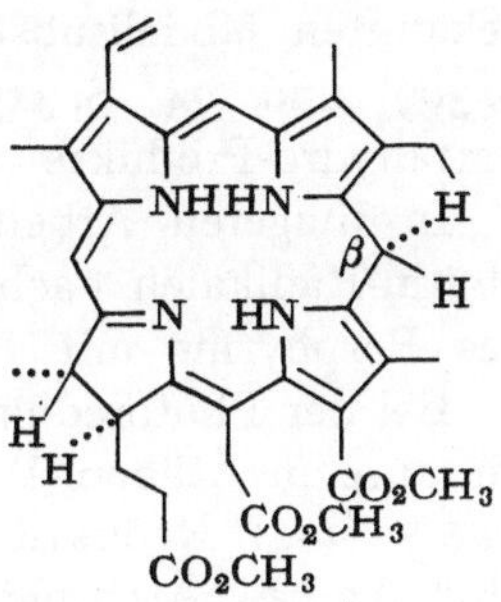   (77)

Ganz analoge Zusammenhänge bestehen bei der Chlorin-Reduktion. Während aus Verbindungen, die keinen γ-Substituenten tragen, ein Gemisch aller vier Chlorin-phlorine entsteht, wird bei γ-substituierten Chlorinen nur eine — in der Regel die β-Brücke — zur Phlorinbildung herangezogen (77). Ursache für diese Bevorzugung ist die geringere sterische Verdichtung dieser Reduktionsprodukte (130). Aus Pyrophäophorbiden entstehen — ohne Borsäure-Zusatz — sowohl α- als auch β-Chlorin-phlorine (129).

Darüber hinaus ist das elektrochemische Verhalten des Eisen-Protoporphyrin-Komplexes das Thema spezieller Untersuchungen gewesen (154).

Photochemische Reduktion. Interessante Untersuchungen zur Photoreduktion von Porphyrinen unternahm MAUZERALL (166—168). Bei der photochemischen Reduktion von Uroporphyrin und einigen verwandten Porphyrinen beobachtete er, daß, analog der chemischen, auch bei der photochemischen Reduktion diskrete Dihydro- und Tetrahydrostufen durchlaufen werden. Die dritte, der Leuko-Verbindung entsprechende Hexahydrostufe, wird nur mit speziellen Bedingungen erreicht.

Unter Belichtung reduzierten Äthylendiamintetraessigsäure, N,N,N,N-Tetramethyläthylendiamin, δ-Aminolävulinsäure, Na-dithionit und einige andere Reduktionsmittel in bestimmten pH-Bereichen nur bis zur Tetrahydrostufe. Nach Oxidation konnte das Porphyrin in hoher Ausbeute zurückgewonnen werden. Dagegen ließ die Einwirkung von Ascorbinsäure die Lösung farblos werden, erzeugte also vermutlich das Porphyrinogen. Während meist beide Reduktionsprodukte nebeneinander vorlagen, gelang es, Bedingungen auszuarbeiten, unter denen wahrscheinlich die Dihydro- und die Tetrahydro-Verbindung selektiv entstehen, und so quantitative Elektronenspektren der Dihydro- und Tetrahydro-Produkte zu erhalten. Auf dieser Grundlage konnte der Hydrierungsgrad durch spektralphotometrische Titration mit geeigneten Oxidationsmitteln mit guter Genauigkeit bestätigt werden, wobei das Porphyrin praktisch quantitativ zurückgewonnen wurde.

Gestützt auf den Vergleich seiner Versuchsergebnisse, insbesondere der Spektren, spektralen Änderungen in Abhängigkeit vom pH-Wert und der Lage der pK-Werte der Protonierungsreaktion, mit denen von bekannten Modellsubstanzen, schlug Mauzerall die Strukturen (58, S. 309) und (74, S. 315) als konjugierte Säuren für die Dihydro- und Tetrahydro-Produkte vor.

In jüngeren Arbeiten konnte Mauzerall die Existenz von Porphyrin-Radikalen nachweisen, die er durch photochemische Reduktion des Porphyrins mit Hilfe des Phlorins erzeugte (169, 170).

Bei der Photoreduktion von Äthylchlorophyllid a („Chl") mit Ascorbinsäure in Äthanol-Pyridin-Lösungen beobachteten Seely und Folkmanis (188) Reduktionsprodukte, die nach Reoxidation Äthylchlorophyllid a bzw. Äthylphäophorbid a („Phä") ergaben. Für die Struktur der Reduktionsprodukte „ChlH$_2$" (λ_{max} 522 mμ), „PhäH$_2$ 620" und „PhäH$_2$ 525" (λ_{max} 620, respektive 525 mμ) wurden auf Grund der Elektronenspektren in Verbindung mit stereochemischen Überlegungen die Formeln (78) und (79) vorgeschlagen.

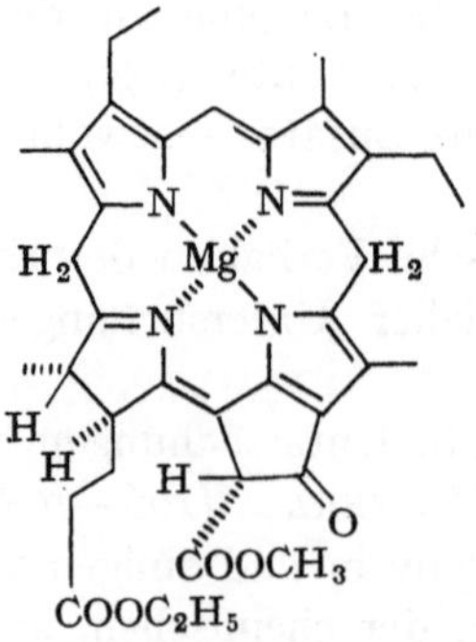

(78) 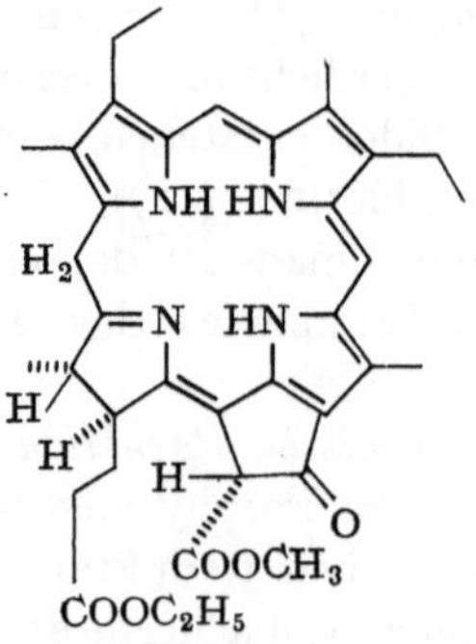   (79)

Literaturverzeichnis: SS. 345—355

„ChlH$_2$“ — das Zwischenprodukt der Krasnowski-Reaktion — ist nach dem Elektronenspektrum mit der Struktur eines β,δ-Dihydrochlorin-magnesiums (78) vereinbar. Daß „PhäH$_2$ 620“ als Chlorinphlorin (79) angesprochen wird, erscheint indessen fraglich, da keines der bekannten Chlorin-phlorine bei 620 mμ absorbiert (Chlorin-phlorin des Methylphäophorbids a: λ_{max} 480 mμ (*130*). „PhäH$_2$ 525“ wird als das metallfreie Analogon zu (78) aufgefaßt.

d) Prototrope Umlagerungen

Schon FISCHER hat prototrope Umlagerungen von Hydroporphyrinen gleicher Oxidationsstufe beobachtet und gedeutet (*88, 86*), z. B. (75a) →
→ (73a).

$$\text{(75a)} \xrightarrow{\text{H}^{\oplus}} \text{(73a)}$$

R^1 oder R^2 = CO · R oder C≡N; R^3 = Alkyl oder substit. Alkyl

Danach handelt es sich um eine säurekatalysierte, doppelte Allylverschiebung unter Ausbildung eines neuen aromatischen Pyrrolringes. Diese Umlagerung dürfte der lange bekannten, zunächst paradox erscheinenden Überführung von Chlorinen in Porphyrine mit Reduktionsmitteln wie Jodwasserstoff oder katalytisch erregtem Wasserstoff in Eisessig (*62*) oder Eisen in Ameisensäure (*79, 138*) zugrundeliegen. Der Prozeß ist an das Vorhandensein einer Oxo- oder Nitrilfunktion in 2- oder 6-Stellung gebunden. Vinylgruppen sind nicht erforderlich, werden aber meistens hydriert. Der Übergang von der Chlorin- in die Porphyrinreihe findet also auf der Hexahydrostufe statt; die Reoxidation des Porphinogens (73a) gibt ein Porphin. Die hier formulierte Wasserstoffaddition hat gegenüber der von SEELY (*185*) diskutierten HJ-Addition einen größeren Anwendungsbereich.

Prototrope Umlagerungen lagen auch auf dem Wege der WOODWARDschen Chlorophyll-Synthese (S. 291). Durch neuere Ergebnisse (*38, 120, 121*) wird ferner nahegelegt, daß die von WOODWARD postulierte Phlorin-Chlorin-Umlagerung (*214*) sich eher als α,γ-Dihydroporphin-Chlorin-Umlagerung (in Form gewisser Metallchelate) verwirklichen läßt. Diese gelingt offenbar auch photochemisch (*189*) und man sollte prüfen, ob ein β,δ-Dihydro-chlorin-magnesium, etwa (78), in ein Bacteriochlorophyll-Derivat zu überführen sei.

e) Dehydrierung

Während brückenhydrierte Porphyrine in der Regel durch Luft oder Jod rasch dehydriert werden, muß man bei peripher hydrierten Porphyrinen stärkere Oxidationsmittel einsetzen. Allerdings nimmt auch hier die Oxidationsanfälligkeit mit steigendem Hydrierungsgrad zu und ist von der speziellen Struktur abhängig.

So wurden in Cyclohexan an Luft und diffusem Licht folgende Halblebensdauern gefunden: *trans*-Octaäthylchlorin 73 Tage, b-Tetrahydro-octaäthylporphin 26 Tage, *cis*-Octaäthylchlorin 16 Tage und a-Tetrahydro-octaäthylporphin 4 Tage (*122*). Danach ist das *trans*-Chlorin am beständigsten.

Allgemein sinkt die Stabilität der Chlorine mit der Einführung eines Zentralmetalls (*49, 50*), steigt jedoch mit zunehmender sterischer Verdichtung, so daß Metallkomplexe nativer Chlorine luftstabil sind (*199, 164*).

Dehydrierung von Chlorinen ohne γ-Substituenten zu Porphinen ist mit folgenden Reagenzien bewerkstelligt worden: Alkali in der Hitze (*71, 72*), Silber(I)-Ionen und Kupfer(II)-Ionen. a-Tetrahydro-porphine ließen sich selektiv zu Chlorinen oxidieren mit Eisen-III-chlorid (*70*) und Trikaliumoctacyano-molybdat(V) (*182*).

Chinone hohen Redoxpotentials, etwa 2,3-Dichlor-4,5-dicyan-*p*-benzochinon („DDC") (oder 3,4,5,6-Tetrachlor-*o*-benzochinon) sind von Linstead und Eisner in die Chlorophyll-Chemie eingeführt worden (*50, 47*) und haben in der Folgezeit wiederholt Anwendung gefunden (*101, 117, 138, 184, 185, 195*).

Die Wirkung der sterischen Verdichtung am Ring III und IV zeigt sich eindrucksvoll am spezifisch bei Raumtemperatur herbeigeführten Übergang des Bacteriochlorin e_6-TME (Trimethylester) (**80**) in 2-Acetyl-2-desvinyl-chlorin e_6-TME (**82**), der seinerseits erst bei 80° zum Porphyrin dehydriert wird (*101*). Im gleichen Sinne ist die im Vergleich zum Chlorin e_6-TME (**47**, S. 304) sehr viel rascher erfolgende Dehydrierung des Rhodochlorin-trimethylesters zu deuten (*138*).

(80)    (81)

Chinone ohne elektronenziehende Substituenten bewirken nur bei Belichtung eine schnelle Dehydrierung (*187, 184, 195, 186*). Der Überführung von Chlorinen in Porphine über die Hexahydro-porphine ist schon gedacht worden (S. 319).

Literaturverzeichnis: SS. 345—355

2. Addition von Sauerstoff

a) Photo-oxidation der Chlorin-phlorine und Folgereaktionen

Die Chlorin-phlorine, die sich vom Chlorin e_6-TME (47), vom 2-Desvinyl-2-acetylchlorin e_6-TME (82) und vom 2-Desvinyl-2-acetylpyromethylphäophorbid a (81) ableiten, z. B. (77) und (83), gelöst in einem

Formelübersicht 6. Elektrolytische Phlorinbildung mit Photo-oxidation

geeigneten nucleophilen Lösungsmittel unter Borsäurezusatz, ergeben bei der Bestrahlung mit sichtbarem Licht unter Sauerstoff Produkte, die das Gerüst des Bacteriochlorins (84) enthalten („Bacterioprodukte") (*130*). Die Substituenten R^1 und R^2 an der aufgehobenen 3,4-Doppelbindung wurden unter Heranziehung physikalischer Verfahren identifiziert. Während R^2 stets OH ist, ist R^1 durch den nucleophilen Lösungsmittel-Rest gegeben, d. h. bei der Photooxidation in Benzol-Methanol ist $R^1 = OCH_3$, bei der in Dioxan-Wasser ist $R^1 = OH$ (*Formelüber-*

sicht 6). Das entstandene, chromatographisch trennbare Diastereomeren-Gemisch (**84a** und **b**) enthält die neu eingetretenen Substituenten in *trans*-Stellung.

Die Elektronenspektren des natürlichen Bacteriochlorinesters und der (partial-)synthetischen Bacterioprodukte sind in *Abb. 1* wiedergegeben.

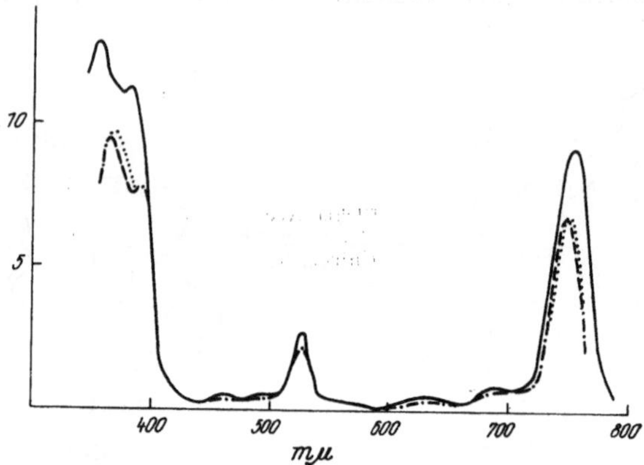

Abb. 1. Elektronenspektren des Bacteriochlorin e_6-TME (**80**) (———) und der Bacterioprodukte (**84a**) (......) und (**84b**) (— · — · —)

Die Massenspektren von (**84a** und **b**) sind so ähnlich, daß ihre Isomerie sicher ist. Die Massenzahl des Molekül-Ions ist 702, die Molmasse entspricht der Summenformel $C_{38}H_{46}N_4O_9$ (**84**). Verglichen mit dem eingesetzten Chlorin $C_{37}H_{42}N_4O_7$ (**83**) ergibt sich die Zunahme von CH_4O_2, so daß die Reaktion schematisch folgenden Verlauf nehmen dürfte:

$$C_{37}H_{42}N_4O_7 \xrightarrow[\text{Elektrolyse}]{+2H^{\oplus}+2e} C_{37}H_{44}N_4O_7 \xrightarrow[\text{Photooxydation}]{+O_2}$$

2-Acetylchlorin e_6-TME (**82**)   Chlorinphlorin (**83**)

Chlorin–phlorin   Peroxid   Epoxid   Bacterioprodukte (**84**)

Die Stellung der Substituenten an den C-Atomen 3 und 4 konnte durch massenspektrometrische Analyse geklärt werden, und zwar am Beispiel des 2-Desvinyl-2-acetyl-pyromethylphäophorbids a (**81**): Photooxidation mit $^{18}O_2$ ergibt einen um 2 Masseneinheiten erhöhten Molekularpeak (85). Der markierte Sauerstoff in der OH-Gruppe verbleibt *nach* der Methanol-Abspaltung im (M-32)-Fragment: (85). Bestrahlt man in C-monodeuteriertem Methanol (CH_2DOH), so erhöht sich die Massenzahl des Molekül-Ions um 1 Einheit (86). Jetzt ist ein intensives (M-33)-Fragment charakteristisch, das der Abspaltung von CH_2DOH entspricht: (86).

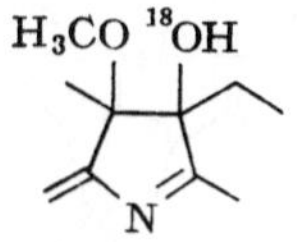

$$H_3CO\ ^{18}OH \qquad\qquad H_2DCO\ OH$$

$$(85) \qquad\qquad\qquad (86)$$

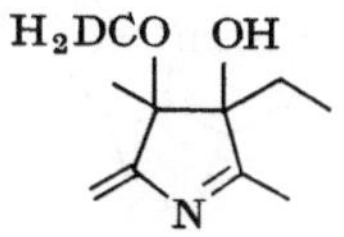

Auch die strukturelle Zuordnung der Substituenten OCH_3 und OH zu den Kohlenstoffatomen 3 bzw. 4 kann massenspektroskopisch bestimmt werden, gemäß nachstehender Fragmentierungsfolge:

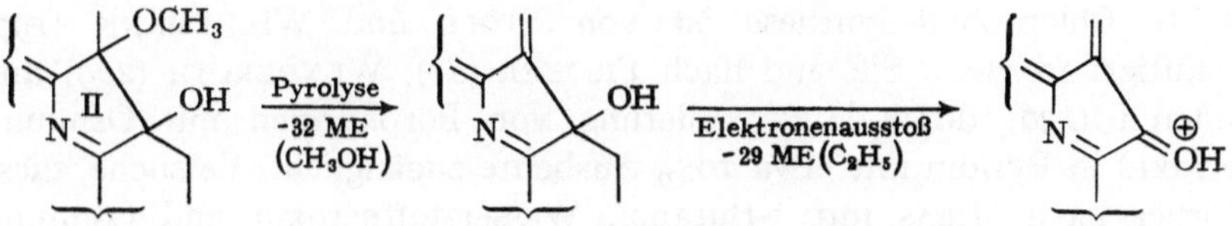

Die Photo-oxidation des Phlorins vom Chlorin e_6-TME (77) in Dioxan-Wasser ergibt so das bereits erwähnte, für die Darstellung des Rhodin g_7-TME wichtige „Bacteriodiol" (siehe S. 305).

Die Bacterioprodukte sind stark säureempfindlich und gehen bei Säureeinwirkung eine unter Rückbildung des Chlorin-Systems ablaufende Umlagerungsreaktion ein, die für $R^1 = OCH_3$ (84) die Funktionalisierung des Substituenten am C-Atom 4 ermöglicht:

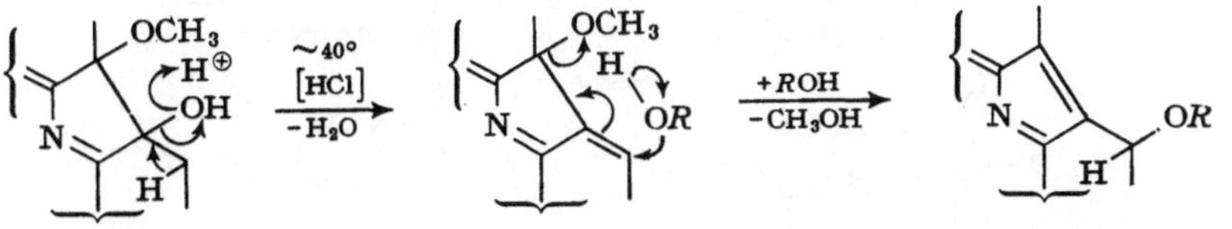

$$(84) \qquad\qquad\qquad (81a)\ R = CH_3$$
$$(81b)\ R = C_2H_5$$
$$(81c)\ R = H$$

Wird die Säure-Behandlung von (81 c) bei 90° durchgeführt, so kommt es zur Wasserabspaltung und man erhält in der 4-Stellung eine Vinyl-Gruppe (87). Ist $R = H$ (81 c), so kann die freie OH-Gruppe zu Carbonyl oxidiert werden und man erhält die 4-Acetyl-Verbindung (88). Für $R^1 = OH$ werden bei Behandlung von (84) mit Säure beide Hydroxyl-Gruppen mit nahezu gleicher Geschwindigkeit angegriffen, so daß auch die Funktionalisierung des Substituenten am C-Atom 3 möglich ist (siehe S. 304).

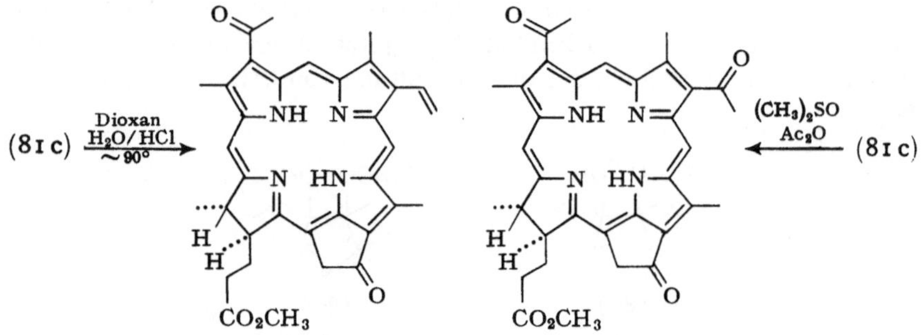

b) Hydroxylierung des Porphyrinsystems

Die Bedeutung von 1,2-Dihydroxy-chlorinen (89) für eine früher geplante Chlorophyll-Synthese ist von Stoll und Wiedemann (*197*) diskutiert worden. Sie sind nach Fischer (*81*), Wenderoth (*209*) und anderen (*136*) durch Hydroxylierung von Porphyrinen mit Osmiumtetroxid in Pyridin mit etwa 10% Ausbeute zugänglich. Versuche, diese Körper nach Milas mit t-Butanol, Wasserstoffperoxid und Osmiumtetroxid oder nach Prevost mit Perbenzoesäure zu gewinnen, waren erfolglos. Heute sind Dihydroxy-chlorine interessant, weil sie sich in starken Säuren, nach Art einer Pinakolinumlagerung in Oxo-Verbindungen verwandeln (*136*, *152*), in die sogenannten Gemini-porphyrinketone (90), die uns auf S. 337 noch beschäftigen werden.

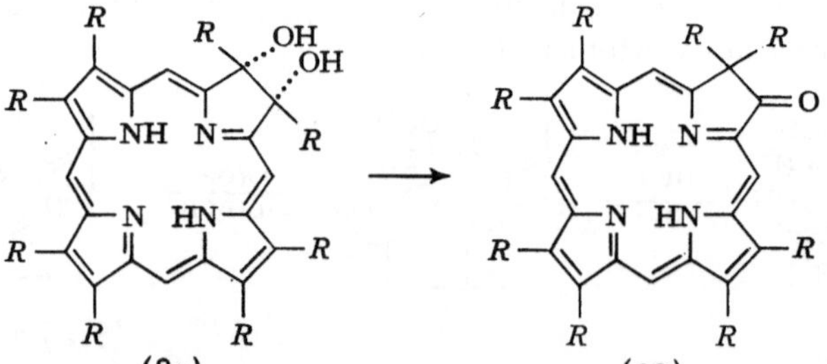

3. Addition von Kohlenstoff an Porphyrindoppelbindungen

Kenner und Mitarb. (*40*) haben durch Einwirkung von Phosphoroxidchlorid auf Acetamido-äthylporphyrine, etwa (91), in einer abgewandelten intramolekularen Vilsmeier-Acylierung das Gemini-methylen-porphyrin (92) erhalten, das zum Chlorin hydriert werden kann.

$$(91) \qquad (92)$$

Die Umsetzung von Octaäthylporphinato-zinndihydroxid (OÄP)Sn-$(OH)_2$ mit Aluminiumbromid in Chloroform ergab ein Chlorinchelat, für das die Formel (93) wahrscheinlich gemacht werden konnte (*123*). (Siehe ferner S. 312.)

$$(93)$$

VI. Substitutionsreaktionen

1. Elektrophile Substitution

Der aromatische Charakter des Porphinsystems äußert sich darin, daß in meso- und peri-Stellung durch elektrophile Substitution Deuterium, Halogene, Nitrogruppen oder Acylreste eingeführt werden können. Schon H. FISCHER und Mitarb. haben Halogenierungen, Nitrierungen und Acylierungen durchgeführt (*61, 64, 69, 73, 75, 78, 80, 84, 87, 89*). Da im Lichte neuerer physikalischer Methoden eine gründlichere strukturelle Sicherung der Reaktionsprodukte allgemein wünschenswert erschien, wurden die Arbeiten von vielen Seiten wieder aufgenommen, und zwar hauptsächlich an vier Substrattypen, nämlich symmetrischen Octa-alkylporphinen (94) und unsymmetrischen Chlorinen natürlicher Herkunft (96) mit unsubstituierten meso-C-Atomen sowie Deuteroporphyrin

IX-DME (95a) (Dimethylester) und 2-Desvinyl-isochlorin e_4-DME (97a). Die letzten beiden unsymmetrischen Stoffe haben freie meso- und peri-Stellungen, deren unterschiedliche Neigung zur elektrophilen Substitution hier gut untersucht werden kann. Wichtigstes Mittel zur Konstitutionsaufklärung ist stets die ^{1}H-Resonanzspektroskopie (siehe S. 306).

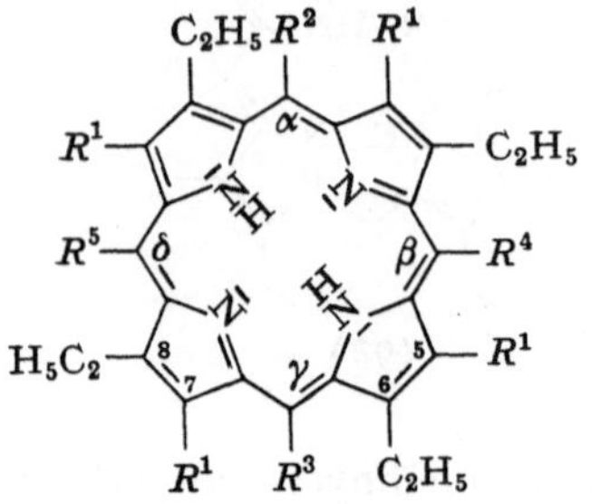

(94)

(94a) $R^1 = C_2H_5$; $R^{2,3,4,5} = H$

(94b) $R^1 = C_2H_5$; $R^{2,3} = CH_3$; $R^{4,5} = H$

(94c) $R^1 = CH_3$; $R^2 = NH_2$; $R^{3,4,5} = H$

(94d) $R^1 = CH_3$; $R^{2,3,4,5} = H$

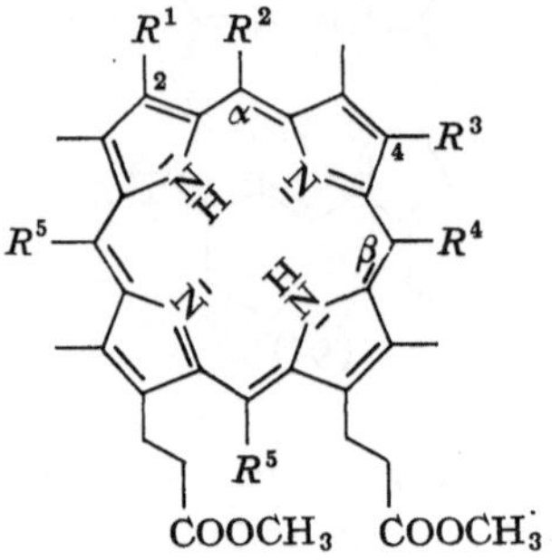

(95)

(95a) $R^{1,2,3,4,5} = H$

(95b) $R^{1,3} = C_2H_5$; $R^{2,4,5} = H$

(95c) $R^1 = CH_3CO$; $R^{2,3,4,5} = H$

(95d) $R^3 = CH_3CO$; $R^{1,2,4,5} = H$

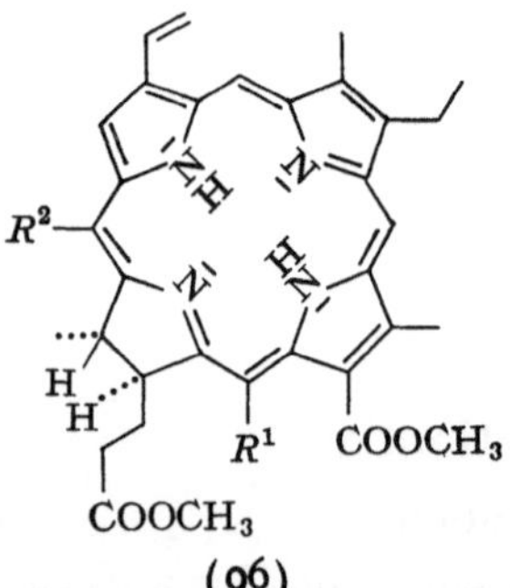

(96)

(47) $R^1 = CH_2COOH_3$; $R^2 = H$

(96a) $R^1 = H$; $R^2 = H$

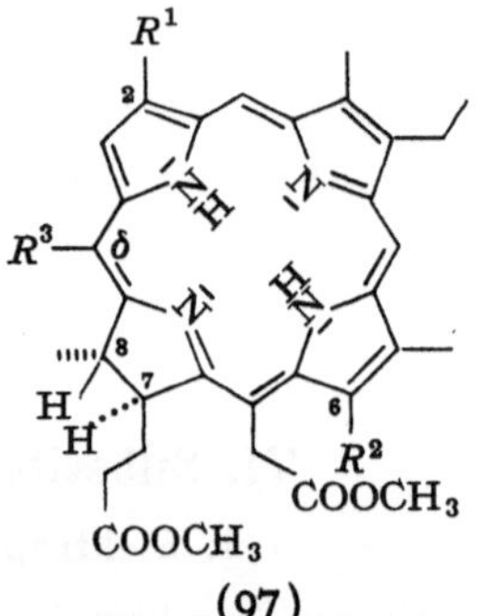

(97)

(97a) $R^{1,2,3} = H$

(97b) $R^{1,3} = H$; $R^2 = CH_3CO$

(97c) $R^{2,3} = H$; $R^1 = CH_3CO$

(97d) $R^3 = H$; $R^{1,2} = CH_3CO$

(97e) $R^1 = C_2H_5$; $R^2 = COOCH_3$; $R^3 = H$

(97f) $R^1 = C_2H_5$; $R^2 = CH_3CO$; $R^3 = H$

Kürzlich wurde auch ein Angriff auf ein endo-C-Atom beobachtet. Bei der Einwirkung von Äthoxycarbonylnitren auf Porphine fand Einschiebung des Nitrens zwischen ein meso- und endo-C-Atom statt. Das Ringerweiterungs-Produkt erlitt beim Erhitzen Ringverengung unter Ausbildung eines meso-Äthoxycarbonylamino-porphins (*106*).

Literaturverzeichnis: SS. 345—355

a) Deuterierung

Die einfachste elektrophile Substitution, der Austausch von Wasserstoff gegen Deuterium mittels deuteronen-spendender Säuren, zeigt bei Porphinen, Chlorinen und Tetrahydroporphinen einen unterschiedlichen Verlauf. Einige Regelmäßigkeiten lassen sich aus *Tabelle 1* entnehmen. Mit zunehmendem Hydrierungsgrad wächst die Elektronendichte an den austauschbereiten meso-C-Atomen (S. 307), und dementsprechend werden die zum Austausch erforderlichen Reaktionsbedingungen immer milder.

Die stark elektronenschiebende Aminogruppe im α-Amino-ätioporphyrin I (94c) bewirkt eine mindestens ebenso große, wenn nicht größere Beschleunigung (*152*). Umgekehrt vermögen elektronenanziehende Acetylreste je nach Lage und Anzahl die Deuterierung zunehmend zu hemmen (*132, 131*); die Austauschgewindigkeit sinkt bei 2-Desvinyl-isochlorin e_4-DME (97a) und seinen Acetylderivaten (97b)—(97d) in der Reihenfolge (97a) > (97b) > (97c) > (97d). Bemerkenswert ist, daß hier stets die δ-Protonen ersetzt werden, während C-2 und C-6 unangetastet bleiben.

Auch sterische Effekte scheinen eine Rolle zu spielen. So ist beim α,γ-Decamethylporphin (*135*) und beim α,γ-Dimethyl-octaäthylporphin (94b) der Austausch der β,δ-Protonen schon bei Raumtemperatur feststellbar und läßt sich bei 55° in wenigen Stunden vervollständigen. Der Übergang von C-β und C-δ in die sp³-Hybridisierung des σ-Übergangs-Komplexes der Deuterierung sollte nach den Erfahrungen an Chlorin-phlorinen (*130*) auch C-α und C-γ von Raumspannung entlasten. Natürlich kann der induktive Effekt der CH_3-Gruppen erleichternd wirken, jedoch sind solche Störungen in die bisher vorliegenden MO-Berechnungen (siehe S. 307) nicht einbezogen.

b) Halogenierung

Im Gegensatz zur Deuterierung ist das Reaktionsgeschehen bei der Halogenierung weniger übersichtlich. In der Regel wird ein Gemisch von Wasserstoffperoxid und Halogenwasserstoffsäure in wäßrig-organischer Phase eingesetzt, wenn man zu meso-Mono- und -Dihalogen-Derivaten kommen will, während ein solches Gemisch in Eisessig glatt zu meso-Tetrachlorporphinen führt. Mit dem letztgenannten Ergebnis wurde Sulfurylchlorid in Chloroform angewendet. Einige Beispiele sind in *Tabelle 2* dargestellt (dort Literaturhinweise). Bei den Chlorinen erfolgt die Halogenierung ausschließlich in γ,δ-Stellung; auch ein γ-Brom-octaäthylchlorin ist zugänglich. Dieses konnte mit Dichlor-dicyan-benzochinon (siehe S. 320) zu meso-Brom-octaäthylporphin dehydriert werden, während die direkte Bromierung an Octaalkylporphinen noch nicht aufgeklärt ist (*17*). Deuteroporphyrin IX (95a, S. 326) läßt sich dagegen glatt in 2,4-Dibrom-deuteroporphyrin überführen.

Auffällig ist die leichte Weitersubstitution: Ohne Schwierigkeiten werden vier Halogenatome eingebaut. Dafür wird die abnehmende Basizität der substituierten Porphine oder ein anderer Reaktionsmechanismus (z. B. Additions-

Tabelle 1. Deuterierungsreaktionen (Austausch H—D)

Eingesetzte Substanz	Formel	Produkt					Agens	Temp.	Zeit[a]	Literatur
		R^1	R^2	R^3	R^4	R^5				
Octaäthylporphin	(94a)	C_2H_5	D	D	D	D	CF_3COOD	110°	72 Stunden[b]	(18)
							D_2SO_4	20°	einige Stunden	
α,γ-Dimethyloctaäthylporphin ...	(94b)	C_2H_5	CH_3	CH_3	D	D	CF_3COOD	55°	5 Stunden	(121)
meso-Amino-ätioporphyrin I	(94c)	CH_3	NH_2	D	H	H	CF_3COOD	20°	20 Minuten	(151)
		CH_3	NH_2	D	D	D	CF_3COOD	100°	1 Stunde	
Octaäthylchlorin (7,8-Dihydro-94a)		C_2H_5	H	D	H	D	CF_3COOD	50°	200 Minuten[b]	(18)
		C_2H_5	D	D	D	D	CF_3COOD	100°	640 Minuten[b]	
Chlorophyll a (Austausch des δ-Protons)	(24)	—	—	—	—	—	CD_3OD/THF	38°	100 Stunden	(45)
Chlorin e_6-TME	(47)	E^e	D	—	—	—	CH_3COOD	80°	2 Stunden	(216)
Rhodochlorin-DME	(96a)	D	D	—	—	—	CH_3COOD	80°	4 Stunden	(216)
2-Desvinyl-isochlorin e_4-DME....	(97a)	H	H	D	—	—	CH_3COOD	65°	2,5 Stunden	(131)
6-Acetyl-2-desvinyl-isochlorin e_4-DME	(97b)	H	CH_3CO	D	—	—	CH_3COOD	85°	2,5 Stunden	(132)
2-Acetyl-2-desvinyl-isochlorin e_4-DME	(97c)	CH_3CO	H	D	—	—	CH_3COOD	85°	4 Stunden[c]	
2,6-Diacetyl-2-desvinylisochlorin e_4-DME	(97d)	CH_3CO	CH_3CO	D	—	—	CH_3COOD	85°	8 Stunden[d]	
a-Tetrahydro-octaäthylchlorin (5,6,7,8-Tetrahydro-94a)		C_2H_5	H	D	H	H	CF_3COOD	50°	9 Minuten	(18)
			D	D	D	D	CF_3COOD	50°	37 Minuten	

[a] Zeit bis zum vollständigen Austausch;   [b] Halbwertzeit, durch kinetische Messung ermittelt;   [c] Austausch etwa 40%;   [d] Austausch etwa 60%;   [e] E = CH_2COOCH_3; TME = Trimethylester;   DME = Dimethylester.

Tabelle 2. Halogenierung von Porphyrinen

Substanz	Formel	Produkt					Bedingungen	Ausbeute	Literatur
		R^1	R^2	R^3	R^4	R^5			
Octaäthylporphin	(94a)	C_2H_5	Cl	H	H	H	$HCl/H_2O_2/THF$	24%	(17)
		C_2H_5	Cl	Cl	H	H		20%	
		C_2H_5	Cl	Cl	Cl	Cl	$CH_3COOH/HCl/H_2O_2$	43%	
Octaäthylchlorin (7,8-Di-hydro-94a)		C_2H_5	H	Cl	H	H	$HCl/H_2O_2/THF$	21%	(17)
		C_2H_5	H	Cl	H	Cl		22%	
		C_2H_5	H	Br	H	H	$HBr/H_2O_2/THF$	40%	
Ätioporphyrin I	(94d)	CH_3	Cl	H	H	H	$POCl_3/(CH_3)_2N \cdot CHO$	16%	(153)
Mesoporphyrin IX	(95b)	C_2H_5	Cl	C_2H_5	Cl	Cl	$SO_2Cl_2/CHCl_3$	44%	(141)
„Propylporphin"[b]			Cl	Cl	Cl	Cl	$SO_2Cl_2/CHCl_3$	10%	(141)
(Propylporphin)Cu[a]			Cl	Cl	Cl	Cl	$SO_2Cl_2/CHCl_3$	35%	(141)
Deuteroporphyrin IX	(95a)	Br	H	Br	H	H	N-Brompyridiniumperbromid	45%	(32)
Mesochlorin e_6-TME	(97e)	C_2H_5	$COOCH_3$	Cl	—	—	$HCl/H_2O_2/Dioxan$	49%	(131)
6-Acetyl-isochlorin e_4-DME ...	(97f)	C_2H_5	$COCH_3$	Cl	—	—	$HCl/H_2O_2/Dioxan$	48%	(131)

[a] Hier verläuft die Reaktion unter Entmetallierung; [b] 1,3,5,8-Tetramethyl-2,4-diäthyl-6,7-di(n-propyl)-porphin, R^2 bis R^5 wie in (94); DME, TME siehe Tabelle 1.

Eliminierungs-Mechanismus) verantwortlich gemacht (*18*). Zumindest bleibt die Frage nach dem Mechanismus der Halogenierung schon bei der Vielfalt der Agenzien offen. Die klassischen Halogenierungskatalysatoren, wie $FeCl_3$, verwendet man ungern, da sie in das Porphyrinsystem unter Chelatbildung eingebaut werden. Bei einer nachfolgenden Entmetallierung im sauren Milieu besteht die Gefahr, daß auch das Halogen verdrängt wird (*131, 78*).

Die Halogen-porphine haben charakteristische Elektronenspektren und bilden Metallkomplexe.

c) Nitrierung

Die Nitrierung des Porphyringerüstes ist neuerdings von Bonnett (*19*) und von Johnson (*152*) näher untersucht worden. Abgesehen von der glatten Darstellung des Mononitro-octaäthylporphins (siehe *Tabelle 3*) entstehen Gemische, in denen je nach Reaktionsbedingungen bestimmte Produkte überwiegen. Auch hier werden Chlorine ausschließlich an C-γ und C-δ nitriert. Die Nitrierung über die Mononitrostufe hinaus ist nicht schwierig, allerdings wurde kein Tetranitro-Derivat erhalten. Überraschend ist die schon auf Fischer (*75*) zurückgehende Feststellung, daß Deuteroporphyrin IX an C-α und C-β substituiert wird, während die Bromierung in 2- und 4-Stellung stattfindet.

Durch Nitrierung mit Kupfer(II)-nitrat in Acetanhydrid wurde aus Ätioporphyrin I direkt Nitro-ätioporphinato-kupfer bereitet (*151*). α-Nitro-octaäthylporphin wurde mit $SnCl_2$/HCl in 46%iger Ausbeute zu α-Amino-octaäthylporphin reduziert (*19*). Auch α-Amino-ätioporphyrin I ist so erhältlich und wurde in das Kupferchelat sowie einige andere Derivate überführt (*151—153*). Mit salpetriger Säure entstand ein nicht näher charakterisiertes Nitroprodukt, wahrscheinlich α-Amino-γ-nitro-ätioporphyrin I.

d) Acylierung

Die Acetylierung des Deuterohämins IX (S. 302) bildete den letzten Aufbauschritt in H. Fischers Totalsynthese des Hämins. Wegen der Eignung von Formyl- und Acetylresten als Anknüpfungspunkte für weitere Synthesen ist ihre Einführung von aktuellem Interesse. Sie geschieht zweckmäßig durch Einwirkung von Säurechloriden, -amiden und -anhydriden auf Metallporphyrine mit Kupfer(II), Nickel(II) und Eisen(III) als Zentralmetallen. Porphyrine ohne Zentralmetall sind für die Acylierung ungeeignet, eigenartigerweise auch Kobalt(II)-chelate, schließlich Zinn(IV)-chelate. Als Acylierungs-Katalysatoren dienen Lewis-Säuren, die das Zentralmetall nicht verdrängen. Die *Tabellen 4* und *5* bringen neuere Beispiele. Zu den schon von H. Fischer benutzten Reagenzien Acetanhydrid und (Dichlormethyl)-äthyläther mit Zinnhalogeniden als Katalysatoren gesellte sich das Vilsmeier-Reagens aus Dimethylformamid und Phosphoroxidchlorid (*125, 153*). Übersichtlich ver-

Tabelle 3. Nitrierung des Porphyrinsystems

Substanz	Formel	Hauptprodukte					Bedingungen	Temp.	Zeit	Ausbeute	Literatur
		R^1	R^2	R^3	R^4	R^5					
Octaäthylporphin ..	(94a)	C_2H_5	NO_2	H	H	H	HNO_3/CH_3COOH	$0°$	1,5 Minuten	92%	(19)
Octamethylporphin statt Ät: CH_3, sonst wie	(94a)	CH_3	NO_2	H	H	H	konz. HNO_3	$22°$	45 Sekunden	34%	(19)
Octaäthylporphin ..	(94a)	C_2H_5	NO_2	H	NO_2	H	konz. HNO_3	$22°$	12 Minuten	13%	(19)
		C_2H_5	NO_2	NO_2	H	H				26%	
		C_2H_5	NO_2	NO_2	NO_2	H	HNO_3/H_2SO_4	$0°$	1 Minute	22%	
Octaäthylchlorin		C_2H_5	H	NO_2	H	H	$(NO_2)(BF_4)$	$26°$	10 Minuten	60%	(17)
(7,8-Dihydro-94a)		C_2H_5	H	NO_2	H	NO_2	in Sulfolan	$31°$	2 Stunden	44%	(19)
Ätioporphyrin I	(94d)	CH_3	NO_2	H	H	H	63% H_2SO_4 + 25% HNO_3	$12°$	30 Minuten	55%	(151)
Ätioporphyrin I	(94d)	CH_3	NO_2	NO_2	NO_2	H	konz. HNO_3	$20°$	36 Stunden	13%	(152)
Deuteroporphyrin IX-dimethylester	(95a)	H	NO_2	H	NO_2	H	HNO_3/H_2SO_4 $(\sim 3 + 100)$	$0°$	7 Minuten	27%	(32)

Tabelle 4. Acylierung des Porphyrinsystems

Substanz	Produkt					Bedingungen	Zeit Temp.	Ausbeute	Literatur
	R^1	R^2	R^3	R^4	R^5				
Octaäthylporphin-Cu s. (94)	C_2H_5	CHO	H	H	H	$(CH_3)_2N\text{-}CHO/POCl_3$	20 Minuten 55°	77%	(125)
Ätioporphin I-Ni s. (94)	CH_3	CHO	H	H	H	$(CH_3)_2N\text{-}CHO/POCl_3$	20 Minuten 55°	84%	(153)
Deuteroporphyrin IX-DME-Cu s. (95)	$COCH_3$	H	H	H	H	$C_6H_6/(CH_3CO)_2O/SnCl_4$	1 Minute 0°	31%	(119)
	H	H	$COCH_3$	H	H				
	$COCH_3$	H	$COCH_3$	H	H	$CH_2Cl_2/(CH_3CO)_2O/SnCl_4$	1,5 Minuten 0°	58%	(119)
Deuteroporphyrin IX-FeCl s. (95)	COC_2H_5	H	COC_2H_5	H	H	$(C_2H_5CO)_2O/SnCl_4$	12 Minuten 0°	37%	(32)
1,4,5,8-Tetramethyl-2-carbäthoxy-porphin-Cu (98a)	—CHO	H	—	—	—	$Cl_2CHOC_2H_5/SnCl_4$	2 Minuten 0°	63%	(177)
2-Desvinyl-isochlorin e_4-DME-Cu (99)	H	$COCH_3$	—	—	—	$(CH_3CO)_2O/SnCl_2$	60 Minuten 0°	68%	(132)
	$COCH_3$	H	—	—	—	$(CH_3CO)_2O/SnCl_2$; H_2O	3 Minuten 100°	94%	(132)
	$COCH_3$	$COCH_3$	—	—	—	$(CH_3CO)_2O/BF_3\text{-}O(C_2H_5)_2$	5 Minuten 100°	90%	(132)
Phylloporphyrin-FeCl (34)	—CHO an C-6					$Cl_2CHOC_2H_5/SnBr_4$	10 Minuten	40—50%	(80)

läuft die Reaktion am Octaäthylporphin-kupfer und Ätioporphyrin I-nickel, da hier nur ein Monosubstitutionsprodukt auftreten kann (Tabelle 4).

Kam bei diesen peripher vollständig substituierten Porphinen als Ort der Substitution nur die meso-Position in Frage, so waren beim Deuteroporphyrin mehrere Mono- und Disubstitutionsprodukte zu erwarten. Es wurden aber nur die in Tabelle 5 aufgeführten Kombinationen der an C-2, C-4, C-α und C-β formylierten Deuteroporphyrine gefunden, die sich nach Abspaltung des Kupfers durch Säulenchromatographie an feinem Kieselgel trennen ließen. Mit zunehmender Temperatur wurden auch Diformylprodukte beobachtet. 2- und 4-Monoformyl-deuteroporphyrin IX-DME wurden nach der Trennung in gleicher Ausbeute erhalten. Auch bei anderen Reaktionen wurde eine gleiche Reaktivität der Ringe I und II des Deuteroporphyrins beobachtet (*119, 32*).

Tabelle 5. Vilsmeier-Formylierung des (Deuteroporphyrin IX-DME)Cu
(*119*)

Reaktionsbedingungen			Ausbeute (in % der Theorie nach Entmetallierung)				Ausgangs-material %
			Monoformylprodukt		Diformylprodukt		
Zeit	Temp.	Lösm.	$(\alpha\sim + \beta\sim)$	$(2\sim + 4\sim)$	$2,\beta\sim$	$2,4\sim$ *	
2 Stunden	0°	$CHCl_3$	21	27	—	—	10
2 Stunden	30°	$CHCl_3$	35	22	—	—	11
2 Stunden	60°	$CHCl_3$	16	6	6	10	10
2 Stunden	80°	$Cl(CH_2)_2Cl$	19	5	7	10	5

* $\sim$ = Formyldeuteroporphyrin.

Die Acetylierung des Deuteroporphyrin-kupfers mit Acetanhydrid/$SnCl_4$ in Benzol milderte die Reaktionsbedingungen derart, daß neben etwas Diacetyl-Verbindung nur zwei Monoacetyl-Produkte im Verhältnis 1 : 1 anfielen (s. Tabelle 4). 2-Acetyl- und 4-Acetyl-deuteroporphyrin IX-DME (**95c**) und (**95d**, S. 326) konnten nach der Entmetallierung chromatographisch an feinem Kieselgel getrennt und durch ihre NMR-Spektren strukturell gesichert werden. (**95d**) dürfte sich durch Reduktion der Oxogruppe und Dehydratisierung im Hochvakuum in Pemptoporphyrin (**41a**, S. 333) überführen lassen.

Nach all diesen Befunden nimmt es nicht wunder, daß FISCHER bei der Gewinnung von Monoformyl-deuteroporphyrin ein Gemisch der 2- und 4-Formylprodukte erhielt, wie er selbst erkannt hat (*91*).

Eine der Acylierung nahestehende Reaktion ist die zum 2,4-Bis(hydroxymethyl)-deuteroporphyrin IX führende Umsetzung des Deuterohämins mit Chlormethylmethyläther $ClCH_2OCH_3$ und Zinntetrachlorid (*82*).

Die dirigierende Wirkung einer Carbäthoxygruppe wurde am (1,4,5,8-Tetra-methyl-2-carbäthoxy-porphin)kupfer (98a) von Ponomarev und Mitarb. (*177*) studiert. Hier wurde das entlegene C-6 bevorzugt substituiert, es entstand haupt-sächlich (98b) neben geringen Mengen von (98c).

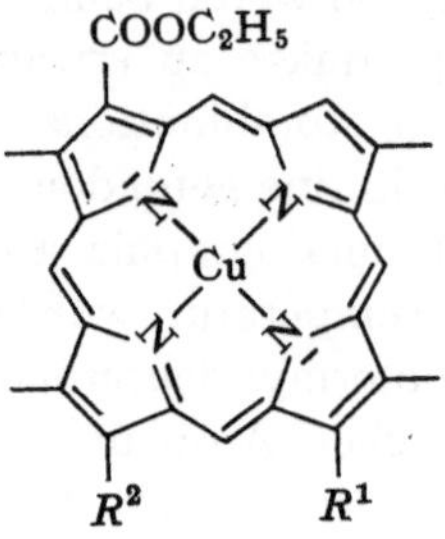

COOC$_2$H$_5$

R^1

Cu

Cu

H

H

R^2 R^1

R^2
COOCH$_3$

COOCH$_3$

(98)

(99)

(98a) $R^{1,2}$ = H

(98b) R^1 = CHO; R^2 = H

(98c) R^1 = H; R^2 = CHO

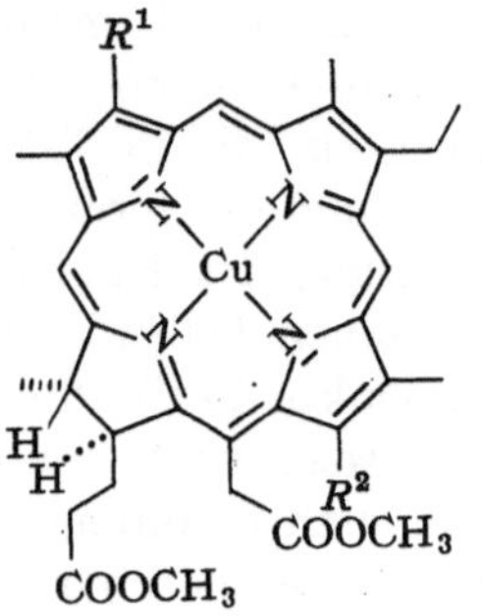

(99a) $R^{1,2}$ = H

(99b) R^1 = H; R^2 = COCH$_3$

(99c) R^1 = COCH$_3$; R^2 = H

(99d) $R^{1,2}$ = COCH$_3$

Die schon betonte, im Vergleich zum Porphyrinsystem größere Neigung des Chlorinsystems (99) zu elektrophilen Substitutionen äußert sich ein-mal darin, daß schwächere Lewis-Säuren, etwa SnCl$_2$, die Acetylierung bereits herbeiführen (*65*), zum anderen in der Weise, daß gewisse Acetyl-gruppen auch wieder durch Protonen verdrängt werden können. Die Acetylierungs- und Entacetylierungsgleichgewichte zwischen (99a), (99b), (99c) und (99d) unter verschiedenen Bedingungen (Temperatur, Zeit, Konzentration und Wassergehalt) wurden ermittelt und hierbei optimale Bedingungen für die Gewinnung jedes der drei Acetylchlorine festgestellt (Tabelle 4, S. 332). Die Substitution an C-6 ist reversibel, an C-2 jedoch nicht (*116, 132*).

Meso-Formylgruppen sind nicht sehr haftfest, was bei Umsetzungen an meso-Formylporphyrinen bisweilen stört (*125, 139*).

2. Nucleophiler Angriff

Nach Pullman (*178*) ist am Porphyrinsystem die Lokalisierungs-energie L (S. 307) für eine nucleophile Substitution in den meso-Posi-tionen kleiner als für eine elektrophile Substitution. Es ist sonderbar, daß ein nucleophiler Angriff auf das Porphinsystem selten diskutiert wird, obwohl er hiernach begünstigt wäre. Tatsächlich hat Woodward (*214*) eine nucleophile Addition nach vorausgegangener Protonisierung eines Porphyrins (100) beobachtet.

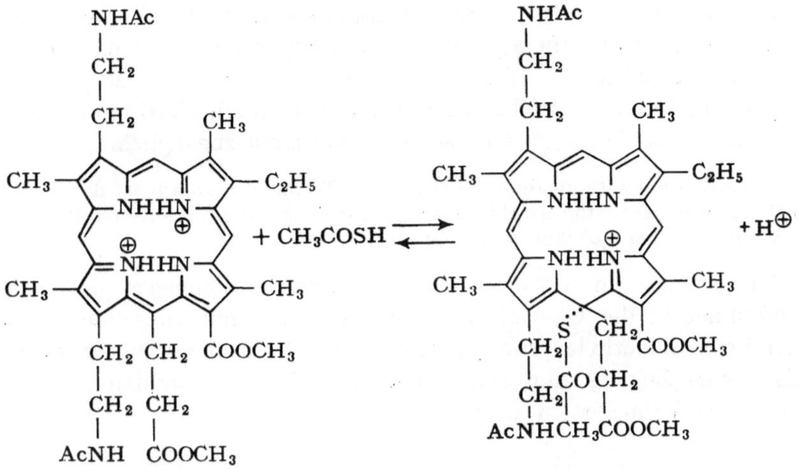

(100)

VII. Einführung von Oxofunktionen

1. Xanthoporphinogene

Die $\alpha,\beta,\gamma,\delta$-Tetraoxoporphinogene wurden zuerst von FISCHER und seinen Mitarbeitern bei Oxydationsversuchen mit Bleidioxid dargestellt (90). Es gelang, in den erhaltenen Reaktionsprodukten, die wegen ihrer intensiv gelben Farbe als Xanthoporphinogene bezeichnet wurden, einen

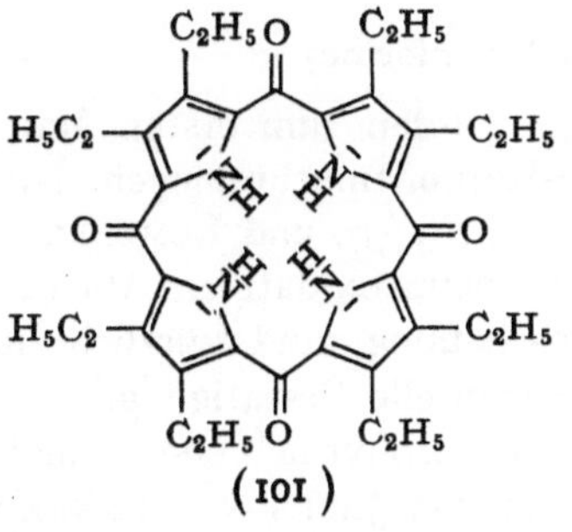

(101) 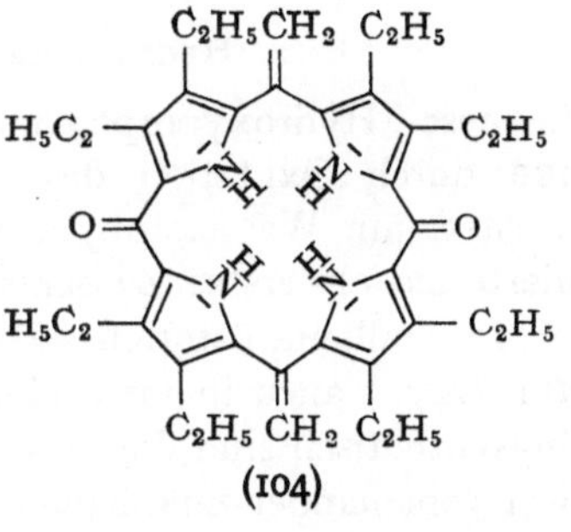 (104)

Mehrgehalt von vier Sauerstoffatomen pro Molekül festzustellen; jedoch versagten alle klassischen Methoden weiterer Strukturbestimmung, speziell alle Keton-Nachweisreaktionen. Die Reaktion wurde neuerdings untersucht (124). Mit Hilfe hauptsächlich von NMR- und Massenspektren konnte den Xanthoporphinogenen eindeutig die meso-Tetraoxoporphyrinogen-Struktur zugeordnet werden, z. B. entstand aus Octaäthylporphin das meso-Tetraoxo-octaäthylporphinogen (101).

Die Reduktion des $\alpha,\beta,\gamma,\delta$-Tetraoxo-octaäthylporphinogens mit Zink/Eisessig führt zum α,γ-Dioxo-octaäthylporphinogen (**102**), das bei der Luftoxidation in α,β,γ,N-Tetrahydro-α,γ-dioxo-octaäthylporphin (**103**) übergeht. Das gleiche Diketon ist auch durch Natriumborhydrid-Reduktion des Xanthoporphinogens in Methanol zugänglich.

Reduktion mit Lithiumaluminiumhydrid in Tetrahydrofuran führt das Xantho-porphinogen ins Octaäthylporphin über. Behandlung mit Bromwasserstoff/Eisessig im Bombenrohr bei 150° läßt das Oxophlorin (**106**) entstehen.

Ausgehend vom $\alpha,\beta,\gamma,\delta$-Tetraoxo-octaäthylporphinogen gelang mit Lithiummethyl die Überführung von zwei „gegenüberliegenden" Oxo-Funktionen in Methylengruppen (**104**). Diese Reaktion eröffnet möglicherweise einen Zugang zum ringsynthetisch schwer darstellbaren α,γ-Dimethylporphyrin-System (*135*).

(102)   (103)

2. Hydroxyporphine (Oxophlorine)

Die meso-,,Hydroxyporphyrine" (**105**) wurden zum ersten Mal von Fischer durch Oxidation des Pyridin-Koprohämochromogen I-tetra-methylesters mit Wasserstoffperoxid erhalten (*163*), und Lemberg (*162*) postulierte sie als erstes Zwischenprodukt beim oxidativen Abbau von Hämin zu Gallenfarbstoffen. Diese Anschauung fand kürzlich durch Kenner (*147a*) auch in vitro eine experimentelle Bestätigung.

Ringsynthetisch sind die meso-,,Hydroxyporphyrine" erst neuerdings auf zwei voneinander unabhängigen Wegen von Jackson, Kenner und Mitarb. (*145*) sowie von Clezy und Nichol (*37*) dargestellt worden. Sie entstehen auch bei der partiellen Reduktion des Xanthoporphino-gens (**101**) mit HBr-Acetanhydrid unter Druck (*124*). Die meisten bekannten meso-,,Hydroxyporphyrine" liegen in der tautomeren Keto-form vor, so daß sie korrekterweise als Oxophlorine (**106**) zu bezeichnen sind. Lediglich ihre Metallkomplexe leiten sich von der Hydroxyporphy-rin-Struktur (**105**) ab. Es ist bemerkenswert, daß einige Oxophlorine unter gewissen Bedingungen paramagnetisch sind (*36, 147, 142*).

Literaturverzeichnis: SS. 345—355

Die Darstellung eines „echten", metallfreien meso-Hydroxyporphyrins ist möglich, wenn man die meso-Hydroxygruppe durch intramolekulare Wasserstoff-brückenbindung fixiert. Es konnte gezeigt werden (*126*), daß sich eine peri-ständige Oxo-Gruppe hierfür eignet.

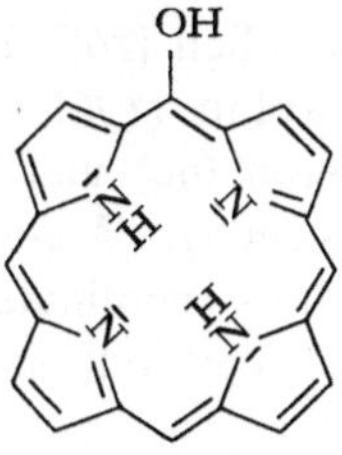

(105) 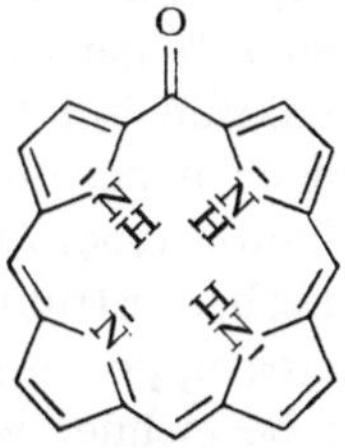   (106)

Geht man vom Geminiporphyrin-monoketon (**90**) aus, so ergibt die Behandlung mit Bleitetraacetat ein Gemisch dreier isomerer meso-Acetoxy-Verbindungen, die sich durch saure Hydrolyse in die entsprechenden meso-„Hydroxyporphyrine" umwandeln lassen. Mit Hilfe physikalischer Befunde, insbesondere durch Vergleich der NMR-Spektren der isomeren Reaktionsprodukte mit dem des Geminiporphyrin-monoketons, konnten die Strukturen festgelegt werden. Während zwei Produkte als β- und γ-Oxophlorine identifiziert wurden, erwies sich das dritte als δ-Hydroxy-gemini-porphyrinmonoketon (**107**).

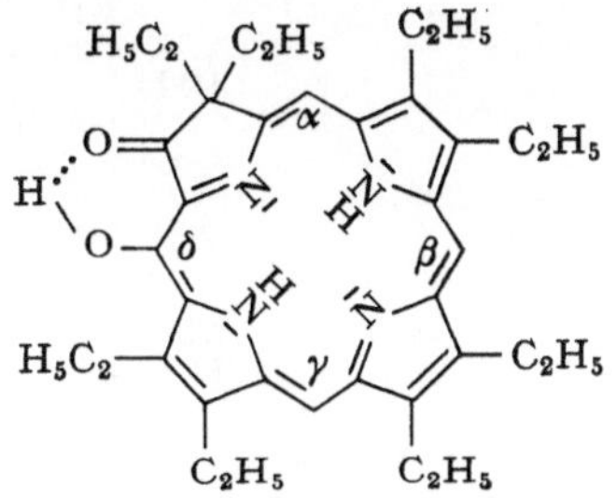

3. Geminiporphyrin-ketone

Die Einführung von Oxogruppen in die peri-Stellungen des Porphyrinsystems geht auf Reaktionen von FISCHER und Mitarb.

(107)

zurück. Bei der Einwirkung von konz. Schwefelsäure und Wasserstoffper-oxid auf Porphyrine erhielten sie Verbindungen, die um ein Sauerstoffatom reicher waren als die Ausgangsprodukte und die von ihnen für Epoxide gehalten wurden (*63*). Zu den gleichen, als Epoxide angesprochenen Ver-bindungen gelangten FISCHER und Mitarb. (*91*), wenn sie Porphyrine, bei-spielsweise mit Osmiumtetroxid, hydroxylierten und die Hydroxy-porphyrine mit konz. Schwefelsäure einer Wasserabspaltung unter-warfen (*81*). Diese Reaktionen konnten am Beispiel des Octaäthyl-porphins (**94a**, S. 326) aufgeklärt, und die Struktur der Anhydrokörper als Ketohalbchlorine (**90**) ermittelt werden (*136*). Demnach erfolgt zu-nächst Hydroxylierung, entweder mittels Wasserstoffperoxid oder Osmiumtetroxid. Aus dem Hydroxylierungsprodukt (**89**) entsteht durch Pinakolumlagerung ein Keton (**90**), das durch zwei geminale Äthylgruppen

und eine im Fünfring des Pyrrolkerns stehende Carbonylgruppe ausgezeichnet ist (siehe S. 324): (94 a) → (89) → (90).

Bonnett und Johnson griffen die von Fischer erstmals durchgeführte Wasserstoffperoxid-Oxidation auf und konnten das Ketohalb-
chlorin (90) in bis zu 20%iger Ausbeute herstellen (16). Die erneute Anwendung dieser Reaktion auf Octaäthylporphin (136) ergab unter bestimmten Bedingungen neben dem Monoketon (90) vier Diketone, z. B.
(108), sowie ein Triketon (109) (Formelübersicht 7). Wegen ihres gemeinsamen charakteristischen Strukturelements — geminale Alkylgruppen
benachbart zur Ketogruppe — sollen diese Verbindungen als Gemini-
porphin-polyketone bezeichnet werden.

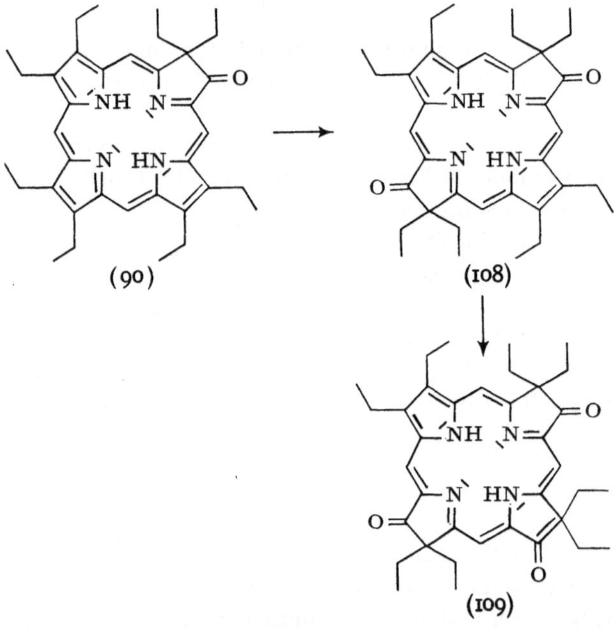

Formelübersicht 7.   Geminiporphin-ketone

VIII. Abwandlung von Seitenketten

1. Abwandlung der Vinylgruppe

a) Entfernung der Vinylgruppe

Eine wichtige Reaktion der Vinylgruppe ist, scheinbar paradoxerweise,
ihre Entfernung aus dem Molekül. Hierdurch wird eine freie Position
zur Einführung von Seitenketten durch elektrophile Substitution geschaffen. Die übliche Methode zur Abspaltung ist die Resorcin-Schmelze,

bei der auch Formyl-, Hydroxymethyl- und Hydroxyäthylgruppen entfernt werden. Für die Desvinylierung diskutieren JACKSON, KENNER et al. (*30*), die ein entsprechendes Additionsprodukt (**110**) isolieren konnten, den folgenden Mechanismus:

b) Reduktion der Vinylgruppe

Die Reduktion der Vinylgruppe tritt als Nebenreaktion auf, wenn Porphyrine hydriert werden (siehe S. 308). Die Umwandlung zur Äthylgruppe wird ebenfalls beobachtet, wenn Abkömmlinge des Chlorophylls, z. B. Chlorin e_6-TME (Trimethylester) und Rhodin g_7-TME mit Diboran (B_2H_6) umgesetzt werden (*122*). Daneben ermöglichte diese Reaktion den Zugang zu Produkten, die auf anderem Wege nur schwer zugänglich sind. So wurden der 2''-Alkohol (**111**), dessen Acetat und dessen (Thiomethyl)-methyläther gewonnen *(Formelübersicht 8)*.

Formelübersicht 8. Hydroborierung

c) Oxidation der Vinylgruppe

Oxidationsreaktionen an Vinylgruppen sind in jüngerer Zeit — abgesehen von der Kaliumpermanganat-Oxidation (*32*) — besonders mit OsO_4 und photochemisch durchgeführt worden. Bei der Umsetzung von Protoporphyrin IX-dimethylester (**41c**, S. 302) mit OsO_4 (*196*) wurde 2',2'',4',4''-Tetrahydroxy-mesoporphyrin IX-dimethylester erhalten, ohne daß bei Vermeidung eines Überschusses von OsO_4 das Porphyrinsystem angegriffen wurde. Durch Perjodat-Spaltung des Glykols konnte dieses in 2,4-Diformyl-deuteroporphyrin IX-dimethylester (95e; $R^{1,3} = CHO$; $R^{2,4,5} = H$; S. 326) überführt werden. Dieser Reaktion kommt einige Bedeutung zu. Sie liefert auf eindeutige Weise eine Vergleichssubstanz für den Identitätsbeweis eines durch Vilsmeier-Reaktion (siehe S. 333) aus Deuteroporphyrin gewonnenen Diformylproduktes. Die Photooxidation von Vinylgruppen mit molekularem Sauerstoff wurde erneut untersucht (*118*) und die zwei isomeren Photoprotoporphyrine, z. B. (**42**, S. 303), gewonnen. Interessant erscheint die Beobachtung, daß die ein diamagnetisches Metallatom enthaltenden Zink- und Magnesiumkomplexe des Protoporphyrins unter den gleichen Bedingungen eine Photooxidation eingingen, nicht aber die Kupfer- und Eisenkomplexe.

2. Abwandlung der Formylgruppe

Die Formylgruppe kommt, wie die Vinylgruppe, in natürlichen Porphyrinen und Chlorinen vor, z. B. als Substituent an C-2 [Spirographis-Porphyrin (**41 b**, S. 302)], C-3 [Chlorophyll b (**24**, S. 295)] und C-8 [Porphyrin a (**45**, S. 303)]. Außerdem existieren, wie auf S. 333 ausgeführt, Methoden, Formylsubstituenten an den Makrocyclus anzufügen. Ihre Umwandlungen haben daher großes präparatives Interesse.

a) Abspaltung der Formylgruppe

Entfernt wird die Formylgruppe, ebenso wie die Vinylgruppe, durch die Resorcin-Schmelze. Enthält ein Porphyrin zwei Formylgruppen in chemisch verschiedener Stellung, so ist unter Umständen eine selektive partielle Entformylierung möglich. Die meso-Formyl-deuteroporphyrine ließen sich in der Resorcin-Schmelze bei tieferen Temperaturen (150°) entformylieren als die entsprechenden 2- oder 4-Formyl-deuteroporphyrine (160°). So konnte 2,β-Diformyl-deuteroporphyrin partiell zum 2-Formyl-deuteroporphyrin entformyliert werden (*119*).

b) Reduktion der Formylgruppe

Die Reduktion der Formyl- zur Methylgruppe kann üblicherweise mit Hydrazin nach Wolff-Kishner erfolgen (*68*). Die Reduktion der Formylgruppe im α-Formyl-octaäthylporphin (**94e**; $R^1 = C_2H_5$; $R^2 = CHO$; $R^{3,4,5} = H$; S. 326), die einen Zugang zum α-Methyloctaäthylporphin eröffnen würde, erwies sich als schwierig. Die direkte Überführung in die Methylgruppe gelang bisher am metallfreien Porphyrin (**94e**) nur mit Diboran und am Zn-Komplex mit $NaBH_4$ in 30%iger Ausbeute (*139*).

3. Reaktionen am isocyclischen Fünfring

Den Chlorophyllen liegt ein spezielles Chloringerüst zugrunde, das zwischen C-6 und C-γ einen isocyclischen Fünfring enthält und das wegen der Bedeutung der sich davon ableitenden Stoffe einen eigenen Namen hat. Chlorine mit isocyclischem Fünfring, z. B. das Methylphäophorbid a (**112a**), werden auch als Phorbide bezeichnet.

(**112a**) $R^1 = CO_2CH_3$; $R^2 = H$

(**112b**) R^1, $R^2 = H$

(**112c**) R^1, $R^2 = D$

(**112d**) $R^1 = CO_2CH_3$; $R^2 = CH_3O$

(**112e**) $R^1 = CH_3O$; $R^2 = CO_2CH_3$

(**112f**) $R^1 = H$; $R^2 = CH_3O$

(**112g**) $R^1 = CH_3O$; $R^2 = H$

(**112**)

Einige Reaktionen an C-9 und C-10, die nicht Teil des aromatischen Makrocyclus sind, sollen hier erwähnt werden [siehe auch (*185*), dort S. 86].

Die durch Abbau von Chlorophyll gewonnenen Phorbide tragen am C-9 eine Oxogruppe. HOLT (*114*) stellte bei vergleichenden Reduktionsversuchen von carbonylgruppenhaltigen Porphyrinen und Chlorophyll-Derivaten fest, daß die Formylgruppe am C-3 leichter der Reduktion unterliegt als die Carbonylfunktion am C-9.

Im Hinblick auf die Enolisierung der Oxofunktion am C-9 in (**112**) war die Deuterierung des Pyromethylphäophorbids a (**112b**) in siedendem Pyridin mit D_2O aufschlußreich. Dabei zeigte sich, daß durch Enolisierung der Carbonylgruppe am C-9 der Wasserstoff nicht nur am C-10 zu (**112c**), sondern mit geringerer Geschwindigkeit auch am C-5 unter Bildung von (**113**) ausgetauscht wird (*171*) *(Formelübersicht 9)*.

(**112c**) → → (**113**)

Formelübersicht 9. Methyl-Deuterierung

PENNINGTON und Mitarb. (*174*) setzten Chlorophyll a und b mit verschiedenen primären und sekundären Aminen um. Dabei trat, wie schon von FISCHER an magnesiumfreien Chlorophyll-Abkömmlingen beobachtet worden war (*185*), Spaltung des isocyclischen Fünfringes unter Bildung von Chlorin-6-amiden (**114**) ein.

In Chlorophyll-Abkömmlingen trägt C-10 eine Carboxylgruppe. Die leichte Abspaltbarkeit dieser Gruppe, beispielsweise die Umwandlung von Methylphäophorbid a in Pyromethylphäophorbid a (**112b**), ist lange bekannt. Kürzlich stellten Pennington und Mitarb. (*175*) fest, daß auch Chlorophyll selbst sowie eine Reihe Mg-haltiger Chlorophyllabkömmlinge diese Reaktion eingehen.

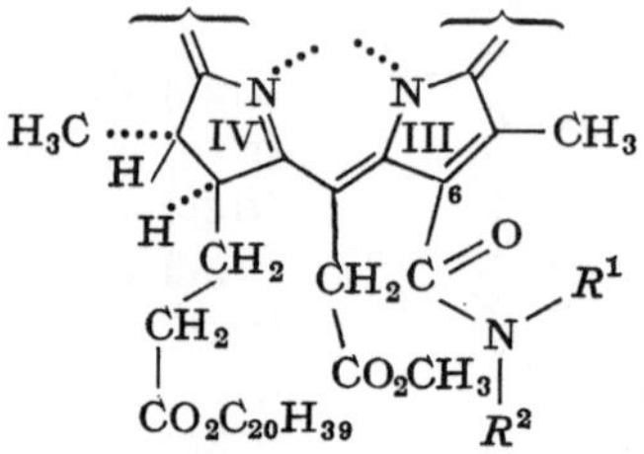

$$(114)$$

Pennington und Mitarb. (*176*) konnten 10-Hydroxy-chlorophylle (**115**) nachweisen, die durch enzymatische Oxidation aus Chlorophyll a und b entstanden. Diese Produkte waren identisch mit einer Substanz, die bei der Allomerisations-Reaktion (*185*) der Chlorophylle in Methanol entsteht, während für eine zweite Substanz dieser Reaktion nach Ansicht der Autoren eine Lactonringbildung (**116**) anzunehmen ist.

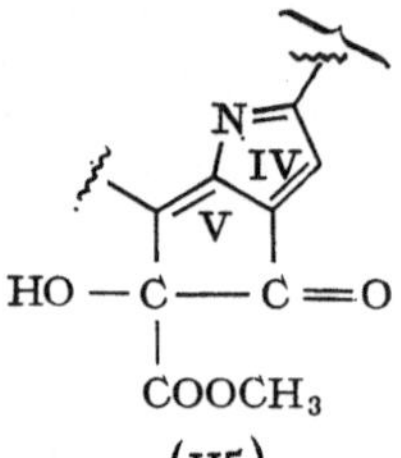

$$(115) \qquad (116)$$

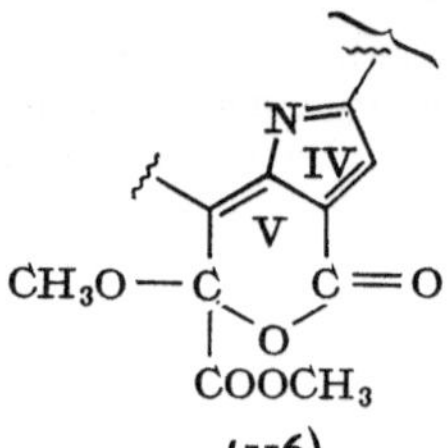

Aus Methylphäophorbid a (**112a**) konnten die am C-10 epimeren 10-Methoxymethylphäophorbide a (**112d**) und (**112e**) dargestellt und chromatographisch getrennt werden. Die Diastereomeren (**112d**) und (**112e**) wurden getrennt pyrolysiert. Durch präparative Schichtchromatographie gelang die Trennung der aus (**112d**) wie (**112e**) erhaltenen, am C-10 epimeren 10-Methoxy-pyromethylphäophorbide a (**112f**) und (**112g**) *(Formelübersicht 10)*. Die Interpretation der ORD- und NMR-Spektren erlaubt unabhängig voneinander die Bestimmung der relativen Konfiguration am C-10 der Diastereomerenpaare. Diese Untersuchungen bestätigen die *trans*-Konfiguration von 10-Methoxycarbonyl-Gruppe und Propionsäure-Seitenkette im Methylphäophorbid a (**112a**) (*211*).

Literaturverzeichnis: SS. 345—355

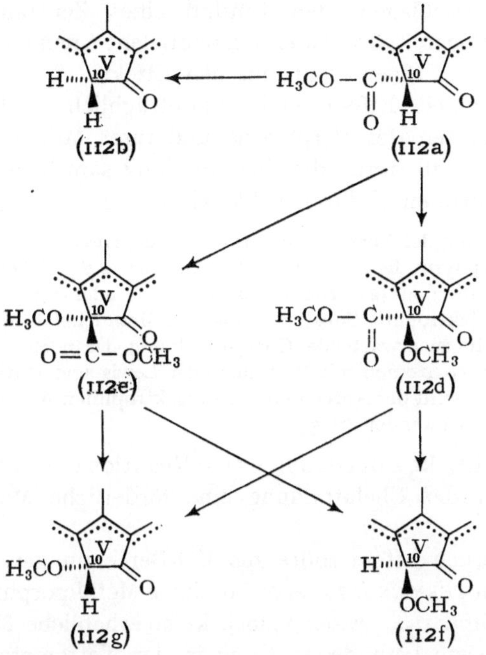

Formelübersicht 10. Diastereomere 10-Methoxy-Verbindungen

IX. Zum Einfluß der Zentralmetalle
auf chemische Reaktionen des Porphyrinsystems

Es war nicht notwendig, in diesem Artikel auf die Koordinations-chemie der Porphyrine einzugehen, da zu diesem Thema eine umfassende Monographie von FALK (*54*) vorliegt. Dagegen sei kurz der Einfluß des Zentralmetalls auf chemische Reaktionen am Porphyrinsystem (siehe S. 308 und 330) diskutiert. Systematische Untersuchungen, die die Abhängigkeit der Reaktivität von speziellen metall-eigenen Parametern erhellen könnten, liegen nicht vor. Sie müßten mindestens eine Periode (z. B. von Ca bis Zn) und zusätzlich noch Gruppen (z. B. Ni/Pd/Pt) an Hand mehrerer solcher Modell-reaktionen erfassen, die das Metall nicht gleichzeitig abspalten. Dabei wäre zudem noch die Wertigkeit des Zentralmetalls M in (**117**) sowie Art, Anzahl und Anordnung der axialen Liganden L_1 und L_2 zu variieren. Schließlich sollte die jeweils in (**117**) vorliegende Fein-struktur (Abweichung der Metallposition aus der Porphyrinebene, Faltung des Moleküls; siehe S. 305) bekannt sein.

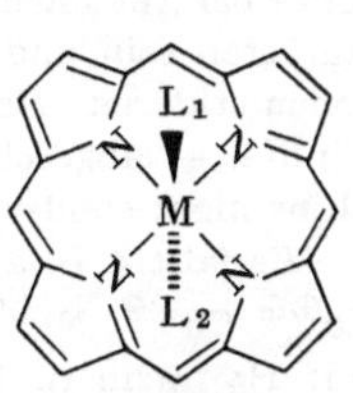

Es wird vorgeschlagen, den Einfluß eines Zentralmetallions auf chemische Reaktionen in drei Effekte aufzuteilen: einen konstitutionellen, einen sterischen und einen elektronischen Effekt.

Der konstitutionelle Effekt betrifft hauptsächlich den Unterschied im chemischen Verhalten der Porphyrine und ihrer Metallchelate und ist dadurch gegeben, daß durch die Chelatbildung sämtliche σ-Elektronenpaare an den zentralen N-Atomen blockiert sind.

Bei der Hydrierung der Porphyrine wirkt sich der Effekt so aus, daß im Gegensatz zum freien Porphyrin die Ausbildung der Phlorine, α,β,γ,N-Tetrahydroporphine und Porphinogene (S. 313) (27) zugunsten einer 1,2- oder α,γ-Hydrierung unterbunden wird. Die elektrophile Substitution ist am Metallchelat erleichtert, weil die am freien Porphyrin zu erwartende Komplexbildung der noch verfügbaren Elektronenpaare an den N-Atomen mit Protonen oder Lewis-aciden Katalysatoren verhindert wird. Damit wird der Aufbau einer den elektrophilen Angriff erschwerenden positiven Ladung abgewendet (S. 330).

Der sterische Effekt kann dadurch eine Reaktion erleichtern, daß durch die spezielle Art der Chelatbildung eine förderliche Molekülgeometrie fixiert wird.

Der elektronische Effekt sollte aus MO-Berechnungen zu entnehmen sein. Die von Gouterman (217, 220) für einige Metallporphine gefundenen Werte für q_{meso} und q_{peri} lassen jedoch keine erhebliche Metallabhängigkeit erkennen. Dem gefundenen Gang in der Nettometalladung Mg > > Zn > Co > Ni > Cu (+ 0,59 > + 0,40 > + 0,34 > + 0,30 > + 0,28) steht eine andere Stabilitätsreihenfolge der Metalloporphyrine (54) gegenüber. Bei anderen Reaktionen sind zu wenig verschiedene Metalle untersucht worden.

Der elektronische Effekt eines Metalls auf den Porphinliganden könnte nach Falk (54) durch eine Metall-Ligand-Rückbindung zwischen den d_{xz}, d_{yz}, (t_{2g})-Metallorbitalen und den π^* (e_g)-Ligandorbitalen zustandekommen. Diese Bindung sollte bei ,,weichen'' (173) Metallionen, wie $Cu^{\oplus\oplus}$, $Ni^{\oplus\oplus}$, $Co^{\oplus\oplus}$, $Pd^{\oplus\oplus}$, $Pt^{\oplus\oplus}$, am stärksten sein und würde die besondere Stabilität der Metalloporphine mit diesen Ionen erklären. Bei ,,harten'' Metallionen sollte dagegen der genannte elektronische Einfluß — abgesehen von einem elektrostatischen — die Molekülorbitale des Porphins nicht sonderlich stören (28).

Es fällt z. B. auf, daß die Metalloporphyrine mit ,,weichen'' Zentralionen, etwa $Co^{\oplus\oplus}$, $Cu^{\oplus\oplus}$, $Ni^{\oplus\oplus}$, weder mit Natrium in Isoamylalkohol (182) noch photochemisch mit Hydrazin in Pyridin (181) zu Dihydroporphinen umgesetzt werden konnten. Auch die reduzierende Alkylierung (S. 312) ist bei diesen Metalloporphyrinen wenig erfolgreich (121), während die genannten Reaktionen von ,,harten'' Zentralionen günstiger beeinflußt werden (,,hart'' sind vergleichsweise $Zn^{\oplus\oplus}$, $Mg^{\oplus\oplus}$, $Sn^{4\oplus}$).

Bisher lassen sich jedoch noch keine widerspruchsfreien Deutungssysteme aufbauen und die Diskussion muß daher skizzenhaft bleiben. In Anbetracht der großen physiologischen Bedeutung der Magnesium-, Eisen- und Kobaltchelate der Tetrapyrrolpigmente (Chlorophyll, Hämin, Vitamin B_{12}) liegt hier für zukünftige Arbeiten ein weites Feld.

Literaturverzeichnis

1. ABRAHAM, R. J., P. A. BURBIDGE, A. H. JACKSON and G. W. KENNER: Concentration Effects in Proton Magnetic Resonance Spectra of Porphyrins. Proc. Chem. Soc. (London) **1963**, 134.

2. ABRAHAM, R. J., P. A. BURBIDGE, A. H. JACKSON and D. B. MACDONALD: The Proton Magnetic Resonance Spetra of Porphyrins. IV. Coproporphyrin Tetramethyl Esters. J. Chem. Soc. (London) **B 1966**, 620.

3. ABRAHAM, R. J., A. H. JACKSON and G. W. KENNER: The Proton Magnetic Resonance Spectra of Porphyrins. I. The Effect of β-Substitution on Proton Chemical Shifts of Porphyrins. J. Chem. Soc. (London) **1961**, 3468.

4. ABRAHAM, R. J., A. H. JACKSON, G. W. KENNER and D. WARBURTON: The Proton Magnetic Resonance Spectra of Porphyrins. III. meso-Substituted Porphyrins. J. Chem. Soc. (London) **1963**, 853.

5. ARCHIBALD, J. L., S. F. MACDONALD and K. B. SHAW: Synthetic Porphyrins Related to Chlorobium Chlorophylls. J. Amer. Chem. Soc. **85**, 644 (1963).

6. ARCHIBALD, J. L., D. M. WALKER, K. B. SHAW, A. MARKOVAC and S. F. MACDONALD: The Synthesis of Porphyrins Derived from Chlorobium Chlorophylls. Canad. J. Chem. **44**, 345 (1966).

7. ARSENAULT, G. P., E. BULLOCK and S. F. MACDONALD: Pyrromethanes and Porphyrins Therefrom. J. Amer. Chem. Soc. **82**, 4384 (1960).

8. BADGER, G. M., R. L. N. HARRIS and R. A. JONES: Porphyrins. IV. Further Preliminary Studies on the Synthesis of Porphyrin a. Austral. J. Chem. **17**, 1002 (1964).

9. BAKER, E. W., T. F. YEN, J. P. DICKIE, R. E. RHODES and L. F. CLARK: Mass Spectrometry of Porphyrins. II. Characterization of Petroporphyrins. J. Amer. Chem. Soc. **89**, 3631 (1967).

10. BALL, R. H., G. D. DOROUGH and M. CALVIN: A Further Study of the Porphyrin-like Products of the Reaction of Benzaldehyde and Pyrrole. J. Amer. Chem. Soc. **68**, 2278 (1946).

11. BAMFIELD, P., R. GRIGG, R. W. KENYON and A. W. JOHNSON: Synthesis of Deuteroporphyrin-IX and Pemptoporphyrin. Chem. Commun. **1967**, 1029.

12. BAMFIELD, P., R. L. N. HARRIS, A. W. JOHNSON, I. T. KAY and K. W. SHELTON: Synthesis of Rhodoporphyrin XV Diethyl Ester and Related Porphyrins. J. Chem. Soc. (London) **C 1966**, 1436.

13. BECKER, E. D. and R. B. BRADLEY: Effects of "Ring Currents" on the NMR Spectra of Porphyrins. J. Chem. Physics **31**, 1413 (1959).

14. BERTELE, E., H. BOOS, J. D. DUNITZ, F. ELSINGER, A. ESCHENMOSER, I. FELNER, H. P. GRIBI, H. GSCHWEND, E. F. MEYER, M. PESARO und R. SCHEFFOLD: Ein synthetischer Zugang zum Corrinsystem. Angew. Chem. **76**, 393 (1964).

15. BOGORAD, L.: The Biosynthesis of Chlorophylls. In: L. P. VERNON and G. R. SEELY (Ed.), The Chlorophylls, p. 481. New York-London: Academic Press. 1966.

16. BONNETT, R., D. DOLPHIN, A. W. JOHNSON, D. OLDFIELD and G. F. STEPHENSON: The Oxidation of Porphyrins with Hydrogen Peroxide in Sulphuric Acid. Proc. Chem. Soc. (London) **1964**, 371.

17. BONNETT, R., I. A. D. GALE and G. F. STEPHENSON: The meso-Reactivity of Porphyrins and Related Compounds. II. Halogenation. J. Chem. Soc. **C 1966**, 1600.

18. — — — The meso-Reactivity of Porphyrins and Related Compounds. III. Deuteriation. J. Chem. Soc. **C 1967**, 1168.

19. BONNETT, R. and G. F. STEPHENSON: The meso-Reactivity of Porphyrins and Related Compounds. I. Nitration. J. Organ. Chem. (USA) **30**, 2791 (1965); Proc. Chem. Soc. (London) **1964**, 79.

346　　　　H. H. INHOFFEN, J. W. BUCHLER und P. JÄGER:

20. BOUCHER, L. J. and J. J. KATZ: The Infrared Spectra of Metalloporphyrins (4000—160 cm^{-1}). J. Amer. Chem. Soc. **89**, 1340 (1967).
21. — — Aggregation of Metallochlorophylls. J. Amer. Chem. Soc. **89**, 4703 (1967).
22. BOUCHER, L. J., H. H. STRAIN and J. J. KATZ: The Far Infrared Spectra of Monomeric and Aggregated Chlorophylls a and b. J. Amer. Chem. Soc. **88**, 1341 (1966).
23. BRIAT, B., D. A. SCHOOLEY, R. RECORDS, E. BUNNENBERG and C. DJERASSI: Magnetic Circular Dichroism Studies. III. Investigation of some Optically Active Chlorins. J. Amer. Chem. Soc. **89**, 6170 (1967).
24. BROCKMANN jr., H.: Kernresonanzspektroskopische Untersuchungen an Chlorinen. Unveröffentlicht.
25. — Zur absoluten Konfiguration des Chlorophylls. Angew. Chem., **80**, 233 (1968).
26. — Zur absoluten Konfiguration des Bacteriochlorophylls. Angew. Chem., **80**, 234 (1968).
27. BUCHLER, J. W.: Systematik und Problematik der Reduktion des Porphyrinsystems. Unveröffentlicht.
28. — Zur Anwendung des Prinzips der harten und weichen Säuren und Basen auf die Metalloporphyrinchemie. In Vorbereitung.
29. BUDZIKIEWICZ, H., F. v. d. HAAR und H. H. INHOFFEN: Zur weiteren Kenntnis des Chlorophylls und des Hämins, X. Zum Aussagewert metastabiler Ionen bei Strukturermittlungen mit Hilfe von Massenspektren: Kombinierte Zerfallsreaktionen bei Chlorinen. Liebigs Ann. Chem. **701**, 23 (1967).
30. BURBIDGE, P. A., G. L. COLLIER, A. H. JACKSON and G. W. KENNER: The Proton Magnetic Resonance Spectra of Porphyrins. V. Syntheses and Spectra of Some meso-Methylated Porphyrins. J. Chem. Soc. (London) **B 1967**, 930.
31. CARR, R. P., P. J. CROOK, A. H. JACKSON and G. W. KENNER: New Syntheses of Protoporphyrin IX. Chem. Commun. **1967**, 1025.
32. CAUGHEY, W. S., J. O. ALBEN, W. Y. FUJIMOTO and J. L. YORK: Substituted Deuteroporphyrins. I. Reactions at the Periphery of the Porphyrin Ring. J. Organ. Chem. (USA) **31**, 2631 (1966).
33. CAUGHEY, W. S., R. M. DEAL, C. WEISS and M. GOUTERMAN: Electronic Spectra of Substituted Metal Deuteroporphyrins. J. Mol. Spectry. **16**, 451 (1965).
34. CAUGHEY, W. S., W. Y. FUJIMOTO and B. P. JOHNSON: Substituted Deuteroporphyrins. II. Substituent Effects on Electronic Spectra, Nitrogen Basicities, and Ligand Affinities. Biochemistry **5**, 3830 (1966).
35. CLACK, D. W. and N. S. HUSH: Successive One-Electron Reduction Potentials of Porphins and Metal Porphins in Dimethylformamide. J. Amer. Chem. Soc. **87**, 4238 (1965).
36. CLEZY, P. S., F. D. LOONEY, A. W. NICHOL and G. A. SMYTHE: The Chemistry of Pyrrolic Compounds. III. The Free Radical Nature of the Oxyporphyrins. Austral. J. Chem. **19**, 1481 (1966).
37. CLEZY, P. S. and A. W. NICHOL: The Chemistry of Pyrrolic Compounds. I. Synthesis of Oxyporphyrins. Austral. J. Chem. **18**, 1835 (1965).
38. CLOSS, G. L. and L. E. CLOSS: Negative Ions of Porphin Metal Complexes. J. Amer. Chem. Soc. **85**, 818 (1963).
39. CLOSS, G. L., J. J. KATZ, F. C. PENNINGTON, M. R. THOMAS and H. H. STRAIN: Nuclear Magnetic Resonance Spectra and Molecular Association of Chlorophylls a and b, Methyl Chlorophyllides, Pheophytins, and Methyl Pheophorbides. J. Amer. Chem. Soc. **85**, 3809 (1963).
40. COLLIER, G. L., A. H. JACKSON and G. W. KENNER: Pyrroles and Related Compounds. IX. Synthesis of two Acetamidoethyl Porphyrins and their Conversion into Vinyl Porphyrins and Chlorins. J. Chem. Soc. (London) **C 1967**, 66.

41. Corwin, A. H. and O. D. Collins: Reduction of Tin Porphyrins to Tin Chlorins. J. Organ. Chem. (USA) **27**, 3060 (1962).

42. Cox, M. T., R. Fletcher, A. H. Jackson, G. W. Kenner and K. M. Smith: Synthesis of Porphyrins through b-Bilenes. Chem. Commun. **1967**, 1141.

43. Dodd, J. W. and N. S. Hush: The Negative Ions of Some Porphin and Phthalocyanine Derivatives, and their Electronic Spectra. J. Chem. Soc. (London) **1964**, 4607.

44. Dorough, G. D. and J. R. Miller: An Attempted Preparation of a Simple Tetrahydroporphine. J. Amer. Chem. Soc. **74**, 6106 (1952).

45. Dougherty, R. C., H. H. Strain and J. J. Katz: Hydrogen Exchange in Chlorophyll and Related Compounds, and Correlation with Molecular Orbital Calculations. J. Amer. Chem. Soc. **87**, 104 (1965).

46. Eichhorn, E. W.: Rotatory Dispersion Studies of Biologically Active Metalloporphyrin Compounds. Tetrahedron **13**, 208 (1961).

47. Eisner, U.: Some Novel Hydroporphyrins. J. Chem. Soc. (London) **1957**, 3461.

48. Eisner, U. and M. J. C. Harding: Some Novel Demetallation Reactions. J. Chem. Soc. (London) **1964**, 4089.

49. Eisner, U. and R. P. Linstead: Chlorophyll and Related Substances. I. The Synthesis of Chlorin. J. Chem. Soc. (London) **1955**, 3742.

50. — — Chlorophyll and Related Substances. II. The Dehydrogenation of Chlorin to Porphin and the Number of Extra Hydrogen Atoms in the Chlorins. J. Chem. Soc. (London) **1955**, 3749.

51. Ellis, J., A. H. Jackson, A. C. Jain and G. W. Kenner: Pyrroles and Related Compounds. III. Syntheses of Porphyrins from Pyrromethanes and Pyrromethenes. J. Chem. Soc. (London) **1964**, 1935.

52. Eschenmoser, A., R. Scheffold, E. Bertele, M. Pesaro and H. Gschwend: Synthetic Corrin Complexes. Proc. Roy. Soc. (London) A **288**, 306 (1965).

53. Evstigneeva, R. P., A. M. Fargali, T. A. Lubkova, I. N. Khandii and N. A. Preobrazhenskii: Synthesis of Isomeric Tetramethyltetrakis(β-diethylaminoethyl)porphyrins and their Derivatives. Chem. Abstr. **65**, 3880 (1966).

54. Falk, J. E.: Porphyrins and Metalloporphyrins. Amsterdam-London-New York: Elsevier. 1964.

55. Felton, R. H. and H. Linschitz: Polarographic Reduction of Porphyrins and Electron Spin Resonance of Porphyrin Anions. J. Amer. Chem. Soc. **88**, 1113 (1966).

56. Felton, R. H., G. M. Sherman and H. Linschitz: Formation of Phase Test Intermediate of Chlorophyll by Electrolytic Reduction. Nature **203**, 637 (1964).

57. Ficken, G. E., R. B. Johns and R. P. Linstead: Chlorophyll and Related Compounds. IV. The Position of the Extra Hydrogens in Chlorophyll. The Oxidation of Pyrophaeophorbide a. J. Chem. Soc. (London) **1956**, 2272.

58. Ficken, G. E., R. P. Linstead, E. Stephen and M. Whalley: Conjugated Macrocycles. XXXI. Catalytic Hydrogenation of Tetrazaporphins, with a Note on its Stereochemical Course. J. Chem. Soc. (London) **1958**, 3879.

59. Filippovich, E. I., V. N. Luzgina, L. I. Korsuntseva, R. P. Evstigneeva und N. A. Preobrazhenskii: Synthese des Mesoporphyrin-dimethylesters. Zhurn. Obshchei Khimii **36**, 1383 (1966) [Chem. Abstr. **66**, 2548d (1967)].

60. Fischer, H.: Synthese des Hämins. Naturwiss. **17**, 611 (1929).

61. Fischer, H. und L. Beer: Über Formyl-pyrro-porphyrin und Formyl-deuteroporphyrin. Z. physiol. Chem. **244**, 31 (1936).

62. Fischer, H. und K. Bub: Über Phäoporphinogen a_5, Phylloerythrinogen und Versuche zur Inaktivierung des Chlorophylls und seiner Derivate. Liebigs Ann. Chem. **530**, 213 (1937).

63. Fischer, H. und H. Eckoldt: Überführung von Porphyrinen in Dioxychlorine durch Einwirkung von Osmiumtetroxyd. Liebigs Ann. Chem. **544**, 138 (1940).

64. Fischer, H. und F. Gerner: Teilsynthese von Meso-methylphäophorbid a und von 9-Oxy-9-desoxo-meso-methylphäophorbid a. Über Chlor-meso-chlorine. Liebigs Ann. Chem. **559**, 77 (1948).

65. Fischer, H., F. Gerner, W. Schmelz und F. Baláž: Über die Einführung von Substituenten in β-Stellung bei Pyrrolen und Pyrrolfarbstoffen. Liebigs Ann. Chem. **557**, 134 (1944).

66. Fischer, H. und H. Gibian: Über die Hydrierung von Vinyl- zu Mesoverbindungen mit Hydrazinhydrat. Liebigs Ann. Chem. **548**, 183 (1941).

67. — — Racemisierung von Chlorophyllderivaten. Liebigs Ann. Chem. **550**, 208 (1942).

68. — — Übergang von der Chlorophyll b- in die a-Reihe. Liebigs Ann. Chem. **552**, 153 (1942).

69. Fischer, H. und P. Halbig: Über Iso-ätioporphyrin, sein Tetrabromderivat, seinen oxydativen und reduktiven Abbau, sowie eine Synthese eines Iso-mesoporphyrins und des Opsopyrrols. Liebigs Ann. Chem. **450**, 151 (1926).

70. Fischer, H. und H. Helberger: Synthese von Chlorinen. Liebigs Ann. Chem. **471**, 285 (1929).

71. Fischer, H. und K. Herrle: Quantitative Dehydrierung von Chlorinkupfersalzen mit Sauerstoff. Liebigs Ann. Chem. **527**, 138 (1937).

72. — — Über Anhydrochlorine, Rhodorhodin und katalytische Reduktion von Porphyrinen zu Chlorinen. Liebigs Ann. Chem. **530**, 230 (1937).

73. Fischer, H., H. Kellermann und F. Baláž: Über die Bromierung der Ester von Meso-isochlorin e_4 und Mesochlorin e_6. Chem. Ber. **75**, 1778 (1942).

74. Fischer, H. und A. Kirstahler: Synthese des Deuterohämins und Deuteroporphyrins. Liebigs Ann. Chem. **466**, 178 (1928).

75. Fischer, H. und W. Klendauer: Über die Chlorierungs- und Nitrierungsreaktion bei Porphyrinen und Chlorinen. Liebigs Ann. Chem. **547**, 123 (1941).

76. Fischer, H. und O. Laubereau: Über die Teilsynthese des Meso-pyrophäophorbids und weitere synthetische Versuche in der Chlorophyllreihe. Liebigs Ann. Chem. **535**, 17 (1938).

77. Fischer, H., H. Mittenzwei und D. B. Hevér: Überführung von Dehydrobacteriophäophorbid a in Chlorophyll a. Liebigs Ann. Chem. **545**, 154 (1940).

78. Fischer, H. und W. Neumann: Über einige Derivate von Ätioporphyrin I. Liebigs Ann. Chem. **494**, 225 (1932).

79. Fischer, H., A. Oestreicher und A. Albert: Über Acetylrhodin g_7 und einige Vinylporphyrine. Liebigs Ann. Chem. **538**, 128 (1939).

80. Fischer, H. und W. Orth: Die Chemie des Pyrrols. 1. Band: Pyrrol und seine Derivate. Mehrkernige Pyrrolsysteme ohne Farbstoffcharakter. 2. Band: Pyrrolfarbstoffe. 1. Hälfte: — Porphyrine — Hämin — Bilirubin und ihre Abkömmlinge. Leipzig: Akad. Verlagsges. 1934 und 1937.

81. Fischer, H. und H. Pfeiffer: Oxydation von Porphyrinen und Chlorinen mit Osmiumtetroxyd. Liebigs Ann. Chem. **556**, 131 (1944).

82. Fischer, H. und H. J. Riedl: Einführung des Oxymethyl-, des Methylmalonsäure- und des Propionsäurerestes in Porphyrine. Liebigs Ann. Chem. **482**, 214 (1930).

83. Fischer, H., H. Röse und E. Bartholomäus: Zur Kenntnis der Porphyrinbildung. 2. Mitt. Über Porphinogen und seine Beziehungen zum Blutfarbstoff und dessen Derivaten. Z. physiol. Chem. **84**, 262 (1913).

84. Fischer, H. und H. Röse: Über Tetrachlormesoporphyrin. Chem. Ber. **46**, 2460 (1913).

85. FISCHER, H. und A. ROTHHAAS: Überführung von Mesoporphyrin IX in eine Porphyrin-monopropionsäure sowie über einige Porphyrinsynthesen. Liebigs Ann. Chem. **484**, 85 (1930).

86. FISCHER, H. und C. G. SCHRÖDER: Zur Konstitution der Verdine und über synthetische Rhodine. Liebigs Ann. Chem. **537**, 250 (1939).

87. FISCHER, H. und A. SCHWARZ: Synthese des 6-Formyl-pyrroporphyrins und des 6-Formylphylloporphyrins. Liebigs Ann. Chem. **512**, 239 (1934).

88. FISCHER, H. und G. SPIELBERGER: Teilsynthese von Äthylchlorophyllid b sowie über 10-Äthoxymethylphäophorbid b. Liebigs Ann. Chem. **515**, 130 (1934).

89. FISCHER, H. und A. STERN: Die Chemie des Pyrrols. 2. Band: Pyrrolfarbstoffe. 2. Hälfte. Chlorophyll a und b sowie Derivate — Chlorophyllabbau — Imidoporphyrine. Leipzig: Akad. Verlagsges. 1940.

90. FISCHER, H. und A. TREIBS: Über Ätioxantho- und Mesoxanthoporphinogen. Liebigs Ann. Chem. **457**, 209 (1927).

91. FISCHER, H. und G. WECKER: Synthese des Spirographisporphyrins. Z. physiol. Chem. **272**, 1 (1941).

92. FISCHER, H. und K. ZEILE: Synthese des Hämatoporphyrins, Protoporphyrins und Hämins. Liebigs Ann. Chem. **468**, 98 (1928).

93. FISCHER, H. und W. ZERWECK: Zur Kenntnis der natürlichen Porphyrine. 7. Mitt. Über Uroporphyrinogen-heptamethylester und eine neue Überführung von Uro- in Koproporphyrin. Z. physiol. Chem. **137**, 242 (1924).

94. FLEISCHER, E. B.: The Structure of Nickel Etioporphyrin-I. J. Amer. Chem. Soc. **85**, 146 (1963).

95. — The Structure of Copper Tetraphenyl-porphine. J. Amer. Chem. Soc. **85**, 1353 (1963).

96. FLEISCHER, E. B., C. K. MILLER and L. E. WEBB: Crystal and Molecular Structures of Some Metal Tetraphenylporphines. J. Amer. Chem. Soc. **86**, 2342 (1964).

97. FLEISCHER, E. B. and A. L. STONE: The Molecular Structure of the Diacid Species of $\alpha,\beta,\gamma,\delta$-Tetra-4-pyridylporphine and $\alpha,\beta,\gamma,\delta$-Tetraphenylporphine. Chem. Commun. **1967**, 332.

98. FLEMING, I.: Absolute Configuration and the Structure of Chlorophyll. Nature **216**, 151 (1967).

99. GLICK, M. D., G. H. COHEN and J. L. HOARD: The Stereochemistry of the Coordination Group in Aquozinc(II) Tetraphenylporphine. J. Amer. Chem. Soc. **89**, 1996 (1967).

100. GOEDHER, J. C.: Visible Absorption and Fluorescence of Chlorophyll and its Aggregates in Solution. In: L. P. VERNON and G. R. SEELY (Ed.), The Chlorophylls, p. 147. New York and London: Academic Press. 1966.

101. GOLDEN, J. H., R. P. LINSTEAD and G. H. WHITHAM: Chlorophyll and Related Compounds. VII. The Structure of Bacteriochlorophyll. J. Chem. Soc. (London) **1958**, 1725.

102. GOUTERMAN, M.: Spectra of Porphyrins. J. Mol. Spectry. **6**, 138 (1961).

103. GOUTERMAN, M., G. WAGNIERE and L. C. SNYDER: Spectra of Porphyrins. II. Four-Orbital Model. J. Mol. Spectry. **11**, 108 (1963).

104. GRASSL, M., G. AUGSBURG, U. COY und F. LYNEN: Zur chemischen Konstitution des Cytohämins. Biochem. Z. **337**, 35 (1963).

105. GRASSL, M., U. COY, R. SEYFFERT und F. LYNEN: Die chemische Konstitution des Cytohämins. Biochem. Z. **338**, 771 (1963).

106. GRIGG, R.: The Reaction of Ethoxycarbonylnitrene with Porphyrins. A Ring Expansion-Contraction Reaction of the Porphyrin Ring. Chem. Commun. **1967**, 1238.

350   H. H. INHOFFEN, J. W. BUCHLER und P. JÄGER:

107. HAMOR, M. J., T. A. HAMOR and J. L. HOARD: The Structure of Crystalline Tetraphenylporphine. The Stereochemical Nature of the Porphine Skeleton. J. Amer. Chem. Soc. **86**, 1938 (1964).

108. HARRIS, R. L. N., A. W. JOHNSON and I. T. KAY: The Synthesis of Porphins and Related Macrocycles. Quart. Rev. (Chem. Soc. London) **1966**, 211.

109. — — — A Stepwise Synthesis of Unsymmetrical Porphyrins. J. Chem. Soc. (London) **C 1966**, 22.

110. HOARD, J. L., G. H. COHEN and M. D. GLICK: The Stereochemistry of the Co-ordination Group in an Iron(III) Derivative of Tetraphenylporphine. J. Amer. Chem. Soc. **89**, 1992 (1967).

111. HOARD, J. L., M. J. HAMOR and T. A. HAMOR: Configuration of the Porphine Skeleton in Unconstrained Porphyrin Molecules. J. Amer. Chem. Soc. **85**, 2334 (1963).

112. HOARD, J. L., M. J. HAMOR, T. A. HAMOR and W. S. CAUGHEY: The Crystal Structure and Molecular Stereochemistry of Methoxyiron(III) Mesoporphyrin-IX Dimethyl Ester. J. Amer. Chem. Soc. **87**, 2312 (1965).

113. HOLT, A. S.: Recently Characterized Chlorophylls. In: L. P. VERNON and G. R. SEELY (Ed.), The Chlorophylls, p. 111. New York and London: Academic Press. 1966.

114. — Reduction of Chlorophyllides, Chlorophylls and Chlorophyll Derivatives by Sodium Borohydride. Plant Physiol. **34**, 310 (1959).

115. HUSH, N. S. and J. R. ROWLANDS: Hyperfine Structure in the Electron Spin Resonance Spectra of Reduced Porphines. J. Amer. Chem. Soc. **89**, 2976 (1967).

116. INHOFFEN, H. H.: Acetylierungsreaktionen an Chlorinen, zugleich ein Beitrag zur Münchener Chlorophyll-Chemie. Angew. Chem. **76**, 383 (1964).

117. INHOFFEN, H. H. und H. BIERE: Zur weiteren Kenntnis des Chlorophylls und des Hämins, XI. Ein Weg zum Protochlorophyll. Tetrahedron Letters **1966**, 5145.

118. INHOFFEN, H. H., C. BLIESENER und H. BROCKMANN jr.: Zur weiteren Kenntnis des Chlorophylls und des Hämins, VIII. Umwandlung von Protoporphyrin IX über Photoprotoporphyrin in Spirographis- und Isospirographis-porphyrin. Tetrahedron Letters **1966**, 3779.

119. — — — Zur weiteren Kenntnis des Chlorophylls und des Hämins, XIV. Substituierte Deuteroporphyrine. Tetrahedron Letters **1967**, 727.

120. INHOFFEN, H. H. und J. W. BUCHLER: Einwirkung von Diisobutylaluminium-hydrid auf Porphyrine. Unveröffentlicht.

121. — — Reduzierende Methylierung des Porphyrinsystems. Unveröffentlicht.

122. INHOFFEN, H. H., J. W. BUCHLER und R. THOMAS: Beiträge zur chemischen Reduktion des Porphyrinsystems. Unveröffentlicht.

123. INHOFFEN, H. H. und J.-H. FUHRHOP: Über die Addition von Chloroform an eine periphere Doppelbindung des Octaäthylporphinato-zinn(IV)-dihydro-xids. Unveröffentlicht.

124. INHOFFEN, H. H., J.-H. FUHRHOP und F. v. d. HAAR: Zur weiteren Kenntnis des Chlorophylls und des Hämins. VII. Über Meso-oxo-porphinogene. Liebigs Ann. Chem. **700**, 92 (1966).

125. INHOFFEN, H. H., J.-H. FUHRHOP, H. VOIGT und H. BROCKMANN jr.: Zur weiteren Kenntnis des Chlorophylls und des Hämins, VI. Formylierung der meso-Kohlenstoffatome von alkylsubstituierten Porphyrinen. Liebigs Ann. Chem. **695**, 133 (1966).

126. INHOFFEN, H. H. und A. GOSSAUER: Darstellung von meso-oxidierten Gemini-porphyrin-ketonen. Unveröffentlicht.

127. INHOFFEN, H. H. und P. JÄGER: Zur weiteren Kenntnis des Chlorophylls und des Hämins, II. Elektrochemische Reduktionen an Porphyrinen (I). Tetrahedron Letters **1964**, 1317.

128. — — Zur weiteren Kenntnis des Chlorophylls und des Hämins, III. Elektrochemische Reduktionen an Porphyrinen und Chlorinen (II). Tetrahedron Letters **1965**, 3387.

129. INHOFFEN, H. H., P. JÄGER und R. MÄHLHOP: Eine Partialsynthese von Rhodin g_7-TME aus Chlorin e_6-TME. Unveröffentlicht.

130. INHOFFEN, H. H., P. JÄGER, R. MÄHLHOP und C.-D. MENGLER: Zur weiteren Kenntnis des Chlorophylls und des Hämins, XII. Elektrochemische Reduktionen an Porphyrinen und Chlorinen (IV). Liebigs Ann. Chem. **704**, 188 (1967).

131. INHOFFEN, H. H. und G. JECKEL: Zur Konstitution des 6-Acetyl-2-desvinyl-isochlorin e_4-dimethylesters. Unveröffentlicht.

132. INHOFFEN, H. H., G. KLOTMANN und G. JECKEL: Zur weiteren Kenntnis des Chlorophylls und des Hämins, V. Acetylierungs- und Entacetylierungsreaktionen an Chlorinen. Liebigs Ann. Chem. **695**, 112 (1966).

133. INHOFFEN, H. H. und R. MÄHLHOP: Zur weiteren Kenntnis des Chlorophylls und des Hämins, IX. Elektrochemische Reduktionen an Porphyrinen und Chlorinen (III). Tetrahedron Letters **1966**, 4283.

134. INHOFFEN, H. H. und C.-D. MENGLER: Katalytische Hydrierung des Pyro-methylphäophorbids a. Unveröffentlicht.

135. INHOFFEN, H. H. und N. MÜLLER: Einführung von meso-Methylgruppen in Octaäthylporphin. Unveröffentlicht.

136. INHOFFEN, H. H. und W. NOLTE: Zur weiteren Kenntnis des Chlorophylls und des Hämins, XV. Umwandlung des Octaäthylporphins in Octaäthyl-gemini-porphyrinpolyketone. Tetrahedron Letters **1967**, 2185.

137. INHOFFEN, H. H., H. PARNEMANN and R. G. FOSTER: Reduction Products of the Sn^{IV}-Complex of Octaethylporphine. Illinois Inst. Techn. NMR Newsletter **65**, 45 (1964).

138. INHOFFEN, H. H. und R. SAMBLEBE: Zur Hydrierung des Chloroporphyrin e_6-trimethylesters. Unveröffentlicht.

139. INHOFFEN, H. H. und O. SOMAYA: Darstellung des meso-Methyl-octaäthyl-porphins. Unveröffentlicht.

140. INHOFFEN, H. H. und J. ULLRICH: Beitrag zur Porphyrin-Synthese von Mac-Donald. Unveröffentlicht.

141. INHOFFEN, H. H. und H. VOIGT: Zur Chlorierung des Octaäthylporphins. Unveröffentlicht.

142. JACKSON, A. H. and G. W. KENNER: New Syntheses of Porphyrins and Related Tetrapyrroles. Nature **215**, 1126 (1967).

143. JACKSON, A. H., G. W. KENNER, H. BUDZIKIEWICZ, C. DJERASSI and J. M. WILSON: Pyrroles and Related Compounds. X. Mass Spectrometry in Structural and Stereochemical Problems. XC. Mass Spectra of Linear Di-, Tri- and Tetrapyrrolic Compounds. Tetrahedron **23**, 603 (1967).

144. JACKSON, A. H., G. W. KENNER, G. McGILLIVRAY and G. S. SACH: Two New Porphyrin Syntheses. J. Amer. Chem. Soc. **87**, 676 (1965).

145. JACKSON, A. H., G. W. KENNER, G. McGILLIVRAY and K. M. SMITH: Pyrroles and Related Compounds. XIII. Porphyrin Synthesis through b-Oxobilanes and Oxophlorins (Oxyporphyrins). J. Chem. Soc. (London) **C 1968**, 294.

146. JACKSON, A. H., G. W. KENNER and G. S. SACH: Pyrroles and Related Compounds. XII. Stepwise Synthesis of Porphyrins through a-Oxo-bilanes. J. Chem. Soc. (London) **C 1967**, 2045.

352   H. H. Inhoffen, J. W. Buchler und P. Jäger:

147. Jackson, A. H., G. W. Kenner and K. M. Smith: Oxyporphyrins. J. Amer. Chem. Soc. **88**, 4539 (1966).
147 a. — — — Pyrroles and Related Compounds. XIV. The Structure and Transformations of Oxophlorins (Oxyporphyrins). J. Chem. Soc. (London) **C 1968**, 302.
148. Jackson, A. H., G. W. Kenner, K. M. Smith, R. T. Aplin, H. Budzikiewicz and C. Djerassi: Pyrroles and Related Compounds. VIII. Mass Spectrometry in Structural and Stereochemical Problems. LXXVI. The Mass Spectra of Porphyrins. Tetrahedron **21**, 2913 (1965).
149. Jackson, A. H., G. W. Kenner and J. Wass: Rational Syntheses of Chlorocruoroporphyrin (Spirographis Porphyrin) and Pemptoporphyrin. Chem. Commun. **1967**, 1027.
150. Johnson, A. W.: Macrocyclic Tetrapyrrolic Pigments. Chem. in Brit. **1966**, 253.
151. Johnson, A. W. and D. Oldfield: Meso-nitro- and Amino-aetioporphyrins. Tetrahedron Letters **1964**, 1549.
152. — — The Nitration and Hydroxylation of Aetioporphyrin I. J. Chem. Soc. (London) **1965**, 4303.
153. — — meso-Substitution Products of Aetioporphyrin I. J. Chem. Soc. (London) **C 1966**, 794.
154. Jordan, J. and T. M. Bednarski: Polarography of Hemin. Evidence for a Two Electron Transfer. J. Amer. Chem. Soc. **86**, 5690 (1964).
155. Katz, J. J., R. C. Dougherty and L. J. Boucher: Infrared and Nuclear Magnetic Resonance Spectroscopy of Chlorophyll. In: L. P. Vernon and G. R. Seely (Ed.), The Chlorophylls, p. 186. New York and London: Academic Press. 1966.
156. Katz, J. J., R. C. Dougherty, H. L. Crespi and H. H. Strain: Nuclear Magnetic Resonance Studies of Plant Biosynthesis. A Bacteriochlorophyll Isotope Mirror Experiment. J. Amer. Chem. Soc. **88**, 2856 (1966).
157. Katz, J. J., R. C. Dougherty, F. C. Pennington, H. H. Strain and G. L. Closs: Hydrogen Exchange at Methine and C-10 Positions in Chlorophyll. J. Amer. Chem. Soc. **85**, 4049 (1963).
158. Katz, J. J., M. R. Thomas and H. H. Strain: Site of Exchangeable Hydrogen in Chlorophyll a from Proton Magnetic Resonance Measurements on Deuteriochlorophyll a. J. Amer. Chem. Soc. **84**, 3587 (1962).
159. Ke, B.: Optical Rotatory Dispersion of Chlorophyll-Containing Particles from Green Plants and Photosynthetic Bacteria. In: L. P. Vernon and G. R. Seely (Ed.), The Chlorophylls, p. 427. New York and London: Academic Press. 1966.
160. Koenig, D. F.: The Structure of α-Chlorohemin. Acta Crystallogr. **18**, 663 (1965).
161. Lascelles, J.: Tetrapyrrole Biosynthesis and its Regulation. New York and Amsterdam: Benjamin. 1964.
162. Lemberg, R.: Chemical Mechanism of Bile Pigment Formation. Rev. Pure Appl. Chem. **6**, 1 (1956).
163. Libowitzky, H. und H. Fischer: Über Iso-oxy-koproporphyrin-I-ester. Z. physiol. Chem. **255**, 209 (1938).
164. Loach, P. A. and M. Calvin: Oxidation States of Manganese Methyl Phaeophorbide a in Aqueous Solution. Nature **202**, 343 (1964).
165. Mason, S. F.: The Infrared Spectra of Heteroaromatic Systems. I. The Porphins. J. Chem. Soc. (London) **1958**, 976.
166. Mauzerall, D.: The Photoreduction of Porphyrins and the Oxidation of Amines by Photo-excited Dyes. J. Amer. Chem. Soc. **82**, 1832 (1960).

167. MAUZERALL, D.: The Photoreduction of Porphyrins: Structure of the Products. J. Amer. Chem. Soc. **84**, 2437 (1962).

168. — The Photoreduction of Uroporphyrin: Effect of pH on the Reduction with EDTA. J. Chem. Phys. **66**, 2531 (1962).

169. MAUZERALL, D. and G. FEHER: A Study of the Photoinduced Porphyrin Free Radical by Electron Spin Resonance. Biochim. Biophys. Acta **79**, 430 (1964).

170. — — Optical Absorption of the Porphyrin Free Radical Formed in a Reversible Photochemical Reaction. Biochim. Biophys. Acta **88**, 658 (1964).

171. MENGLER, C.-D.: Trideuterierung der C-5-Methylgruppe im Methylphäophorbid a. Bestätigung der Zuordnung des C-5-NMR-Signals. Illinois Inst. Techn. NMR-Letters **100**, February 1967.

172. NEILANDS, J. B. and H. TUPPY: Crystalline Synthetic Porphyrin c. Biochim. Biophys. Acta **38**, 351 (1960).

173. PEARSON, R. G.: Hard and Soft Acids and Bases. J. Amer. Chem. Soc. **85**, 3533 (1963).

174. PENNINGTON, F. C., S. D. BOYD, H. HORTON, S. W. TAYLOR, D. G. WULF, J. J. KATZ and H. H. STRAIN: Reactions of Chlorophylls a and b with Amines. Isocyclic Ring Rupture and Formation of Substituted Chlorin-6-amides. J. Amer. Chem. Soc. **89**, 3871 (1967).

175. PENNINGTON, F. C., H. H. STRAIN, W. A. SVEC and J. J. KATZ: Preparation and Properties of Pyrochlorophyll a, Methyl Pyrochlorophyllide a, Pyropheophytin a, and Methyl Pyropheophorbide a Derived from Chlorophyll by Decarbomethoxylation. J. Amer. Chem. Soc. **86**, 1418 (1964).

176. — — — — Preparations and Properties of 10-Hydroxychlorophylls a and b. J. Amer. Chem. Soc. **89**, 3875 (1967).

177. PONOMAREV, G. V., R. P. EVSTIGNEEVA, V. N. STROMNOV und N. A. PREOBRAZHENSKII: Synthese formylsubstituierter Porphyrine. Khim. heterocycl. Soed. **1966**, 628 [Chem. Abstr. **66**, 37905n (1967)].

178. PULLMAN, A. E.: On the Specific Reactivity of Chlorins Toward an Electrophilic Attack on the Methene Bridges. J. Amer. Chem. Soc. **85**, 366 (1963).

179. RICCI, A., S. PINAMONTI e V. BELLAVITA: Comportamento polarografico ed oscillografico di alcune porfirine. Ricerca Sci. **30**, 2497 (1960) [Chem. Abstr. **60**, 14127 (1964)].

180. SANO, S., T. SHINGU, J. M. FRENCH and E. THONGER: The Chemical Structure of Pemptoporphyrin. Biochem. J. **97**, 250 (1965).

181. SAVEL'EV, D. A., A. N. SIDOROV, R. P. EVSTIGNEEVA und G. V. PONOMAREV: Dunkel- und photochemische Reduktion von Metallderivaten aus der Porphin-Serie. Doklady Akad. Nauk. SSSR **167**, 135 (1966) [Chem. Abstr. **64**, 18947 (1966)].

182. SCHLESINGER, W., A. H. CORWIN and L. J. SARGENT: Synthetic Chlorins and Dihydrochlorins. J. Amer. Chem. Soc. **72**, 2867 (1950).

183. SCHWARTZ, S., M. H. BERG, I. BOSSENMEYER and H. DINSMORE: Determination of Porphyrins in Biological Materials. Methods Biochem. Anal. **8**, 221 (1960).

184. SEELY, G. R.: Hypochlorophyll. J. Amer. Chem. Soc. **88**, 3417 (1966).

185. — The Structure and Chemistry of Functional Groups. In: L. P. VERNON and G. R. SEELY (Ed.), The Chlorophylls, p. 67. New York and London: Academic Press. 1966.

186. — Photochemistry of Chlorophylls in Vitro. In: L. P. VERNON und G. R. SEELY (Ed.), The Chlorophylls, p. 523. New York and London: Academic Press. 1966.

187. SEELY, G. R. and M. CALVIN: Photochemical Studies of the Porphyrins. III. Photoreduction of a Porphyrin by Benzoin. J. Chem. Phys. **23**, 1068 (1955).

188. Seely, G. R. and A. Folkmanis: Photoreduction of Ethyl Chlorophyllide a by Ascorbic Acid in Ethanol-Pyridine Solutions. J. Amer. Chem. Soc. **86**, 2763 (1964).

189. Seely, G. R. and K. Talmadge: Photoreduction of Zinc Porphyrin by Ascorbic Acid. Photochem. Photobiol. **3**, 195 (1964).

190. Sidorov, A. N.: Spektroskopisches Studium der Hydrierung von Substanzen aus der Porphinreihe. Biofizika **10**, 226 (1965).

191. — Spectroscopic Studies on Reversible Reactions in Compounds of the Porphin Series. Russ. Chem. Rev. **35**, 153 (1966).

192. Sidorov, A. N. und V. E. Kholmogorov: Absorption Spectrum of Negative Pyridine Ions and their Reaction with Tetraphenylporphyrin and Magnesium Phthalocyanine. Doklady Akad. Nauk. SSSR **170** (5), 1202 (1966) [Chem. Abstr. **66**, 34907d (1967)].

193. — — Spectral Study of Conversions of Zinc Etioporphyrin in its Reaction with Sodium and Pyridin Negative Ions. Doklady Akad. Nauk. SSSR **170** (6), 1433 (1966) [Chem. Abstr. **66**, 60484h (1967)].

194. Silvers, S. J. and A. Tulinsky: The Crystal and Molecular Structure of Triclinic Tetraphenylporphyrin. J. Amer. Chem. Soc. **89**, 3331 (1967).

195. Smith, J. R. L. and M. Calvin: Studies on the Chemical and Photochemical Oxidation of Bacteriochlorophyll. J. Amer. Chem. Soc. **88**, 4500 (1966).

196. Sparatore, F. and D. Mauzerall: Osmium Tetroxide Oxidation of Protoporphyrin IX and Synthesis of Deuteroporphyrin IX 2,4-Diacrylic Acid. J. Organ. Chem. (USA) **25**, 1073 (1960).

197. Stoll, A. und E. Wiedemann: Chlorophyll. Fortschr. chem. Forsch. **2**, 538 (1952).

198. Strell, M., A. Kalojanoff und H. Koller: Teilsynthese des Grundkörpers von Chlorophyll a, des Phäophorbids a. Angew. Chem. **72**, 169 (1960).

199. Strell, M. und F. Zuther: Reaktionen in der Chlorophyllreihe, 5. Mitt. Veränderungen von Phorbiden und Chlorinen durch komplexbildende Metalle. Liebigs Ann. Chem. **612**, 264 (1958).

200. Tarlton, E. J., S. F. MacDonald and E. Baltazzi: Uroporphyrin 3. J. Amer. Chem. Soc. **82**, 4389 (1960).

201. Treibs, A. und R. Schmidt: Synthetische Arbeiten auf dem Chlorophyllgebiet. Synthese des 2-Desäthyl-phylloporphyrins. Liebigs Ann. Chem. **577**, 105 (1952).

202. Treibs, A. und E. Wiedemann: Über Chlorophyll. Liebigs Ann. Chem. **466**, 264 (1928).

203. — — Über den Abbau des Chlorophylls durch Alkali. Liebigs Ann. Chem. **471**, 146 (1929).

204. Urry, D. W.: Model Systems for Interacting Heme Moieties. I. The Heme Undecapeptide of Cytochrome c. J. Amer. Chem. Soc. **89**, 4190 (1967).

205. Urry, D. W. and J. W. Pettegrew: Model Systems for Interacting Heme Moieties. II. The Ferriheme Octapeptide of Cytochrome c. J. Amer. Chem. Soc. **89**, 5276 (1967).

206. Webb, L. E. and E. B. Fleischer: The Structure of Porphine. J. Amer. Chem. Soc. **87**, 667 (1965).

207. — — Crystal Structure of Porphine. J. Chem. Phys. **43**, 3100 (1965).

208. Weiss, C., H. Kobayashi and M. Gouterman: Spectra of Porphyrins. III. Selfconsistent Molecular Orbital Calculations of Porphyrin and Related Ring Systems. J. Mol. Spectry. **16**, 415 (1965).

209. Wenderoth, H.: Überführung von Porphyrinen in Isochlorine. Liebigs Ann. Chem. **558**, 53 (1947).

210. Wolf, H.: Zur weiteren Kenntnis des Chlorophylls und des Hämins, IV. Rotationsdispersions- und Zirkulardichroismus-Spektren von Chlorinen. Liebigs Ann. Chem. **695**, 98 (1966).

211. Wolf, H., H. Brockmann jr., H. Biere und H. H. Inhoffen: Zur weiteren Kenntnis des Chlorophylls und des Hämins, XIII. Darstellung der diastereomeren 10-Methoxy-(pyro)-methyl-phäophorbide a und Bestimmung der relativen Konfiguration am C-Atom 10. Liebigs Ann. Chem. **704**, 208 (1967).

212. Woodward, R. B.: Totalsynthese des Chlorophylls. Angew. Chem. **72**, 651 (1960).

213. — The Total Synthesis of Chlorophyll. Pure Appl. Chem. **2**, 383 (1961).

214. — Fundamental Studies in the Chemistry of Macrocyclic Systems Related to Chlorophyll. Industrie chim. belge **1962**, No. 11, 1293.

215. Woodward, R. B., W. A. Ayer, J. M. Beaton, F. Bickelhaupt, R. Bonnett, P. Buchschacher, G. L. Closs, H. Dutler, J. Hannah, F. P. Hauck, S. Itô, A. Langemann, E. Le Goff, W. Leimgruber, W. Lwowski, J. Sauer, Z. Valenta and H. Volz: The Total Synthesis of Chlorophyll. J. Amer. Chem. Soc. **82**, 3800 (1960).

216. Woodward, R. B. and V. Škarić: A New Aspect of the Chemistry of Chlorins. J. Amer. Chem. Soc. **83**, 4676 (1961).

217. Zerner, M. and M. Gouterman: Porphyrins. IV. Extended Hückel Calculations on Transition Metal Complexes. Theoret. Chim. Acta **4**, 44 (1966).

218. — — Porphyrins. V. Extended Hückel Calculations on Vanadyl ($VO^{2\oplus}$) and Vanadium (II) Complexes. Inorgan. Chem. **5**, 1699 (1966).

219. — — Porphyrins. VI. Extended Hückel Calculations on the Scandium Complex. Inorgan. Chem. **5**, 1707 (1966).

220. — — Porphyrins. X. Extended Hückel Calculations on Alkaline Earth Complexes. Theoret. Chim. Acta **8**, 26 (1967).

221. Zerner, M., M. Gouterman and H. Kobayashi: Porphyrins. VIII. Extended Hückel Calculations on Iron Complexes. Theoret. Chim. Acta **6**, 363 (1966).

(Eingelaufen am 9. Januar 1968)

Methoden und Ergebnisse der Sequenzanalyse von Ribonucleinsäuren

Von DIETER DÜTTING, Tübingen

Mit 24 Abbildungen

Inhaltsübersicht

Die experimentellen Arbeiten an den Serin-Transfer-RNA's wurden im Laboratorium von Professor H. G. ZACHAU *(Institut für Genetik der Universität Köln) ausgeführt, dem ich an dieser Stelle für die großzügige Unterstützung in allen Phasen der gemeinsamen Arbeit danken möchte. Für die Unterstützung der vorliegenden Arbeit bin ich Professor* A. GIERER *zu Dank verpflichtet. Professor* F. SANGER, *Drs.* J. T. MADISON *und* U. L. RAJBHANDARY *stellten mir dankenswerterweise noch unveröffentlichte Versuchsergebnisse zur Verfügung.*

Abkürzungen, Symbole, Definitionen

(siehe auch Abbildungen 2, 3 und 7, S. 361, 364, 377)

Für weitere Definitionen siehe Text und Glossary in: J. D. WATSON: Molecular Biology of the Gene. New York: Benjamin. 1965.

Anticodon: Sequenz von 3 benachbarten Nucleotiden einer tRNA, die sich in antiparalleler Weise mit dem Codon einer mRNA paaren läßt.

Cistron oder *Gen:* Abschnitt auf dem Doppelstrang der DNA, der eine Einheit der Information und Funktion darstellt.

Codon: Eine Sequenz von 3 benachbarten Nucleotiden einer mRNA, die 1 Aminosäure (oder Kettenabbruch) codiert.

DEAE-: Diäthylaminoäthyl-.

Deoxyribonucleosid: der Pentose-Zucker des ($\rightarrow$) Nucleosids ist β-D-Deoxyribose.

DNA: Deoxyribonucleinsäure.

EDTA: Äthylendiamintetraessigsäure.

Gen: siehe Cistron.

Genetische Information: Die in der Nucleotidsequenz einer DNA oder mRNA enthaltene Information.

Komplementäre Basenpaarung: Wasserstoffbrückenbindung zwischen Adenin und Thymin (oder Uracil) sowie Guanin und Cytosin in einer Nucleinsäure-Doppelhelix.

Mutation: Veränderung der Basensequenz eines Gens:
1. Punktmutation, Umwandlung einer Base in eine andere (es gibt Transitionen und Transversionen);
2. Deletion, Verlust eines oder mehrerer Nucleotide;
3. Insertion, Einfügung eines oder mehrerer Nucleotide;
4. Inversion, Umkehrung einer Nucleotidsequenz.

Nucleosid: Besteht aus einem Pentose-Zucker, mit dem in N-glycosidischer Bindung eine Purin- oder Pyrimidin-Base verknüpft ist.

Nucleotid: Grundbaustein der Nucleinsäuren, Phosphorsäureester eines Nucleosids.

p und *—:* bezeichnen eine Phosphatgruppe, und zwar links von der Abkürzung eines Nucleosids ein 5′-Phosphat (z. B. pA oder —A = Adenosin-5′-phosphat), rechts davon ein 3′-Phosphat (z. B. Gp oder G— = Guanosin-3′-phosphat).

p!: 2′,3′-Cyclophosphat, z. B. Ip! = Inosin-2′,3′-cyclophosphat.

PDE: Phosphodiesterase, spaltet vom 3′- oder 5′-Ende der Polynucleotide schrittweise Mononucleotide ab.

PME: Phosphomonoesterase, spaltet endständige Phosphatgruppen von Polynucleotiden ab bzw. überführt Mononucleotide in Nucleoside; im allgemeinen wird die alkalische Phosphomonoesterase aus *Escherichia coli* verwendet.

Polynucleotid: Eine lineare Sequenz von Nucleotiden, deren Pentose-Einheiten durch 3′,5′-Phosphodiesterbrücken verbunden sind.

PP: Pyrophosphat.

Ribonucleosid: Der Pentose-Zucker des (→) Nucleosids ist β-D-Ribose. In Abb. 7 (S. 377) sind die in den tRNA's aus Hefe gefundenen Ribonucleoside mit ihren Abkürzungen zusammengestellt. Die Art der Verknüpfung von Base und Ribose (R) ist in Abb. 3 (S. 364) zu sehen. Zusätzliche Abkürzungen: U*, Gemisch von U und UH_2; rT und T (in RNA), Ribothymidin.

RNA: Ribonucleinsäure.

mRNA: Messenger-Ribonucleinsäure, Matrizen-RNA, die zur Nucleotidsequenz eines Gens oder Gen-Komplexes komplementär ist und die Synthese einer oder mehrerer Aminosäuresequenzen dirigiert.

sRNA: Lösliche Ribonucleinsäure, Gemisch verschiedener tRNA's.

tRNA: Transfer-Ribonucleinsäure, bi-funktionelle RNA mit einer Akzeptor-Funktion (Bindung einer bestimmten Aminosäure am 3′-Hydroxy-Ende des 3′-terminalen Adenosins) und einer Transfer-Funktion (Wechselwirkung mit dem für die übertragene Aminosäure spezifischen Codon der mRNA im aktiven Proteinsynthese-Komplex).

RNase: Ribonuclease, Enzym, das bestimmte Phosphodiesterbindungen von Ribopolynucleotiden spaltet (vgl. Abb. 3, S. 364).

T1-RNase: Ribonuclease T1 aus Takadiastase.

T2-RNase: Ribonuclease T2 aus Takadiastase.

I. Einleitung

Eines der erstaunlichsten Phänomene der Biologie ist die Tatsache, daß die Kontinuität der Lebewesen, d. h. ihre Fähigkeit zur Reproduktion, letztes Endes auf dem Wechselspiel zweier Typen von linearen Makromolekülen in den Zellen beruht, den Nucleinsäuren und Proteinen. Die für diese Kontinuität verantwortliche genetische Information einer Zelle ist als eindimensionale 4-Buchstaben-Schrift in den *Nucleotidsequenzen* der *Deoxyribonucleinsäuren* (DNA's) verschlüsselt und wird im Verlauf der Proteinsynthese in die eindimensionale 20-Buchstaben-Schrift der *Aminosäuresequenzen* von *Proteinmolekülen* übersetzt.

Abschnitte auf dem Doppelstrang der DNA, die eine Einheit der Information und Funktion darstellen, nennt man *Gene* oder *Cistren*. Die Nucleotidsequenz eines Gens spezifiziert die Aminosäuresequenz einer Polypeptidkette und ist zu dieser colinear. Um mit PERUTZ (*73a*) zu sprechen: "The entire edifice of molecular biology is based on the hypothesis of the sequence of bases in a nucleic acid forming a code which determines the sequence of amino acids in a polypeptide chain. Moreover, the two sequences are thought to be co-linear, so that the order of loci on a genetic fine-structure map is the same as the order of amino acids controlled by those loci in the corresponding polypeptide chain". Die Faltung der linearen Aminosäureketten in die dreidimensionalen Strukturen der Proteine ist ein automatischer, keine zusätzliche Information benötigender Vorgang.

Überträger der Information von der DNA zu den Proteinen ist eine lineare, einzelsträngige Ribonucleinsäure-Kopie, die auf Grund komplementärer Basenpaarung (vgl. Abb. 2, S. 361) an einem Strang der

Literaturverzeichnis: SS. 414—421

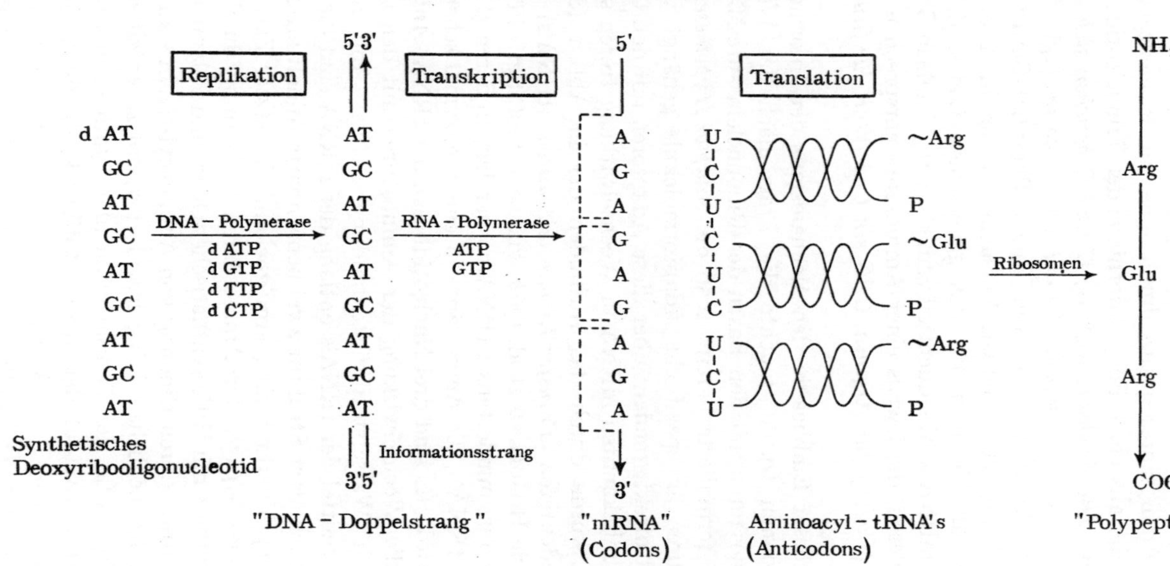

Die senkrechten Pfeile geben die Wachstumsrichtung der Nucleotid - bzw. Aminosäureketten und damit die Richtung der Replikation, Transkription und Translation an.

Abb. 1. Fluß der genetischen Information von der DNA über RNA zum Protein

DNA enzymatisch synthetisiert wird, die sogenannte *Messenger-Ribo-nucleinsäure* (mRNA). *Abb. 1* zeigt diesen Vorgang der Informations-übertragung an einem Modellbeispiel (*48*), nämlich der in vitro-Bio-synthese eines Arginin-Glutaminsäure-Copolypeptides. Dirigiert wird die Synthese dieser Aminosäuresequenz durch eine im Reagenzglas her-gestellte mRNA, die aus einer Folge alternierender Trinucleotide, AGA und GAG, besteht. *Die Buchstaben A, G, U und C beziehen sich auf die vier Standardbasen Adenin, Guanin, Uracil und Cytosin* (vgl. Abb. 7, S. 377), deren Anordnung am linearen Ribose-Phosphat-Gerüst einer RNA in Abb. 3 (S. 364) zu sehen ist. Versuche wie der in Abb. 1 dar-gestellte beweisen die Triplettnatur des genetischen Codes, d. h. die Basensequenz der mRNA wird sequentiell vom 5'- zum 3'-Ende in Gruppen von 3 Basen abgelesen, die jeweils einer Aminosäure entsprechen. *AGA ist ein Code-Wort (Codon) für Arginin, GAG ein Code-Wort für Glutamin-säure.*

Schon bevor dieser Einblick in den Mechanismus der Informations-übertragung gewonnen wurde, tauchte die Frage nach der Wechsel-wirkung der Aminosäuren mit den Basen der Ribonucleinsäure-Matrizen auf. CRICK (*20a*) formulierte bereits 1956 die *Adaptor-Hypothese*, nach der die Aminosäuren an spezifische Adaptormoleküle gehängt werden müssen, also nicht direkt, sondern über diese Adaptoren mit den Codons der damals noch unbekannten mRNA in Wechselwirkung treten sollten. Diese Hypothese wurde durch das Auffinden der in Abb. 1 (S. 359) schematisch angedeuteten *Transfer-Ribonucleinsäuren* (tRNA's) durch HOLLEY (*37a*) sowie HOAGLAND et al. (*36b*) glänzend bestätigt. Für jede der 20 Aminosäuren ist mindestens 1 tRNA vorhanden, an deren 3'-Ende, —CpCpA in allen tRNA's, die Aminosäure in einem enzymatischen Vor-gang angehängt wird. Es sind drei Nucleotidbasen der tRNA, *Anticodon* genannt, welche die Wechselwirkung der Aminosäuren mit den *Codons* der mRNA durch Wasserstoffbrückenbindungen (vgl. Abb. 2) ver-mitteln. Transportmittel der tRNA's entlang der mRNA sind die *Ribo-somen*, in deren ungeklärter Struktur zwei benachbarte Anheftungsstellen für tRNA's die Peptidverknüpfung ermöglichen. In letzter Zeit ist es gelungen, die *Codon-Tripletts* der Aminosäuren vollständig durch zell-freie Polypeptidsynthese mit Hilfe synthetischer Oligo- und Polynucleotide als „mRNA" (*48*) sowie durch Bindung von Aminoacyl-tRNA's an Ribo-somen mittels der 64 möglichen Trinucleotide (*67*, *20*) zu entziffern.

Während die Spezifität der Transkription (der Synthese von mRNA an den Genen der DNA) auf der Wasserstoffbrückenbindung komple-mentärer Standard-Basen, T : A, A : U, G : C, C : G, beruht (vgl. Abb. 2), liegt der Spezifität der *Translation* (d. h. der richtigen Aminosäure-verknüpfung an der mRNA) die Wechselwirkung des Anticodons einer tRNA mit einem Codon der mRNA zugrunde. Hierbei scheinen außer

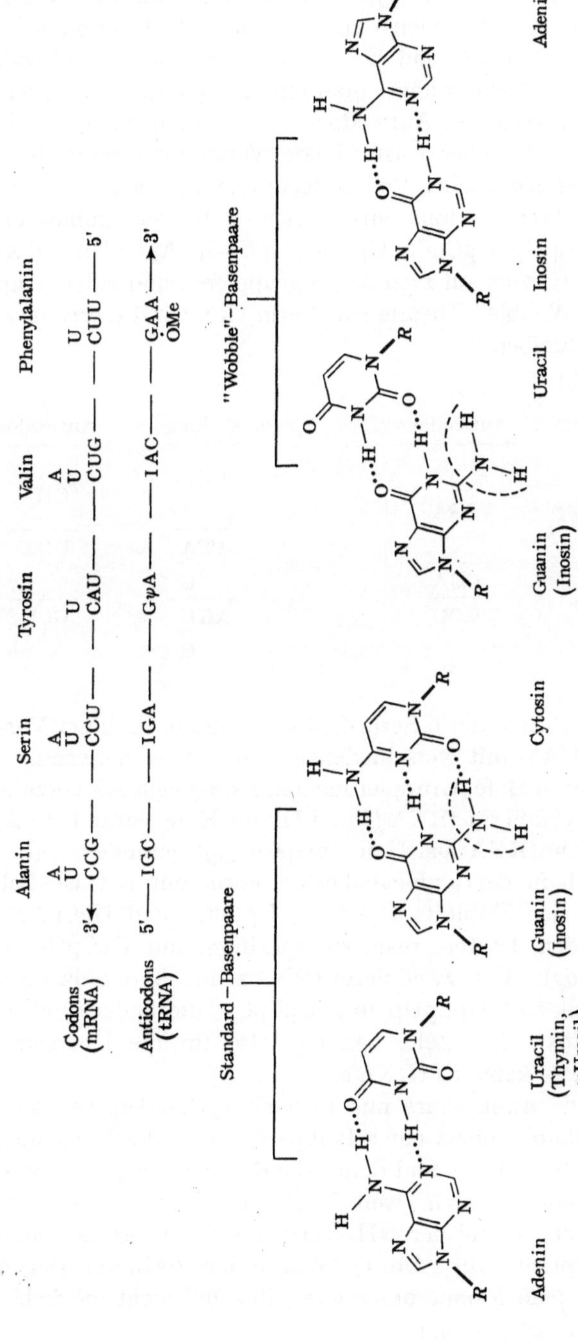

Abb. 2. Wechselwirkung komplementärer Nucleotidbasen bei der Translation. Die Anticodons der 5 tRNA's wurden in analogen Positionen ihrer Nucleotidsequenzen nachgewiesen (vgl. Abb. 20, S. 404)

den Standard-Basenpaaren auch sogenannte „Wobble"-Basenpaare *(Abb. 2)* eine Rolle zu spielen *(19)*. Der „Wobble" beruht in der Paarung von I (1. Position des Anticodons) mit C, U und A (3. Position der Codons), von G (Anticodon) mit C und U (Codon) bzw. von U (Anticodon) mit G und A (Codon). Dabei nimmt man eine antiparallele Wechselwirkung an, bei der die 1. Base des Anticodons (5'-terminal) mit der 3. Base des Codons (3'-terminal) gepaart ist. Diese Vorstellung trägt der starken Degeneration des genetischen Codes Rechnung, in welchem einer Aminosäure häufig mehrere Codons entsprechen, z. B. der Aminosäure Serin die 6 Codons UpCpUp, UpCpCp, UpCpA, UpCpGp, ApGpUp und ApGpCp. Man kann diese Codons auf zwei Arten gruppieren und jeder Gruppe entsprechend der „Wobble"-Theorie *eine* Serin-tRNA mit *einem* spezifischen Anticodon zuschreiben:

1. Codons →	Anticodons ←	2. Codons →	Anticodons ←
UCU		UCU	
C	IGA	C	GGA
A		UCA	UGA
UCG	CGA	G	
AGU	GCU	AGU	GCU
C		C	

Für die Translation der 6 Serin-Codons wären nach dieser Vorstellung nur 3 Serin-tRNA's mit verschiedenen Anticodons notwendig. Gruppierung 1 scheint in Hefe, Gruppierung 2 in *Escherichia coli* vorzukommen. In der Sequenz von Serin-tRNA (I und II) aus Hefe wurde IpGpAp *(113)*, in der von Alanin-tRNA aus Hefe IpGpCp *(39)* gefunden (vgl. Abb. 2). Erstere ließ sich in der aminoacylierten Form nur mit UC-Polymeren *(117)* bzw. mit den Tripletts UpCpUp, UpCpCp und UpCpAp *(67)* an Ribosomen binden, letztere reagierte spezifisch mit GpCpUp, GpCpCp und GpCpAp *(67)*. Von zwei Serin-tRNA's aus *E. coli* dagegen zeigte die eine Bindung mit UpCpUp und UpCpCp, die andere mit UpCpAp und UpCpGp *(95)*. Die Zelle benötigt also für die Übersetzung der 64 Codon-Tripletts keine 64 tRNA's.

Wäre für jede Aminosäure nur 1 Codon vorhanden, so könnte man aus der Aminosäuresequenz einer Polypeptidkette die Nucleotidsequenz der entsprechenden mRNA und damit die des zugehörigen Gens ableiten. Dies wäre deshalb möglich, weil Nucleotid- und Aminosäuresequenz colinear sind *(111)*, wobei der NH_2-Terminus des Proteins dem 5'-Ende der mRNA entspricht (vgl. Abb. 1). Wegen der erwähnten Degeneration des Codes ist es jedoch ohne besondere „Regeln" nicht möglich, vorherzusagen, welches Codon einer Codon-Gruppe jeweils in der mRNA für eine bestimmte Aminosäure verwendet wird. Da solche „Regeln" bisher nicht gefunden wurden, kann man vorläufig nur über die direkte Sequenzanalyse in die Primärstruktur einer mRNA eindringen (eine Ausnahme wird in Kap. VI erwähnt, S. 412).

Wegen der Länge und Labilität von mRNA-Molekülen liegt heute ihre direkte Sequenzanalyse in den meisten Fällen noch außerhalb des experimentell Möglichen. Eine der kleinsten mRNA's ist die etwa 3300 Nucleotide lange RNA der Phagen f2 und MS2, auf der mindestens drei Gene untergebracht sind. Das Mantelprotein des f2-Phagen besteht aus identischen Polypeptidketten von 129 Aminosäuren, die kürzlich in ihrer Sequenz aufgeklärt wurden (*108*). 129 Aminosäuren entsprechen etwa einer Sequenz von 390 Nucleotiden, die nur einen kleinen Teil der gesamten RNA ausmacht. Arbeiten an der RNA dieser Phagen sind in verschiedenen Laboratorien im Gange, haben aber bisher erst die Struktur von Oligonucleotiden der vollständigen Spaltung mit Pankreas-RNase und Endgruppen geliefert.

Günstigere Objekte sind die aus etwa 80 Nucleotiden aufgebauten tRNA's und die aus 120 Nucleotiden bestehende 5S-Ribosomale RNA. An diesen RNA's wurden die meisten der zur Zeit zur Verfügung stehenden Methoden der Sequenzanalyse von Ribonucleinsäuren entwickelt. Wir wollen uns daher im folgenden hauptsächlich mit diesen Molekülen befassen.

Für zusätzliche Ergebnisse der Sequenzanalyse, die im folgenden nicht behandelt werden können, sei auf eine Zusammenfassung von RajBhandary und Stuart (*77*) verwiesen, in der weitere Literaturangaben zu finden sind.

Die Arbeiten an der Nucleotidsequenz von tRNA's gipfelten Anfang 1965 in der vollständigen Strukturaufklärung der Alanin-tRNA aus Bäckerhefe durch Holley und Mitarbeiter (*39*). In der Folgezeit wurden die Arbeiten an der Struktur von vier weiteren tRNA's aus Hefe, Serin-tRNA I und II (*113*), Tyrosin-tRNA (*56*) und Phenylalanin-tRNA (*76, 78*) sowie von 5S'-Ribosomaler RNA aus *E. coli* (*13*) abgeschlossen. Größere Teilsequenzen liegen von Valin-tRNA aus Hefe (*6*) vor.

Während tRNA's außer den vier Standard-Nucleosiden A, G, C und U zusätzlich „seltene" Nucleoside enthalten (*35*), besteht die 5S-Ribosomale RNA nur aus den Standard-Nucleosiden (*13*). Die biologische Funktion der letzteren ist noch unbekannt.

Für eine erfolgreiche Sequenzanalyse müssen verschiedene Voraussetzungen erfüllt sein. Zunächst muß sich die RNA hochgradig reinigen lassen, wobei Spuren von spaltenden Enzymen (Nucleasen) ausgeschaltet werden müssen. Die Anreicherung einer bestimmten tRNA aus dem Gemisch der gesamten sRNA läßt sich weitgehend durch Bestimmung

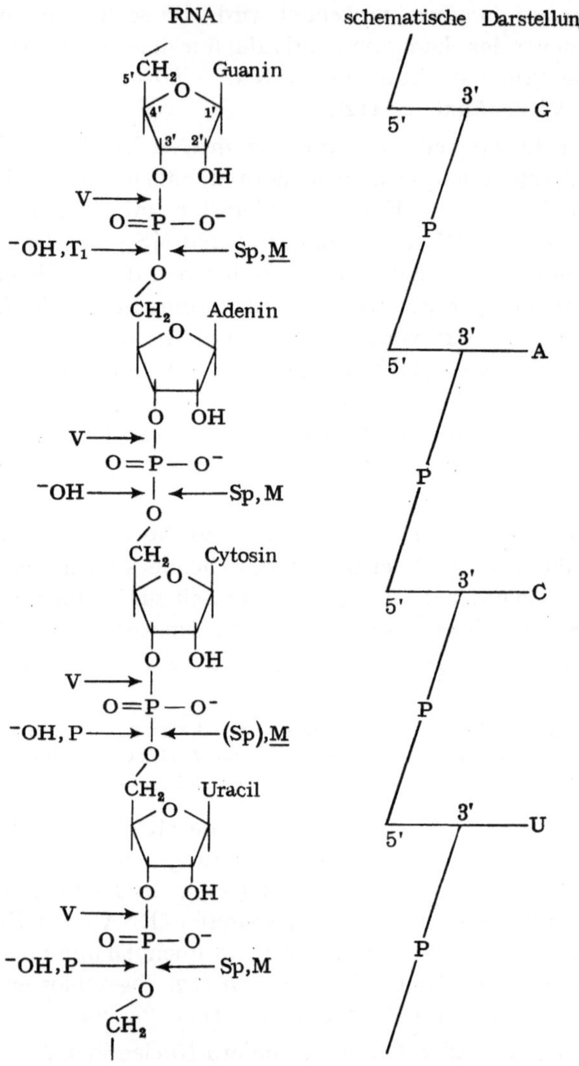

Abb. 3. Ausschnitt aus der Struktur einer RNA. Dargestellt ist eine Sequenz von 4 Nucleotiden, in abgekürzter Schreibweise: —GpApCpUp—. Die Pfeile an den 4 Phosphodiesterbindungen zeigen die Spaltstellen der hydrolytischen Enzyme an (Bezifferung: vgl. Text, S. 365)

ihrer Aminosäure-Akzeptor-Aktivität verfolgen, die auf der enzymatischen Aktivierung der zugeordneten radioaktiven Aminosäure duch ATP und ihrer anschließenden Veresterung mit dem CpCpA-Ende der tRNA

Literaturverzeichnis: SS. 414—421

beruht (*8*). Hierbei wird die Radioaktivität von einer säurelöslichen Form in eine säureunlösliche, die RNA, überführt. Dieser Test sagt nicht unbedingt etwas über die Intaktheit der RNA aus, da die partielle Spaltung von sRNA (*68*) und Serin-tRNA (*24*) mit T1-RNase in Gegenwart von Mg^{++}, welches die 3-dimensionale Struktur der tRNA stabilisiert, nur zu teilweisem Verlust der Akzeptoraktivität führt.

Ein weiteres gutes Reinheitskriterium sind die Mengen der Oligonucleotide, die man bei vollständiger Spaltung der RNA mit einer Endonuclease erhält. Eine weitgehend reine RNA sollte etwa 1 Mol Oligonucleotid per Mol RNA geben oder ein geradzahliges Vielfaches hiervon.

Wie die Sequenzanalyse von Polypeptiden spezifisch spaltende Peptidasen erfordert, so sind für die Analyse von RNA's Nucleasen notwendig, die bestimmte Internucleotid-Bindungen der RNA spalten. Man unterscheidet *Endonucleasen*, welche die RNA-Ketten im Innern zu Oligonucleotiden mit terminalen 3′-Phosphatgruppen spalten, und *Exonucleasen*, welche RNA- und DNA-Ketten schrittweise vom 3′- oder 5′-Ende her unter Freisetzung von Mononucleotiden abbauen. Zum anderen gibt es Enzyme, die sowohl Endo- wie Exonuclease-Aktivität besitzen und nach einer raschen endonucleolytischen Anfangsphase die entstandenen Oligonucleotide exonucleolytisch weiter abbauen, wobei die Reaktion auf dem Dinucleotid-Niveau stehenbleiben kann (*Micrococcus*-Nuclease, M). Die wichtigsten Endonucleasen sind die *Pankreas-RNase* (P), welche spezifisch CpX- und UpX-Bindungen der RNA spaltet, und die *T1-RNase* (T1), welche GpX-Bindungen der RNA hydrolysiert (*29*). Als Exonucleasen werden hauptsächlich die *Schlangengift-Phosphodiesterase* (V), welche RNA schrittweise vom nicht phosphorylierten 3′-Ende unter Freisetzung von 5′-Mononucleotiden (pA, pG, pC, pU) abbaut, und *Milz-Phosphodiesterase* (Sp) verwendet, welche vom dephosphorylierten 5′-Ende her 3′-Mononucleotide (Ap, Gp, Cp, Up) freisetzt. In *Abb. 3* sind an dem Ausschnitt einer RNA-Kette die Angriffsstellen der verschiedenen Enzyme durch Pfeile angedeutet. Die unspezifische Spaltung mit *Alkali* (⁻OH) verläuft, wie die Spaltung mit Pankreas- und T1-RNase, über ein intermediäres 2′,3′-Cyclophosphat und ist daher nur bei RNA-Molekülen, nicht bei DNA, möglich. Mit Endonucleasen wie der Pankreas- oder T1-RNase kann man RNA in kleinere, aus der Spezifität des Enzyms resultierende Oligonucleotide spalten (*vollständige Spaltung*) oder sie unter kontrollierten Bedingungen in größere Fragmente zerlegen (*partielle Spaltung*). Ein Fragment der partiellen Spaltung besteht aus zwei oder mehreren Oligonucleotiden der vollständigen Spaltung, und kann daher durch vollständige Nachspaltung mit einer der beiden RNasen und Identifizierung der kleineren Oligonucleotide analysiert werden.

Aus je einem Oligonucleotid der vollständigen Spaltung mit Pankreas-RNase und mit T1-RNase läßt sich ein längeres Oligonucleotid rekonstruieren, wenn in beiden Spaltprodukten ein „seltenes" Nucleotid oder ein ausgezeichnetes Di-, Tri- oder Tetranucleotid bei der Analyse gefunden wurde, das in der gesamten tRNA nur einmal vorhanden ist. In einem solchen Fall gibt eine *eindeutige Überlappung* zwischen den beiden Oligonucleotiden die verlängerte Sequenz (vgl. Abb. 10, S. 383). Zum Beispiel:

Pankreas-RNase-Spaltprodukt	T1-RNase-Spaltprodukt
ApGpGp*rTp*	*rTp*ψpCpGp

Überlappende Sequenz

ApGpGp*rTp*ψpCpGp

Dieses Verfahren ist nur begrenzt anwendbar (*25*), vor allem wenn — wie in der 5S-Ribosomalen RNA — keine seltenen Nucleotide vorhanden sind.

Zwei Fragmente der partiellen Spaltung können einander mit einem oder mehreren der Oligonucleotide überlappen, die bei der vollständigen Spaltung der beiden Fragmente entstehen (vgl. Tab. 1, S. 394). In diesem Fall kann eine längere Nucleotidsequenz der RNA rekonstruiert werden. Zum Beispiel:

Fragment 1

ApiPApApψpCpUpUpUpOMeUpGpGpGpCpUpUpUpGp*CpCpCpGp*

Fragment 2

*CpCpCpGp*MeCpGpCpApGpGpTpψpCpGp

Überlappende Sequenz

— — — — — CpUpUpUpGp*CpCpCpGp*MeCpGp — — — —

Literaturverzeichnis: SS. 414—421

Mit diesem Prinzip der Rekonstruktion überlappender Sequenzen war es möglich, die gesamte Nucleotidsequenz mehrerer tRNA's aufzuklären. Die Voraussetzung für den Erfolg dieser Arbeiten ist in drei Entwicklungen zu sehen, wenn man von der Reinigung der RNA's und der Entdeckung der Enzyme absieht:

1. der Analysenmethodik, mit der die Struktur der kleineren Oligonucleotide aufgeklärt wurde;

2. der Entdeckung der hochspezifischen partiellen Spaltung mit Pankreas- und T1-RNase (*73, 2, 52*);

3. der Ausarbeitung chromatographischer Trennmethoden. Bei der Trennung der großen Fragmente kam der Säulenchromatographie an DEAE-Cellulose und DEAE-Sephadex in 7 M Harnstoff (*105*) besondere Bedeutung zu.

II. Reinigung von Transfer-RNA's und 5 S-Ribosomaler RNA
1. Transfer-RNA's

Lösliche oder sRNA, die ein Gemisch vieler tRNA's darstellt, zeigt bei der Zonensedimentation im Sucrose-Gradienten einen Sedimentationskoeffizienten von etwa 4 S und keinerlei Andeutung einer Trennung in Einzelfraktionen. Sie wird aus Hefe oder *E. coli* durch direkte, kalte Phenolextraktion der Zellen isoliert, aus der wäßrigen Phase mit Äthanol ausgefällt, durch Extraktion mit 1 M NaCl von ribosomaler RNA weitgehend befreit und zweckmäßig an einer DEAE-Cellulose-Säule durch Elution mit M NaCl weiter gereinigt (*38, 120, 83*). *E. coli*-sRNA, die ohne Extraktion mit M NaCl hergestellt worden war, zeigte bei der Gelfiltration an Sephadex G 100 etwa 25% andere RNA-Komponenten, die als Ribosomale RNA, vor allem 5 S-Ribosomale RNA identifiziert werden konnten (*90*).

Aus der sRNA lassen sich einzelne tRNA's durch *Gegenstromverteilung* in komplexen zweiphasigen Lösungsmittelsystemen oder *Säulenchromatographie* herausfraktionieren. Für präparative Zwecke geeignet ist die Kombination verschiedener Gegenstromverteilungssysteme: z. B. konnte Serin-tRNA durch zwei Verteilungen von sRNA aus Brauereihefe im Tri-n-butylamin-System (*118*) und eine anschließende ausgedehnte Verteilung in einem Ammoniumsulfat-enthaltenden System (*22*) in guter Ausbeute mit einer maximalen Akzeptoraktivität von etwa 40 nMol ^{14}C-Serin per Fraktion isoliert werden (*114*). Dabei wurden zwei Serin-tRNA-Fraktionen, I und II, teilweise getrennt. Abbau von Randmaterial beider Fraktionen mit T1-RNase zeigte Unterschiede im Oligonucleotidmuster, die auf drei einfache Nucleotidaustausche in internen Positionen der RNA's zurückgeführt werden konnten. Die vermutlichen Anticodon-Regionen, —*ψ*—U—*I*—*G*—*A*—iPA—A—*ψ*—, waren in beiden Molekülen gleich (*117*).

Alanin-, Tyrosin- und Phenylalanin-tRNA aus Bäckerhefe wurden in einem stark Phosphat-haltigen Verteilungssystem (*3*) gereinigt und zeigten nach der Re-Verteilung eine weitere Art von Multiplizität: Es wurden zwei Phenylalanin- und zwei Tyrosin-tRNA's (*78, 56*) isoliert, die sich jeweils dadurch unterschieden, daß die rascher wandernde Nebenkomponente II das vollständige —CpCpA Ende trug, während die Sequenz der langsamer wandernden Hauptkomponente mit —CpC endete. Der gleiche Effekt wurde bei Serin-tRNA aus Bäckerhefe gefunden, die in eine Spezies mit der 3'-terminalen Sequenz —GpCpCpA und eine zweite mit —GpCpC als Ende zerlegt werden konnte (*49*). Serin-tRNA I und II aus Brauereihefe (*114*) endeten dagegen beide mit —GpCpCpA.

Allein aus dem Auffinden multipler tRNA's für eine Aminosäure auf distinkte, in ihren Codonspezifitäten verschiedene Moleküle zu schließen, ist auf Grund dieser Befunde nicht zulässig.

Die einzelnen Gegenstromverteilungssysteme und ihre Anwendung auf die Fraktionierung von sRNA sind vor kurzem ausführlich beschrieben worden (*21*) und sollen daher hier nicht weiter behandelt werden. Durch Flotation des Cetyl-trimethyl-ammoniumbromidsalzes der tRNA in einem Wasser-Äther-System (*6*) ließ sich die RNA-Ausbeute bei der Aufarbeitung der großen Volumina stark salzhaltiger Verteilungssysteme wesentlich verbessern (vgl. *114*).

Hochgereinigte tRNA's konnten auch durch Säulenchromatographie gewonnen werden. Zum Beispiel Phenylalanin-tRNA (*46*) und eine Leucin-tRNA (*47*) aus *E. coli* durch:

1. Umgekehrte Phasen-Chromatographie an Säulen, deren Füllung aus 1 Teil organischer Phase (4% Dimethyldilaurylammoniumchlorid in Isoamylacetat) gemischt mit 2 Teilen Trägermaterial (hydrophobe Diatomeen-Erde) bestand. Chromatographiert wurden bei 37° bis zu 4,2 g tRNA mit einem linearen NaCl-Gradienten in 0,01 M MgCl$_2$, 0,01 M Na-acetat, pH 4,5 (gesättigt mit Isoamylacetat). 2. Fraktionen mit hoch-angereicherter RNA wurden an Bio-Gel P2-Polyacrylamidgel-Säulen entsalzt und 3. an Bio-Gel P 100-Polyacrylamidgel-Säulen durch Filtration in 0,4 M NaCl, 0,01 M MgCl$_2$, 0,05 M Tris, pH 7, von höhermolekularen RNA-Verunreinigungen (ribosomaler RNA) befreit.

Ein anderes vielversprechendes Säulenverfahren ist die Chromatographie von sRNA an benzoylierter DEAE-Cellulose (*102, 34a*). Substitution der Hydroxy-Gruppen von DEAE-Cellulose durch Benzoyl-Reste erhöht ihre nicht-ionische Wechselwirkung mit Polynucleotiden und führt zu Trennungen auf Grund von Basenzusammensetzung und Sekundärstruktur der tRNA's. Nach einer Vorfraktionierung mit linearen NaCl-Gradienten in 0,01 M MgCl$_2$ wurden einzelne Akzeptor-Fraktionen bei pH 4,0 oder 3,5 in An- oder Abwesenheit von Mg^{++}

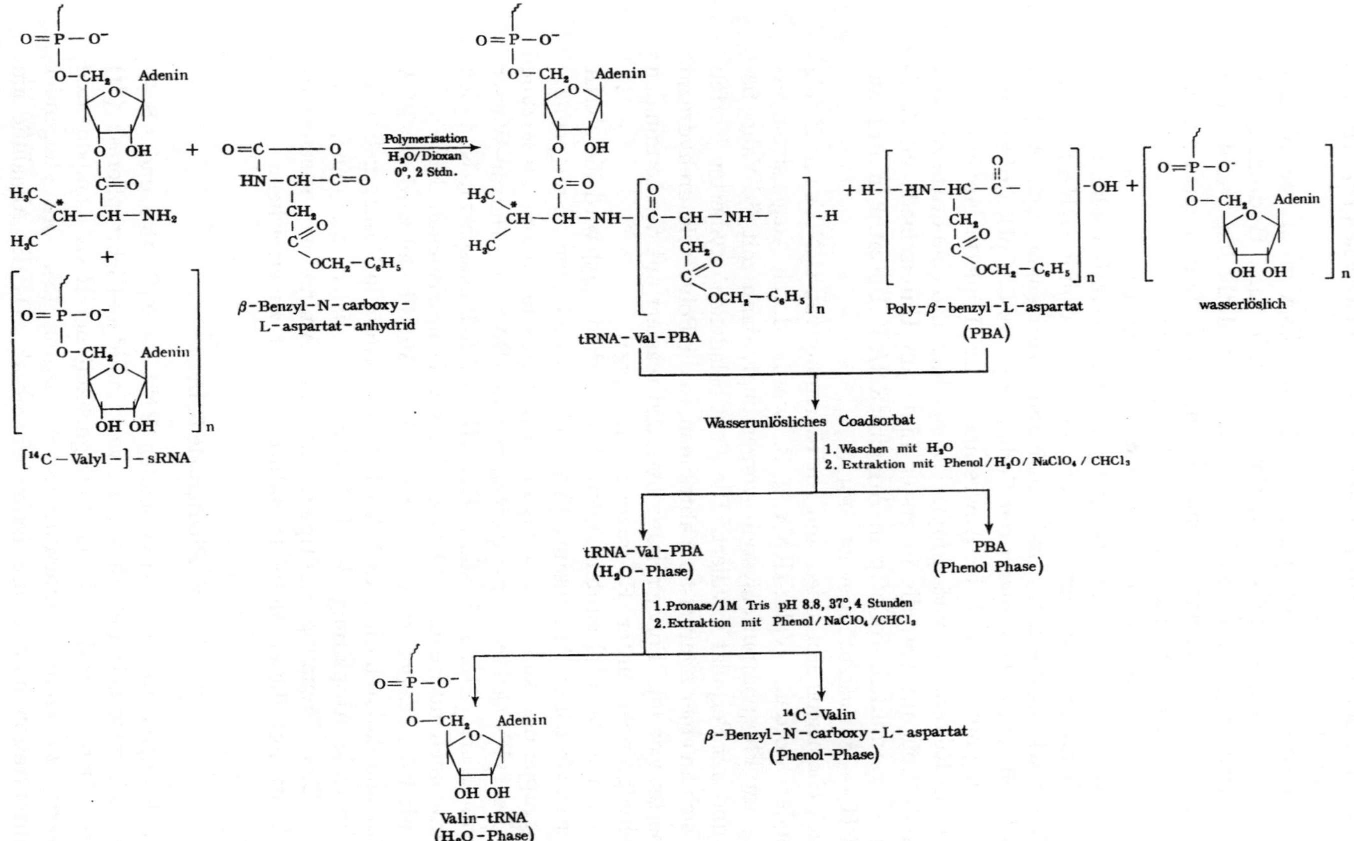

Abb. 4. Isolierung einer einzelnen tRNA aus dem tRNA-Gemisch nach KATCHALSKI et al. (45)

rechromatomatographiert und dadurch die entsprechenden tRNA's hoch angereichert.

Eine ausgezeichnete Trennung von *E. coli*-sRNA in 14 verschiedene aminosäurespezifische tRNA-Fraktionen wurde an Hydroxyapatit-Säulen erzielt (*72*). In Verbindung mit anderen Methoden (wie der umgekehrten Phasen-Chromatographie) ist dieses Verfahren auch für die gleichzeitige Reinigung mehrerer tRNA's geeignet.

Durch sukzessive Chromatographien an DEAE-Cellulose bei pH 7,5 und an DEAE-Sephadex A 50 bei pH 4,5 mit Harnstoff-Gradienten bzw. NaCl-Gradienten in 7 M Harnstoff (*17*) gelang es, Arginin-tRNA I und II aus Hefe stark anzureichern und voneinander zu trennen (*95*). Arginin-tRNA I ließ sich in Gegenwart der Codon-Tripletts CpGpU, CpGpC und CpGpA, Arginin-tRNA II in Anwesenheit der Tripletts ApGpA und ApGpG an Ribosomen binden (*95*). Es liegt hier also eine dritte Art von tRNA-Multiplizität vor, die in erster Linie auf Unterschiede im Anticodon — vermutlich IpCpGp in Arginin-tRNA I, UpCpUp in Arginin-tRNA II — zurückzuführen ist (*19*).

Auch chemische Methoden sind zur Reinigung von tRNA's mit Erfolg verwendet worden. Valin-tRNA z. B. wurde durch Gegenstromverteilung im Phosphatpuffersystem vorgereinigt, dann mit ^{14}C-Valin beladen und mit Perjodat oxydiert. Die nicht beladenen oxydierten Ketten ließen sich an einer Agargelsäule abtrennen, in die Polyacrylsäurehydrazid eingebettet war (*6*). Polyacrylsäurehydrazid reagiert mit den terminalen Dialdehydgruppen unter Hydrazonbildung (*75*).

In einer anderen eleganten chemischen Methode (*45*) wird die sRNA mit der gewünschten Aminosäure (AA) beladen und dann an die primäre Aminogruppe des Aminoacylendes Polyasparaginsäure durch Umsetzung mit β-Benzyl-N-carboxy-L-aspartat-anhydrid (BAA) in Dioxan-Wasser anpolymerisiert (*Abb. 4*, S. 369). Der tRNA-AA-Polyaspartat-Komplex fällt zusammen mit freiem PBA aus, während die unbeladenen tRNA's, die nicht mit BAA reagierten, in Lösung bleiben. Mit Phenol wird das PBA entfernt und dadurch die tRNA-AA-Polyaminosäure in Lösung gebracht. Anschließende Abspaltung der Polyaminosäure liefert die reine, freie tRNA. Eine Trennung multipler, für eine Aminosäure spezifischer tRNA's ist mit dieser Methode natürlich nicht zu erreichen.

2. Ribosomale RNA

Die Komponenten der Ribosomalen RNA — 23 S, 16 S und 5 S — lassen sich ausgezeichnet durch Polyacrylamidgel-Elektrophorese (*81*) trennen. Dabei verhält sich die Beweglichkeit der RNA-Komponenten umgekehrt zu ihren Sedimentationskoeffizienten (S). Mit steigender Gelkonzentration nimmt ihre Beweglichkeit ab. Da die Abnahme am

Literaturverzeichnis: SS. 414—421

größten für die Komponenten mit höherem Molekulargewicht ist, werden 23S- und 16S-RNA im 2,6%-Gel weiter voneinander getrennt als im 2,2%-Gel (*53*). Die beste Trennung von tRNA's (4S) und 5S-RNA wird in 7,5%-Gelen erzielt, welche die 16S- und 23S-Komponenten vollkommen ausschließen (*53*).

Für die Isolierung von RNA aus dem Gel wurde von HINDLEY (vgl. *12*, *36 a*) eine Methode entwickelt, in der nach dem Anfärben und Entfärben der RNA die ausgeschnittene Gel-Scheibe mit festem Harnstoff homogenisiert und die entstehende Paste dann mit einem alkalischen Puffer (0,1 M Tris, 0,01 M EDTA, 1 M NaCl) extrahiert wird.

Die Fraktionierung durch Polyacrylamidgel-Elektrophorese ist im wesentlichen eine Molekularfiltration, bei der die Porengröße des Gels durch die Acrylamid (und Bisacrylamid)-Konzentration beliebig variiert werden kann. Änderungen in der Sekundär-Struktur der RNA beeinflussen ihre Beweglichkeit, da ein kompaktes Molekül besser durch eine bestimmte Porengröße geht als ein entfaltetes. Diese Methode, mit der ^{32}P-markierte 5S-RNA für die Sequenzanalyse gereinigt wurde (*12*), gibt in vielen Fällen bessere Trennungen als die Zonensedimentation im Sucrose-Gradienten (*53*).

III. Analyse von Endgruppen und terminalen Sequenzen

Die Sequenzanalyse von Ribonucleinsäuren bedient sich drei prinzipieller Methoden: 1. der Bestimmung von Endgruppen und terminalen Sequenzen; 2. der vollständigen Spaltung mit Endonucleasen und Analyse der Mono- und Oligonucleotide und 3. der partiellen Spaltung mit Endonucleasen und vollständigen Weiterspaltung der großen Fragmente. Die vollständige und partielle Spaltung mit Exonucleasen wird hauptsächlich zum Analysieren der unter 2. anfallenden Oligonucleotide verwendet und daher dort besprochen (S. 382). Alle drei Methoden basieren auf chromatographischen Trennverfahren, die an Hand von speziellen Beispielen erläutert werden sollen.

Transfer-RNA's und 5S-Ribosomale RNA sind nur am 5'-Ende, nicht am 3'-Ende phosphoryliert. Bei der alkalischen Hydrolyse oder der vollständigen Spaltung mit T2-RNase (*87*) erhält man daher neben Mononucleotiden und (eventuell) ribosemethylierten Di- oder Trinucleotiden das *5'-Ende* als *Nucleosiddiphosphat* (pGp bei den meisten tRNA's, pUp bei 5S-Ribosomaler RNA), das *3'-Ende* als *Nucleosid* (A bei allen tRNA's, U bei 5S-Ribosomaler RNA). Nucleoside und Nucleosiddiphosphate lassen sich von Mononucleotiden leicht an DEAE-Cellulose-Säulen oder durch Papierelektrophorese bei pH 2,7—3,5 abtrennen.

In analoger Weise zeichnen sich die terminalen Sequenzen, die bei der vollständigen Spaltung der RNA mit Pankreas- oder T1-RNase an-fallen, durch ein 5'-terminales Phosphat oder ein 3'-terminales Nucleosid aus. Zum Beispiel gab die T1-RNase-Spaltung von 5 S-Ribosomaler RNA pUpGp und CpApU (*12*), die von Tyrosin-tRNA pCpUpCpUpCpGp und ApCpC (*56*) als 5'- bzw. 3'-Ende.

Durch partiellen Abbau mit T1-RNase lassen sich diese Endsequenzen je nach den Spaltungsbedingungen verschieden weit nach rechts oder links verlängern, z. B. im Falle der Tyrosin-tRNA das 5'-Ende zu pCpUpCpUpCpGpUpAp2MeGp, das 3'-Ende zu CpCpCpCpCpGpGpGp ApGpApCpC (*56*). Ebenso konnte das 5'-terminale Pankreas-RNase-Spaltprodukt von Serin-tRNA, pGpGpCp, durch partielle Spaltung mit Pankreas-RNase zu pGpGpCpApApCpUpUp verlängert werden (*24*).

Bei höhermolekularen RNA's, in deren Struktur man von den beiden Enden her durch vollständige und partielle Spaltungen eindringen will, empfiehlt es sich, die Enden vorher radioaktiv zu markieren.

Zur Markierung des *5'-Endes* stehen eine enzymatische und eine chemische Methode zur Verfügung. Nach Abspaltung des 5'-Phosphats von RNA oder DNA mit PME kann man *enzymatisch* die γ-^{32}P-Gruppe von ATP auf das 5'-Hydroxy-Ende der Nucleinsäure übertragen:

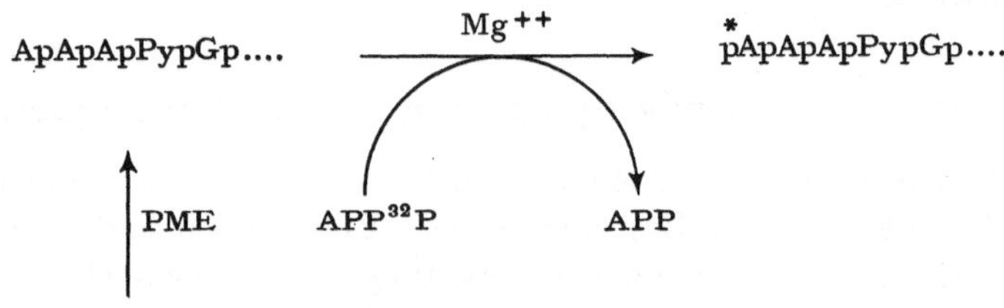

Das diese Reaktion katalysierende Enzym, *Polynucleotidkinase*, wird in *E. coli*-Zellen durch Infektion mit T4- oder T2-Phagen induziert (*82*, *69*). Anwendung dieser Methode auf hochmolekulare RNA setzt die Abwesenheit von RNase-Verunreinigungen im Enzympräparat voraus. Das Enzym wurde zu diesem Zweck aus der T4-infizierten *E. coli*-Mutante *A 19* (RNase 1-negativ) isoliert und zur Phosphorylierung von PME-behandelter 16 S- und 23 S-Ribosomaler RNA verwendet (*101*). 16 S-5' (^{32}P) RNA- und 23 S-5' (^{32}P) -RNA-Fraktionen wurden nach ihrer Re-Isolierung durch Gelfiltration an Sephadex G 200 und Sucrose-Dichtegradienten-Zentrifugation mit T1-RNase bzw. Pankreas-RNase gespalten, und die Oligonucleotide an DEAE-Sephadex A 25-Säulen in 7 M Harnstoff bei pH 8 nach ihrer Kettenlänge fraktioniert. Als radioaktiv markierte Hauptfragmente wurden pApApApPypGp bzw. pApAp ApPyp in den 16 S-RNA-Hydrolysaten, pGp bzw. pGpPupPyp in den 23 S-RNA-Hydrolysaten identifiziert.

Literaturverzeichnis: SS. 414—421

Die *chemische* Markierungsmethode besteht aus der Reaktion des Cetyltrimethylammonium-Salzes der RNA oder DNA mit einem etwa 40—50fachen Überschuß des Tri-n-hexyl-ammoniumsalzes von (^{14}C)-Methyl-phosphoromorpholidat in wasserfreiem Dimethylsulfoxyd bei Raumtemperatur (79):

$$\text{HO}-\overset{\overset{\textstyle O}{\|}}{\underset{\underset{\textstyle O^-}{|}}{P}}-\text{OXpYpZp}\ldots\ldots + {}^{14}\text{CH}_3-\text{O}-\overset{\overset{\textstyle O}{\|}}{\underset{\underset{\textstyle O^-}{|}}{P}}-\text{N}\bigcirc\text{O}\rightarrow$$

$$\rightarrow {}^{14}\text{CH}_3-\text{O}-\overset{\overset{\textstyle O}{\|}}{\underset{\underset{\textstyle O^-}{|}}{P}}-\text{O}-\overset{\overset{\textstyle O}{\|}}{\underset{\underset{\textstyle O^-}{|}}{P}}-\text{OXpYpZp}\ldots + \text{HN}\bigcirc\text{O}$$

Eine derartige pyrophosphorylierte sRNA zeigt die gleiche Akzeptoraktivität und ein ähnliches Elutionsverhalten an DEAE-Cellulose-Säulen wie unbehandelte sRNA und kann daher als intakt betrachtet werden.

Das *3'-Ende* einer RNA läßt sich am besten durch Oxydation der freien *2',3'-cis*-Diol-Gruppierung der terminalen Ribose mit Perjodat und Reduktion der entstandenen Aldehyd-Gruppen mit ^{3}H-Natrium-borhydrid (Abb. 6, S. 375) markieren (78).

Doppelmarkierung der beiden Enden einer hochmolekularen RNA, des 5'-Endes mit ^{32}P (Polynucleotidkinase) und des 3'-Endes mit ^{3}H (Perjodat, NaB^{3}H$_4$), sollte es möglich machen, bei der anschließenden partiellen endonucleolytischen Spaltung definierte Sequenzen von beiden Enden zu isolieren. Da man die Bedingungen der partiellen Spaltung in weiten Grenzen variieren kann, besteht die Aussicht, schrittweise längere Sequenzen von den Enden isolieren zu können. Eine erfolgreiche Verwendung dieser Markierungsmethoden erfordert allerdings eine völlig intakte RNA, da Brüche zu neuen Endgruppen und damit unübersichtlichen Resultaten führen.

Die 3'-terminalen Sequenzen bestimmter tRNA's ließen sich nach radioaktiver Markierung auch ohne vorangehende Fraktionierung von sRNA isolieren. Durch simultane Beladung von Hefe-sRNA mit ^{3}H-Threonin und ^{14}C-Alanin oder ^{3}H-Serin und ^{14}C-Glycin und anschließende Spaltung mit T1-RNase bei *pH 5,4* konnten die terminalen Sequenzen von Threonin- und Alanin-tRNA bzw. Serin- und Glycin-tRNA als Aminoacyl-Oligonucleotide freigesetzt und von den nicht aminoacylierten Oligonucleotiden an Dowex 50-X2-Säulen abgetrennt werden (93). Unter pH 3 sind Aminoacyl-Oligonucleotide wie CpApCpCpA-Aminosäure basisch und werden daher von Kationenaustauschern zurückgehalten, während die normalen sauren Oligonucleotide durch 0,2 M

Ammoniumformiat, pH 2,7, vollständig aus der Säule gewaschen werden. Mit 0,05 M Ammoniumformiat bei pH 3,8 kann man dann z. B. das ^{14}C-Alanin-Oligonucleotid, bei pH 5,2 das ^{3}H-Threonin-Oligonucleotid eluieren. Analyse nach Abspaltung der Aminosäure bei pH 9 ergab folgende Strukturen (*92*): CpCpA (Serin), CpApCpCpA (Glycin und Threonin), UpCpCpApCpCpA (Alanin). Die gleichen terminalen

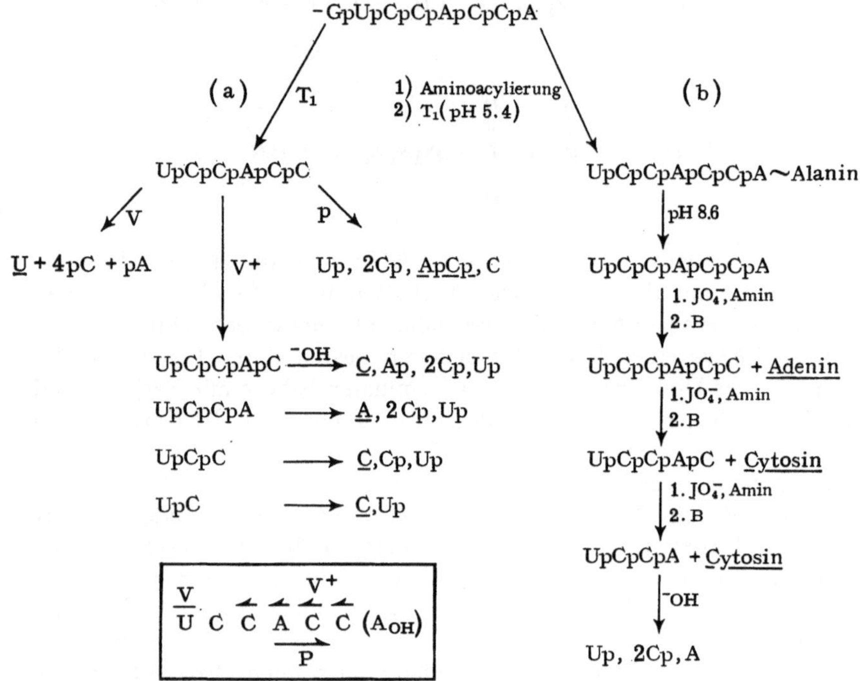

Abb. 5. Analyse der 3′-terminalen Nucleotidsequenz von Alanin-tRNA: a) durch sequentiellen Abbau mit PDE und b) durch sequentiellen Abbau mit Perjodat. Vgl. Abb. 10 und Text, S. 383/382, für die Erklärung der Enzymsymbole

Sequenzen für Serin und Alanin-tRNA wurden aus den entsprechenden hochangereicherten tRNA's isoliert (*114, 41*). In *Abb. 5a* ist die Strukturaufklärung der terminalen Sequenz von Alanin-tRNA durch Holley et al. (*41*) derjenigen von Smith und Herbert (*92*) in *Abb. 5b* gegenübergestellt. Erstere verwendeten den sequentiellen Abbau mit Schlangengift-PDE (*42*) (vgl. Kap. IV, S. 377), letztere den sequentiellen Abbau mit Perjodat (*110, 66*). Der letztgenannte Abbau besteht aus vier Reaktionen: 1. der im Neutralen ablaufenden Oxydation des terminalen Glycols zum Dialdehyd; 2. der aminkatalysierten Spaltung der terminalen Phosphodiesterbindung (β-Eliminierung); 3. der Freisetzung der Purin- oder

Pyrimidinbase und 4. der Abspaltung des terminalen Phosphats mit PME. Einen möglichen Mechanismus für diesen Abbau (*16*) zeigt *Abb. 6*. Bei wiederholter Anwendung auf hochmolekulare RNA verursachen die Amine Kettenbrüche. Durch Verwendung von Anilin als Katalysator der β-Eliminierung (*100*) ließ sich diese Schwierigkeit umgehen.

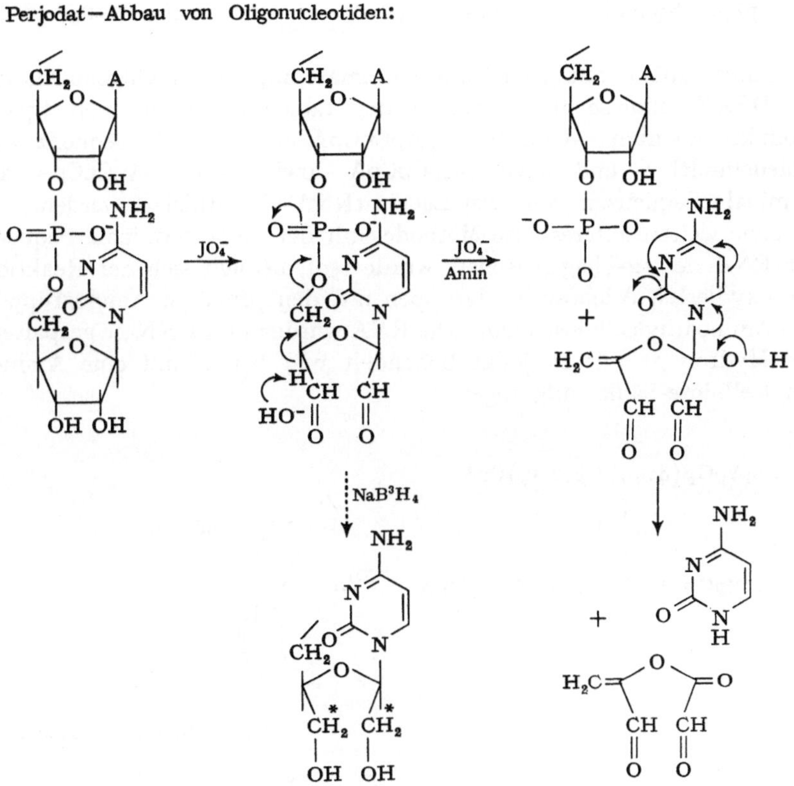

Perjodat—Abbau von Oligonucleotiden:

Abb. 6. Reaktionsmechanismus des schrittweisen Perjodat-Abbaus von Oligonucleotiden (*15*). Die nicht zum Abbau gehörende Reduktion des Dialdehyds mit Tritium-markiertem NaBH$_4$ bei neutralem pH führt zum stabilen radioaktiven Diol

BERG et al. (*9*) analysierten die terminalen Sequenzen von *E. coli* Leucin- und Isoleucin-tRNA, indem sie sRNA mit Leucin oder Isoleucin beluden, die freien Adenosin-Enden der nicht aminoacylierten RNA-Ketten mit Perjodat oxydierten und durch β-Elimination entfernten. Nach Hydrolyse der schützenden Aminoacylester konnten sie die intakten CpCpA-Enden der Leucin- bzw. Isoleucin-tRNA's enzymatisch in Gegenwart von PP$_1$ abspalten. Das diese Reaktion katalysierende Enzym sRNA-Pyrophosphorylase greift Ketten, die mit Cp enden, kaum an.

In der umgekehrten, vom gleichen Enzym katalysierten Reaktion wurde dann mit (α-^{32}P)CTP das —CpCpA der Leucin- oder Isoleucin-tRNA's durch zwei Äquivalente ^{32}pC ersetzt:

$$pZp\ldots XpCpCpA \atop + \atop pZp\ldots XpCpCp \quad \xrightarrow[\text{Enz.}]{PP_i} \quad {pZp\ldots X \atop + \atop pZp\ldots XpCpCp} \quad \xrightarrow[\text{Enz.}]{(\alpha\text{-}^{32}P)\,CTP} \quad {pZp\ldots Xp\overset{*}{C}p\overset{*}{C} \atop + \atop pZp\ldots XpCpCp}$$

Durch Abbau mit T1-RNase, Chromatographie der Oligonucleotide an DEAE-Cellulose und Analyse der radioaktiv markierten Spaltprodukte konnten —GpCp(UpC)pApCpCpA als terminale Sequenz von Isoleucin-tRNA und —GpCpApCpCpA sowie —GpUpApCpCpA als terminale Sequenzen von zwei Leucin-tRNA's identifiziert werden.

Eine weitere interessante Methode, mit der die 3'-terminale Sequenz der RNA des f2-Phagen isoliert wurde (50), bedient sich der Reaktion der oxydierten Adenosin-Endgruppen mit den primären Aminogruppen von Aminoäthylcellulose (120). Die RNA wurde mit T1-RNase gespalten, das Hydrolysat mit Perjodat behandelt und bei 4° auf eine Aminoäthylcellulose-Säule aufgezogen:

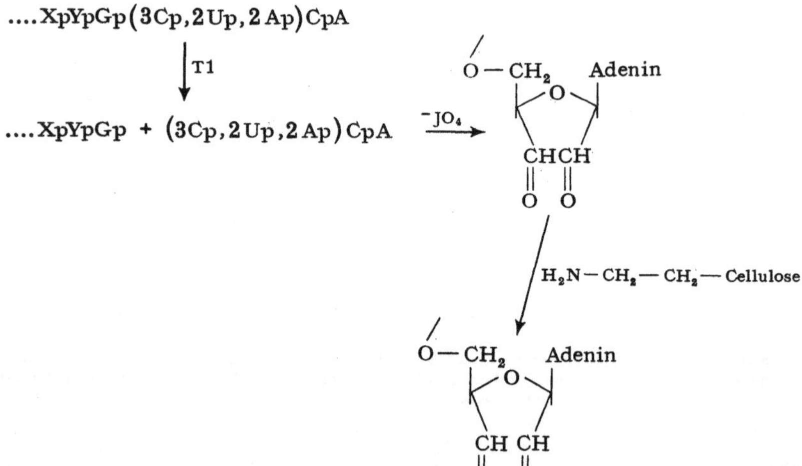

$$\ldots XpYpGp(3Cp, 2Up, 2Ap)CpA$$
$$\Big\downarrow T1$$
$$\ldots XpYpGp + (3Cp, 2Up, 2Ap)CpA \xrightarrow{^-JO_4}$$

Mit 4 M NaCl ließen sich alle internen nichtoxydierten Oligonucleotide (…XpYpGp) bei 4° heraus-eluieren. Nach Waschen der Säule mit KHCO$_3$ und H$_2$O konnte das terminale Oligonucleotid bei 45° mit 0,5 M n-Propylammoniumbikarbonat durch β-Elimination aus der Schiff-Base freigesetzt und als (3Cp, 2Up, 2Ap)Cp gewonnen werden.

Literaturverzeichnis: SS. 414—421

IV. Vollständige enzymatische Spaltung
von Ribonucleinsäuren und Analyse der Oligonucleotide

1. Transfer-Ribonucleinsäuren

a) Spaltweise der Ribonucleasen

Die Spezifität der gebräuchlichsten Endonucleasen, Pankreas- und
T1-RNase, erstreckt sich auch auf die „seltenen" Nucleotide, deren

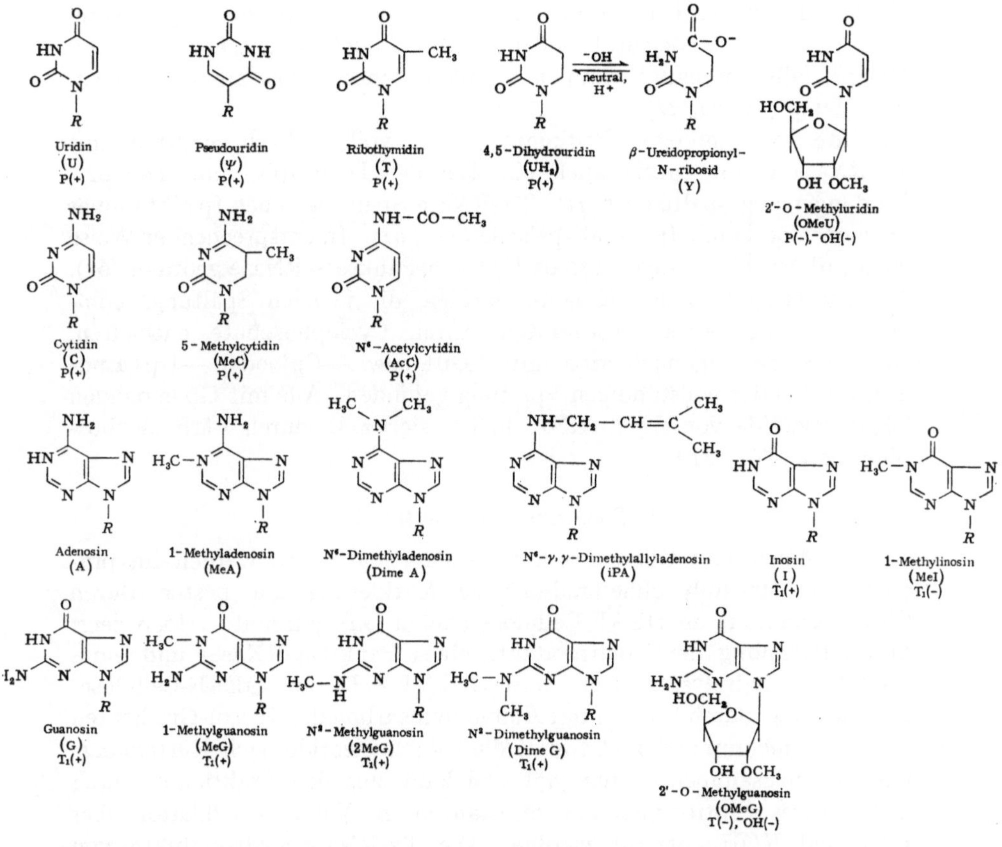

Uridin (U) P(+) — Pseudouridin (Ψ) P(+) — Ribothymidin (T) P(+) — 4,5-Dihydrouridin (UH₂) P(+) — β-Ureidopropionyl-N-ribosid (Y) — 2'-O-Methyluridin (OMeU) P(-), ⁻OH(-)

Cytidin (C) P(+) — 5-Methylcytidin (MeC) P(+) — N⁴-Acetylcytidin (AcC) P(+)

Adenosin (A) — 1-Methyladenosin (MeA) — N⁶-Dimethyladenosin (Dime A) — N⁶-γ,γ-Dimethylallyladenosin (iPA) — Inosin (I) T₁(+) — 1-Methylinosin (MeI) T₁(-)

Guanosin (G) T₁(+) — 1-Methylguanosin (MeG) T₁(+) — N²-Methylguanosin (2MeG) T₁(+) — N²-Dimethylguanosin (Dime G) T₁(+) — 2'-O-Methylguanosin (OMeG) T(-), ⁻OH(-)

Abb. 7. Nucleoside aus Alanin-, Serin- und Tyrosin-tRNA. Abkürzungen in Klammern. P (+), T1 (+) usw., vgl. Text, S. 377. R = Ribose

Nucleosidbausteine für Alanin-, Serin- und Tyrosin-tRNA in *Abb.* 7
zusammengestellt sind. So spaltet T1-RNase neben GpX- auch
DimeGpX-, 2MeGpX-, MeGpX- und IpX-Bindungen [in Abb. 7 symbolisiert durch T1 (+)], Pankreas-RNase neben UpX- und CpX- auch ψpX-,
TpX-, UH₂pX-, MeCpX- und AcCpX-Bindungen [in Abb. 7 symbolisiert
durch P (+)].

Nicht gespalten werden Bindungen, an denen 2'-O-Methylribo-nucleoside wie OMeU und OMeG beteiligt sind, da die 2'-O-Methyl-Gruppe die Bildung des intermediär entstehenden 2',3'-Cyclophosphats (vgl. Kap. I, S. 366) blockiert. Es wurden daher aus T1-RNase-Hydrolysaten von Serin-tRNA nach der Weiterspaltung mit Pankreas-RNase die RNase- und Alkali-stabilen Dinucleotide OMeUpGp und OMeGpGp isoliert (*27, 61*), die durch Exonucleasen weiter abgebaut werden können. Häufig findet man, daß die 2',3'-Cyclophosphatenden von „seltenen" Nucleotiden wesentlich langsamer geöffnet werden als die der entsprechenden Standardnucleotide. So überführt T1-RNase das DimeGp! und MeGp!, falls überhaupt, dann nur außerordentlich langsam in DimeGp und MeGp (*26, 96, 41*).

Unter den milderen Bedingungen der vollständigen Spaltung mit T1-RNase bleibt häufig auch ein Teil der DimeGpX- und MeGpX-Bindungen ungespalten (*61, 41*). T1-RNase-Spaltung neben Ip gibt immer Oligonucleotide mit Ip!- und Ip-Enden (*41, 61*). In entsprechender Weise wird ψp! erheblich langsamer als Up! von Pankreas-RNase geöffnet (*61*). Im Unterschied zu den Mononucleotiden, die je nach Spaltungsbedingungen immer einen bestimmten Anteil Cyclophosphate enthielten, wurden keine Oligonucleotide mit —Gp! bzw. —Cp! oder —Up!-Endgruppen bei der vollständigen Spaltung gefunden. Alle mit Gp endenden Oligonucleotide von Serin-tRNA ließen sich z. B. durch PME dephosphorylieren (*61, 114*).

b) Trennung der Spaltprodukte

Die Mono- und Oligonucleotide beider Spaltungen werden im präparativen Maßstab ohne radioaktive Markierung am besten durch Chromatographie an DEAE-Cellulose-Säulen (*74*) getrennt. *Abb. 8* zeigt die Auftrennung der Spaltprodukte eines Pankreas-RNase- und eines T1-RNase-Hydrolysats von Serin-tRNA I + II an DEAE-Cellulose-Säulen bei 4° durch Elution mit Ammoniumcarbonat (pH 8,6)-Gradienten (*114*). Ammoniumcarbonat trennt die Oligonucleotide nach Kettenlänge und Basenzusammensetzung (*97*) und kann aus den Fraktionen durch mehrmaliges Eintrocknen der Substanzen im Vakuumexsikkator über P_2O_5 und KOH entfernt werden. Die T1-RNase-Spaltprodukte von Serin-tRNA ließen sich am besten nach vorangehender Abspaltung des 3'-terminalen Phosphats durch PME trennen (*27*). Innerhalb einer Gruppe von Substanzen gleicher Kettenlänge wurden zuerst pyrimidin- später purinreiche Oligonucleotide eluiert.

UH$_2$p- sowie Cp-reiche Substanzen kamen besonders früh von der Säule, so UH$_2$pUH$_2$pApApG zusammen mit ApApApG, CpCpCpG zusammen mit UpUpG. Von den purinreichen Pankreas-RNase-Spaltprodukten wurden die Ap-reichen vor den Gp-reichen Oligonucleotiden gleicher Kettenlänge eluiert. OMeUpGpGpGpCp haftete am stärksten an der DEAE-Cellulose und konnte nur nach vorangehender PME-Behandlung des Hydrolysats mit M NaCl oder besser mit M NaCl in 7 M Harnstoff in schlechter Ausbeute von der Säule eluiert werden (*114*).

Bis zum Hexanucleotid ließen sich Gemische von T1-RNase-Spaltprodukten durch Papierelektrophorese in 0,2 M Ammoniumformiat, pH 2,7, 10^{-3} M EDTA weiter auftrennen (*27*). Jedes dieser T1-RNase-Spaltprodukte hatte eine charakteristische Wanderungsgeschwindigkeit (im Vergleich zu Up) und nach der Elution

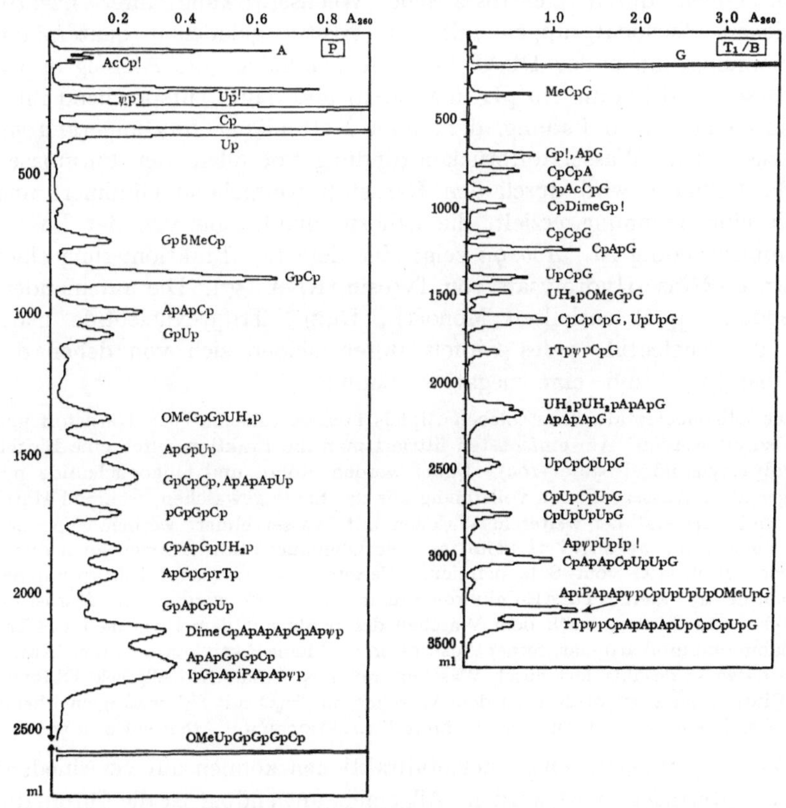

Abb. 8. Pankreas- und T1-RNase-Spaltprodukte von Serin-tRNA I + II (*114*). Die Hydrolysate wurden an DEAE-Cellulose-Säulen (Pankreas-RNase-Hydrolysat: 1 × 40 cm, T1-RNase/PME-Hydrolysat: 1 × 50 cm) mit linearen Ammoniumcarbonat-Gradienten bei 4° chromatographiert. Vgl. Text, S. 378

vom Papier im allgemeinen auch ein charakteristisches UV-Spektrum (*27, 114*). Beide Daten zusammen erlaubten später, die Spaltprodukte der größeren Fragmente ohne weitergehende Analyse zu identifizieren (*24*).

Die elektrophoretische Trennung von Pankreas-RNase-Spaltprodukten gelang nur bis zu den Trinucleotiden. Längere, purinreiche Oligonucleotide gaben im obigen Elektrophorese-System keine distinkten Flecken mehr, auch nicht nach vorangehender Dephosphorylierung. Die nicht oder schlecht getrennten längeren Oligonucleotide beider Spaltungen konnten durch Rechromatographie an DEAE-Cellulose in 7 M Harnstoff — verd. HCl oder — 0,1 M HCOOH mit NaCl-Gradienten getrennt werden (*114*).

Holley und Mitarbeiter (*41*, *119*) trennten die Pankreas- und T1-RNase-Spaltprodukte von Alanin-tRNA zunächst an DEAE-Sephadex A 25-Säulen (0,25 × 225 cm) mit Ammoniumcarbonatgradienten, gingen aber später zur Chromatographie in 7 M Harnstoff, pH 8, mit NaCl-Gradienten über (*2*) (Abb. 15 b, S. 393).

In Gegenwart von 7 M Harnstoff bei pH 7 werden Oligonucleotide hauptsächlich durch elektrostatische Wechselwirkung ihrer negativ geladenen Phosphatgruppen mit den positiv geladenen quaternären Ammoniumgruppen der DEAE-Cellulose gebunden (die Aminogruppen der Basen sind bei diesem pH ungeladen) und daher entsprechend ihrer gesamten negativen Ladung, d. h. ihrer Kettenlänge fraktioniert (*105*). Die zusätzliche Wasserstoffbrücken-Bindung vor allem der Purinbasen an die Cellulose wird durch den Harnstoff weitgehend eliminiert und damit eine Trennung erzielt, die nahezu unabhängig von der Basen-zusammensetzung ist. *Abb. 9a* zeigt eine derartige Fraktionierung eines Pankreas-RNase-Hydrolysats von Tyrosin-tRNA (*57*). Die aufeinander-folgenden Gipfel enthalten Mono(1)-, Di(2)-, Tri(3)-Nucleotide usw., d. h. die Nucleotide jedes Gipfels unterscheiden sich von denen des Nachbargipfels um eine negative Ladung.

Die Oligonucleotide eines solchen Gipfels müssen zunächst von Harnstoff und Salz befreit werden. Am einfachsten filtriert man die Fraktion durch eine Biogel-P 2-Polyacrylamidgel-Säule (*106*). Dabei werden Mono- und Oligonucleotide mit weniger als 1 Wasservolumen vollständig aus der Säule gewaschen, während Harn-stoff und Salz erst bei weiterem Waschen mit Wasser eluiert werden. Von den Pentanucleotiden an aufwärts kann man die Oligonucleotidfraktionen auch durch Dialyse gegen H_2O vom Salz befreien. Weitere Methoden sind Adsorption der Nucleotide an kleine DEAE-Cellulose-Säulen aus stark verdünnten Harnstoff-lösungen und Elution nach dem Waschen der Säule mit dem flüchtigen 1 M Tri-äthylammoniumbikarbonat, ferner Adsorption an kleine Aktivkohle-Säulen (Aktiv-kohle teilweise desaktiviert durch Waschen mit 95% Äthanol — 8% sek. Oktanol) und Elution der Nucleotide nach dem Waschen mit H_2O mit 8% wäßrigem Phenol (*23*). Das Phenol kann dann durch mehrere Extraktionen mit Äther entfernt werden.

Die vorgereinigten Oligonucleotidfraktionen können auf verschiedene Weise weiterfraktioniert werden. Allgemein anwendbar ist die Chromato-graphie an DEAE-Cellulose in 7 M Harnstoff, 0,1 M HCOOH, oder an DEAE-Sephadex A 25 in 7 M Harnstoff, HCl, pH 2,7—3,0 (*84*, *86*, *24*). Bei diesen pH-Werten sind die Aminogruppen von Cytosin und Adenin protoniert und tragen je nach der Basenzusammensetzung in verschie-denem Grade zur Gesamtladung eines Oligonucleotids bei. Gemischte Oligonucleotide gleicher Kettenlänge, wie die Pankreas-RNase-Penta-nucleotide (*5*) von Tyrosin-tRNA (Abb. 9a), können so auf Grund ihrer Basenzusammensetzung getrennt werden (*57*) (Abb. 9b).

Die Dekameren aus T1-RNase-Hydrolysaten von hochmolekularer MS 2-RNA ließen sich auf Grund ihres (Ap + Cp)/Up-Gehaltes außer durch Chromatographie an DEAE-Sephadex, 7 M Harnstoff, 0,06 N HCl, pH 2,7, auch durch Papierelektro-

phorese in 0,02 M Ammoniumformiat, pH 2,7, und anschließende Papierchromatographie in t-Butanol-0,02 M Ammoniumformiat (170 : 230, v/v), pH 3,8, in 5 Fraktionen zerlegen (*86*).

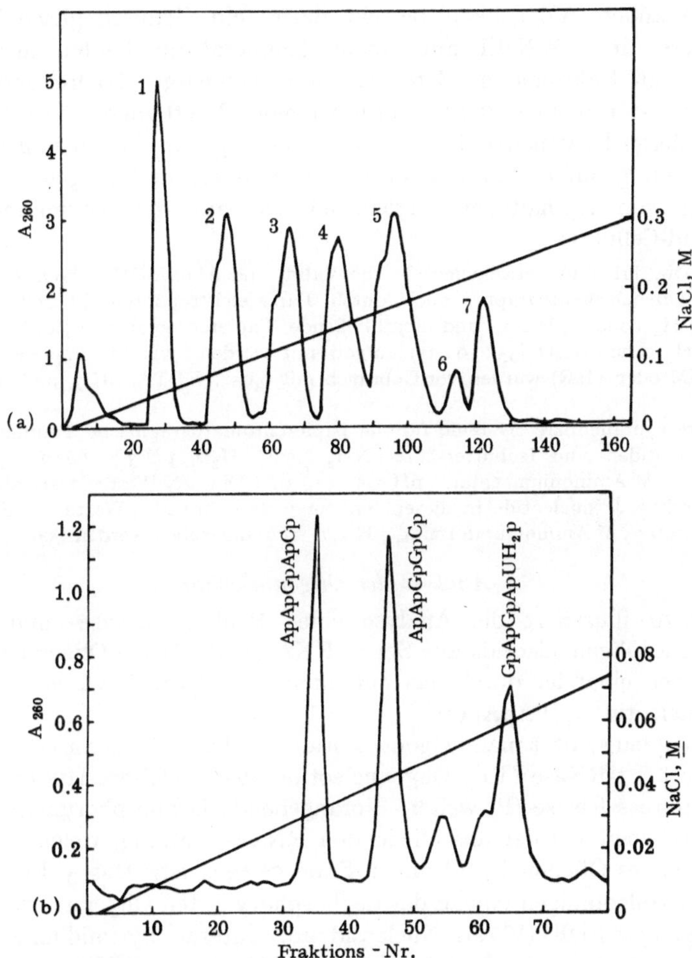

Abb. 9.   Pankreas-RNase-Spaltprodukte von Tyrosin-tRNA (*57*).

a) Trennung an DEAE-Cellulose (0,35 × 25-cm-Säule) in 7 M Harnstoff, pH 7,0,
b) Rechromatographie der Pentanucleotide (*5*) an DEAE-Cellulose (0,35 × 25-cm-Säule) in 7 M Harnstoff,
0,1 M HCOOH. Eluiert wurde mit linearen NaCl-Gradienten. Vgl. Text, S. 380. Mit freundlicher Genehmigung von Dr. J. T. Madison [J. Biol. Chem. **242**, 1318 (1967)]

Weitere Trennungen derartiger nichthomogener Fraktionen waren an DEAE-
Cellulose in 7 M Harnstoff, 0,2 M Ammoniumacetat, pH 4,4, möglich, da DEAE-
Cellulose auch in 7 M Harnstoff eine höhere Affinität für Purine als für Pyrimidine
hat und Cp bei diesem pH teilweise protoniert ist. Bei pH 4,0—4,4 konnten daher
Oligomere fraktioniert werden, die sich nur durch ihren Ap/Cp-Gehalt unterschieden
(*86*).

Auch die zuerst beschriebene Chromatographie an DEAE-Cellulose oder -Sephadex in Ammoniumcarbonat ist zur Auftrennung von Oligonucleotiden gleicher Kettenlänge geeignet (*41*).

Eine andere Möglichkeit besteht darin, ein Gemisch gleich langer Oligomerer in 1 M NaCl mit einem Temperaturgradienten an einer Cellulose zu fraktionieren, deren sterisch begünstigte Hydroxygruppen mit den 5′-terminalen Phosphatgruppen von Polythymidylsäure-Ketten oder anderen Polynucleotiden verestert sind (*34*). In diesem Fall beruht die Trennung auf der in 1 M NaCl stark temperaturabhängigen Basenpaarung von Ap-haltigen Oligonucleotiden mit der Thymidin-Polynucleotid-Cellulose.

Die Di-, Tri- und Tetranucleotide von Valin- und Alanin-tRNA konnten außer durch Säulenchromatographie auch durch Papierelektrophorese in 20% Essigsäure—NH_3 conc., pH 2,7, und anschließende Papierchromatographie in n Propanol-NH_3 conc.—H_2O (55 : 10 : 35) aufgetrennt werden (*43*). Die Papiere (Whatman 3MM oder 3HR) wurden vor Gebrauch mit 0,05% EDTA, pH 7, und H_2O gewaschen.

Weitere brauchbare Systeme für die papierchromatographische Trennung von Oligonucleotiden sind Isobuttersäure—NH_3 conc.—H_2O, pH 3,7 (66 : 1 : 33) und Äthanol—1 M Ammoniumacetat, pH 7,5 (7 : 3). Die R_f-Werte fast aller 64 synthetischen Trinucleotide in diesen Systemen und ihre R_{up}-Werte bei Elektrophorese in 0,05 M Ammoniumformiat, pH 2,7, sind angegeben worden (*54*).

c) Analyse der Oligonucleotide

Abb. 10 illustriert die Analyse eines Pankreas-RNase- und eines T1-RNase-Oligonucleotids aus Serin-tRNA (*114*). Beide Oligonucleotide überlappen einander durch das nur einmal in Serin-tRNA vorhandene Tetranucleotid ApiPApApψp.

Prinzipiell spaltet man Oligonucleotide aus Pankreas-RNase-Hydrolysaten mit T1-RNase (T1), Oligonucleotide aus T1-RNase-Hydrolysaten mit Pankreas-RNase (P) weiter. Vorangehende Dephosphorylierung mit PME (B) zeigt bei der anschließenden RNase-Spaltung, welches Mono- oder kleinere Oligonucleotid am 3′-Ende gelegen ist. Das 5′-Ende fällt bei dem vollständigen Abbau des dephosphorylierten Oligonucleotids mit Schlangengift-PDE (V) als Nucleosid an. Interne Pyrimidinnucleotid-Sequenzen können durch den partiellen Abbau mit *Schlangengift-PDE* (V⁺) aufgeklärt werden (*42*). Dieser verläuft asynchron und erlaubt die Isolierung aller Zwischenprodukte des stufenweisen Abbaus vom 3′-Ende her, wird allerdings durch seltene Nucleotide und Purinnucleotid-„cluster" stark verlangsamt. Anschließende Spaltung der Zwischenprodukte mit Pankreas-RNase (P) setzt die 3′-Enden als Nucleoside frei. Diese zeigen den Verlauf der Sequenz an.

Die Trennung der Zwischenprodukte des partiellen PDE-Abbaus (V⁺) von GpGpGpApGpApGpU* aus Alanin-tRNA an DEAE-Cellulose in 7 M Harnstoff, pH 7, mit einem linearen Na-acetat-Gradienten (*42*) zeigt

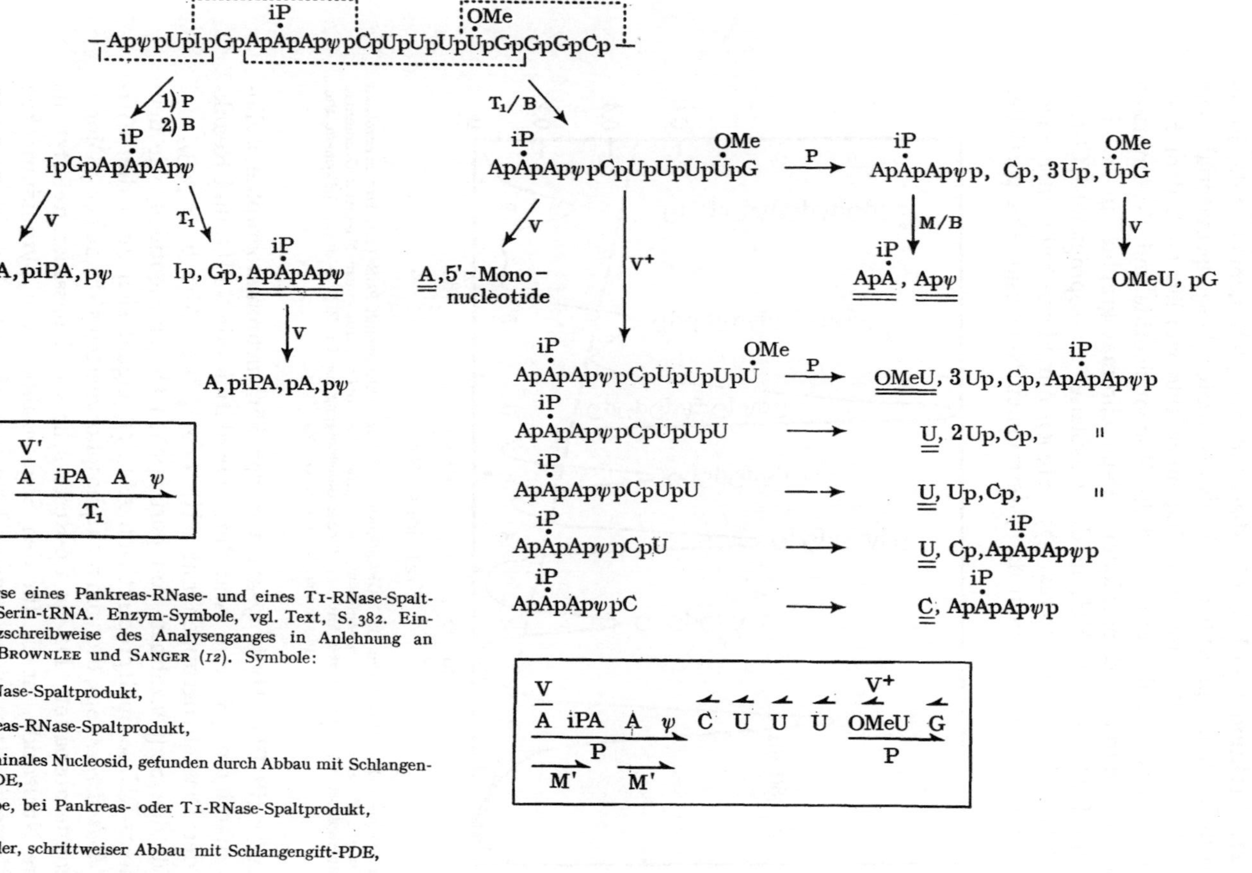

Abb. 10. Analyse eines Pankreas-RNase- und eines T1-RNase-Spaltprodukts aus Serin-tRNA. Enzym-Symbole, vgl. Text, S. 382. Eingerahmt: Kurzschreibweise des Analysenganges in Anlehnung an BROWNLEE und SANGER (12). Symbole:

T1 → T1-RNase-Spaltprodukt,

P → Pankreas-RNase-Spaltprodukt,

V → 5'-terminales Nucleosid, gefunden durch Abbau mit Schlangengift-PDE,

V' → dasselbe, bei Pankreas- oder T1-RNase-Spaltprodukt,

V+ → partieller, schrittweiser Abbau mit Schlangengift-PDE,

M → Micrococcus-Nuclease-Abbauprodukt,

M' → dasselbe, bei Pankreas- oder T1-RNase-Spaltprodukt,

Sp+ → partieller, schrittweiser Abbau mit Milz-PDE,

S S → Fragmente einer sekundären Spaltung mit T1-RNase

Abb. 11. Fraktion 2 enthält die abgespaltenen 5'-Mononucleotide, Fraktion 9 nicht dephosphoryliertes, PDE-resistentes Ausgangsmaterial. Die in Fraktion 3—7 vorhandenen Oligonucleotide wurden nach dem Entsalzen mit Alkali hydrolysiert und die Mononucleotide und 3'-terminalen Nucleoside durch 2-dimensionale Papierchromatographie in 1. Isopropanol-H_2O (65 : 35), NH_3 in der Gasphase, 2. Isopropanol-HCl-H_2O (340 : 82 : 500) aufgetrennt (*42, 57*). Stehen nur kleine Substanzmengen zur Verfügung, so kann man die 3'-terminalen Nucleoside des partiellen

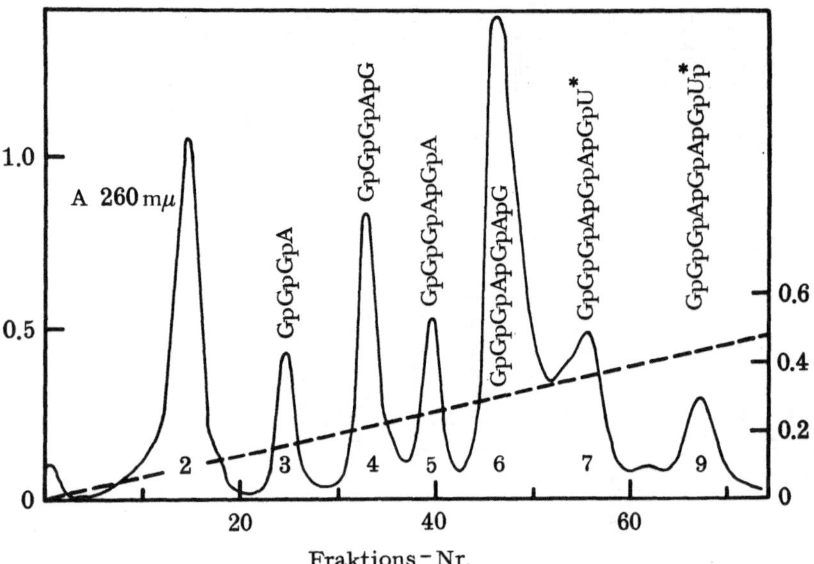

Abb. 11. Sequentieller Abbau von GpGpGpApGpApGpU* mit Schlangengift-PDE (*42*). Das Hydrolysat wurde an DEAE-Cellulose (0,35 × 30-cm-Säule) in 7 M Harnstoff, pH 7,5, mit einem Na-acetat-Gradienten chromatographiert. Vgl. Text, S. 382. Mit freundlicher Genehmigung von Dr. R. W. Holley [Biochem. Biophys. Res. Commun. **17**, 389 (1964)]

Schlangengift-PDE-Hydrolysats vor der Säulenchromatographie radioaktiv markieren (*78*), und zwar durch Oxydation mit $NaJO_4$ und Reduktion der Dialdehydgruppen mit (^{3}H)-$NaBH_4$ (vgl. Abb. 6, S. 375).

Für den Sequenzabbau von purinreichen Oligonucleotiden, wie dem Pankreas-RNase-Spaltprodukt von Abb. 11, eignet sich auch die *Milzphosphodiesterase* (Sp), die vom nichtphosphorylierten 5'-Ende her 3'-Mononucleotide abspaltet. Da sie Cp-Reste sehr viel langsamer freisetzt als andere Nucleotide, ist sie für den Sequenzabbau von pyrimidinreichen T 1-RNase-Spaltprodukten weniger brauchbar (*89, 12*). Die 5'-terminalen Nucleoside der Zwischenprodukte können nach dem Dephosphorylieren der 3'-Enden durch vollständigen Schlangengift-PDE-Abbau identifiziert werden (vgl. Abb. 13, S. 388).

Literaturverzeichnis: SS. 414—421

Kürzere Oligonucleotide (Tetra- und Pentanucleotide) mit Pyrimidinnucleotid- bzw. Purinnucleotid-Sequenzen werden von Schlangengift-PDE unter partiellen Abbaubedingungen häufig zu schnell bzw. zu langsam gespalten, so daß man keine Zwischenprodukte fassen kann. In solchen Fällen hilft der Abbau mit *Micrococcus*-Nuclease (M) (*119*) oder Polynucleotid-Phosphorylase (K) weiter. Zum Beispiel ließen sich rTpψpCpG und CpUpCpUpG aus Serin-tRNA mit *Micrococcus*-Nuclease unter kontrollierten Bedingungen in rTpψp + CpG bzw. CpUp + Cp + UpG spalten (*30*). Die interne Sequenz von ApiPApApψp (Abb. 10) ergab sich aus dem vollständigen Abbau mit Schlangengift-PDE (A als 5′-Terminus) (*114*) und der Spaltung mit *Micrococcus*-Nuclease (in Gegenwart von PME) in ApiPA und Apψ (*30*). Der partielle Schlangengift-PDE-Abbau der dephosphorylierten 5′-terminalen Sequenz von Tyrosin-tRNA, pCpUpCpUpCpGp, blieb auf der Tetranucleotidstufe, Cp(Up, Cp)U, stehen (*57*). Hier führte die partielle, ebenfalls asynchron verlaufende Phosphorolyse mit *Polynucleotid*-Phosphorylase (K+) weiter (*57, 58a*), bei der durch P$_i$ 5′-Nucleosiddiphosphate vom nichtphosphorylierten 3′-Ende her freigesetzt werden (*91*):

$$\text{CpUpCpUpCpG} \xrightarrow[\text{(K+)}]{\text{P}_i} \text{(CpUpCpUpC)} + \text{ppG}$$
$$\text{CpUpCpU} + \text{ppC}$$
$$\text{CpUpC} + \text{ppU}$$

Die partielle Phosphorolyse endete also erst auf dem Trinucleotid-Niveau und lieferte damit die vollständige Sequenz des Hexanucleotids. Die gleiche Beobachtung wurde bei der partiellen Phosphorolyse von GpGpGpApGpApC aus Tyrosin-tRNA gemacht, bei der GpGpG als 5′-Terminus zurückblieb.

Die Spaltprodukte der Oligonucleotide wurden bei Serin-tRNA im allgemeinen zuerst an kleinen DEAE-Cellulose-Säulen mit Ammoniumcarbonat-Gradienten fraktioniert (*61*). Gemische von Nucleotiden und Nucleosiden lassen sich durch Papierelektrophorese in 0,05 M Ammoniumformiat, pH 3,5, oder in 20% Essigsäure—NH$_3$, pH 3,0 (*43*) trennen. Zu diesem Zweck eignet sich auch die erwähnte zweidimensionale Papierchromatographie (*42*). Von den Standard-Nucleotiden können die meisten „seltenen" Nucleotide nach der PME-Behandlung als Nucleoside durch Papierchromatographie in n Butanol-H$_2$O-NH$_3$ conc. (86 : 14 : 5) abgetrennt werden (*61, 30*).

Ein spezielles Problem stellte die Analyse der UH$_2$p-enthaltenden Oligonucleotide dar. Bei der Spaltung von Oligonucleotiden der T1-RNase-Hydrolysate bzw. Pankreas-RNase-Hydrolysate war aufgefallen, daß im ersteren Fall Schlangengift-PDE keine bei A$_{260}$ meßbaren 5′-terminalen Nucleoside produzierte, während im letzteren Fall bei der Nachspaltung mit T1-RNase keine meßbaren 3′-terminalen Pyp-Enden freigesetzt wurden. Nachbehandlung zweier angeblicher „Dinucleotide"

aus dem T1-RNase-Hydrolysat von Alanin-tRNA, „CpGp" und „ApGp",
mit Pankreas-RNase zeigte bei der Chromatographie an DEAE-Sephadex
und Bestimmung des organischen Phosphats in den Eluatfraktionen einen
A_{820}-Gipfel, der keinem A_{260}-Gipfel entsprach (*58*). Das Material dieser
Fraktion gab auf dem Papier nach dem Besprühen mit 0,5 M NaOH
eine positive Farbreaktion mit einem salzsauren *p*-Dimethylamino-
benzaldehyd-Reagens, die charakteristisch für freie Ureido-Gruppen ist
(*31*). Die bekannte Alkalilabilität der Dihydropyrimidine (*18*) lenkte den
Verdacht auf UH_2p, das bei 260 mμ eine vernachlässigbare Absorption
zeigt und durch verd. Alkali in die ringoffene Form Yp umgewandelt
werden kann (Abb. 7, S. 377). Mischchromatographie des fraglichen
Materials nach PME-Behandlung mit authentischem UH_2 in verschiedenen Systemen bestätigte den Verdacht (*58*). Die fraglichen „Dinucleotide" stellten sich also als Trinucleotide mit der Struktur UH_2pCpGp und
UH_2pApGp heraus. UH_2p wurde anschließend in anderen tRNA's aus
Hefe gefunden (*25, 56, 78, 6*) und ^{32}P-markiertes Yp in Alkalihydrolysaten
von ^{32}P-markierten Oligonucleotiden nachgewiesen (*89*).

Während sich die qualitative Bestimmung von UH_2 auf den erwähnten
Farbtest stützt, ist eine semiquantitative Bestimmung außer durch
Phosphatanalyse auch durch unmittelbare Messung von A_{230} nach Zugabe von 0,1 M NaOH möglich (*5*). Das nur im Alkalischen zwischen 230
und 235 mμ vorhandene Absorptionsmaximum von UH_2 nimmt wegen
der Umwandlung $UH_2 \rightarrow Y$ rasch ab. Division des anfänglichen A_{230}-
Wertes durch den millimolaren Extinktionskoeffizienten 8,2 gibt die
ungefähre UH_2-Menge. Fehler treten dann auf, wenn unter den Bedingungen der Chromatographie bereits ein Teil des UH_2 in Y umgewandelt
worden ist.

Auf die Isolierung und Charakterisierung der anderen „seltenen"
Nucleoside (Abb. 7, S. 377) soll hier nicht eingegangen werden. Eines
der ungewöhnlichsten ist das Cytokinin-Aktivität besitzende N^6-γ,γ-
Dimethylallyladenosin (iPA) aus Serin-tRNA, dessen Struktur durch
Massenspektrometrie und Kernresonanzspektroskopie aufgeklärt werden
konnte (*10*). Unabhängig davon wurde seine Struktur durch Synthese
und Vergleich des synthetisierten Materials mit iPA aus Hefe-sRNA
und Serin-tRNA bestätigt (*36*). Wahrscheinlich ist das ursprünglich als
N^6-DimeA bezeichnete Nucleosid in Tyrosin-tRNA ebenfalls iPA (*55*).

2. 5 S-Ribosomale RNA

a) Trennung der Oligonucleotide

SANGER et al. (*89*) entwickelten eine zweidimensionale Papierelektro-
phoresetechnik, mit der es ihnen gelang, die ^{32}P-markierten Oligonucleotide aus Pankreas- und T1-RNase-Hydrolysaten von radioaktiver

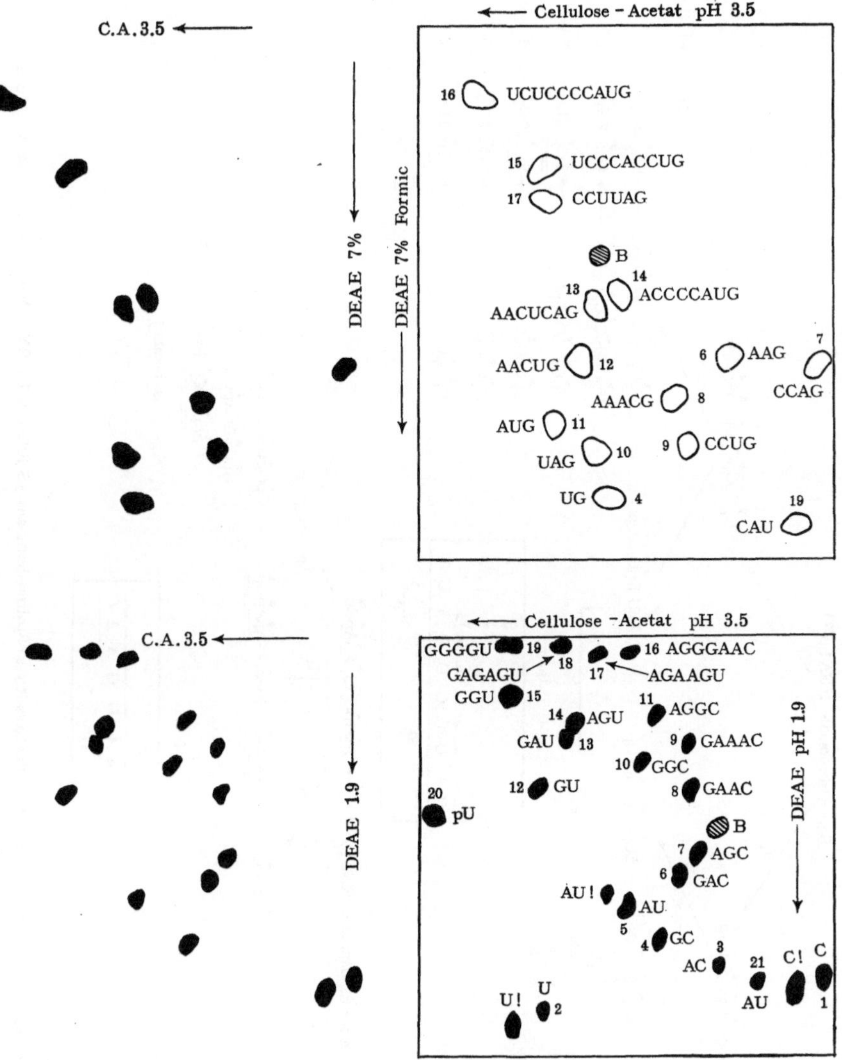

Abb. 12. Elektrophorese-„Fingerprints" der T1-RNase/PME-Spaltprodukte (oben) und Pankreas-RNase-Spaltprodukte (unten) von 5 S-Ribosomaler RNA (*12*). Vgl. Text, S. 387/389. Mit freundlicher Genehmigung von Dr. F. Sanger [J. Mol. Biol. **23**, 337 (1967)]

5 S-Ribosomaler RNA nahezu vollständig aufzutrennen (*12*) und durch Radioautographie sichtbar zu machen *(Abb. 12)*. Nach Elektrophorese in 0,5% Pyridin—5% Essigsäure (v/v), pH 3,5, auf Celluloseacetat-Streifen wurden die Substanzen durch ein Stempelverfahren auf DEAE-Cellulose-Papier übertragen und entweder in 7% Ameisensäure (T1-

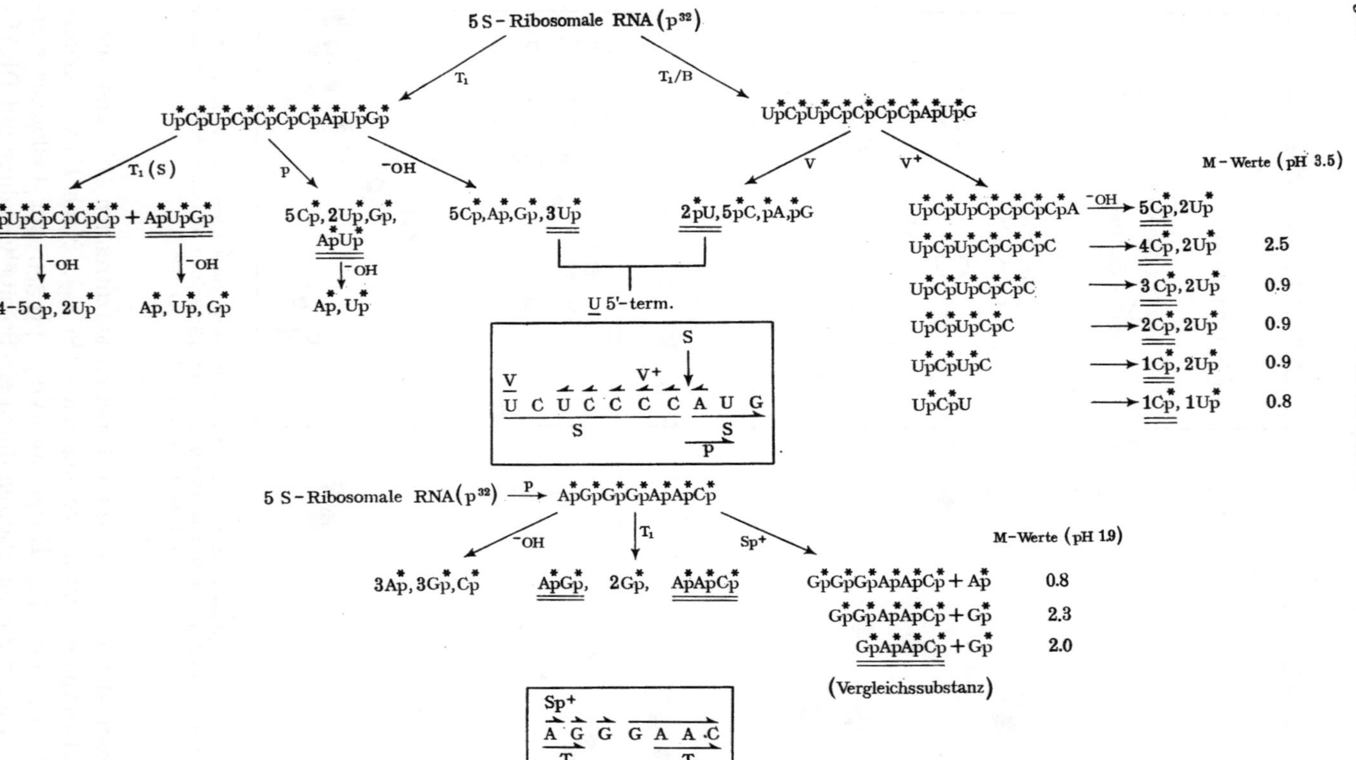

Abb. 13. Analyse eines ³²P-markierten T₁-RNase- und eines Pankreas-RNase-Spaltprodukts aus 5 S-Ribosomaler RNA. Vgl. Text und Abb. 10, S. 383

RNase/PME-Spaltprodukte) oder bei pH 1,9 in 2,5% Ameisensäure—8,7% Essigsäure (v/v) (Pankreas-RNase-Spaltprodukte) durch Elektrophorese aufgetrennt. Die hervorragende Trennung in der 2. Dimension kommt durch die Verbindung von Ionenaustausch-Chromatographie mit Elektrophorese zustande (vgl. auch Abb. 14, S. 390).

b) Analyse der ³²P-markierten Oligonucleotide

Die vom DEAE-Papier eluierten Oligonucleotide wurden durch vollständigen Abbau mit Alkali, Schlangengift-PDE, Weiterspaltung mit Pankreas- oder T1-RNase sowie durch partiellen Abbau mit Exonucleasen allein auf Grund der elektrophoretischen Beweglichkeit der Abbauprodukte analysiert. Die zur Verfügung stehenden geringen radioaktiven Substanzmengen schließen die üblichen UV-spektroskopischen Methoden aus. *Abb. 13* zeigt die Analyse eines Dekanucleotids aus T1-RNase-Hydrolysaten und eines Heptanucleotids aus Pankreas-RNase-Hydrolysaten.

Im folgenden seien einige Besonderheiten der Analysenmethodik erwähnt:

1. Das 5'-terminale Nucleotid ist dasjenige, das im alkalischen Hydrolysat vorhanden ist, dagegen im Schlangengift-PDE-Hydrolysat fehlt, da es als Nucleosid auf den Radioautogrammen unsichtbar ist. 2. PypA-Bindungen von T1-RNase-Oligonucleotiden lassen sich durch Nachinkubation mit T1-RNase unspezifisch spalten. Diese sekundäre Spaltung (S) führte zu zwei Bruchstücken (Abb. 13), die sich durch Elektrophorese auf DEAE-Cellulose-Papier bei pH 1,9 trennen ließen *(Abb. 14a)* und die Strukturaufklärung der längeren Oligonucleotide erleichterten. Die Brauchbarkeit dieser Sekundärspaltung sei noch an der Analyse zweier weiterer Oligonucleotide aus 5 S-Ribosomaler RNA (a) und Serin-tRNA (b) *(114)* in der von BROWNLEE und SANGER *(12)* vorgeschlagenen Kurzschreibweise erläutert:

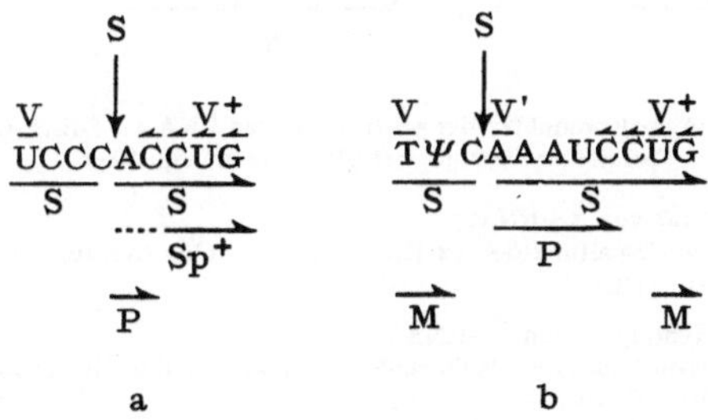

3. Die vollständige Sequenz der T1-RNase-Oligonucleotide ergab sich aus dem partiellen Abbau mit Schlangengift-PDE. Die schrittweise kleiner werdenden Spaltprodukte wurden durch Elektrophorese auf

DEAE-Papier bei pH 3,5 getrennt *(Abb. 14c)* und mit Alkali analysiert. Das unsichtbar bleibende 3′-terminale Nucleosid entspricht dem fehlenden Nucleotid des um ein Glied kürzeren Abbauprodukts.

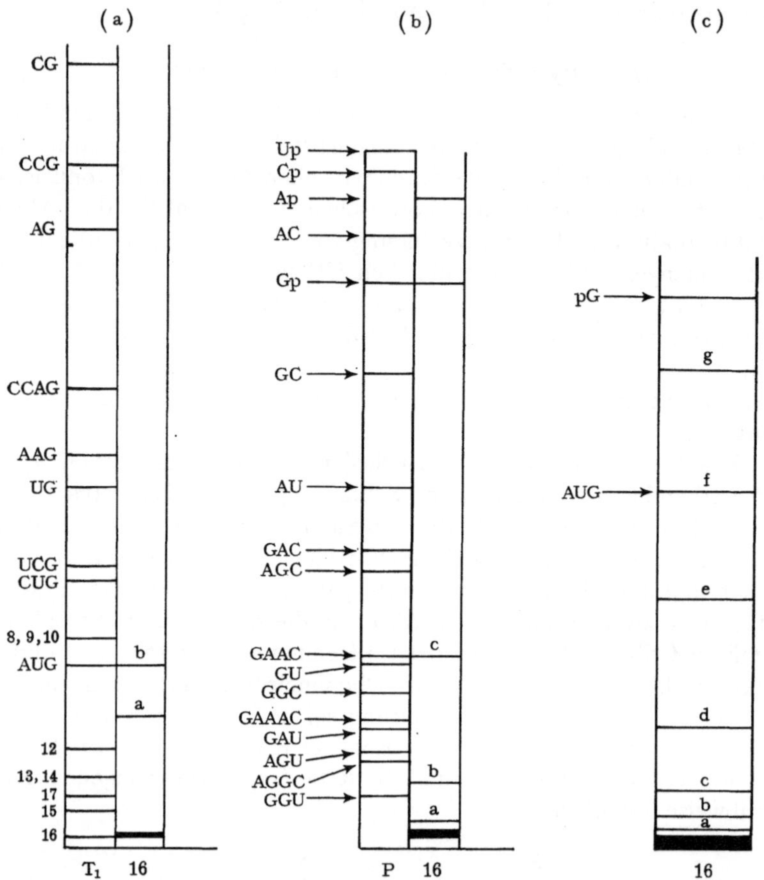

Abb. 14. Elektrophorese von Spaltprodukten der 5 S-Ribosomalen RNA auf DEAE-Cellulose-Papier (a und b) bei pH 1,9 bzw. (c) bei pH 3,5 *(12)*. Vgl. Text.

(a) T1: T1-RNase-Hydrolysat von 5 S-RNA.
    16: Produkte der Sekundärspaltung des T1-RNase-Spaltprodukts (Nr. 16) von Abb. 12 mit T1-RNase.
       a: UCUCCCC,   b: AUG.

(b) P: Pankreas-RNase-Hydrolysat von 5 S-RNA.
    16: Produkte des partiellen Abbaus des Pankreas-RNase-Spaltprodukts (Nr. 16) von Abb. 12 mit Milz-PDE.
       a: GGGAAC,   b: GGAAC,   c: GAAC.

(c) 16: Produkte des partiellen Abbaus des T1-RNase-Spaltprodukts (Nr. 16) von Abb. 12 mit Schlangengift-PDE.
       a: UCUCCCCA, b: UCUCCCC, c: UCUCCC,
       d: UCUCC,   e: UCUC,   g: UCU.

    Die dicke Linie in der Nähe des Starts zeigt die Position des unveränderten Oligonucleotids an.

Literaturverzeichnis: SS. 414—421

Es war möglich, die Struktur von Oligonucleotiden direkt aus der elektrophoretischen Beweglichkeit der Produkte des Sequenzabbaus abzuleiten. Bezeichnet man mit y den Abstand eines Oligonucleotids von der Auftragsstelle auf dem DEAE-Papier, mit x den Abstand zwischen dem Oligonucleotid und seinem ersten Abbauprodukt, so kann man einen Wert $M = x/y$ definieren, der für jedes abgespaltene 5′-Mononucleotid in definierten Grenzen schwankt. So findet man bei der pH 3,5-Elektrophorese von Schlangengift-PDE-Abbauprodukten auf DEAE-Papier:

Abgespaltenes Nucleotid M-Wert

$$pC \quad \ldots\ldots\ldots\ldots \quad 0,6\text{—}1,2$$
$$pA \quad \ldots\ldots\ldots\ldots \quad 2,1\text{—}2,9$$
$$pU \quad \ldots\ldots\ldots\ldots \quad 1,7\text{—}1,9$$
$$pG \quad \ldots\ldots\ldots\ldots \quad 2,6\text{—}4,4$$

4. Die vollständige Struktur der längeren Pankreas-RNase-Oligonucleotide wurde außer durch die üblichen Spaltungen durch partiellen schrittweisen Abbau mit Milz-PDE und elektrophoretische Trennung der Abbauprodukte auf DEAE-Papier bei pH 1,9 ermittelt *(Abb. 14b)*. Mittels der M-Werte kann man auch hier wieder die Struktur der abgespaltenen Mononucleotide aus der elektrophoretischen Beweglichkeit der einzelnen Zwischenprodukte des Abbaus ermitteln.

3. Quantitative Aspekte der vollständigen Spaltungen

Die quantitativen Mengen von Oligonucleotiden in einem Hydrolysat können aus den bei pH 7 oder 8,6 gemessenen A_{260}-Werten der getrennten Produkte mittels millimolarer Extinktionskoeffizienten (ε_{mM}) berechnet werden. Zum Beispiel wurde für GpUp (Abb. 8a, S. 379) $A_{260} = 17,7$ — bezogen auf 25 mg ungespaltene Serin-tRNA — gefunden, angenommen, daß 1 mg tRNA 25 A_{260}-Einheiten bei pH 7 entspricht. Der ε_{mM}-Wert von GpUp ergibt sich durch Addition der ε_{mM}-Werte von Gp (11,7) und Up (9,6) bei 260 mμ und pH 8,6 abzüglich 5% Hyperchromizität zu 20,3 (*98*). Daraus folgen $17,7/20,3 = 0,87$ μMol GpUp per μMol tRNA. In analoger Weise wurden die Mengen der anderen Oligonucleotide mittels abgeschätzter Hyperchromizitäten (*98, 62*) bestimmt. Für einmal in der RNA vorhandene Oligonucleotide schwanken die gewonnenen μMol-Werte um 0,87 μMol per μMol tRNA. Setzt man daher die Menge GpUp gleich 1, so kann man die Mengen aller anderen Pankreas-RNase-Spaltprodukte in Mol/Mol GpUp angeben (*114*). Da der ε_{260}-Wert der Serin-tRNA nicht bekannt ist und außerdem 23—24 A_{260}-Einheiten per mg gereinigter tRNA den Verhältnissen besser gerecht werden dürften, liefert das beschriebene Verfahren die größtmögliche Genauigkeit. Oligonucleotid-Analysen werden ebenfalls aus den A_{260}-Werten der Spaltprodukte mittels der veröffentlichten ε_{260}-Werte der Mononucleotide (*7, 41*) und anderer Spaltprodukte berechnet.

Im Falle der ^{32}P-markierten Oligonucleotide von 5 S-ribosomaler RNA (*12*) werden die counts/min in der Papierregion bestimmt, die das Oligonucleotid enthält. Counts/min dividiert durch die Zahl der Phosphatreste des Oligonucleotids bezogen auf ApGp = 2,0 in T 1-RNase-Hydrolysaten bzw. ApUp = 2,0 in Pankreas-RNase-Hydrolysaten gibt die relativen molaren Mengen der Nucleotide.

V. Partielle Spaltung von Ribonucleinsäuren und Analyse der Oligonucleotidfragmente

1. Partielle enzymatische Spaltung von Transfer-Ribonucleinsäuren

Wie bereits erwähnt wurde, lassen sich aus den Oligonucleotiden der vollständigen Pankreas- und T1-RNase-Spaltung dann längere Sequenzen rekonstruieren, wenn durch einmal vorhandene „seltene" Nucleotide oder ausgezeichnete Spaltprodukte eindeutige Überlappungen hergestellt werden können. Für die Einordnung der übrigen Mono- und Oligonucleotide in die endgültige Sequenz sind zusätzliche Methoden erforderlich. Holley et al. (*40*) diskutierten 1963 zwei Möglichkeiten:

1. Baut man die gesamte tRNA schrittweise vom 3′- bzw. 5′-Ende her mit einer asynchron spaltenden Exonuclease ab und entfernt das Enzym, z. B. nach 50%iger Spaltung, dann sollte man nach der anschließenden Hydrolyse der restlichen RNA mit Pankreas- oder T1-RNase einen vom 3′- zum 5′-Ende bzw. vom 5′- zum 3′-Ende kontinuierlich ansteigenden Mengengradienten der Oligonucleotide erhalten. Beim Angriff vom 3′-Ende her sollten Oligonucleotide in der Nähe des 3′-Endes in geringster, solche in der Nähe des 5′-Endes in größter Menge gefunden werden. Eine solche Methode ist allerdings nur sinnvoll, wenn die verschiedenen 5′-Mononucleotide (bei Schlangengift-PDE-Abbau vom 3′-Ende) bzw. 3′-Mononucleotide (bei Milz-PDE-Abbau vom 5′-Ende) mit etwa gleichen Geschwindigkeiten von der Exonuclease freigesetzt werden. Diese Forderung war beim partiellen Schlangengift-PDE-Abbau von Serin-tRNA I + II (*115*), offensichtlich wegen der stabilen Sekundärstruktur in den Akzeptor-Hälften der RNA's und wegen seltener Nucleotide (Schlangengift-PDE spaltet z. B. rTpψp wesentlich langsamer als andere PypPyp-Bindungen), nicht erfüllt. Zudem wurde der Gradient durch Oligonucleotide gestört, die a) zweimal in beiden Serin-tRNA's (wie CpApG), b) einmal in I, aber zweimal in II (wie ApG), c) einmal in I *oder* II (wie rTpψpCpG in II) vorkamen. Nur für die drei nach der PDE-Behandlung in kleinster Menge vorhandenen 3′-terminalen Oligonucleotide beider Serin-tRNA's, UpUpG, UpCpG und CpCpA, wurde ein vom UpUpG über UpCpG zum CpCpA abfallender eindeutiger Gradient gefunden (*115*).

2. Als zweite Möglichkeit bot sich die partielle endonucleolytische Spaltung der tRNA in große Oligonucleotidfragmente an, die dann durch vollständige Spaltung in die bereits bekannten kleineren Oligonucleotide weiter zerlegt werden können. In diesem Fall erhält man eine Überlappung zwischen zwei Fragmenten, wenn eines oder mehrere der bekannten Oligonucleotide in beiden Fragmenten vorhanden sind. Voraussetzung für den Erfolg dieser Methode ist eine definierte dreidimensionale Struktur der RNA bei niedriger Temperatur, in der bestimmte Phos-

phodiesterbindungen dem enzymatischen Angriff leichter zugänglich sind als andere.

a) Partielle Spaltung mit T1-RNase

Nachdem sich gezeigt hatte, daß Mg^{++} die Inaktivierung der Akzeptoraktivität bestimmter tRNA's hemmt, wenn man sRNA in seiner Gegenwart mit *B. subtilis*-RNase oder T1-RNase inkubiert (*68*), gelang es

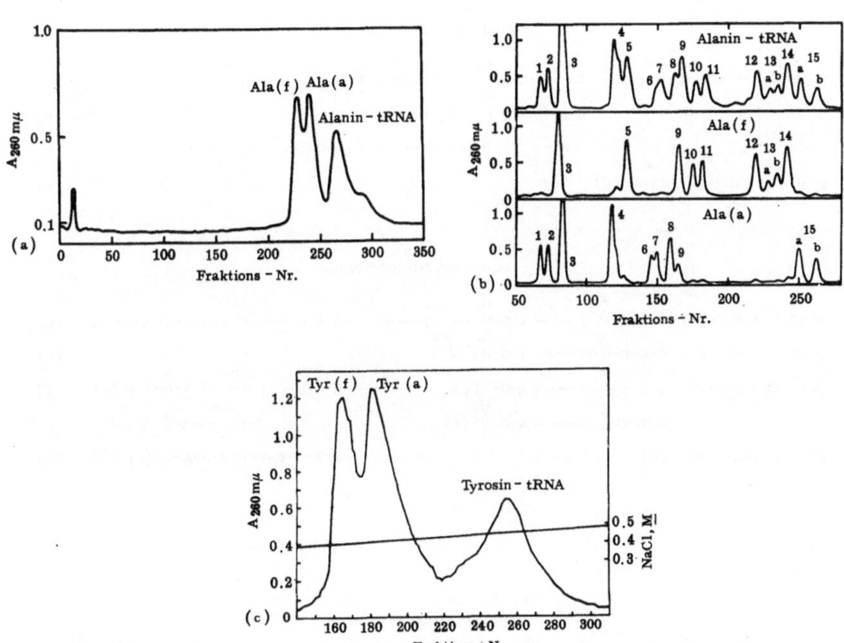

Abb. 15. Chromatographische Trennung der Hälften von Alanin- und Tyrosin-tRNA.

(a) Chromatographie des T1-RNase-Partialhydrolysats von Alanin-tRNA (vgl. Text, S. 393) an DEAE-Cellulose (0,5 × 100-cm-Säule) in 7 M Harnstoff, pH 8,0 (*73*).

(b) Oligonucleotidmuster der Hälften von Alanin-tRNA im Vergleich zu dem von ganzer Alanin-tRNA. Chromatographie an DEAE-Cellulose (0,35 × 220-cm-Säulen) in 7 M Harnstoff, pH 8,0 (*73*). Eluiert wurde mit linearen NaCl-Gradienten.

(c) Chromatographie des T1-RNase-Partialhydrolysats von Tyrosin-tRNA (vgl. Text) an DEAE-Cellulose (0,35 × 120-cm-Säule) bei 55° in 7 M Harnstoff, pH 7 (*59*).

Mit freundlicher Genehmigung von Drs. R. W. HOLLEY und J. T. MADISON [Proc. Nat. Acad. Sci. (USA) **53**, 543 (1965) und J. Biol. Chem. **242**, 1324 (1967)].

PENSWICK und HOLLEY (*73*), Alanin-tRNA (5 mg) zu mehr als 50% bei 0° (4 min) in 0,02 M MgCl$_2$—0,02 M Tris, pH 7,5 (1 ml), mit einer kleinen Menge T1-RNase (25 Einheiten) in zwei etwa gleich lange Stücke zu spalten und diese durch Chromatographie an DEAE-Cellulose in 7 M Harnstoff, pH 8, weitgehend zu trennen *(Abb. 15a)*. Nach der Rechromatographie wurden die Hälften vollständig mit T1-RNase abgebaut. *Abb. 15b* zeigt die Oligonucleotidmuster der 5'-Phosphat-Hälfte Ala(a) und der Akzeptor-Hälfte Ala(f) (Tabelle 1, s. unten) im Vergleich zu dem

Tabelle 1. Fragmente der partiellen Spaltung von Alanin- und Tyrosin-
Pankreas-
Die Fragmente sind so untereinander angeordnet, daß gleiche Positionen der drei anderen RNA's, in denen sie nicht vorkom-

Alanin-tRNA, linke Hälfte

```
          Me                      H₂                    H₂            Dime
pG-G-G-C-G-U-G-U-Ġ-G-C-G-C-G-U-A-G-Ü-C-G-G ———— Ü-A-G-C-G-C-Ġ-C-U-C-C-C-U-U-I-G- (a)
                                 H₂                    H₂            Dime
                C-G-U-A-G-Ü-C-G-G ———— Ü-A-G-C-G-C-Ġ-C-U-C-C-C-U-U-I-G- (b)
                                 H₂                    H₂            Dime
                     Ü-C-G-G ———— Ü-A-G-C-G-C-Ġ-C-U-C-C-C-U-U-I-  (c)
          Me                                           Dime
pG-G-G-C-G-U-G-U-Ġ-G- (d)                 C-Ġ-C-U-C-C-C-U-U-I-  (e)
```

Serin-tRNA I+II, linke Hälfte

```
                  Ac                H₂   OMe    H₂H₂              Dime
pG-G-C-A-A-C-U-U-G-G-C-Ċ-G ——— A-G-Ü — Ġ-G ——— U-U-A-A-G-G-C-Ġ-A-A-A-G-A-ψ-U-I-  (a)
                  Ac                H₂   OMe
pG-G-C-A-A-C-U-U-G-G-C-Ċ-G ——— A-G-Ü — Ġ-G-  (b)
                  Ac                        H₂H₂              Dime
pG-G-C-A-A-C-U-U-G-G-C-Ċ-G ——— A-G-  (c)    U-U-A-A-G-G-C-Ġ-A-A-A-G-A-ψ-U-I-  (d)
                  Ac                H₂                  Dime
            G-G-C-Ċ-G ——— A-G-Ü- (f)        C-Ġ-A-A-A-G-A-ψ-U-I-  (e)

pG-G-C-A-A-C-U-U- (g)            A-A-G-G-C-G-A-A-A-G-A-ψ-U-I-  (h)
```

Tyrosin-tRNA, linke Hälfte

```
               2Me                  H₂H₂ OMe H₂H₂H₂                Dime
pC-U-C-U-C-G-G-U-A-Ġ-C-C-A ———— A-G-Ü-Ü-Ġ-G-Ü-Ü-Ü-A-A-G-G-C-Ġ-C-A-A-G-A-C-U-G-  (a)
               2Me
pC-U-C-U-C-G-G-U-A-Ġ-C-C-A ———— A-G-  (b)
                                         H₂H₂H₂                Dime
                             U-Ü-Ü-A-A-G-G-C-Ġ-C-A-A-G-A-C-U-G-  (c)
               2Me                                 Dime
pC-U-C-U-C-G-G-U-A-Ġ- (d)              C-Ġ-C-A-A-G-  (e)
```

von ganzer Alanin-tRNA. Durch die Länge der verwendeten DEAE-Cellulose-Säulen (0,35 × 220 cm), in 7 M Harnstoff, pH 8, wurden fast alle Oligonucleotide aufgetrennt.

Wie Alanin-tRNA konnten auch Tyrosin- (*59*) und Valin-tRNA (*6*) in Hälften gespalten werden. Für beide tRNA's waren stärkere Spaltungsbedingungen notwendig (Tyrosin-tRNA, 15 mg, 150 Einheiten T1-RNase, 4 ml 0,01 M MgCl₂—0,02 M Tris, pH 7,5, 0°, 30 min). Die Hälften von Tyrosin-tRNA *(Tabelle 1)*, Tyr(a) und Tyr(f), wurden an DEAE-Cellulose bei 55° in 7 M Harnstoff, pH 7, getrennt *(59)*. Bei dieser Temperatur ist der Trenneffekt größer als bei Raumtemperatur *(Abb. 15c*, S. 393).

tRNA mit T1-RNase und von Serin-tRNA I + II mit T1-RNase und RNase

Moleküle untereinander fallen. Insertionen von Nucleotiden in einer RNA sind in den men, durch verlängerte Striche ausgeglichen

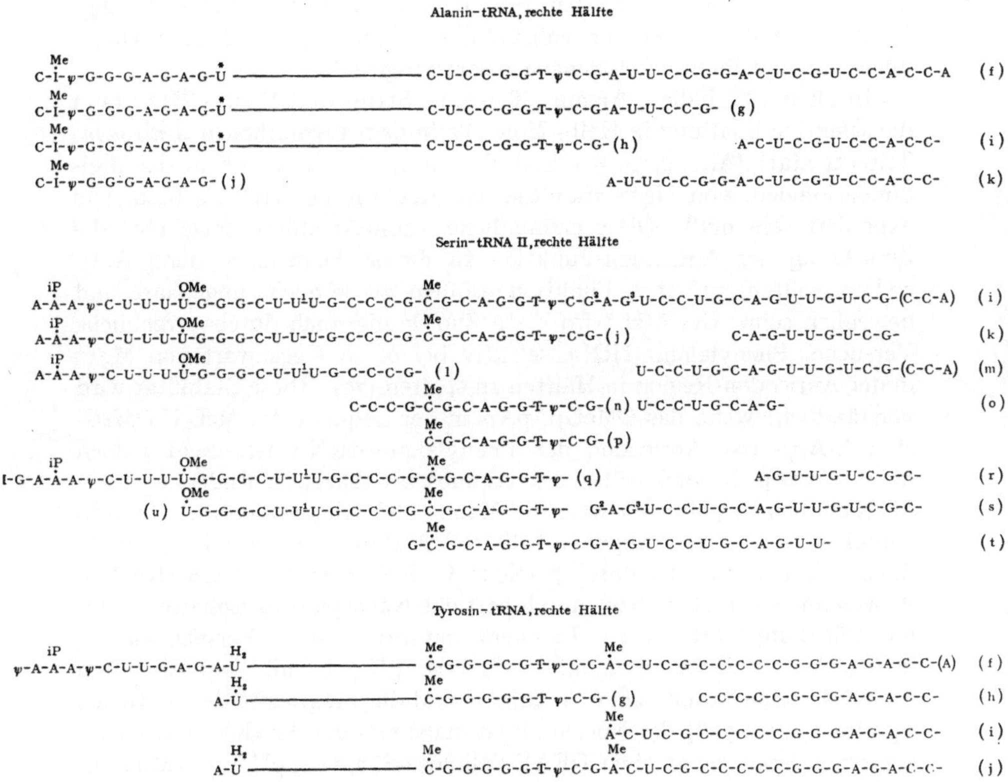

Alanin-tRNA, rechte Hälfte

```
        Me
C-I-ψ-G-G-G-A-G-A-G-U ————————————— C-U-C-C-G-G-T-ψ-C-G-A-U-U-C-C-G-G-A-C-U-C-G-U-C-C-A-C-C-A    (f)
        Me
C-I-ψ-G-G-G-A-G-A-G-U ————————————— C-U-C-C-G-G-T-ψ-C-G-A-U-U-C-C-G-  (g)
        Me
C-I-ψ-G-G-G-A-G-A-G-U ————————————— C-U-C-C-G-G-T-ψ-C-G- (h)              A-C-U-C-G-U-C-C-A-C-C-    (i)
        Me
C-I-ψ-G-G-G-A-G-A-G- (j)                                      A-U-U-C-C-G-G-A-C-U-C-G-U-C-C-A-C-C-    (k)
```

Serin-tRNA II, rechte Hälfte

```
   iP              OMe                         Me
A-A-A-ψ-C-U-U-U-U-G-G-G-C-U-U[1]U-G-C-C-C-G-C-G-C-A-G-G-T-ψ-C-G[2]A-G[3]U-C-C-U-G-C-A-G-U-U-G-U-C-G-(C-C-A)   (i)
   iP              OMe                         Me
A-A-A-ψ-C-U-U-U-U-G-G-G-C-U-U-U-G-C-C-C-G-C-G-C-A-G-G-T-ψ-C-G- (j)                  C-A-G-U-U-G-U-C-G-         (k)
   iP              OMe
A-A-A-ψ-C-U-U-U-U-G-G-G-C-U-U[1]U-G-C-C-C-G- (l)                    U-C-C-U-G-C-A-G-U-U-G-U-C-G-(C-C-A)       (m)
                                    Me
                        C-C-C-G-C-G-C-A-G-G-T-ψ-C-G- (n) U-C-C-U-G-C-A-G-                                     (o)
                                    Me
                        C-G-C-A-G-G-T-ψ-C-G- (p)
     iP              OMe                     Me
I-G-A-A-A-ψ-C-U-U-U-U-G-G-G-C-U-U[1]U-G-C-C-C-G-C-G-C-A-G-G-T-ψ- (q)              A-G-U-U-G-U-C-G-C-            (r)
                  OMe                       Me
         (u) U-G-G-G-C-U-U[1]U-G-C-C-C-G-C-G-C-A-G-G-T-ψ-       G[2]A-G[3]U-C-C-U-G-C-A-G-U-U-G-U-C-G-C-        (s)
                                    Me
                        G-C-G-C-A-G-G-T-ψ-C-G-A-G-U-C-C-U-G-C-A-G-U-U-                                         (t)
```

Tyrosin-tRNA, rechte Hälfte

```
 iP                  H2                      Me                Me
ψ-A-A-A-ψ-C-U-U-G-A-G-A-U ————————————— C-G-G-G-C-G-T-ψ-C-G-A-C-U-C-G-C-C-C-C-C-G-G-G-A-G-A-C-C-(A)    (f)
                    H2                      Me
                   A-U ————————————— C-G-G-G-G-G-T-ψ-C-G- (g)        C-C-C-C-C-G-G-G-A-G-A-C-C-         (h)
                                                                 Me
                                                                A-C-U-C-G-C-C-C-C-C-G-G-G-A-G-A-C-C-    (i)
                    H2                      Me                Me
                   A-U ————————————— C-G-G-G-G-G-T-ψ-C-G-A-C-U-C-G-C-C-C-C-C-G-G-G-A-G-A-C-C-           (j)
```

[1] Serin-tRNA I = C   [2],[3] Serin-tRNA I = A

Aus T1-RNase-Partialhydrolysaten von Serin-tRNA I + II, die in Abwesenheit von Mg++ (z. B. Serin-tRNA, 22 mg, 5000 Einheiten T1-RNase, 0,2 M Tris, pH 7,5, 0°, 30 min) hergestellt waren *(24)*, konnten drei große Fragmente isoliert werden, die der gemeinsamen 5′-Phosphathälfte von Serin-tRNA I + II und den beiden Akzeptor-Hälften von Serin-tRNA I bzw. II entsprechen (Tabelle 1). Die Akzeptorhälfte von Serin-tRNA ist wegen der beiden Gp-Reste im rT-ψ-C-„loop" (Abb. 20,

S. 404) labiler gegenüber T1-RNase als diejenige von Serin-tRNA I. Beiden fehlte das 3′-terminale CpCpA im Einklang mit den Sekundärstrukturmodellen, in denen die letzten 4 Nucleotide ungepaart sind und bei Serin-tRNA (GpCpCpA) der T1-RNase eine ungeschützte GpCp-Bindung darbieten. *Abb. 16* zeigt die Isolierung der 5′-Phosphat-Hälfte, Ser (a), und der Akzeptor-Hälfte von Serin-tRNA II, Ser II (i), aus dem oben erwähnten Hydrolysat durch DEAE-Cellulose-Chromatographie bei 50° (pH 7,5) und zweimalige Rechromatographie der letzten Fraktion an DEAE-Sephadex A 25 bei pH 3,0, jeweils in 7 M Harnstoff (*24*). Darunter sind die bei der vollständigen Spaltung anfallenden Oligonucleotide mit ihren nMol-Mengen zusammengestellt.

In allen vier Fällen, Alanin-, Tyrosin-, Serin- und Valin-tRNA, fand die selektive Spaltung in Halb- Moleküle in dem vermutlichen *Anticodon-*Triplett statt (Abb. 2, S. 361 und Tabelle 1), das demnach in der dreidimensionalen, von Mg++ stabilisierten Struktur der tRNA's besonders exponiert sein muß. Diese erstaunliche Labilität stützt umgekehrt die Zuordnung der Anticodon-Funktion zu diesen Sequenzen, denn Anticodons sollten, um ihre Funktion erfüllen zu können, ungepaart und beweglich sein. Gestützt wird diese Zuordnung auch durch vergebliche Versuche, Phenylalanin-tRNA selektiv bei 0° in Gegenwart von Mg++ in der Anticodon-Region in Hälften zu spalten (*76*). Diese Stabilität wird verständlich, wenn das OMeGpApAp in der Sequenz A-OMeC-U-*OMeG-A-A*-X-A-ψ- das Anticodon der Phenylalanin-tRNA ist, nicht jedoch das GpApAp in der Nähe des Aminosäure-Akzeptor-Endes.

Nachdem es gelungen war, die Mono- und Oligonucleotide der vollständigen RNase-Spaltungen auf die tRNA-Hälften zu verteilen, konnte die *Struktur der Hälften* durch partielle T1-RNase-Spaltung der tRNA in Abwesenheit von Mg++ bzw. durch partielle Nachspaltung isolierter Halbmoleküle aufgeklärt werden. In einem charakteristischen Versuch wurden 30 mg Alanin-tRNA in 15 ml 0,2 M Tris, pH 7,5, mit 6750 Einheiten T1-RNase eine Stunde bei 0° inkubiert und die Fragmente der partiellen Spaltung (etwa 70% des Nucleotidmaterials) von den Produkten der vollständigen Spaltung an einer DEAE-Cellulose-Säule bei pH 8 in 7 M Harnstoff abgetrennt. Homogene Fragmente wurden bei der Rechromatographie einzelner Fraktionen an DEAE-Cellulose bei 55° (pH 8) in 7 M Harnstoff gewonnen (*2*).

Unter ähnlichen Bedingungen (45 mg RNA, 6500 Einheiten T1-RNase in 4,5 ml 0,2 M Tris, pH 7,5, 45 min, 0°) ließ sich auch Serin-tRNA I + II zu 60—70% *partiell* mit T1-RNase spalten (*Abb. 17a*, S. 398, Fraktionen 6—11).

Da in diesem Fall analoge Oligonucleotidfragmente gleicher Kettenlänge (aus Serin-tRNA I bzw. II) zu erwarten waren, wurde zwischen die Vortrennung an DEAE-Cellulose (pH 7,5, 7 M Harnstoff) (Abb. 17a,

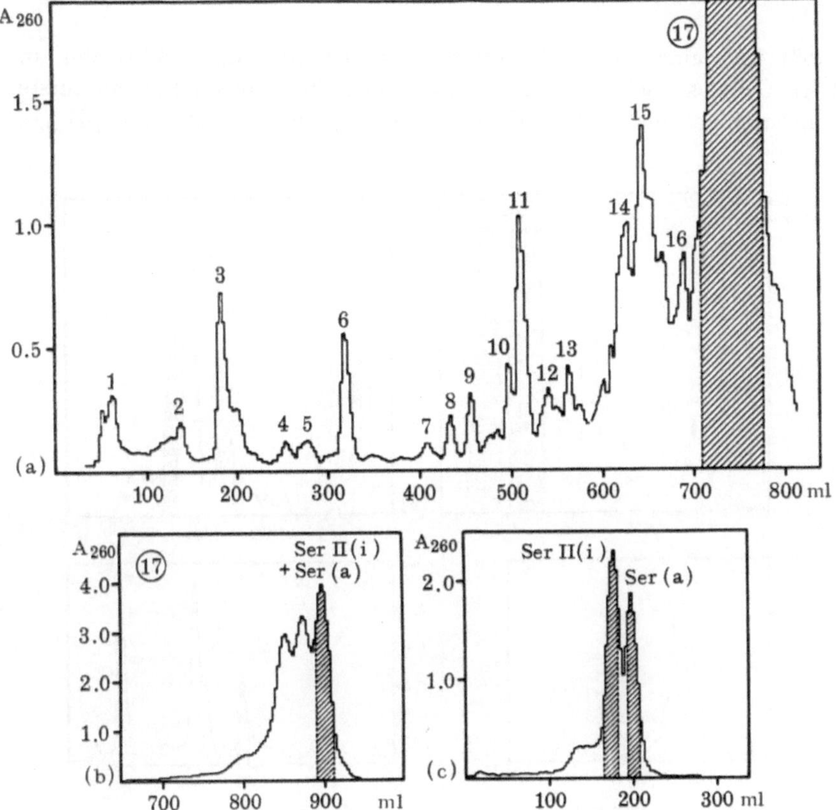

Vollständige Spaltung mit T_1/B:

Ser(a)		Ser II(i)	
Spaltprodukt	nMol	Spaltprodukt	nMol
G + Gp!	4 X 55	G + Gp!	3 X 68
CpApApCpUpUpG	48	ApiPApApψpCpUpUpUpOMeUpG	68
CpAcCpG	41	CpUpUpUpG	50
ApG	59	CpCpCpG	68
UH₂pOMeGpG	56	MeCpG	57
UH₂pUH₂pApApG	43	CpApG	2 X 62
CpDimeGp!	36	TpψpCpG	68
ApApApG	44	ApG	66
ApψpUpI	56	UpCpCpUpG	69
		UpUpG	65
		UpCpG	67

Abb. 16. Isolierung der Hälften von Serin-tRNA II (*115*).

(a) Chromatographie eines T 1-RNase-Partialhydrolysats von Serin-tRNA I + II (vgl. Text, S. 396) an DEAE-Cellulose (0,4 × 210-cm-Säule) bei 50° in 7 M Harnstoff, pH. 7,5.
(b) Rechromatographie von Fraktion 17 an DEAE-Sephadex A 25 (0,4 × 210-cm-Säule) in 7 M Harnstoff HCl, pH 3,0.
(c) Rechromatographie der schraffierten letzten Fraktion von (b) an DEAE-Sephadex A 25 (0,4 × 210-cm-Säule) in 7 M Harnstoff, HCl, pH 3,0. Eluiert wurde mit linearen NaCl-Gradienten.

S. 398) und eine zweite Rechromatographie einzelner Fraktionen an DEAE-Cellulose bei 55° (nur in Ausnahmefällen notwendig) eine erste Rechromatographie an DEAE-Sephadex A 25 in 7 M Harnstoff, pH 3,0,

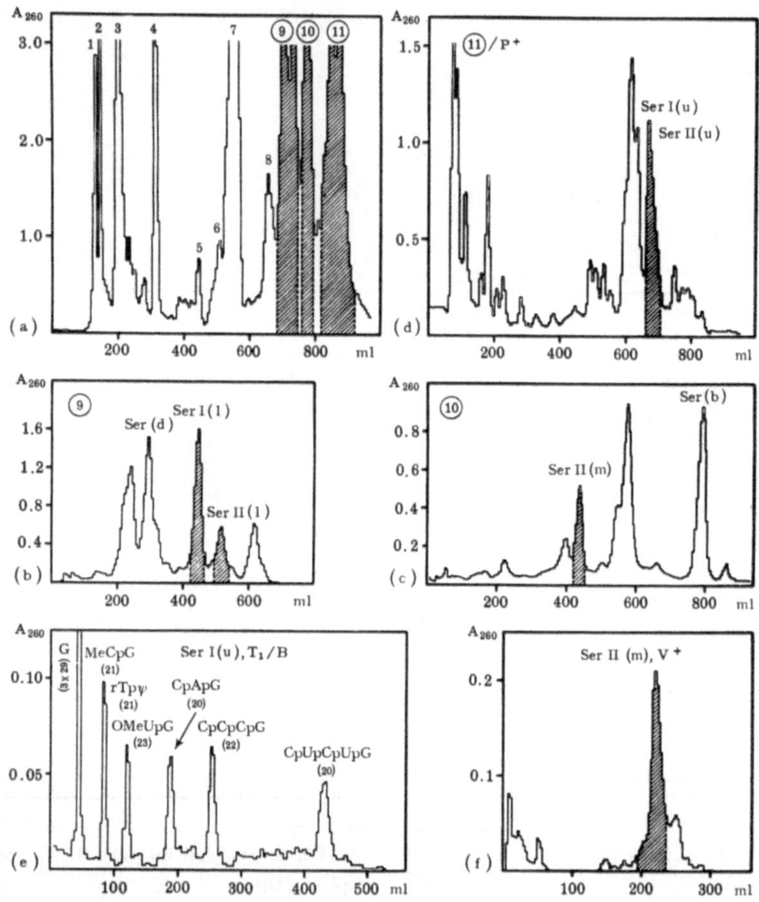

Abb. 17. Isolierung homogener Oligonucleotidfragmente aus einem T1-RNase-Partialhydrolysat von Serin-tRNA I + II (vgl. 24)

(a) Chromatographie des Hydrolysats (vgl. Text, S. 396) an DEAE-Cellulose (0,8 × 210-cm-Säule) in 7 M Harnstoff, pH 7,5.

(b) und (c) Rechromatographie der Fraktionen 9 und 10 an DEAE-Sephadex A 25 (0,68 × 210-cm-Säulen) in 7 M Harnstoff, HCl, pH 3,0 (115).

(d) Chromatographie des partiellen Pankreas-RNase-Hydrolysats von Fraktion 11 an DEAE-Cellulose (0,68 × 210-cm-Säule) bei 55° in 7 M Harnstoff, pH 7,5.

(e) Chromatographie der T1-RNase-Spaltprodukte von Fragment Ser I (u) bei 4° an DEAE-Cellulose (0,8 × × 25-cm-Säule), linearer Ammoniumcarbonat-Gradient. nMol-Mengen der Spaltprodukte in Klammern.

(f) Chromatographie des partiellen Schlangengift-PDE-Hydrolysats von Fragment Ser II (m) an DEAE-Cellulose (1 × 8-cm-Säule) bei 4° in 7 M Harnstoff, pH 7,5 (115).
Eluiert wurde in allen Fällen mit linearen NaCl-Gradienten.

Literaturverzeichnis: SS. 414—421

geschaltet *(24)*. Durch diese konnten u. a. die Fragmente Ser I (l) und Ser II (l) getrennt werden *(Abb. 17 b)*, die sich nur durch einen Cp/Up-Austausch bei einer Kettenlänge von 21 Nucleotiden unterschieden (Tabelle 1, S. 394).

Nicht alle Fragmente, die zur Rekonstruktion der Serin-tRNA-Moleküle notwendig waren, wurden durch ihre partielle Spaltung mit T1-RNase gewonnen. Wichtige zusätzliche Fragmente, wie Ser II (n) und Ser II (p) (Tabelle 1), brachte die partielle Nachspaltung von Fraktion 11 (Abb. 17 a) mit T1-RNase. In diesem Fall wurde das Hydrolysat nach der üblichen Phenolextraktion zum Entfernen der T1-RNase direkt an DEAE-Cellulose (0,68 × 210-cm-Säule) bei pH 7,5 und 55° chromatographiert *(24)*.

Die partielle T1-RNase-Spaltung von Tyrosin-tRNA (6 mg RNA, 1300 Einheiten T1-RNase, 2 ml 0,1 M Tris, pH 7,5, 60 min, 0°) lieferte nur vier Partialstücke in hinreichender Ausbeute *(59)*: Tyr (d), Tyr (g), Tyr (h) und Tyr (i) (Tabelle 1). Fragment Tyr (j) wurde aus einem bei Raumtemperatur hergestellten T1-RNase-Partialhydrolysat (0,01 M Mg^{++}, 15 min) isoliert. Da große Teile der 5'-Phosphathälfte vollständig, nicht partiell, gespalten worden waren, wurde die isolierte Hälfte (2,5 mg) partiell mit T1-RNase (125 Einheiten, 2 ml 0,1 M Tris, pH 7,5, 5 min, 0°) nachgespalten. Chromatographie an DEAE-Cellulose (0,35 × 240-cm-Säule) in 7 M Harnstoff, pH 6,0, mit einem linearen Na-acetat-Gradienten, lieferte die Fragmente Tyr (b) und Tyr (c).

b) Partielle Spaltung mit Pankreas-RNase

Da ähnlich wie bei Tyrosin-tRNA auch die G-reiche 5'-Phosphat-Hälfte von Serin-tRNA in höherem Grade als die Akzeptor-Hälften von T1-RNase gespalten wurde, war es von Interesse, die partielle Spaltung von Serin-tRNA mit Pankreas-RNase zu untersuchen *(24)*. LITT und INGRAM *(52)* hatten bereits durch Gelfiltration an Sephadex G 75 gezeigt, daß Pankreas-RNase bei 0° in Gegenwart von Mg^{++} sRNA bevorzugt in große Fragmente spaltet und daraus geschlossen, daß Mg^{++} durch Stabilisierung der Sekundärstruktur die in helicalen Bereichen liegenden PypX-Bindungen schützt. Serin-tRNA I + II (580 A$_{260}$-Einheiten) wurde in 4 ml 0,1 M Tris-acetat, pH 7,5, 0,01 M Mg-acetat bei 0° 30 min mit Pankreas-RNase (∼0,1 A$_{280}$-Einheiten) inkubiert (vgl. *52*), dann aber die RNase nicht mit Bentonit, sondern mit einem wesentlich wirksameren, gereinigten Hektorit (Na-Mg-Lithofluorosilikat, „Macaloid") *(99)* bei pH 5,7 entfernt. Anschließend wurde das Hydrolysat von Mg^{++} befreit und an DEAE-Cellulose bei pH 7,5 in 7 M Harnstoff chromatographiert *(24) (Abb. 18 a)*. Fraktionen 8—12, zusammen mehr als 70% des Nucleotidmaterials, enthielten Oligonucleotidfragmente der partiellen

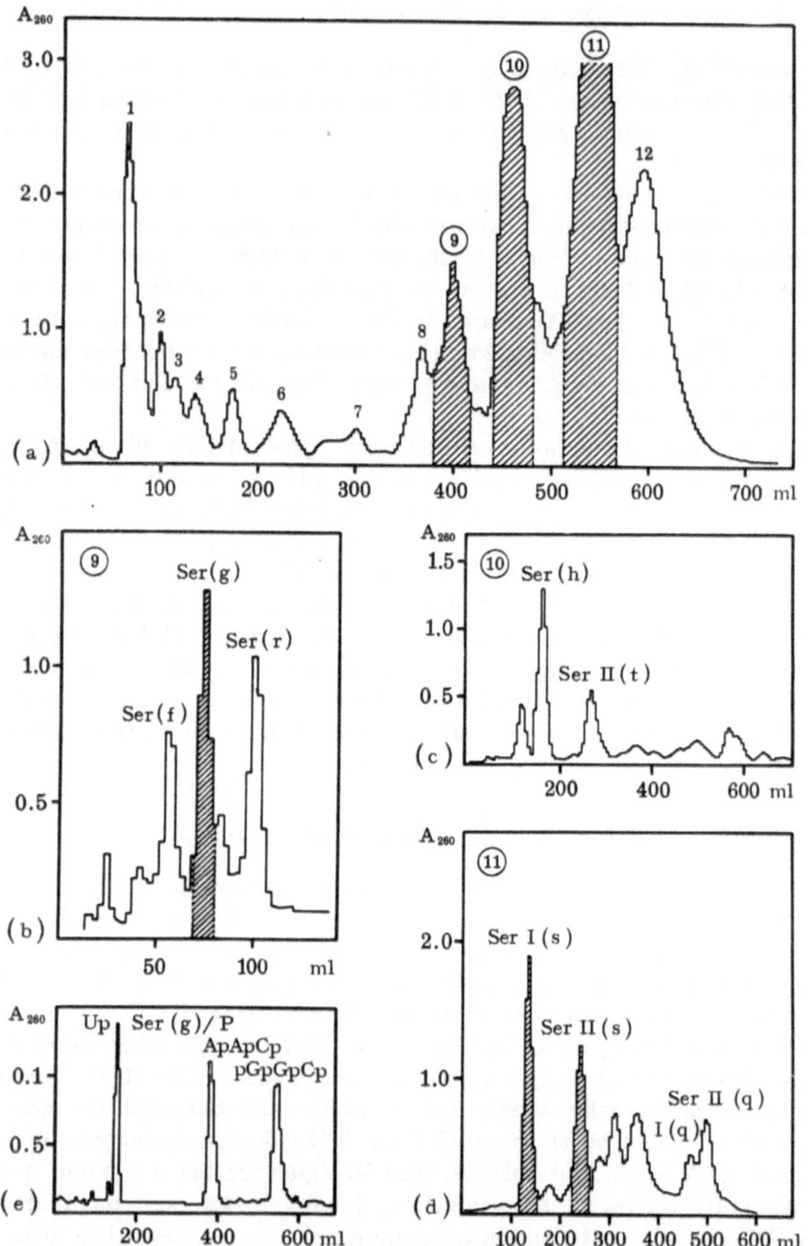

Abb. 18. Isolierung homogener Oligonucleotidfragmente aus einem Pankreas-RNase-Partialhydrolysat von Serin-tRNA I + II (24).

(a) Chromatographie des Hydrolysats (vgl. Text, S. 399) an DEAE-Cellulose (0,68 × 210-cm-Säule) in 7 M Harnstoff, pM 7,5.

(b) Rechromatographie von Fraktion 9 an DEAE-Sephadex A 25 (0,4 × 210-cm-Säule) in 7 M Harnstoff, HCl, pH 3,0.

(c) und (d) Rechromatographie der Fraktionen 10 und 11 an DEAE-Sephadex A 25 (0,68 × 210-cm-Säulen) in 7 M Harnstoff, HCl, pH 3,0.

(e) Chromatographie der Pankreas-RNase-Spaltprodukte von Fragment Ser (g) bei 4° an DEAE-Cellulose (0,8 × 25-cm-Säule), linearer Ammoniumcarbonat-Gradient.

Literaturverzeichnis: SS. 414—421

Spaltung. Die Rechromatographie der schraffierten Fraktionen 9, 10 und 11 an DEAE-Sephadex A 25 in 7 M Harnstoff, bei pH 3,0 (*Abb. 18b bis d*) lieferte eine Serie von homogenen Oligonucleotidfragmenten, die durch vollständige Spaltung mit Pankreas-RNase oder T1-RNase und Trennung der Spaltprodukte an DEAE-Cellulose-Ammoniumcarbonat-Säulen analysiert werden konnten. Als Beispiel sei auf die Isolierung (Abb. 18b) und Analyse (Abb. 18e) des 5′-terminalen Fragmentes Ser (g) hingewiesen.

Mehrere analoge Oligonucleotidfragmente wurden isoliert. Ser II (s) und Ser I (s) (Abb. 18d) ergänzen die T1-RNase-Spaltprodukte Ser II (l) und Ser I (l) insofern, als sie die beiden anderen der insgesamt drei Basenaustausche enthalten, die den Unterschied zwischen den beiden Serin-tRNA's ausmachen (Tabelle 1, S. 394). Nur teilweise ließen sich die Fragmente Ser I (q) und Ser II (q) trennen (Abb. 18d), die beide am 5′-Ende mit dem Anticodon IpGpAp— beginnen und am 3′-Ende mit —rTpψp enden. Sie unterscheiden sich in nur einem Pyrimidin (C/U = Austausch) und zeigen direkt, daß beide Serin-tRNA's das gleiche Anticodon besitzen (*117*). Zusammen mit den Fragmenten Ser I (u) und Ser II (u) bestätigen sie zudem die Struktur des 5′-terminalen Dinucleotids in den Oligonucleotiden rTpψpCpGp (Serin-tRNA II) und rTpψpCpApApApUpCpCpUpGp (Serin-tRNA I), die sich aus *Micrococcus*-Nuclease-Spaltungen der beiden Oligonucleotide ergab (*30*).

Ser I (u) und Ser II (u) entstanden bei der partiellen Nachspaltung von Fraktion 11 des T1-RNase-Partialhydrolysats (Fragmente Ser (i), Ser (j), usw.) (Abb. 17a, S. 398) mit Pankreas-RNase (in Gegenwart von Mg++) und wurden teilweise durch Chromatographie an DEAE-Cellulose bei pH 7,5 und 55° (Abb. 17d) und Rechromatographie der schraffierten Fraktion bei pH 3 an DEAE-Sephadex getrennt. Bei der anschließenden vollständigen Spaltung von Ser I (u) mit T1-RNase/PME (Abb. 17e) wurde u. a. rTpψ freigesetzt.

Bemerkenswert an den Ser(u)-Fragmenten ist die Beständigkeit ihrer Pyrimidinsequenzen bei der partiellen Nachspaltung längerer Fragmente der Akzeptorhälfte mit Pankreas-RNase. Die gleichen Pyrimidinsequenzen hemmen den partiellen Schlangengift-PDE-Abbau der Ser (l)-Fragmente (*24*) sowie der ganzen Serin-tRNA (*115*). Das Sekundärstrukturmodell der Serin-tRNA (Abb. 20, S. 404), in dem diese Sequenzen in einen durch vier GC-Paare stabilisierten „loop" einbezogen sind (S-Region), macht ihre Stabilität verständlich und läßt vermuten, daß auch die isolierten Fragmente entsprechende helicale Bereiche besitzen:

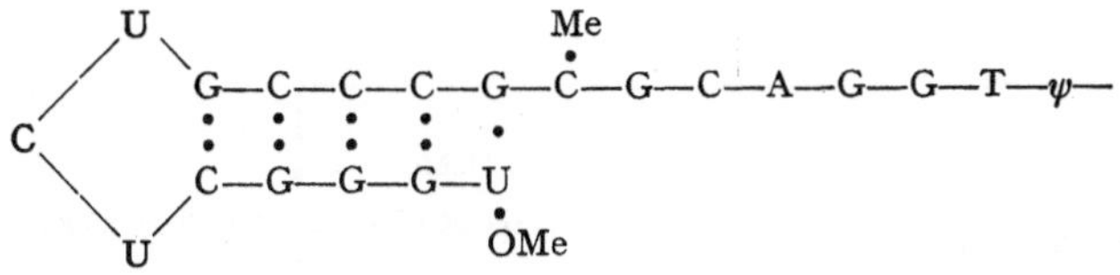

```
        U                    Me
         \                    •
   C      G—C—C—C—G—C—G—C—A—G—G—T—ψ—          Ser I (u)
          C—G—G—G—U
         /          •
        U          OMe


        U
         \
   C      G—C—C—C—G—
          C—G—G—G—U—U—U—U—C—ψ—A—A—A          Ser I (l)
         /          •               •
        U          OMe             iP
```

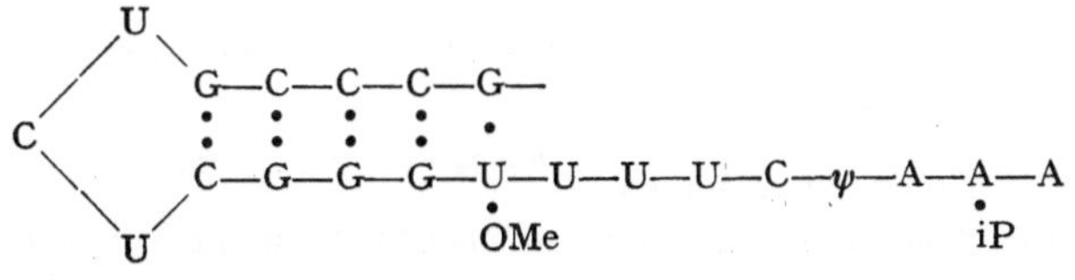

c) 3'-Terminale Oligonucleotide in großen Fragmenten

Im Unterschied zu Ser (l) wurde Fragment Ser II (m) leicht von PDE angegriffen. In Ser II (m) war die gegenseitige Anordnung der end-

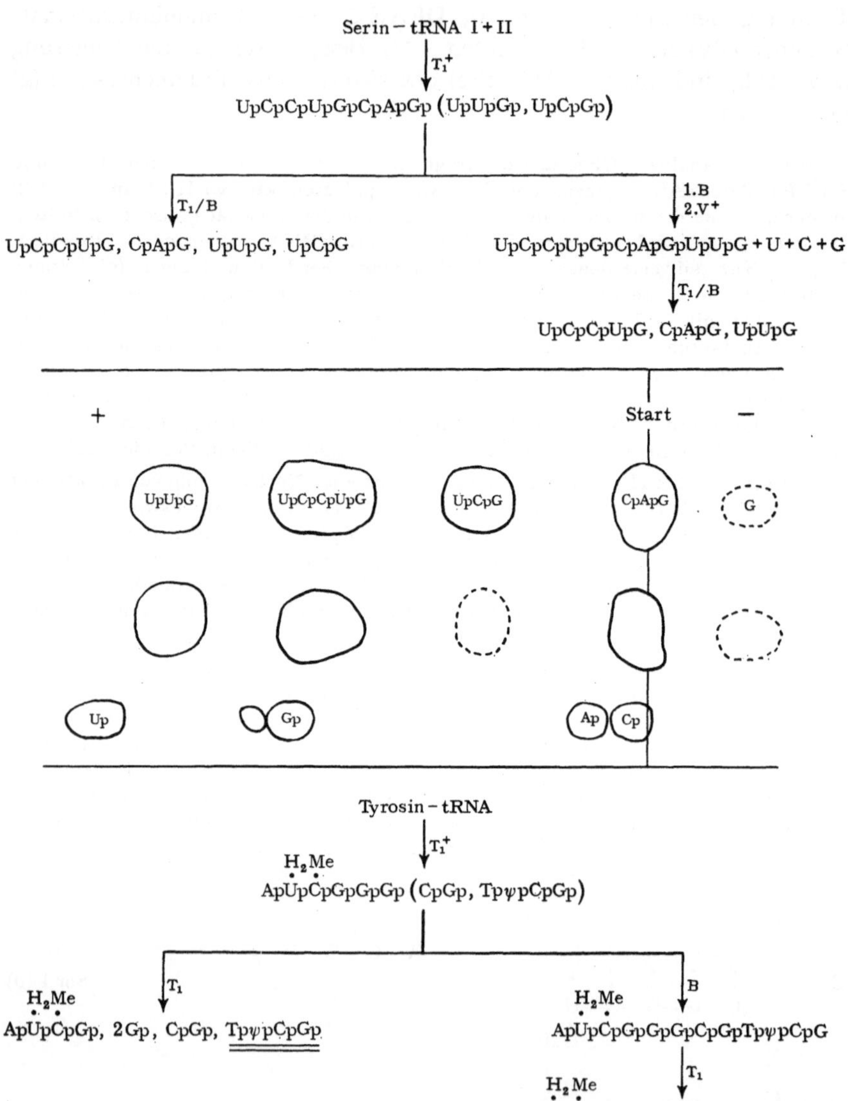

Abb. 19. Methoden zur Identifizierung 3'-terminaler Oligonucleotide in Fragmenten der partiellen Spaltung mit T1-RNase. Vgl. Text
Die Papierelektrophorese wurde in 0,2 M Ammoniumformiat, pH 2,7, 10⁻³ M EDTA, bei 30 V/cm ausgeführt

Literaturverzeichnis: SS. 414—421

ständigen Trinucleotide UpUpGp und UpCpGp unbekannt. Abb. 17c (S. 398) zeigt seine Isolierung aus Fraktion 10 des T1-RNase-Partialhydrolysats durch Rechromatographie an DEAE-Sephadex bei pH 3 (schraffierte Fraktion). Ein Teil des Fragments wurde vollständig mit T1-RNase/PME gespalten, der andere Teil mit PME zur Entfernung des 3'-terminalen Phosphats vorinkubiert und dann kurzzeitig (10 min, Raumtemperatur) mit Schlangengift-PDE behandelt *(Abb. 19)*. Die Chromatographie des Inkubationsgemisches an einer kleinen DEAE-Cellulose-Säule in 7 M Harnstoff, pH 7,5, ist in Abb. 17f zu sehen (S. 398). Der schraffierte Gipfel wurde isoliert und vollständig mit T1-RNase/PME gespalten. Papierelektrophorese (pH 2,7) der beiden T1-RNase-Hydrolysate (Abb. 19, Mitte) zeigte, daß aus dem PDE-behandelten Hydrolysat UpCpG nahezu vollständig verschwunden war. Seine 3'-terminale Position in Ser (II) m ist dadurch gesichert *(24)*.

Eine andere Methode, das 3'-terminale Oligonucleotid eines größeren Fragmentes zu bestimmen, wurde bei der Strukturaufklärung von Tyrosin-tRNA verwendet (Abb. 19). Das Gemisch der Fragmente Tyr (d) (aus der 5'-Phosphat-Hälfte) und Tyr (g) (aus der Akzeptor-Hälfte) wurde zuerst mit PME, dann nach Entfernen der PME, mit T1-RNase inkubiert. Bei der anschließenden DEAE-Cellulose-Chromatographie wurde neben rTpψpCpGp auch rTpψpCpG gefunden, das daher in Fragment Tyr (g) 3'-terminal sein muß *(59)*. Die unvollständige Entfernung des 3'-Phosphats durch PME zeigt, daß ein Teil von Tyr (g) mit Gp! endete.

Bei der PME-Behandlung von Ser II (m) trat diese Schwierigkeit nicht auf, da dieses Fragment durch DEAE-Sephadex-Chromatographie bei pH 3 (Raumtemperatur, 2—3 Tage) isoliert worden war, also unter Bedingungen die Cyclophosphatenden öffnen. Auch die Oligonucleotidfragmente der partiellen Spaltung mit Pankreas-RNase enthielten nach der Chromatographie bei pH 3,0 keine Cyclophosphatenden. So wurde z. B. nach der vollständigen Spaltung von Ser (q) und Ser (u) mit T1-RNase/PME nur rTpψ, nicht aber rTpψp! gefunden (Abb. 17e, S. 398).

Cyclophosphatendgruppen von Oligonucleotidfragmenten können auch enzymatisch mit einer Cyclophosphodiesterase aus *E. coli (1)*, die gleichzeitig 3'-Nucleotidase-Aktivität besitzt, entfernt werden. Phosphatfreie 3'-Enden sind vor allem für die radioaktive Markierung mit der Perjodat-NaB^{3}H$_4$-Methode *(78)* wichtig.

d) Partielle Spaltung mit der „sauren" Ribonuclease aus Milz

Mit der „sauren" Ribonuclease aus Milz *(9a)* wurde 5 S-Ribosomale RNA partiell gespalten. Die entstandenen großen Fragmente endeten im wesentlichen mit —Ap *(88)*, zeigen also eine A-Spezifität dieses Enzyms unter den Bedingungen der partiellen Spaltung an.

2. Sekundärstruktur von Transfer-Ribonucleinsäuren

Die linearen Sequenzen von Alanin-, Serin-, Tyrosin- und Phenylalanin-tRNA, die mit den geschilderten Methoden aufgeklärt wurden,

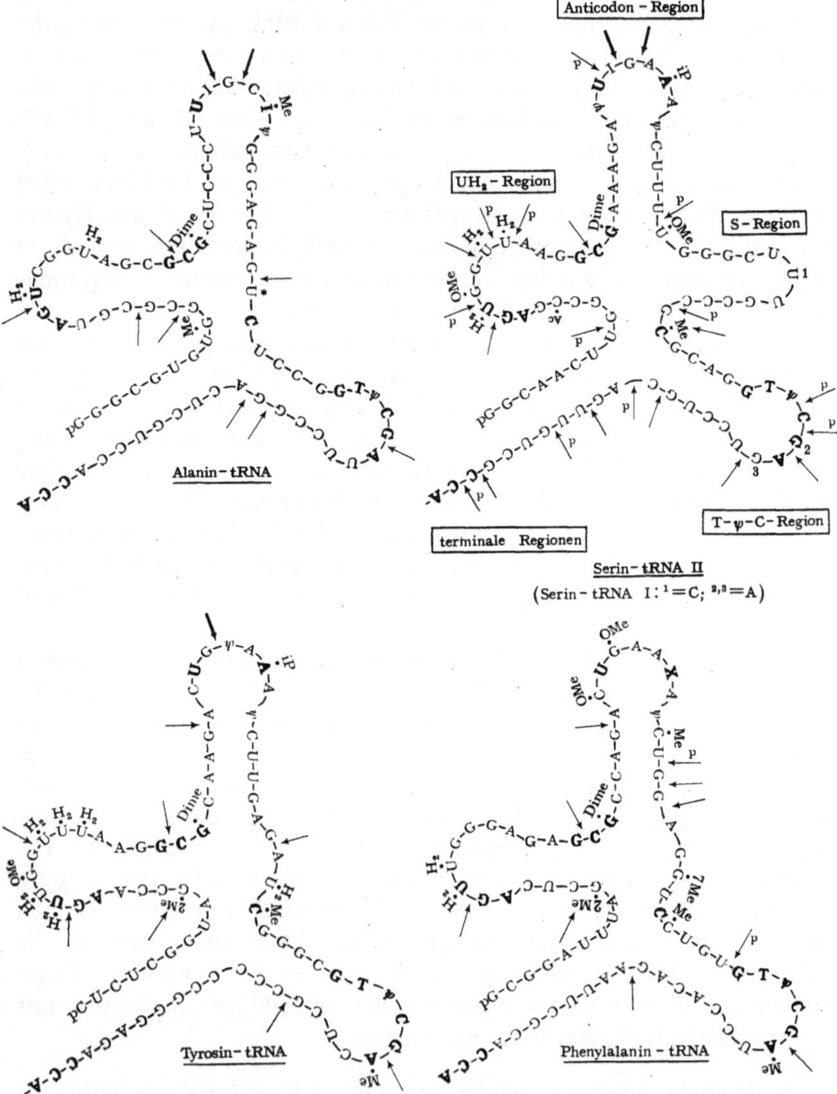

Abb. 20. Sekundärstruktur-Modelle von Alanin-, Serin-, Tyrosin- und Phenylalanin-tRNA. Vgl. Text

lassen sich durch maximale Basenpaarung (entsprechend Abb. 2, S. 361)
in sehr ähnliche „Kleeblatt"-Conformationen falten, die in *Abb. 20* dargestellt sind (*39, 114, 56, 76*).

Literaturverzeichnis: SS. 414—421

Optische Rotationsdispersions(ORD)-Messungen an Lösungen von Alanin- und Tyrosin-tRNA (pH 6,8, 0,15 M KCl, + oder —Mg^{++}) zeigten, daß bei niedriger Temperatur die dreidimensionalen Strukturen dieser tRNA's durch intramolekulare H-Brückenbindungen — hauptsächlich GC-Paare — *und* „stacking"-Wechselwirkungen zwischen benachbarten Basen der linearen Sequenz zusammengehalten werden (*107*). Der Gehalt an GC-Paaren, der aus den ORD-Kurven abgeschätzt wurde, ist in Übereinstimmung mit den Modellen (Abb. 20), bei Alanin-tRNA größer als bei Tyrosin-tRNA.

Außer durch ORD-Kurven (*107*) und Röntgenstrukturdaten (*32*) werden die Sekundärstrukturmodelle von Abb. 20 u. a. durch die Befunde gestützt, die bei der partiellen Spaltung von Serin-tRNA mit Pankreas-RNase (*24*) sowie bei der T1-RNase-Spaltung von Alanin- und Tyrosin-tRNA in Gegenwart von Mg^{++} bei 0° (*73, 59*) gewonnen wurden. Die mit P bezeichneten Pfeile am Modell der Serin-tRNA (Abb. 20) zeigen auf die Bindungen, die von Pankreas-RNase in der anfänglichen partiellen Spaltungsphase in Gegenwart von Mg^{++} hydrolysiert werden. Diese exponierten Bindungen liegen fast ausschließlich in den ungepaarten Bereichen der „Kleeblatt"-Conformation (*116*).

Wie erwähnt, kann die T1-RNase-Spaltung in *Gegenwart* von Mg^{++} so dirigiert werden, daß zunächst nur die am stärksten exponierten Anti-codon-Sequenzen (dicke Pfeile) angegriffen werden. Die dünnen unbe-zifferten Pfeile zeigen auf die Bindungen, deren Spaltung mit T1-RNase in *Abwesenheit* von Mg^{++} zu Partialstücken führt. Da der stabilisierende Einfluß von Mg^{++} unter diesen Bedingungen fortfällt, ist verständlich, daß auch Spaltung in helicalen Bereichen beobachtet wurde. In Überein-stimmung mit dieser Vorstellung konnte aus Tyrosin-tRNA das lange Fragment Tyr (j) nur in Gegenwart, nicht aber in Abwesenheit von Mg^{++} isoliert werden (*59*).

Die *Sekundärstrukturen* von Abb. 20 können durch Aufeinander-faltung bestimmter „loops" in *Tertiärstrukturmodelle* der tRNA's über-führt werden, aus denen vor allem die Anticodon-„loops" herausragen. So kann man den rT—ψ—C-„loop" auf den UH$_2$-„loop" klappen (*37, 14*) und dabei außer Mg^{++}-Brücken zusätzliche Basenpaare, wie

$$\diagdown\text{C—G—A}\diagup \qquad (\text{rT}—\psi—\text{C-Region})$$

$$\diagup\text{G—C—UH}_2\diagdown \qquad (\text{UH}_2\text{-Region})$$

in Alanin-tRNA, annehmen. Eine wahrscheinliche Conformation der Anticodon-„loops" als stereochemische Basis für die „Wobble"-Hypo-these (*19*) wurde kürzlich diskutiert (*32*). In dieser Conformation sind fünf Basen des Anticodon-„loops", z. B. I—G—A—iPA—A— in Serin-tRNA, auf den fünf gepaarten Basen des helicalen Anticodon-Armes, —ψ—C—U—U—U— in Serin-tRNA, durch „stacking" helical angeordnet. Die Modifizierung der Purinbase rechts vom Anticodon (Abb. 20), d. h. bei Serin-tRNA die Isopentenyl-Seitenkette des iPA, scheint zu verhindern, daß sich das Anticodon-Triplett in dieser Conformation um eine Base verschiebt, z. B. statt I—G—A in Serin-tRNA G—A—(A) als Anticodon fungiert.

Für die notwendige Diskriminierung der tRNA's durch die Aminoacyl-RNA-Synthetasen könnten im Rahmen der nativen tertiären Conformationen die strukturellen Unterschiede vor allem der S- und UH_2-Regionen von Bedeutung sein (*115, 56*): Die in verschiedenem Grade „offenen" UH_2-Regionen zeichnen sich durch Menge und Anordnung der UH_2's in sonst purinreichen Sequenzen aus, die als konstantes Element —A—G—UH_2— enthalten. Es ist anzunehmen, daß die UH_2's die „stacking"-Wechselwirkungen benachbarter Basen unterbrechen und dadurch diesen Teil der Moleküle für eine Wechselwirkung mit Proteinen geeigneter machen.

Die aus den Sekundärstrukturmodellen ersichtliche größere Labilität des UH_2-„loops" von Tyrosin-tRNA im Vergleich zu dem von Alanin-tRNA wird durch Partialspaltungen mit T1-RNase unter vergleichbaren (Mg-freien) Bedingungen bestätigt: Es konnten keine in den UH_2-„loop" von Tyrosin-tRNA hineinreichenden Partialstücke isoliert werden (*59*). Die beiden Pfeile am UH_2-„loop" von Tyrosin-tRNA (Abb. 20) bezeichnen Spaltstellen, die bei der partiellen Nachspaltung der isolierten 5'-Phosphathälfte beobachtet wurden.

3. Partielle enzymatische Spaltung von 5 S-Ribosomaler und hochmolekularer RNA

Die Oligonucleotidfragmente der partiellen enzymatischen Spaltungen von ^{32}P-markierter *5S-Ribosomaler RNA* aus *E. coli* wurden auf DEAE-Cellulose-Papier durch eine „displacement" Chromatographie getrennt. Dabei entwickelt man das Chromatogramm mit einem konzentrierten Gemisch nichtmarkierter Oligonucleotide, die das Papier in distinkten Fronten hinabwandern und die mitlaufenden radioaktiven Fragmente etwa entsprechend ihrer Kettenlänge auftrennen (*88*). Mit den auf diese Weise gereinigten Fragmenten konnte die vollständige Nucleotidsequenz der 5 S-Ribosomalen RNA trotz des Fehlens „seltener" Nucleotide aufgeklärt werden (*13*).

Abb. 21 zeigt ein mögliches, weitgehend „offenes" Sekundärstrukturmodell der 5 S-Ribosomalen RNA mit drei helicalen Regionen, die außer den Standard-Basenpaaren auch G—U-Paare enthalten. Die Sequenzen in diesen durch Basenpaarung ausgezeichneten Bereichen waren besonders resistent gegen Spaltung mit RNasen. Wie bei den tRNA's

(Abb. 20, S. 404) lassen sich die beiden Enden des Moleküls miteinander paaren. Selbstverständlich können auch Conformationen mit wesentlich mehr Basenpaaren konstruiert werden (*13a*).

Die *16S*- und *23S-Komponenten* der *Ribosomalen RNA* stellen wegen ihrer Länge (1600 bzw. 3200 Nucleotide) ein schwieriges Problem für die Sequenzanalyse dar. In den „fingerprints" ihrer Pankreas- und T1-RNase-Spaltprodukte (*89*) und in ihren 5'-Termini (*101*) wurden Sequenz-

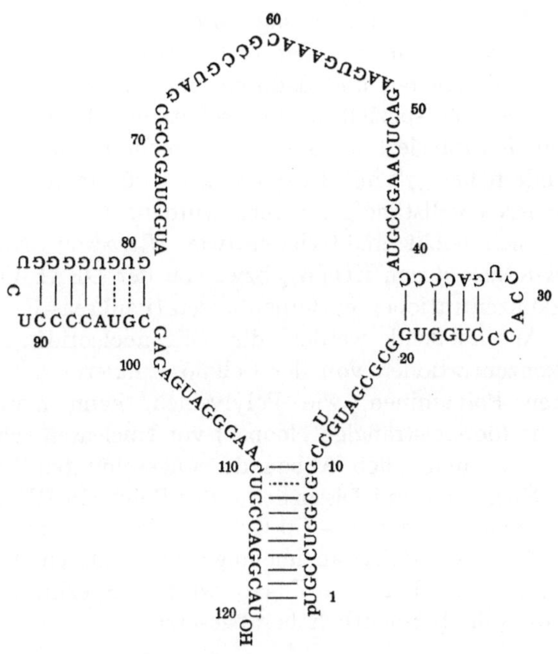

Abb. 21. Sekundärstruktur-Modell der beiden hauptsächlichen 5S-Ribosomalen RNA's von *E. coli* MRE 600 (*13*). Die beiden RNA's unterscheiden sich durch einen G/U-Austausch in Position 13. Standard-Basenpaare sind durch ausgezogene Linien, G—U-Paare durch unterbrochene Linien gekennzeichnet

unterschiede nachgewiesen. Die 23S-Komponente wird außerordentlich leicht — möglicherweise durch eine endogene Ribosomale Nuclease — in zwei Hälften gespalten, und scheint daher eine besonders exponierte labile Phosphodiesterbindung zu besitzen (*63*).

Die partielle Spaltung mit Pankreas- und T1-RNase bei 0° wurde am Gemisch der beiden Komponenten untersucht und der Grad der Spaltung bei zunehmender Enzymmenge durch Polyacrylamidgel-Elektrophorese in 5%-Gelen verfolgt (*60*). Man erhält dabei in An- und Abwesenheit von Mg^{++} eine begrenzte Zahl distinkter Banden (bis zu 12) im Sedimentations- und damit Molekulargewichtsbereich von 4—15 S, die für eine definierte dreidimensionale Struktur der beiden RNA's sprechen. Mit steigender Enzymkonzentration bei der Inkubation verschwinden die langsamer wandernden Banden, während die rascher wandernden an Intensität gewinnen, d. h. die wenigen großen Fragmente der ersten Abbauphase werden, wie bei den tRNA's, in kleinere Fragmente weitergespalten.

Eine noch nicht voll ausgeschöpfte Methode für die Fraktionierung größerer RNA-Fragmente dürfte die Säulenchromatographie an benzoylierter DEAE-Cellulose (*102*) sein.

Hochmolekulare RNA's mit wenig ausgeprägter Sekundärstruktur, wie *MS2-RNA*, kann man vor der endonucleolytischen Spaltung an DEAE-Cellulose adsorbieren und dadurch eine Art künstlicher Tertiärstruktur erzeugen, die speziell in ungeschützten Bereichen gespalten wird (*85*). Man hydrolysiert in diesem Fall nicht bei 0° in Gegenwart von Mg^{++}, sondern bei 37° mit Enzymmengen, die in Abwesenheit des Adsorbens die RNA vollständig abbauen würden. Geeignet sind Endonucleasen, die sich leicht inaktivieren (wie *Micrococcus*-Nuclease und *Bacillus cereus*-RNase durch EDTA), bzw. von der DEAE-Cellulose mit niedrigen Salzkonzentrationen entfernen lassen (Pankreas-RNase, *B. subtilis*-RNase). Anschließend werden die Oligonucleotidfragmente mit höheren Salzkonzentrationen von der Cellulose eluiert.

Mit anderen Polyaminen, wie Polylysinen, kann man spezifisch helicale Bereiche (doppelsträngige „loops") vor Nucleasen schützen (*94*). Die Fragmente, die man nach Abbau der ungeschützten Bereiche mit unspezifischen Nucleasen und Dissoziieren der Polylysin-RNA-Komplexe erhält, zeigen einen hohen Gp + Cp-Gehalt, wie man ihn in stark gepaarten Bereichen der RNA anzunehmen hat. Durch Hitzedenaturierung der RNA vor der Polylysin-Zugabe wird die Spezifität des Schutzeffekts von Polylysin beträchtlich herabgesetzt.

4. Partielle Spaltung nach chemischer Modifizierung der RNA

Gilham (*33*) hat ein wasserlösliches Carbodiimid, 1-Cyclohexyl-3-(2-morpholinyl-4-äthyl)-carbodiimid-metho-*p*-toluolsulfonat (CMC), beschrieben, das sich an den N1-Stickstoff von Up und Gp sowie an die beiden N-Atome von ψp in RNA anlagert *(Abb. 22)*. Pankreas-RNase ist nicht in der Lage, neben dem so modifizierten Up und ψp ($\overline{U}$p bzw. $\overline{\psi}$p) zu spalten, so daß man nur Oligonucleotide mit Cp-Enden erhalten sollte.

Abb. 23 zeigt die DEAE-Cellulose-Chromatographien von Pankreas-RNase-Hydrolysaten, (a) unbehandelter Hefe-sRNA und (b) mit CMC modifizierter Hefe-sRNA (*28*). Für die Modifizierung wurden Bedingungen verwendet, die bei Ribosomaler RNA aus Weizenkeimen nach Pankreas-RNase-Spaltung nur Trinucleotide mit Cp-Enden gegeben hatten (*51*). Die Bedingungen der Pankreas-RNase-Spaltung waren in beiden Versuchen gleich und so gewählt, daß auch im Mononucleotidbereich kaum Cyclophosphate zu erwarten waren. Wie deutlich zu sehen ist, entsteht bei der Spaltung der modifizierten RNA kein freies Up und ψp; außerdem ist der Gehalt an längeren Oligonucleotiden im Hydro-

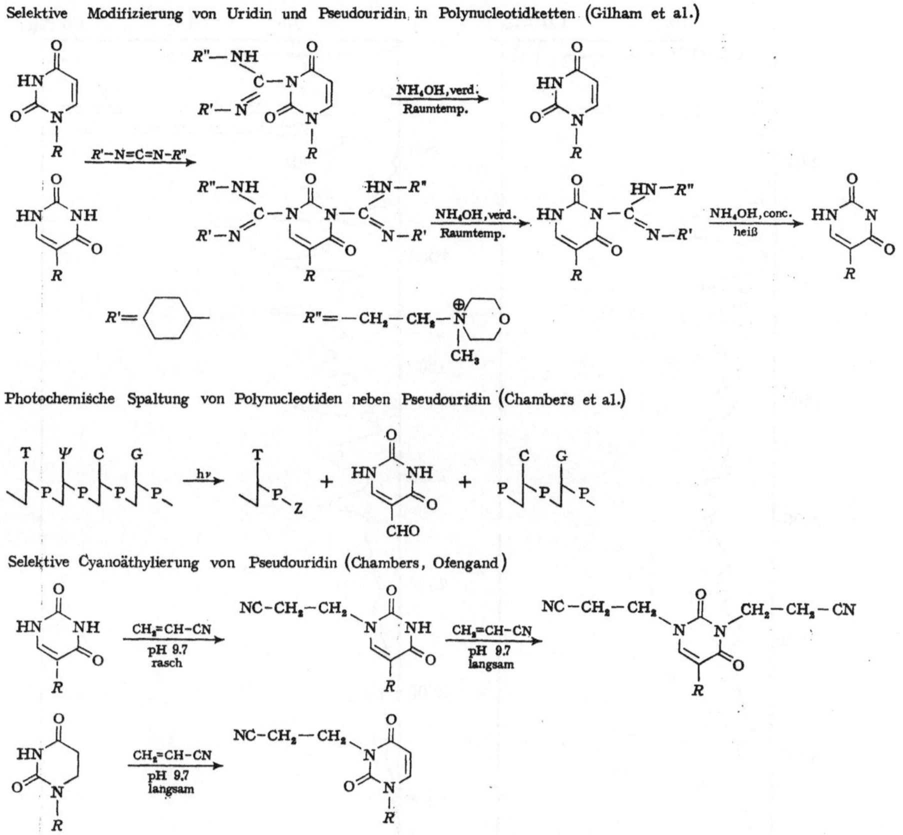

Selektive Modifizierung von Uridin und Pseudouridin in Polynucleotidketten (Gilham et al.)

Photochemische Spaltung von Polynucleotiden neben Pseudouridin (Chambers et al.)

Selektive Cyanoäthylierung von Pseudouridin (Chambers, Ofengand)

Abb. 22. Chemische Modifizierung von Nucleotidbasen in Ribonucleinsäuren. Vgl. Text, S. 408

lysat dieser RNA merklich höher. Mit Erfolg wurde diese Methode bei der Strukturaufklärung von 5 S-Ribosomaler RNA verwendet (*88*).

Schlangengift- und Milz-Phosphodiesterase sind nicht in der Lage, modifizierte Dinucleosidphosphate vom Typ Cp$\overline{\text{U}}$ und Cp$\overline{\psi}$ zu spalten (*64*). Da eine der beiden CMC-Gruppen des ψ gegenüber verd. Ammoniak bei Raumtemperatur, das die CMC-Reste von U und G entfernt, resistent ist (Abb. 22), sollte es möglich sein, den exonucleolytischen Abbau einer tRNA am ersten modifizierten ψ selektiv zu blockieren. CMC reagiert auch mit denaturierter DNA (wahrscheinlich mit den T- und G-Resten),

jedoch kaum mit nativer DNA. Die Reaktion mit sRNA verläuft in
Gegenwart von Mg^{++} langsamer als in seiner Abwesenheit (4).

Etwa 2 Mol CMC-^{14}C-Methoiodid ließen sich an Alanin-tRNA aus Hefe bei 38°
in 0,02 M Mg^{++} innerhalb der ersten 60 Sekunden addieren. Anschließend ver-

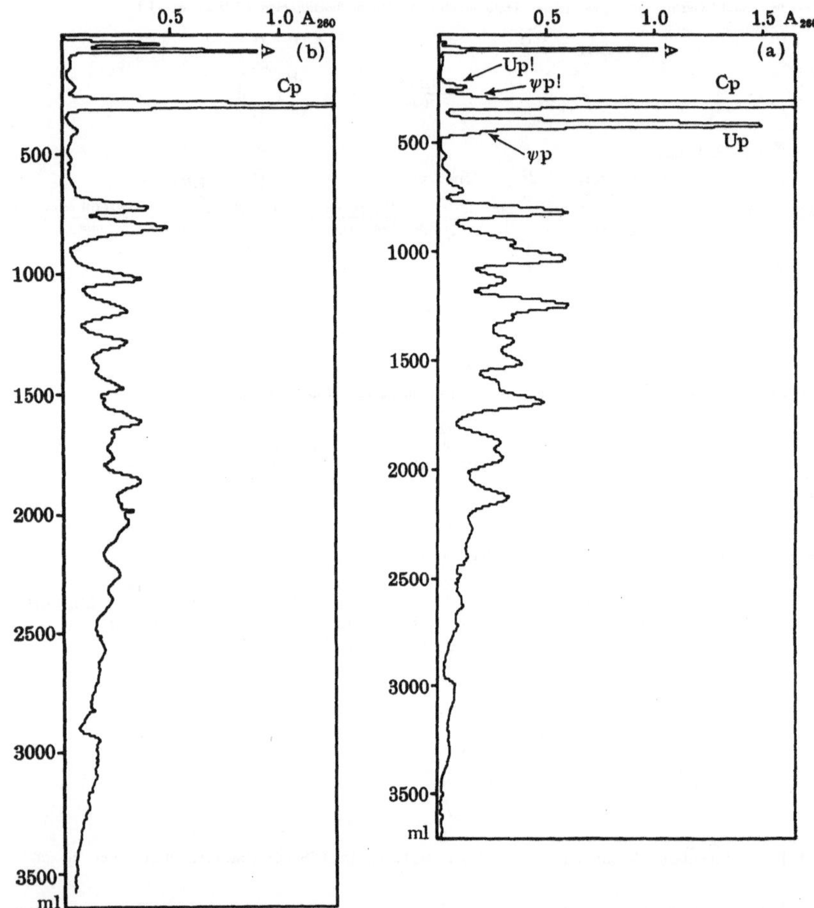

Abb. 23. Unterschied im Pankreas-RNase-Hydrolysat von nichtmodifizierter (a) und mit CMC-modifizierter
Hefe-sRNA (b). Vgl. Text, S. 408. Chromatographie an DEAE-Cellulose (1 × 50-cm-Säulen) mit linearen
Ammoniumcarbonat-Gradienten

langsamte sich die Reaktion stark. Spaltung der modifizierten Alanin-tRNA mit
T1- und Pankreas-RNase und Analyse der radioaktiven Spaltprodukte zeigte, daß
zunächst die UpUpIpGpCpMeIpψpGp-Region reagiert hatte, nicht aber die ebenfalls
in einem „loop" gelegene TpψpCpGpApUp-Sequenz (11a) (vgl. Abb. 20, S. 404).
Die Anticodon-Region muß also in der dreidimensionalen Struktur der Alanin-
tRNA besonders exponiert sein.

Literaturverzeichnis: SS. 414—421

Diese Befunde lassen vermuten, daß CMC bevorzugt mit Basen in nichthelicalen Bereichen von RNA und DNA reagiert.

Ein anderes interessantes Reagens, mit dem man nahezu spezifisch ψ-Reste in tRNA modifizieren kann, ist Acrylonitril (*15*).

Führt man die Cyanoäthylierung (Abb. 22) bei 37° in Dimethylsulfoxyd aus, das 1 M Dimethylaminoäthanol und verschiedene Mengen Wasser enthält, so kann man mit der Wasserkonzentration die Zahl der ψp-Reste kontrollieren, die in den tRNA-Molekülen in der anfänglichen raschen Reaktionsphase cyanoäthyliert werden (*80*): in 25% Wasser 1,2 Reste, in 35% Wasser 2 Reste und in 50% Wasser 3 Reste (d. h. alle ψp's) pro tRNA-Molekül. In der anschließenden langsameren Phase reagieren auch U, I und rT, während ψ dicyanoäthyliert wird (Abb. 22). Da mit der graduellen Cyanoäthylierung von ψ-Resten in tRNA's ein gradueller Verlust an Sekundärstruktur verbunden ist (wie sich aus Schmelz- und ORD-Kurven ergibt) und gleichzeitig die Aminosäure-Akzeptoraktivität über 30% (1. ψp) und 20% (2. ψp) auf 15% Restaktivität (3. ψp cyanoäthyliert) absinkt, scheint eine mögliche Funktion der ψp's in den tRNA's die Aufrechterhaltung der Sekundärstruktur zu sein.

Zunächst reagiert das am stärksten exponierte ψp (wahrscheinlich dasjenige, das sich in Alanin-, Serin-, Tyrosin-, Phenylalanin- und Valin-tRNA aus Hefe im Anticodon-,,loop" befindet; vgl. Abb. 20, S. 404), dann die restlichen weniger exponierten ψp's (vor allem das ψp im rT—ψ—C-,,loop"). Im Einklang hiermit wurde gefunden (*112*), daß das ψp der rTpψpCpGp-Sequenzen von Hefe-sRNA bei der Cyanoäthylierung von nur 1 ψp in Gegenwart von 0,5 M NaCl *nicht*, in Abwesenheit des die Sekundärstruktur stabilisierenden NaCl jedoch fast *vollständig* modifiziert war, also das 2. oder 3. der cyanoäthylierten ψp's sein muß. Aus diesem und den CMC-Versuchen kann man schließen, daß sich die rTpψpCpGp-Sequenz in der voll ausgebildeten dreidimensionalen Struktur der tRNA's nicht in exponierter Lage befindet, sondern, wie in Kap. V, S. 405 diskutiert, an der komplementären Basenpaarung im Innern der tRNA beteiligt ist.

Neben der selektiven Modifizierung von ψp ist auch eine selektive Spaltung von RNA-Ketten an den ψp-Resten beschrieben worden (*104*), bei der photochemisch ψp in 5-Formyluracil umgewandelt wird. Diese Spaltung wurde an der rTpψpCpGp-Sequenz erprobt (Abb. 22, S. 409) und dann auf Alanin-tRNA angewandt. Tatsächlich wurden bei der anschließenden Gelfiltration der photochemisch gespaltenen Alanin-tRNA (Sephadex G-100, 56°) vier teilweise getrennte Fraktionen isoliert, die in ihrer Größe den möglichen Oligonucleotidfragmenten bei selektiver Spaltung am ψp zu entsprechen scheinen.

Bromierung von Alanin-tRNA mit N-Bromsuccinimid bei 25° und pH 7,5 in wäßriger, Mg^{++}-haltiger Lösung (N-Bromsuccinimid : RNA, 50 : 1) führte zum Einbau von etwa vier Brom-Atomen pro RNA-Molekül. Da T1-RNase neben bromierten Nucleotiden nicht spaltet, konnten die Positionen der Brom-Atome in der RNA aus dem veränderten Chromatographie-Muster nach T1-RNase-Spaltung abgeleitet werden (*65*). Bromiert wurde bevorzugt am Anticodon-,,loop", in der UH$_2$-Region und an der Aminosäure-Akzeptor-Sequenz, erst anschließend am rT—ψ—C—,,loop" und in anderen Einzelstrag-Regionen. Diese Versuche sprechen ebenfalls für eine ,,Kleeblatt"-Conformation der Alanin-tRNA.

VI. Schlußbemerkungen

Im Verlauf der Strukturaufklärung von mehreren tRNA's und 5 S-Ribosomaler RNA wurden Methoden der Sequenzanalyse entwickelt, mit denen es möglich sein sollte, auch in die Struktur größerer RNA-Moleküle einzudringen. Wie das Beispiel der 5 S-Ribosomalen RNA zeigt, in der keine „seltenen" Nucleotide die Rekonstruktion von überlappenden Sequenzen erleichtern, wird dabei die partielle Spaltung mit Endonucleasen und die enzymatische Spaltung nach chemischer Modifizierung bestimmter Nucleotide eine besondere Rolle spielen.

Als besonders günstiges Objekt für zukünftige Arbeiten bietet sich die an sich homogene *RNA der Phagen MS2 (f2) und Qβ* an. Mit kleinen Mengen Pankreas-RNase gelang es bereits, $Q\beta$-RNA bei 0° in zwei Fragmente zu spalten, von denen eines 68%, das andere 32% der etwa 3300 Nucleotide enthielt (*4a*). Die Orientierung der beiden Fragmente in der RNA konnte durch spezifische Markierung des 3'-Endes *vor* der partiellen Spaltung festgelegt werden. Zu diesem Zweck wurde das 3'-terminale Nucleosid der $Q\beta$-RNA mit Perjodat oxydiert und der entstandene terminale Dialdehyd mit ^{3}H-Isonicotinsäurehydrazid gekuppelt (vgl. das Schema auf S. 375). Anschließende Fragmentierung zeigte, daß das große Fragment das radioaktiv markierte 3'-Ende der RNA enthielt, das kleine Fragment also von dem Teil der RNA mit dem 5'-Ende herrühren mußte.

Dieses Beispiel demonstriert, wie man sich die Sequenzanalyse von größeren RNA-Molekülen zu denken hat: Zunächst Spaltung der RNA in wenige große Fragmente, deren Orientierung in der Gesamt-RNA durch Endgruppenmarkierung festgestellt werden kann; anschließend Analyse einzelner Fragmente mit den konventionellen Methoden.

Da die $Q\beta$- und MS2-RNA die gesamte genetische Information (3 oder 4 Gene) der zugehörigen Phagen enthält, hat man hier also die Möglichkeit, in die Struktur von Genen selbst einzudringen. Von besonderem Interesse ist dabei die Sequenz, die den Übergang von einem Gen zum nächsten Gen signalisiert. Eine terminale Lage der Hüllprotein-Gene in diesen RNA's würde die Sequenzanalyse entsprechender terminaler Nucleotidfragmente wahrscheinlich erleichtern, da z. B. die gesamte Aminosäuresequenz des f2-Hüllproteins bereits bekannt ist (*108*).

An eine *Sequenzanalyse von DNA-Fragmenten* ist wegen des Fehlens spezifischer Nucleasen für die gezielte enzymatische Spaltung und wegen der Länge der Moleküle vorläufig nicht zu denken. Wegweisend sind hier die elektronenmikroskopischen Untersuchungen von BEER und Mitarbeitern (*63a*) an einzelnen DNA-Strängen, in denen bestimmte Nucleotide mit sichtbaren Markierungen versehen worden waren. So addiert sich 2-Diazo-*p*-benzoldisulfonsäure unter bestimmten Bedingungen be-

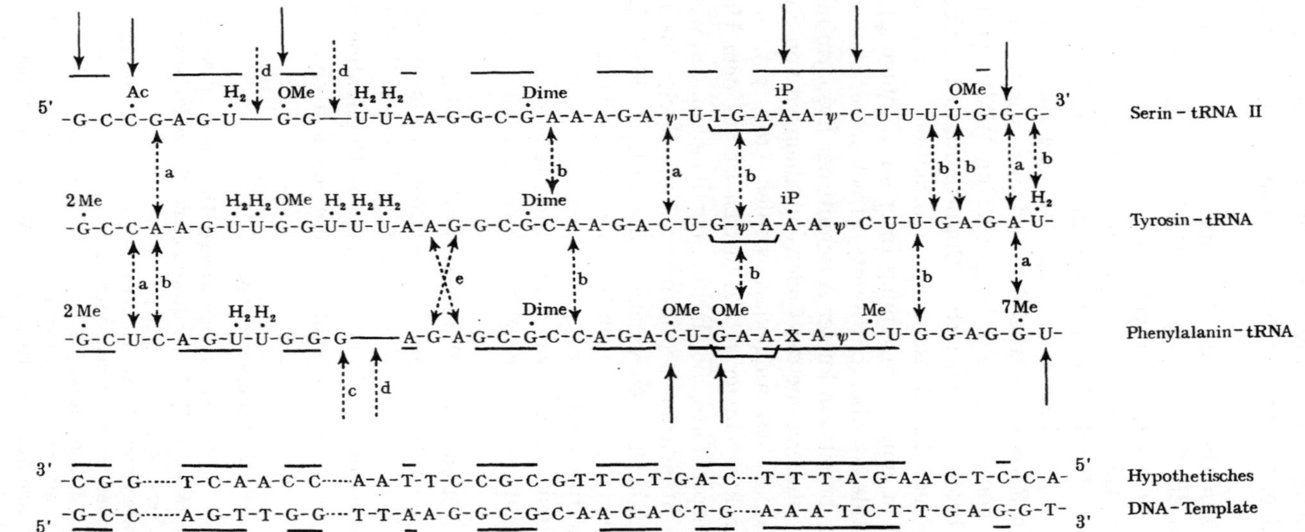

Abb. 24. Ausschnitte aus den Sequenzen von Serin-, Tyrosin- und Phenylalanin-tRNA und einem hypothetischen gemeinsamen DNA-Template. Eingerahmte Bereiche sind in allen drei Molekülen gleich, wenn man von der sekundären enzymatischen Modifizierung einzelner Basen absieht. Ausgezogene Pfeile: vgl. Text, S. 414. Gestrichelte Pfeile: Mutationen, die zur funktionellen Differenzierung der entsprechenden, durch Duplikation entstandenen DNA-Cistren führten

a: Transition, b: Transversion, c: Insertion, d: Deletion, e: Inversion

Auch für Translokationen sind Anhaltspunkte vorhanden (Abb. 20, S. 404): z. B. G—G—G—A—G—A—G—U— (Alanin-tRNA), G—G—G—A—G—A—C— (Tyrosin-tRNA), G—G—G—A—G—A—G—C— (Phenylalanin-tRNA) in verschiedenen Regionen der Moleküle vorzugt an Gp-Reste und kann nach Ausbreitung der DNA auf einem Kohle-Film und Färbung mit Uranylacetat im Elektronenmikroskop sichtbar gemacht werden, da jede der Sulfonsäuregruppen Uranylionen-Komplexe bindet.

Im übrigen bleibt zunächst nur die Hoffnung, auf dem Umweg über ihre mRNA-Kopien in die Feinstruktur der DNA-Gene (Cistren) selbst einzudringen. Am aussichtsreichsten erscheint die Kombination von Mutationsanalyse eines Gens mit Strukturuntersuchungen an der zugehörigen mRNA zu sein. Für die Isolierung spezifischer mRNA's kommen DNA-RNA-Hybridisierungstechniken in Frage *(70)*, die aber wegen der Labilität bakterieller mRNA noch erheblich verfeinert werden müssen. Die Mutationsanalyse könnte auf dem Vergleich von Wildtypproteinen mit den Proteinen Proflavin-induzierter Doppelmutanten aufgebaut werden, in denen die Phase der Ablesung der Information — z. B. durch Deletion (Verlust) und in einiger Entfernung Insertion (Einfügung) eines oder mehrerer Nucleotide — über bestimmte Strecken verschoben ist. Auf diese Weise konnte bereits ein kurzer Abschnitt der Lysozym-mRNA des T4-Phagen unter Zuhilfenahme der Codon-Zuordnungen von NIREN-BERG *(67)* und KHORANA *(48)* aufgeklärt werden *(103, 71)*.

Einblick in die Struktur einiger tRNA-Cistren ist auf dem Umweg über ihre Transkriptionsprodukte, die tRNA's, bereits gewonnen worden. Die Umwandlung von Standardnucleosiden in seltene Nucleoside findet mit größter Wahrscheinlichkeit enzymatisch am fertigen RNA-Strang statt *(11, 109)*. Das zeigt sich u. a. auch bei einem Vergleich der bereits bekannten tRNA-Strukturen *(Abb. 24)*, in denen in bestimmten Positionen (ausgezogene Pfeile in Abb. 24) Standardnucleoside *und* die von ihnen abgeleiteten modifizierten „seltenen" Nucleoside zu finden sind *(116)*.

Die großen Ähnlichkeiten in der Primärstruktur von Serin-, Tyrosin- und Phenylalanin-tRNA aus Hefe legen den Gedanken an eine molekulare Evolution der zugehörigen tRNA-Cistren nahe *(44)*, in deren Verlauf ein hypothetisches tRNA-Cistron durch eine Serie von Duplikationen und anschließende Mutationen in die Cistren der individuellen tRNA's umgewandelt wurde.

Literaturverzeichnis

1. ANRAKU, Y.: A New Cyclic Phosphodiesterase Having a 3′-Nucleotidase Activity from *Escherichia coli* B. I. Purification and Some Properties of the Enzyme. J. Biol. Chem. **239**, 3412 (1964).

2. APGAR, J., G. A. EVERETT and R. W. HOLLEY: Analyses of Large Oligonucleotide Fragments Obtained from a Yeast Alanine Transfer Ribonucleic Acid by Partial Digestion with Ribonuclease T1. J. Biol. Chem. **241**, 1206 (1966).

3. APGAR, J., R. W. HOLLEY and S. H. MERRILL: Purification of Alanine-, Valine-, Histidine-, and Tyrosine-Acceptor Ribonucleic Acids from Yeast. J. Biol. Chem. **237**, 796 (1962).

4. AUGUSTI-TOCCO, G. and G. L. BROWN: Reaction of N-Cyclohexyl,N'-β(4-Methylmorpholinium) Ethyl Carbodiimide Iodide with Nucleic Acids and Polynucleotides. Nature **206**, 683 (1965).

4a. BASSEL, B. A., Jr. and S. SPIEGELMAN: Specific Cleavage of Qβ-RNA and Identification of the Fragment Carrying the 3'-OH Terminus. Proc. Nat. Acad. Sci. (USA) **58**, 1155 (1967).

5. BATT, R. D., J. K. MARTIN, J. McT. PLOESER and J. MURRAY: Chemistry of Dihydropyrimidines. Ultraviolet Spectra and Alkaline Decomposition. J. Amer. Chem. Soc. **76**, 3663 (1954).

6. BAYEV, A., T. VENKSTERN, A. MIRZABEKOV, A. KRUTILINA, V. AXELROD, L. LI and V. ENGELHARDT: Primary Structure of the Valine Transfer RNA. Partial Reconstruction of its Molecule. In: D. Shugar (ed.), Properties and Function of Genetic Elements, p. 287. Sympos. Federat. Europ. Biochem. Soc., Warsaw 1966. New York: Academic Press. 1967.

7. BEAVEN, G. H., E. R. HOLIDAY and E. A. JOHNSON: Optical Properties of Nucleic Acids and Their Components. In: E. Chargaff and J. N. Davidson (ed.), The Nucleic Acids, Vol. I, p. 493. New York: Academic Press. 1955.

8. BERG, P., F. H. BERGMANN, E. J. OFENGAND and M. DIECKMANN: The Enzymic Synthesis of Amino Acyl Derivatives of Ribonucleic Acid. I. J. Biol. Chem. **236**, 1726 (1961).

9. BERG, P., U. LAGERKVIST and M. DIECKMANN: The Enzymic Synthesis of Amino Acyl Derivatives of Ribonucleic Acid. VI. Nucleotide Sequences Adjacent to the ... pCpCpA End Groups of Isoleucine- and Leucine-Specific Chains. J. Mol. Biol. **5**, 159 (1962).

9a. BERNARDI, G.: Spleen Acid Ribonuclease. In: G. L. Cantoni and D. R. Davies (ed.), Procedures in Nucleic Acid Research, p. 37. New York: Harper and Row. 1966.

10. BIEMANN, K., S. TSUNAKAWA, J. SONNENBICHLER, H. FELDMANN, D. DÜTTING und H. G. ZACHAU: Struktur eines ungewöhnlichen Nucleosids aus serinspezifischer Transfer-Ribonucleinsäure. Angew. Chem. **78**, 600 (1966); Angew. Chem. internat. edit. **5**, 590 (1966).

11. BOREK, E.: The Methylation of Transfer RNA: Mechanism and Function. Cold Spring Harbor Sympos. Quant. Biol. **28**, 139 (1963).

11a. BROSTOFF, S. W. and V. M. INGRAM: Chemical Modification of Yeast Alanine-tRNA with a Radioactive Carbodiimide. Science **158**, 666 (1967).

12. BROWNLEE, G. G. and F. SANGER: Nucleotide Sequences from the Low Molecular Weight Ribosomal RNA of *Escherichia coli*. J. Mol. Biol. **23**, 337 (1967).

13. BROWNLEE, G. G., F. SANGER and B. G. BARRELL: Nucleotide Sequence of 5 S-ribosomal RNA from *Escherichia coli*. Nature **215**, 735 (1967).

13a. CANTOR, C. R.: Possible Conformations of 5 S Ribosomal RNA. Nature **216**, 513 (1967).

14. CANTOR, C. R., S. R. JASKUNAS and J. TINOCO, Jr.: Optical Properties of Ribonucleic Acids Predicted from Oligomers. J. Mol. Biol. **20**, 39 (1966).

15. CHAMBERS, R. W.: The Chemistry of Pseudouridine. IV. Cyanoethylation. Biochemistry **4**, 219 (1965).

16. — The Chemistry of Pseudouridine. Progr. Nucleic Acid Res. and Mol. Biol. **5**, 368 (1966).

17. CHERAYIL, J. D. and R. M. BOCK: A Column Chromatographic Procedure for the Fractionation of sRNA. Biochemistry **4**, 1174 (1965).

18. COHN, W. E. and D. G. DOHERTY: The Catalytic Hydrogenation of Pyrimidine Nucleosides and Nucleotides and the Isolation of their Ribose and Respective Ribose Phosphates. J. Amer. Chem. Soc. **78**, 2863 (1956).

19. CRICK, F. H. C.: Codon-Anticodon Pairing. The Wobble Hypothesis. J. Mol. Biol. **19**, 548 (1966).

20. — The Genetic Code. III. Scient. American **215**, No. 4, 55 (1966).

20a. — Discussion. In: E. M. Crook et al. (ed.), The Structure of Nucleic Acids and their Role in Protein Synthesis. Biochem. Soc. Sympos. No. 14, p. 25. Cambridge: Univ. Press. 1957.

21. DOCTOR, B. P.: Fractionation of RNA's by Countercurrent Distribution. In: L. Grossman and K. Moldave (ed.), Methods in Enzymology-Nucleic Acids, Vol. 12, p. 644. New York: Academic Press. 1967.

22. DOCTOR, B. P. and C. M. CONNELLY: Separation of Yeast Amino Acid-Acceptor Ribonucleic Acids by Countercurrent Distribution in Modified Kirby's System. Biochem. Biophys. Res. Commun. **6**, 201 (1961).

23. DUTTA, S. K., A. S. JONES and M. STACEY: The Separation of Desoxypentosenucleic Acids and Pentosenucleic Acids. Biochim. Biophys. Acta **10**, 613 (1953).

24. DÜTTING, D., H. FELDMANN and H. G. ZACHAU: Partial Digestions of Serine Transfer Ribonucleic Acids with Pancreatic and T 1-Ribonucleases. Z. physiol. Chem. **347**, 249 (1966).

25. DÜTTING, D., W. KARAU, F. MELCHERS und H. G. ZACHAU: Nucleotidsequenzen in Serinspezifischen Transfer-Ribonucleinsäuren. Biochim. Biophys. Acta **108**, 194 (1965).

26. DÜTTING, D. und H. G. ZACHAU: Ribonucleinsäurespaltung neben N^2-Dimethyl-guanylsäure-Resten durch T 1-Ribonuclease. Z. physiol. Chem. **336**, 132 (1964).

27. — — Spaltung einer Serinspezifischen Transfer-Ribonucleinsäure-Fraktion mit T 1-Ribonuclease. Biochim. Biophys. Acta **91**, 573 (1964).

28. — — unveröffentlicht.

29. EGAMI, F., K. TAKAHASHI and T. UCHIDA: Ribonucleases in Taka-Diastase: Properties, Chemical Nature, and Applications. Progr. Nucleic Acid Res. and Mol. Biol. **3**, 59 (1964).

30. FELDMANN, H., D. DÜTTING and H. G. ZACHAU: Analyses of Some Oligonucleotide Sequences and Odd Nucleotides from Serine Transfer Ribonucleic Acids. Z. physiol. Chem. **347**, 236 (1966).

31. FINK, R. M., R. E. CLINE, C. McGAUGHEY and K. FINK: Chromatography of Pyrimidine Reduction Products. Analyt. Chemistry **28**, 4 (1956).

32. FULLER, W. and A. HODGSON: Conformation of the Anticodon Loop in tRNA. Nature **215**, 817 (1967).

33. GILHAM, P. T.: An Addition Reaction Specific for Uridine and Guanosine Nucleotides and its Application to the Modification of Ribonuclease Action. J. Amer. Chem. Soc. **84**, 687 (1962).

34. GILHAM, P. T. and W. E. ROBINSON: The Use of Polynucleotide-Cellulose in Sequence Studies of Nucleic Acids. J. Amer. Chem. Soc. **86**, 4985 (1964).

34a. GILLAM, I., S. MILLWARD, D. BLEW, M. V. TIGERSTROM, E. WIMMER and G. M. TENER: The Separation of Soluble Ribonucleic Acids on Benzoylated DEAE-Cellulose. Biochemistry **6**, 3043 (1967).

35. HALL, R. H.: A General Procedure for the Isolation of "Minor" Nucleosides from Ribonucleic Acid Hydrolysates. Biochemistry **4**, 661 (1965).

36. HALL, R. H., M. J. ROBINS, L. STASIUK and R. THEDFORD: Isolation of N(6)-(γ,γ-Dimethylallyl)adenosine from Soluble Ribonucleic Acid. J. Amer. Chem. Soc. **88**, 2614 (1966).

36a. HINDLEY, J.: Fractionation of ^{32}P-labelled Ribonucleic Acid on Polyacrylamide Gels and their Characterization by Fingerprinting. J. Mol. Biol. **30**, 125 (1967).

36 b. HOAGLAND, M. B., P. C. ZAMECNIK and M. L. STEPHENSON: Intermediate Reactions in Protein Synthesis. Biochim. Biophys. Acta **24**, 216 (1957).

37. HOLLEY, R. W.: The Nucleotide Sequence of a Nucleic Acid. Scient. American **214**, No. 2, 30 (1966).

37 a. — An Alanine-dependent, Ribonuclease-inhibited Conversion of AMP to ATP, and its Possible Relationship to Protein Synthesis. J. Amer. Chem. Soc. **79**, 658 (1957).

38. HOLLEY, R. W., J. APGAR, B. P. DOCTOR, J. FARROW, M. A. MARINI and S. H. MERRILL: A Simplified Procedure for the Preparation of Tyrosine- and Valine-Acceptor Fractions of Yeast "Soluble Ribonucleic Acid". J. Biol. Chem. **236**, 200 (1961).

39. HOLLEY, R. W., J. APGAR, G. A. EVERETT, J. T. MADISON, M. MARQUISEE, S. H. MERRILL, J. R. PENSWICK and A. ZAMIR: Structure of a Ribonucleic Acid. Science **147**, 1462 (1965).

40. HOLLEY, R. W., J. APGAR, G. A. EVERETT, J. T. MADISON, S. H. MERRILL and A. ZAMIR: Chemistry of Amino Acid-Specific Ribonucleic Acids. Cold Spring Harbor Sympos. Quant. Biol. **28**, 117 (1963).

41. HOLLEY, R. W., G. A. EVERETT, J. T. MADISON and A. ZAMIR: Nucleotide Sequences in the Yeast Alanine Transfer Ribonucleic Acid. J. Biol. Chem. **240**, 2122 (1965).

42. HOLLEY, R. W., J. T. MADISON and A. ZAMIR: A New Method for Sequence Determination of Large Oligonucleotides. Biochem. Biophys. Res. Commun. **17**, 389 (1964).

43. INGRAM, V. M. and J. A. SJÖQUIST: Studies on the Structure of Purified Alanine and Valine Transfer RNA from Yeast. Cold Spring Harbor Sympos. Quant. Biol. **28**, 133 (1963).

44. JUKES, T. H.: Indications for a Common Evolutionary Origin Shown in the Primary Structure of Three Transfer RNAs. Biochem. Biophys. Res. Commun. **24**, 744 (1966).

45. KATCHALSKI, E., S. YANKOFSKY, A. NOVOGRODSKY, Y. GALENTER and U. Z. LITTAUER: A Chemical Method for the Purification of Amino Acid-specific Transfer RNA's. Biochim. Biophys. Acta **123**, 641 (1966).

46. KELMERS, A. D.: Preparation of Highly Purified Phenylalanine Transfer Ribonucleic Acid. J. Biol. Chem. **241**, 3540 (1966).

47. — Preparation of a Highly Purified Leucine Transfer Ribonucleic Acid. Biochem. Biophys. Res. Commun. **25**, 562 (1966).

48. KHORANA, H. G.: Polynucleotide Synthesis and the Genetic Code. Federat. Proc. (Amer. Soc. Exp. Biol.) **24**, 1473 (1965).

49. LEBOWITZ, P., P. L. IPATA, M. H. MAKMAN, H. H. RICHARDS and G. L. CANTONI: Resolution of Cytidine- and Adenosine-Terminal Transfer Ribonucleic Acids. Biochemistry **5**, 3617 (1966).

50. LEE, J. C. and P. T. GILHAM: A Method for the Determination of Nucleotide Sequences near the Terminals of Ribonucleic Acids of Large Molecular Weight. J. Amer. Chem. Soc. **88**, 5685 (1966).

51. LEE, J. C., N. W. Y. HO and P. T. GILHAM: Preparation of Ribotrinucleotides Containing Terminal Cytidine. Biochim. Biophys. Acta **95**, 503 (1965).

52. LITT, M. and V. M. INGRAM: Chemical Studies on Amino Acid Acceptor Ribonucleic Acids. II. Attempts at Partial Digestion of Yeast Amino Acid Acceptor Ribonucleic Acid with Pancreatic Ribonuclease. Biochemistry **3**, 560 (1964).

53. LOENING, U. E.: The Fractionation of High-Molecular-Weight Ribonucleic Acid by Polyacrylamid-Gel Electrophoresis. Biochem. J. **102**, 251 (1967).

54. LOHRMANN, R., D. SÖLL, H. HAYATSU, E. OHTSUKA and H. G. KHORANA:
Studies on Polynucleotides. 51. Syntheses of the 64 Possible Ribotrinucleotides
Derived from the Four Major Ribomononucleotides. J. Amer. Chem. Soc.
88, 819 (1966).

55. MADISON, J. T.: persönliche Mitteilung.

56. MADISON, J. T., G. A. EVERETT and H. KUNG: Nucleotide Sequence of a Yeast
Tyrosine Transfer RNA. Science **153**, 531 (1966).

57. — — — Oligonucleotides from Yeast Tyrosine Transfer Ribonucleic Acid.
J. Biol. Chem. **242**, 1318 (1967).

58. MADISON, J. T. and R. W. HOLLEY: The Presence of 5,6-Dihydrouridylic Acid
in Yeast "Soluble" Ribonucleic Acid. Biochem. Biophys. Res. Commun. **18**,
153 (1965).

58a. MADISON, J. T., R. W. HOLLEY, J. S. POUCHER and P. H. CONNETT: Use of
Polynucleotide Phosphorylase in Sequence Determination of Oligonucleotides.
Biochim. Biophys. Acta **145**, 825 (1967).

59. MADISON, J. T. and H. KUNG: Large Oligonucleotides Isolated from Yeast
Tyrosine Transfer Ribonucleic Acid After Partial Digestion with Ribonuclease
T1. J. Biol. Chem. **242**, 1324 (1967).

60. McPHIE, P., J. HOUNSELL and W. B. GRATZER: The Specific Cleavage of
Yeast Ribosomal Ribonucleic Acid with Nucleases. Biochemistry **5**, 988 (1966).

61. MELCHERS, F., D. DÜTTING und H. G. ZACHAU: Enzymatische Spaltungen von
Serin-tRNA-Fraktionen. Biochim. Biophys. Acta **108**, 182 (1965).

62. MICHELSON, A. M.: The Chemistry of Nucleosides and Nucleotides, p. 444.
New York: Academic Press. 1963.

63. MIDGLEY, J. E. M.: Effects of Different Extraction Procedures on the Molecular Characteristics ot Bacterial Ribosomal Ribonucleic Acid. Biochim.
Biophys. Acta **95**, 232 (1965).

63a. MOUDRIANAKIS, E. N. and M. BEER: Base Sequence Determination in Nucleic
Acids with the Electron Microscope. III. Chemistry and Microscopy of Guanine-
labeled DNA. Proc. Nat. Acad. Sci. (USA) **53**, 564 (1965).

64. NAYLOR, R., N. W. Y. Ho and P. T. GILHAM: Selective Chemical Modifications
of Uridine and Pseudouridine in Polynucleotides and Their Effect on the Speci-
ficities of Ribonuclease and Phosphodiesterases. J. Amer. Chem. Soc. **87**,
4209 (1965).

65. NELSON, J. A. and R. W. HOLLEY: Bromination of Yeast Alanine Transfer RNA.
Federat. Proc. (Amer. Soc. Exp. Biol.) **26**, 733 (1967).

66. NEU, H. C. and L. A. HEPPEL: Nucleotide Sequence Analysis of Polyribonucleotides by Means of Periodate Oxidation Followed by Cleavage with an
Amine. J. Biol. Chem. **239**, 2927 (1964).

67. NIRENBERG, M., T. CASKEY, R. MARSHALL, R. BRIMACOMBE, D. KELLOG,
B. (P.) DOCTOR, D. HATFIELD, J. LEVIN, F. ROTTMAN, S. PESTKA, M. WILCOX
and F. ANDERSON: The RNA Code and Protein Synthesis. Cold Spring Harbor
Sympos. Quant. Biol. **31**, 11 (1966).

68. NISHIMURA, S. and G. D. NOVELLI: Resistance of sRNA to Ribonucleases in
the Presence of Mg-Ion. Biochem. Biophys. Res. Commun. **11**, 161 (1963).

69. NOVOGRODSKY, A. and J. HURWITZ: The Enzymatic Phosphorylation of Ribonucleic Acid and Deoxyribonucleic Acid. J. Biol. Chem. **241**, 2923 (1966).

70. NYGAARD, A. P. and B. D. HALL: Formation and Properties of RNA-DNA
Complexes. J. Mol. Biol. **9**, 125 (1964).

71. OKADA, Y., E. TERZAGHI, G. STREISINGER, J. EMRICH, M. INOUYE and A.
TSUGITA: A Frame-shift Mutation Involving the Addition of Two Base Pairs in
the Lysozyme Gene of Phage T4. Proc. Nat. Acad. Sci. (USA) **56**, 1692 (1966).

72. PEARSON, R. L. and A. D. KELMERS: Separation of Transfer Ribonucleic Acids by Hydroxyapatite Columns. J. Biol. Chem. **241**, 767 (1966).

73. PENSWICK, J. R. and R. W. HOLLEY: Specific Cleavage of the Yeast Alanine RNA into Two Large Fragments. Proc. Nat. Acad. Sci. (USA) **53**, 543 (1965).

73a. PERUTZ, M. F.: Proteins and Nucleic Acids. Structure and Function. 8th Weizmann Memorial Lecture Series, April 1961, p. 141. Amsterdam: Elsevier. 1962.

74. PETERSON, E. A. and H. A. SOBER: Chromatography of Proteins. I. Cellulose Ion-exchange Adsorbents. J. Amer. Chem. Soc. **78**, 751 (1956).

75. PORTATIUS, H. v., P. DOTY and M. L. STEPHENSON: Separation of L-Valine Acceptor "Soluble Ribonucleic Acid" by Specific Reaction with Polyacrylic Acid Hydrazide. J. Amer. Chem. Soc. **83**, 3351 (1961).

76. RAJBHANDARY, U. L., S. H. CHANG, A. STUART, R. D. FAULKNER, R. M. HOSKINSON and H. G. KHORANA: Studies on Polynucleotides. LXVIII. The Primary Structure of Yeast Phenylalanine Transfer RNA. Proc. Nat. Acad. Sci. (USA) **57**, 751 (1967).

77. RAJBHANDARY, U. L. and A. STUART: Nucleic Acids — Sequence Analysis. Annu. Rev. Biochem. **35**, II, 759 (1966).

78. RAJBHANDARY, U. L., A. STUART, R. D. FAULKNER, S. H. CHANG and H. G. KHORANA: Nucleotide Sequence Studies on Yeast Phenylalanine sRNA. Cold Spring Harbor Sympos. Quant. Biol. **31**, 425 (1966).

79. RAJBHANDARY, U. L., R. J. YOUNG and H. G. KHORANA: Studies on Polynucleotides. XXXII. The Labeling of End Groups in Polynucleotide Chains: The Selective Phosphorylation of Phosphomonoester Groups in Amino Acid Acceptor Ribonucleic Acids. J. Biol. Chem. **239**, 3875 (1964).

80. RAKE, A. V. and G. M. TENER: Effect of Cyanoethylation of Yeast Transfer Ribonucleic Acid on its Amino Acid Acceptor Activity. Biochemistry **5**, 3992 (1966).

81. RICHARDS, E. G., J. A. COLL and W. B. GRATZER: Disc Electrophoresis of RNA in Polyacrylamide Gels. Analyt. Biochem. **12**, 452 (1965).

82. RICHARDSON, C. C.: Phosphorylation of Nucleic Acid by an Enzyme from T4 Bacteriophage-Infected *Escherichia coli*. Proc. Nat. Acad. Sci. (USA) **54**, 158 (1965).

83. RICHARDSON, J. P.: The Binding of RNA-Polymerase to DNA. J. Mol. Biol. **21**, 83 (1966).

84. RUSHIZKY, G. W., E. M. BARTOS and H. A. SOBER: Chromatography of Mixed Oligonucleotides on DEAE-Sephadex. Biochemistry **3**, 626 (1964).

85. RUSHIZKY, G. W., I. H. SKAVENSKI and A. E. GRECO: Preparation of Large Oligonucleotides from High Molecular Weight Ribonucleic Acid. Biochemistry **5**, 3328 (1966).

86. RUSHIZKY, G. W., I. H. SKAVENSKI and H. A. SOBER: Characterization of the Major Compounds Found in Ribonuclease T1 Digests of Ribonucleic Acid. III. Penta- and Higher Oligonucleotides. J. Biol. Chem. **240**, 3984 (1965).

87. RUSHIZKY, G. W. and H. A. SOBER: Studies on the Specificity of Ribonuclease T2. J. Biol. Chem. **238**, 371 (1963).

88. SANGER, F.: persönliche Mitteilung.

89. SANGER, F., G. G. BROWNLEE and B. G. BARRELL: A Twodimensional Fractionation Procedure for Radioactive Nucleotides. J. Mol. Biol. **13**, 373 (1965).

90. SCHLEICH, T. and J. GOLDSTEIN: Gel Filtration Heterogeneity of *Escherichia coli* Soluble RNA. J. Mol. Biol. **15**, 136 (1966).

90a. SCHROEDER, W. A. and R. T. JONES: Some Aspects of the Chemistry and Function of Human and Animal Hemoglobins. Fortschr. Chem. organ. Naturstoffe **23**, 113 (1965).

91. Singer, M. F., R. J. Hilmoe and M. Grunberg-Manago: Studies on the Mechanism of Action of Polynucleotide Phosphorylase. J. Biol. Chem. **235**, 2705 (1960).
92. Smith, C. J. and E. Herbert: Structure of Amino Acyl Oligonucleotides from T 1 Ribonuclease Digests of Soluble Ribonucleic Acid Charged with Serine, Glycine, Threonine, and Alanine. Biochemistry **5**, 1333 (1966).
93. Smith, C. J., P. M. Smith and E. Herbert: Isolation of Amino Acyl Oligonucleotides from T 1 Ribonuclease Digests of Soluble Ribonucleic Acid Charged with Serine, Glycine, Threonine, and Alanine. Biochemistry **5**, 1323 (1966).
94. Sober, H. A., S. F. Schlossman, A. Yaron, S. A. Latt and G. W. Rushizky: Protein-Nucleic Acid Interaction. I. Nuclease-Resistant Polylysine-Ribonucleic Acid Complexes. Biochemistry **5**, 3608 (1966).
95. Söll, D., D. S. Jones, E. Ohtsuka, R. D. Faulkner, R. Lohrmann, H. Hayatsu, H. G. Khorana, J. D. Cherayil, A. Hampel and R. M. Bock: Specificity of sRNA for Recognition of Codons as Studied by the Ribosomal Binding Technique. J. Mol. Biol. **19**, 556 (1966).
96. Staehelin, M.: On the Specificity of RNAase T 1. Biochim. Biophys. Acta **87**, 493 (1964).
97. — Column Chromatography of Oligonucleotides and Polynucleotides. Progr. Nucleic Acid Res. **2**, 169 (1963).
98. — Studies on Nucleotide Sequences in Ribonucleic Acids. II. Spectroscopic Properties of Oligoribonucleotides. Biochim. Biophys. Acta **49**, 20 (1961).
99. Stanley, W. M., Jr. and R. M. Bock: Isolation and Physical Properties of the Ribosomal Ribonucleic Acid of *Escherichia coli.* Biochemistry **4**, 1302 (1965).
100. Steinschneider, A. and H. Fraenkel-Conrat: Studies of Nucleotide Sequences in TMV-RNA. IV. Use of Aniline in Stepwise Degradation. Biochemistry **5**, 2735 (1966).
101. Takanami, M.: Analysis of the 5'-Terminal Nucleotide Sequences of Ribonucleic Acids. I. The 5'-Termini of *Escherichia coli* Ribosomal RNA. J. Mol. Biol. **23**, 135 (1967).
102. Tener, G. M., I. Gillam, M. v. Tigerstrom, S. Millward and E. Wimmer: Purifications of tRNAs on Benzoylated DEAE-Cellulose. Federation Proc. (Amer. Soc. Exp. Biol.) **25**, 519 (1966).
103. Terzaghi, E., Y. Okada, G. Streisinger, J. Emrich, M. Inouye and A. Tsugita: Change of a Sequence of Amino Acids in Phage T 4 Lysozyme by Acridine Induced Mutations. Proc. Nat. Acad. Sci. (USA) **56**, 500 (1966).
104. Tomasz, M. and R. W. Chambers: The Chemistry of Pseudouridine. VII. Selective Cleavage of Polynucleotides Containing Pseudouridylic Acid Residues by a Unique Photochemical Reaction. Biochemistry **5**, 773 (1966).
105. Tomlinson, R. V. and G. M. Tener: The Effect of Urea, Formamide, and Glycols on the Secondary Binding Forces in the Ion Exchange Chromatography of Polynucleotides on DEAE-Cellulose. Biochemistry **2**, 697 (1963).
106. Uziel, M. and W. E. Cohn: Desalting of Nucleotides by Gel Filtration. Biochim. Biophys. Acta **103**, 539 (1965).
107. Vournakis, J. N. and H. A. Scheraga: Optical Rotatory Dispersion Studies of Yeast Alanine and Tyrosine Transfer Ribonucleic Acids. Evidence for Intramolecular Hydrogen Bonding and Discussion of Conformational Aspects. Biochemistry **5**, 2997 (1966).
108. Weber, K., G. Notani, M. Wikler and W. Konigsberg: Amino Acid Sequence of the f 2 Coat Protein. J. Mol. Biol. **20**, 423 (1966).
109. Weiss, S. B. and J. Legault-Demare: Pseudouridine Formation: Evidence for RNA as an Intermediate. Science **149**, 429 (1965).

110. WHITFELD, P. R.: A Method for the Determination of Nucleotide Sequence in Polyribonucleotides. Biochem. J. **58**, 390 (1954).

111. YANOFSKY, C., B. C. CARLTON, J. R. GUEST, D. R. HELINSKI and U. HENNING: On the Colinearity of Gene Structure and Protein Structure. Proc. Nat. Acad. Sci. (USA) **51**, 266 (1964).

112. YOSHIDA, M. and T. UKITA: Pseudouridine Residues Resistant to Cyanoethylation in Yeast Transfer Ribonucleic Acid. Biochim. Biophys. Acta **123**, 214 (1966).

113. ZACHAU, H. G., D. DÜTTING und H. FELDMANN: Nucleotidsequencen zweier serinspezifischer Transfer-Ribonucleinsäuren. Angew. Chem. **78**, 392 (1966); Angew. Chem. internat. edit. **5**, 422 (1966).

114. — — — The Structures of Two Serine Transfer Ribonucleic Acids. Z. physiol. Chem. **347**, 212 (1966).

115. — — — On the Primary Structure of Transfer Ribonucleic Acids. In: D. Shugar (ed.), Properties and Function of Genetic Elements, p. 271. Sympos. Federat. Europ. Biochem. Soc., Warsaw 1966. New York: Academic Press. 1967.

116. ZACHAU, H. G., D. DÜTTING, H. FELDMANN, F. MELCHERS and W. KARAU: Serine Specific Transfer Ribonucleic Acids. Comparison of Nucleotide Sequences and Secondary Structure Models. Cold Spring Harbor Sympos. Quant. Biol. **31**, 417 (1966).

117. ZACHAU, H. G., D. DÜTTING, F. MELCHERS, H. FELDMANN and R. THIEBE: On Serine-Specific Transfer Ribonucleic Acids. In: H. Tuppy (ed.), Ribonucleic Acid Structure and Function, p. 21. Sympos. Federat. Europ. Biochem. Soc., Vienna 1965. Oxford: Pergamon Press. 1966.

118. ZACHAU, H. G., M. TADA, W. B. LAWSON und M. SCHWEIGER: Fraktionierung der löslichen Ribonucleinsäure. Biochim. Biophys. Acta **53**, 221 (1961).

119. ZAMIR, A., R. W. HOLLEY and M. MARQUISEE: Evidence for the Occurrence of a Common Pentanucleotide Sequence in the Structures of Transfer Ribonucleic Acids. J. Biol. Chem. **240**, 1267 (1965).

120. ZUBAY, G.: The Isolation and Fractionation of Soluble Ribonucleic Acid. J. Mol. Biol. **4**, 347 (1962).

(Eingelaufen am 25. September 1967)

Manzsche Buchdruckerei, 1090 Wien